P9-AEZ-687

PHYSICAL PROPERTY DATA
for the
DESIGN ENGINEER

PHYSICAL PROPERTY DATA for the DESIGN ENGINEER

Edited by

C. F. Beaton
Consultant
Peebleshire, Scotland

G. F. Hewitt
Harwell Laboratory
Oxfordshire, England

with contributions by

P. E. LILEY, R. N. MADDOX, S. F. PUGH, M. SCHUNCK,
J. TABOREK AND G. ULRYCH

HEMISPHERE PUBLISHING CORPORATION
A member of the Taylor & Francis Group

New York Washington Philadelphia London

Physical Property Data for the Design Engineer

1 2 3 4 5 6 7 8 9 0 BRBR 8 9 8 7 6 5 4 3 2 1 0 9 8

Cover design by
Debra Eubanks Riffe

Library of Congress Cataloging in Publication Data

Physical property data for the design engineer.

Bibliography: p.
Includes index.
1. Chemicals—Tables. I. Beaton, C. F. II. Hewitt, G. F. (Geoffrey Frederick)
TP200.P49 1988 660.2′0212 88-10969
ISBN 0-89116-739-0

CONTENTS

PREFACE

In process design, often one of the most difficult problems is to find appropriate physical property data for the fluids involved. Even for commonly used fluids, there is often a dearth on such data and the data available is often in inconvenient forms or not readily accessible in the design office. The objective of the present volume is to provide physical data on a wide range of fluids and construction material of relevance in process design. The aim is to present this material in a well-organized and easily-used form. Though no collection of physical data can be wholly comprehensive, the range of substances covered is very wide and, we believe, forms a useful new compilation.

The main source for the information presented in this book is the Heat Exchanger Design Handbook published by Hemisphere Publishing Corporation. The issue of HEDH in 1983 represented the culmination of several years of intensive planning, writing, editing, and production. This major five-part loose-leaf Handbook covers the whole spectrum of design guidance and data for heat transfer and heat exchangers including heat exchanger theory, fluid mechanics and heat transfer, thermal and hydraulic design of heat exchangers, mechanical design of heat exchangers and, in Part 5, physical property data. HEDH has been updated annually and, at the time of writing, preparations were in hand for the fifth supplement.

In considering the physical property material from HEDH, it was realized that a major recompilation of the material would be greatly advantageous. In HEDH, the information was presented in terms of the type of property (saturation properties, subcooled liquid properties, thermodynamic properties, superheated vapor properties etc). Thus, for any given substance, information could be found in four or five different and relatively uncorrelated sections. As experience has grown with the use of the material, it was realized that it would be far better to have all of the information for a given substance grouped together. This has now been done for the present volume and has involved a major recompilation of the material, which we hope the users will find beneficial.

We would like to express our gratitude to the President of Hemisphere Publishing Corporation, Mr. William Begell, whose enthusiasm for this project has been an inspiration. We would also like to thank Mrs. Florence Padgett (Associate Publisher of Hemisphere Publishing Corporation) for her help and encouragement. Finally, we must express our thanks and gratitude to the original contributors to Part 5 of HEDH whose work has been included in this volume. We are particularly grateful to Dr. M. Schunck of BASF Aktiengesellschaft who prepared the original HEDH text material on properties of pure fluids and of solids and to Professor R. M. Maddox of Oklahoma State University who produced the equivalent material on properties and mixtures of fluids and also contributed the data on saturation properties. We are also grateful for the contributions by Professor P. E. Liley of Purdue University and Dr. S. F. Pugh for material on thermal and elastic properties of solids respectively. We have also used material contributed to HEDH by Dr. G. Ulrych of Kraftwerk Union relating to properties of heavy water and material from HEDH on physical property unit conversion charts contributed by Dr. J. Taborek, Consultant. We hope that all these contributors will feel that we have made wise use of their material together with our own in producing what we believe will be a useful addition to the physical property data sources available for the design engineer.

C. F. Beaton
G. F. Hewitt

CONTRIBUTORS

C. F. Beaton	Consultant, Old Howford Cottage, Innerleithen, Peebleshire, Eh44GPS, England
G. F. Hewitt	Thermal Hydraulics Division, Atomic Energy Establishment, Harwell, Oxfordshire OX11 ORA, England
P. E. Liley	Mechanical Engineering Building, Purdue University, West Lafayette, Indiana 79707, United States
R. N. Maddox	School of Chemical Engineering, Oklahoma State University, Stillwater, Oklahoma 74074, United States
S. F. Pugh	Metallurgy Division, Atomic Energy Research Establishment, Harwell Oxfordshire OX11 ORA, England
M. Schunck	BASF Aktiengesellschaft, Abteilung D-DAD/CB, 6700 Ludwigshafen/Rhein, Federal Republic of Germany
G. Ulrych	Kraftwerk Union, Reaktortechnik, Hammerbacherstrasse 12 + 14, Postfach 3220, 8520 Erlangen, Federal Republic of Germany

INTRODUCTION

Clearly, the main purpose of this book is to present actual physical property data which can be used in the design process. However, it is considered important to give some brief introductory material relating to the physical properties and this is done for pure fluids (Chapter 1), for fluid mixtures (Chapter 4) and for solid properties (Chapter 5). Chapter 1 deals first with critical data (covering the correspondence principle, critical temperature, critical pressure, critical volume and the acentric factor). This is followed naturally by specific volume ($p\,\tilde{V}\,T$) correlations and by thermodynamic properties (boiling point, vapour pressure, heat of vaporisation, specific heat capacity etc). This is then followed by a brief introduction to transport properties (viscosity, thermal conductivity and Prandtl number) and the chapter closes with a discussion of surface tension. It should be stressed that the material presented here is not intended to be very detailed and comprehensive; further information will be found in a variety of textbooks which are referenced in the chapter. The aim, as was stated above, is to provide an introduction (for pure fluids) to Chapters 2 and 3.

Chapter 2 (the most important one in this book) contains the physical property data for pure fluids. For the book, the data have been grouped together for each individual substance. The data is of the following principal types:

1. *Liquid physical properties.* Here, physical properties of the liquid phase of various pure substances are given for temperatures below the boiling point.
2. *Saturation properties.* Here, the full range of vapour and liquid thermodynamic and transport properties are tabulated at a variety of points along the saturation curve starting at the normal boiling point. Many design calculations are for boiling and condensing systems where it is particularly useful to have data along the saturation curve. Such data is rarely available in practice and it is believed that these tables make a particularly useful contribution to the available sources.
3. *Transport properties of superheated gases.* Here, transport properties are given for the gaseous form of the substance at temperatures above the normal boiling point and at atmospheric pressure.
4. *Thermodynamic properties of superheated gases.* Here, values of specific volume, enthalpy and entropy are given as a function of temperature and pressure for a number of the more important substances.

In addition to the above, information is given for each substance on chemical formula, molecular weight, normal density, boiling point, and critical parameters. Although not all of the above data is available for each substance, at least some data is available for around 200 substances, covering most of the ones that are of major interest in the process industries.

It is, of course, particularly important to have specific data on water and water physical properties have been the subject of an ongoing international collaborative effort over many years. No book of this kind would be complete without an accurate set of data for water and this is given in Chapter 3. Here, the data have been abstracted from the recent NBS/NRC Steam Tables (also published by Hemisphere Publishing Corporation).

In many industrial applications, mixtures of fluids rather than pure fluids are the norm rather than the exception. Examples here would be mixtures of fluids being reacted in chemical reactors, separated in distillation or solvent extraction columns and the mixtures naturally associated with petroleum exploration, production and refining. Mixtures present very special difficulties in obtaining accurate physical property data. Chapter 4 deals with such mixtures starting with their phase behaviour (which is often very complex) and continuing with their thermodynamic properties, thermophysical properties and interfacial tension. In order to predict the phase behaviour, it is necessary to have data for the constituent binary pairs and this data is also presented in Chapter 4.

The final chapter of the book deals with properties of solids. These are, again, of vital importance in design and Chapter 5 discusses the background for such properties including density, specific heat, thermal conductivity and elastic properties. The chapter also includes specific data on thermal conductivity and elastic properties and concludes with specific data on the common materials of construction.

UNITS AND NOMENCLATURE

SYSTEM INTERNATIONAL (SI) UNITS

With few exceptions, the material in this book conforms to the SI system of units, now almost universally accepted throughout the world. In the SI system, there are seven fundamental and precisely defined basic units from which all others are derived. In the context of the present book, the following five basic quantities are relevant:

Length: This is defined in terms of the meter (m). The meter was originally defined in terms of a prototype held in Paris but this definition was replaced by one in terms of the wavelength of light from a krypton-86 lamp and, more recently, in terms of the velocity of light.
Mass: The standard unit of mass (the kilogram, kg) is related to an international prototype mass.
Time: The standard unit is the second (s), which is defined in terms of the radiation frequency of the cesium-133 atom.
Temperature: This is expressed in terms of the kelvin (K) which is defined in terms of the triple point of water.
Amount of substance: This is referred to as the "mole" (alternatively spelt mol), which is the amount of substance that contains as many elementary entities as there are atoms in 0.012 kg of carbon-12. The elementary entities must be identified when the mole is used (atoms, molecules, ions).

In addition to the base units, there are a number of important derived units with special names. These derived units are given capital letters to indicate their nature (for example, the newton N replaces kg m/s^2. The derived units most commonly used here are shown in Table 1.

Any physical quantity may be regarded as the product of a pure number and the associated units. Thus, for example, a distance of a thousand meters may be written as 1000 × m. Often, the numerical values associated with this rule are inconvenient and it is better to express the physical quantity in terms of units which have a prefix. For the example cited here, the distance can be represented as, for instance, 1 × km where the unit is now the kilometer. The standard prefixes for the SI system are shown in Table 2. Wherever possible, we have used the "preferred" units and avoided the use of those which are given as "accepted" in the above table. Thus, units are expressed in general in ranges of 1000.

Table 1 Most frequently used derived SI units

Quantity	Unit	Symbol	Dimension
Force	newton	N	kg m/s^2
Energy, work, heat (also power × time)	joule	J	N m or kg m^2/s^2 W s = J s/s
Power	watt	W	J/s = kg m^2/s^3
Pressure	pascal	Pa	N/m^2 = kg/m s^2

Table 2 SI prefixes

Factor	Prefix	Symbol	Status
10^{12}	tera	T	Preferred
10^{9}	giga	G	Preferred
10^{6}	mega	M	Preferred
10^{3}	kilo	k	Preferred
10^{2}	hecto	h	Accepted
10	deka	da	Accepted
10^{-1}	deci	d	Accepted
10^{-2}	centi	c	Accepted
10^{-3}	milli	m	Preferred
10^{-6}	micro	μ	Preferred
10^{-9}	nano	n	Preferred
10^{-12}	pico	p	Preferred

SYMBOLS AND UNITS

A wide variety of symbols are used in this book for physical quantities; these symbols are defined locally in each particular area but conform to a general nomenclature which has the following main characteristics:

1. Capital Roman letters (e.g. H,M) denote absolute quantities (for instance enthalpy H, volume V. Exceptions to this rule are time (t) and coordinate distances (x,y,z,r). These exceptions are obvious and fall in with well established practices.
2. Lower-case Roman letters denote the corresponding specific quantities per unit mass, mole, length, area or volume. Examples here are specific enthalphy h (J/kg) and specific volume v (m^3/kg). An allowed exception here is for velocity (u,v,w); the difference between velocity and specific volume is usually clear from the context.
3. Molar quantities are denoted by a tilde (˜) over the symbol. Examples here would be mole fraction in the liquid phase $\tilde{x}$ and molar density $\tilde{\rho}$.
4. Lower-case Greek letters are used for coefficients and physical properties (examples are λ for thermal conductivity and σ for surface tension). An exception (meeting common practice) is the use of c for specific heat capacity.

The subscripts s,ℓ,g and m are used to denote the solid, liquid, and gas phases and the mean values respectively.

Although the symbols are defined locally within the text, Table 3 may be found useful as defining the more common quantities.

CONVERSION FACTORS

Although in an ideal world, there would be no need to be concerned about the use of units other than SI, the fact remains that a knowledge of conversion factors from one set of units to another is indispensable in engineering design. The reasons for this are mainly:

1. Much available data and other information is expressed in alternative units, for instance in terms of pounds, feet, hours, etc. It is important to be able to make accurate conversions to bring such information into line with the SI system.
2. The pace of conversion to the SI system has not been as rapid as many people would have hoped. Thus, in process design, British Imperial units (which should now be properly called US Customary Units!) are still very widely used. Pressures for changes seem to have lessened rather than increased in recent years.

In view of this situation, we have considered it important to present a number of conversion factors here and these are summarized in Table 4. It is clear that, if the units change, then the number in the expression:

$$\text{Physical quantity} = \text{number} \times \text{units}$$

will also change. In going from the units given in the second column to the units given in the fourth column, then the numerical value has to be multiplied by the factor given in the third column. If the units change from those given in

Table 3 Common symbols and their units

Quantity	Symbol	Unit
Acceleration due to gravity*	g	m/s^2
Avogadro or Loschmidt number**	$\tilde{L}$	1/mol
Chemical potential	μ	J/kg
Density	ρ	kg/m^3
Diffusivity	δ	m^2/s
Dynamic (absolute) viscosity	η	kg/ms
Concentration	c	kg/m^3
Enthalpy	H	J
Enthalpy (specific)	h	J/kg
Entropy (specific)	s	J/kgK
Free energy (specific)	$f = u - Ts$	J/kg
Gas constant***	$\tilde{R}$	J/molK
Gibbs function free enthalpy (specific)	$g = h - Ts$	J/kg
Heat quantity	Q	J
Heat flux	$\dot{q}$	W/m^2
Internal energy	U	J
Internal energy (specific)	u	J/kg
Lewis number	$Le = \kappa/\delta = \lambda/\rho c_p \delta$	
Mass	M	kg
Mass flux	$\dot{m}$	kg/m^2s
Mass fraction	x_i, y_i	kg_i/kg
Molar concentration	$\tilde{c}$	mol/m^3
Molar density	$\tilde{\rho}$	mol/m^3
Molar flux	$\dot{n}$	mol/m^2s
Mole fraction	$\tilde{x}_i, \tilde{y}_i$	kg_i/kg
Molecular weight	$\tilde{M}$	g/mol (kg/kmol)
Number of moles	N	mol
Prandtl number	$Pr = \nu = k =$ $\eta c_p/\lambda$	
Pressure	p	Pa (N/m^2)
Specific heat capacity	c_p, c_v	J/kgK
Specific volume	v	m^3/kg
Surface tension	σ	N/m
Temperature	T	K, °C
Thermal conductivity	λ	W/mK
Thermal diffusivity	$\kappa \equiv \lambda/\rho c_p$	m^2/s
Time	t	s
Viscosity (Dynamic absolute)	η	kg/ms
Viscosity (kinematic)	$\nu = \eta/\rho$	m^2/s
Volume	v	m^3

*Standard value = 9.80665 m/s^2
**$\tilde{L}$ = 6.0252 × 10^{23}
***$\tilde{R}$ = 8.314 J/molK = 8314 J/molK

the fourth column to those given in the second column, then the numerical value has to be *divided* by the factor given in the third column.

For temperature the following conversion equations are used:

$$°C = \frac{5}{9}[°F - 32] \qquad °C = (°F - 40)\frac{5}{9} - 40 \qquad \Delta T(°C) = \frac{9}{5}\Delta T(°F) \qquad K = °C + 273.15$$

$$°F = \frac{9}{5}(°C) + 32 \qquad °F = (°C + 40)\frac{9}{5} - 40 \qquad \Delta T(°F) = \frac{5}{9}\Delta T(°C) \qquad °R = °F + 459.67$$

CONVERSION SCALES

Many mistakes are made in unit conversions and as a further aid to avoid these, it is possible to use comparison scales and examples are given in Table 5.

Table 4 Unit conversion

Physical quantity	Given in → / Gives ←	Multiplied by → / Divided by ←	Gives / Given in
Amount of substance	lb_m-mol	453.6	kmol
	g-mol	1.000	mol
	kg-mol	1.000	kmol
	mol	1 000	kmol
Area	ft^2	0.092903	m^2
	in^2	645.16	mm^2
	acre	4 047.0	m^2
Density	lb_m/ft^3	16.0185	kg/m^3
	kg/m^3	0.06243	lb_m/ft^3
	lb_m/U.S. gal	119.7	kg/m^3
Diffusivity	ft^2/s	0.092903	m^2/s
Energy (work) (heat)	Btu[b]	1 055.056	J = N m = W s
	Btu	0.2520	kcal
	Btu	778.28	ft lb_f
	kcal	4 186.8	J
	ft lb_f	1.3558	J
	W h	3 600	J
Enthalpy	Btu/lb_m	2 326	J/kg
	$kcal/kg_m$	4 186.8	J/kg
Force	lb_f	4.44822	N = kg m/s^2
	lb_f	0.45359	kg_f
	kg_f	2.2046	lb_f
	kg_f	9.80665	N
	dyne	0.00001 (exact)	N
Heat flux	Btu/h ft^2	3.1546	W/m^2
	W/m^2	0.317	Btu/h ft^2
	$kcal/cm^2$ s	41.868	W/m^2
Length	ft	0.3048	m
	in	25.4 (exact)	mm
	mil	0.0254	mm
	yard	0.9144	m
	mile (mi)	1 609.3	m
	km	0.621388	mi
Mass	lb_m	0.45359	kg
	kg	2.2046	lb_m
	metric ton	2 204.6	lb_m
	ton (2 000 lb_m)	907.18	kg
Mass flow rate	lb_m/h	0.0001260	kg/s
	kg/s	7 936.51	lb_m/h
	lb_m/s	0.4536	kg/s
	lb_m/min	0.00756	kg/s

Table 4 Unit conversion *(Continued)*

Physical quantity	Given in ⟶ / Gives ⟵	Multiplied by ⟶ / Divided by ⟵	Gives / Given in
Mass flux	lb_m/h ft²	1.356 × 10^{-3}	kg/s m²
	kg/s m²	737.5	lb_m/h ft²
	lb_m/ft² s	4.8824	kg/m² s
Power	Btu/h	0.2931	W = J/s
	W	3.4118	Btu/h
	kcal/h	1.163	W
	ft lb_f/s	1.3558	W
	hp (metric)	735.5	W
	Btu/h	0.2520	kcal/h
	tons refrig.	3 516.9	W
Pressure	lb_f/in² (psi)	6.8948	kN/m² = kPa
	kPa	0.1450	psi
	bar	100	kPa
	lb_f/ft²	0.0479	kPa
	mm Hg (torr)	0.1333	kPa
	in Hg	3.3866	kPa
	mm H_2O	9.8067	Pa
	in H_2O	249.09	Pa
	at (kg_f/cm²)	98.0665	kPa
	atm (normal)	101.325	kPa
Specific heat capacity	Btu/lb_m °F	4 186.8	J/kg K
	kcal/kg °C	4 186.8	J/kg K
Surface tension	dyne/cm	0.001	N/m
	dyne/cm	6.852 × 10^{-5}	lb_f/ft
	lb_f/ft	14.954	N/m
Thermal conductivity	Btu/ft h °F	1.7308	W/m K
	W/m K	0.5778	Btu/ft h °F
	kcal/m h °C	1.163	W/m K
Thermal diffusivity	m²/h	0.0002778	m²/s
	ft²/s	0.092903	m²/s
	ft²/h	25.81 × 10^{-6}	m²/s
Velocity	ft/s[a]	0.3048	m/s
	m/s	3.2808	ft/s
	ft/min	0.00508	m/s
	mi/h	1.6093	km/h
	km/h	0.6214	mi/h
	knots	1.852	km/h
Kinematic viscosity	stoke (St), cm² s	0.0001	m²/s
	centistoke (cSt)	10^{-6}	m²/s
	ft²/s	0.092903	m²/s
Dynamic (absolute) viscosity	centipoise (cP)	0.001	kg/m s
	poise (P)	0.1	Pa s
	cP	1.000	mPa s
	cP	1 000	μPa s
	lb_m/ft h	0.0004134	Pa s
	lb_m/ft h	0.4134	cP
	cP	2.4189	lb_m/ft h
	lb_m/ft s	1.4482	Pa s

Table 4 Unit conversion (*Continued*)

Physical quantity	Given in ⟶ Gives ⟵	Multiplied by ⟶ Divided by ⟵	Gives Given in
Volume	ft^3	0.028317	m^3
	U.S. gal	0.003785	m^3
	U.S. gal	3.785	liter (L)
	L (liter)	0.2642	U.S. gal
	Brit. gal	0.004546	m^3
	U.S. gal	0.13368	ft^3
	barrel (U.S. pet.)	0.15898	m^3
	barrel (U.S. pet.)	42	U.S. gal
Volume flow rate	U.S. gal/min	6.309×10^{-5}	m^3/s
	U.S. bbl/day	0.15899	m^3/day
	U.S. bbl/day	1.84×10^{-6}	m^3/s
	ft^3/s	0.02832	m^3/s
	ft^3/min	0.000472	m^3/s

Table 5 Unit conversion charts

AREA: square meter, m^2; square foot, ft^2

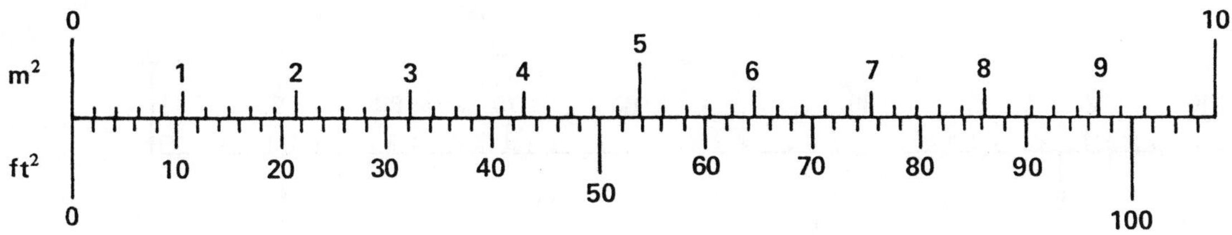

$m^2 \times 10.764 = ft^2$
$ft^2 \times 0.0929 = m^2$

DENSITY: $\frac{kg}{m^3}$; $\frac{lb_m}{ft^3}$

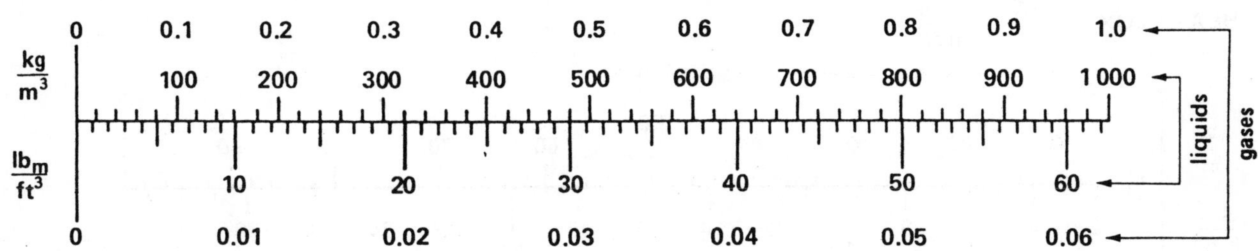

$\frac{kg}{m^3} \times 0.0624 = \frac{lb_m}{ft^3}$

$\frac{lb_m}{ft^3} \times 16.02 = \frac{kg}{m^3}$

ENERGY (WORK): joule, J = N m; Btu

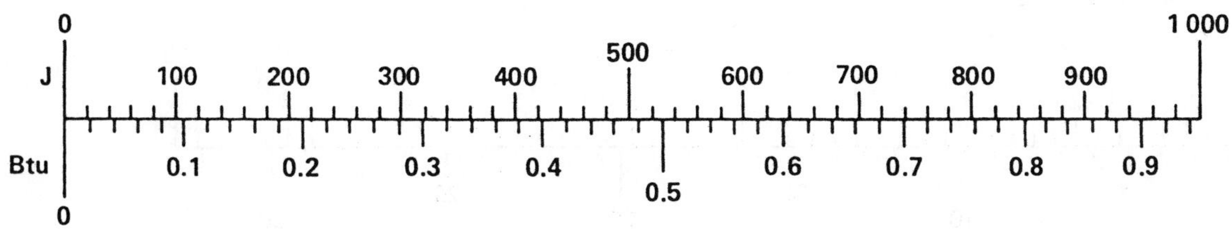

$J \times 0.948(10^{-3}) = Btu$
$Btu \times 1055 = J$

ENTHALPY: $\frac{kJ}{kg}$; $\frac{Btu}{lb_m}$

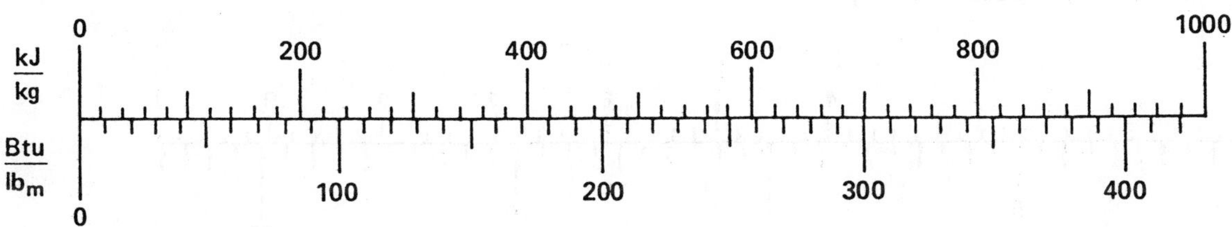

$\frac{kJ}{kg} \times 0.430 = \frac{Btu}{lb_m}$

$\frac{Btu}{lb_m} \times 2.326 = \frac{kJ}{kg}$

Table 5 Unit conversion charts (*Continued*)

FORCE: newton, $N = \frac{kg\ m}{s^2}$; pound-force, lb_f

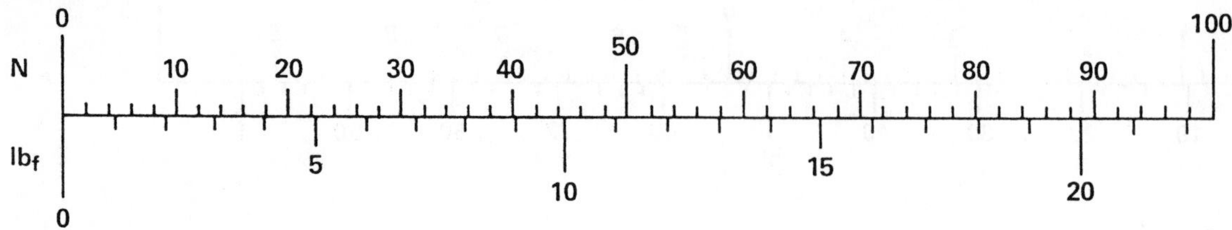

$N \times 0.225 = lb_f$
$lb_f \times 4.448 = N$

HEAT FLUX: $\frac{W}{m^2}$; $\frac{Btu}{h\ ft^2}$

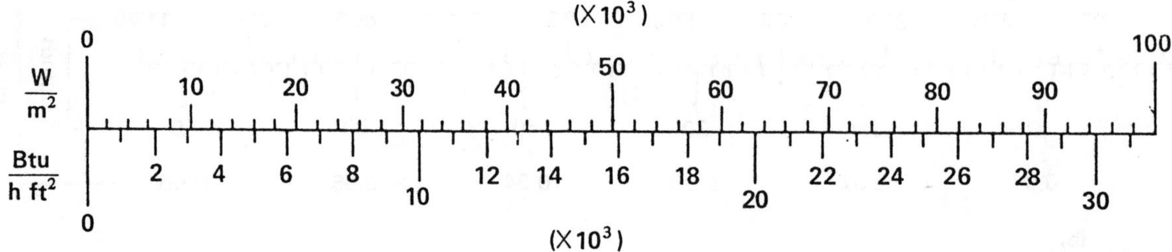

$\frac{W}{m^2} \times 0.316 = \frac{Btu}{h\ ft^2}$
$\frac{Btu}{h\ ft^2} \times 3.16 = \frac{W}{m^2}$

LENGTH: meter, m; foot, ft

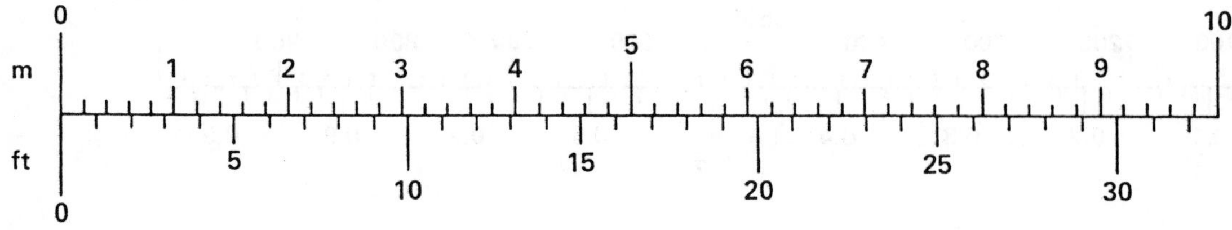

$m \times 3.2803 = ft$
$ft \times 0.3048 = m$

MASS: kilogram, kg; pound-mass, lb_m

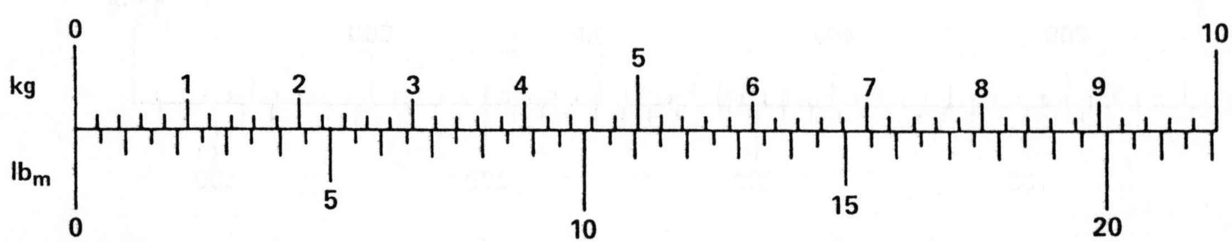

$kg \times 2.205 = lb_m$
$lb_m \times 0.454 = kg$

Table 5 Unit conversion charts (*Continued*)

MASS FLOW RATE: $\frac{kg}{s}$; $\frac{10^3\ lb_m}{h}$

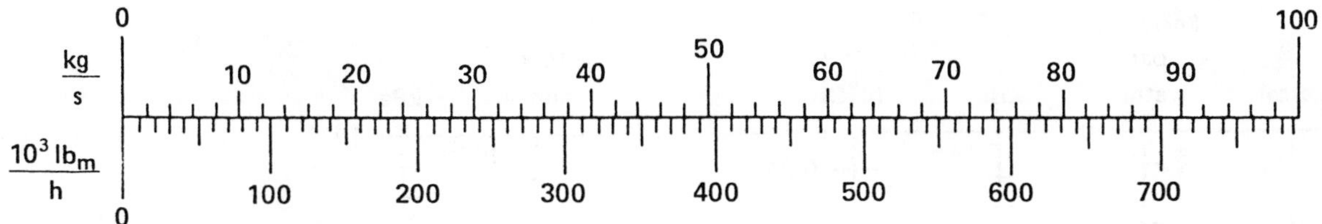

$$\frac{kg}{s} \times 7.937 = \frac{10^3\ lb_m}{h}$$

$$\frac{10^3\ lb_m}{h} \times 0.126 = \frac{kg}{s}$$

MASS VELOCITY: $\frac{kg}{s\ m^2}$; $\frac{10^3\ lb_m}{h\ ft^2}$

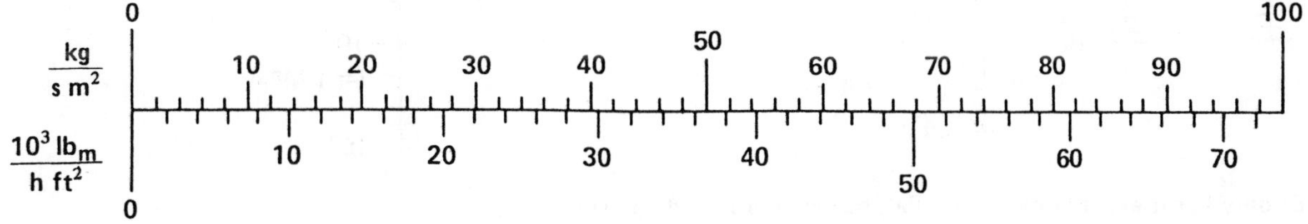

$$\frac{kg}{s\ m^2} \times 0.737 = \frac{10^3\ lb_m}{h\ ft^2}$$

$$\frac{10^3\ lb_m}{h\ ft^2} \times 1.356 = \frac{kg}{s\ m^2}$$

POWER OR ENERGY FLOW RATE: kilowatt, kW $= \frac{kJ}{s}$; $\frac{10^6\ Btu}{h}$

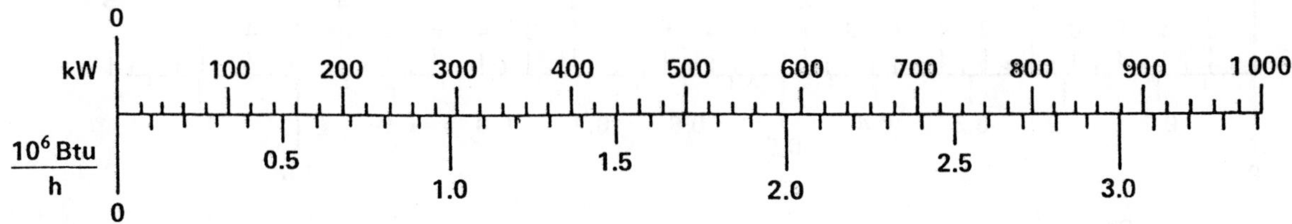

$$kW \times 3.412(10^{-3}) = \frac{10^6\ Btu}{h}$$

$$\frac{10^6\ Btu}{h} \times 293 = kW$$

Table 5 Unit conversion charts (*Continued*)

PRESSURE CONVERSION CHART
(top line = 1 Pa on all charts)

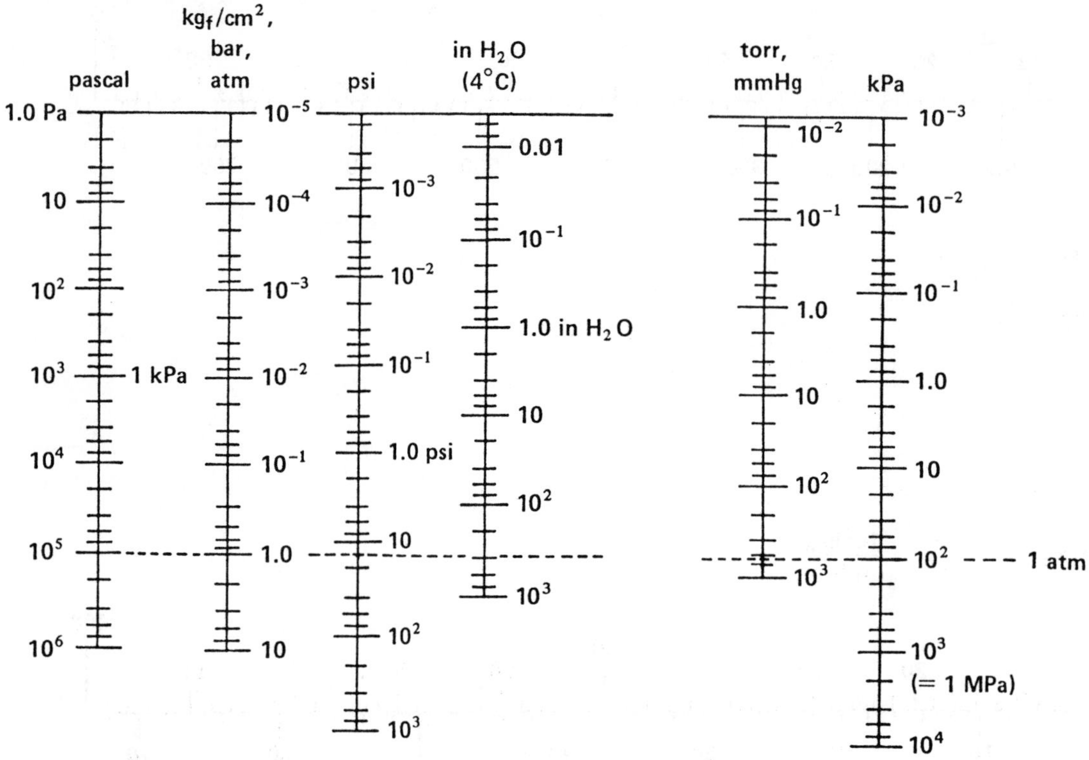

*NOTE: only bar is exact (1 bar = 10^5 Pa); however, atm and kg_f/cm^2 are within 2% accuracy (see list below).

1 atm = 1.01325 X 10^5 Pa
1 psi = 6.89476 X 10^3 Pa
1 in H_2O = 2.49082 X 10^2 Pa
1 kg_f/cm^2 = 9.80665 X 10^4 Pa
1 bar = 10^5 Pa
1 mmHg = 1.33322 X 10^2 Pa

1 Pa = 9.86923 X 10^{-6} atm
1 Pa = 1.45038 X 10^{-4} psi
1 Pa = 4.01474 X 10^{-3} in H_2O
1 Pa = 1.01972 X 10^{-5} kg_f/cm^2
1 Pa = 10^{-5} bar
1 Pa = 7.50064 X 10^{-3} mmHg

SPECIFIC HEAT CAPACITY: $\frac{kJ}{kg\ K}$; $\frac{Btu}{lb_m\ ^\circ F}$

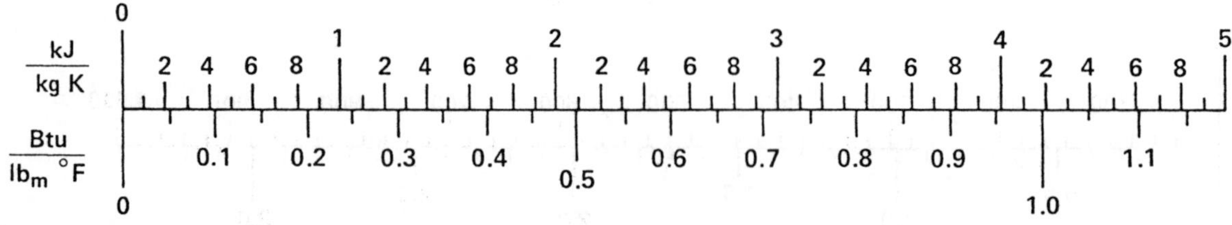

$$\frac{kJ}{kg\ K} \times 0.239 = \frac{Btu}{lb_m\ ^\circ F}$$

$$\frac{Btu}{lb_m\ ^\circ F} \times 4.187 = \frac{kJ}{kg\ K}$$

Table 5 Unit conversion charts ***(Continued)***

TEMPERATURE COMPARISON
(NOT TO SCALE)

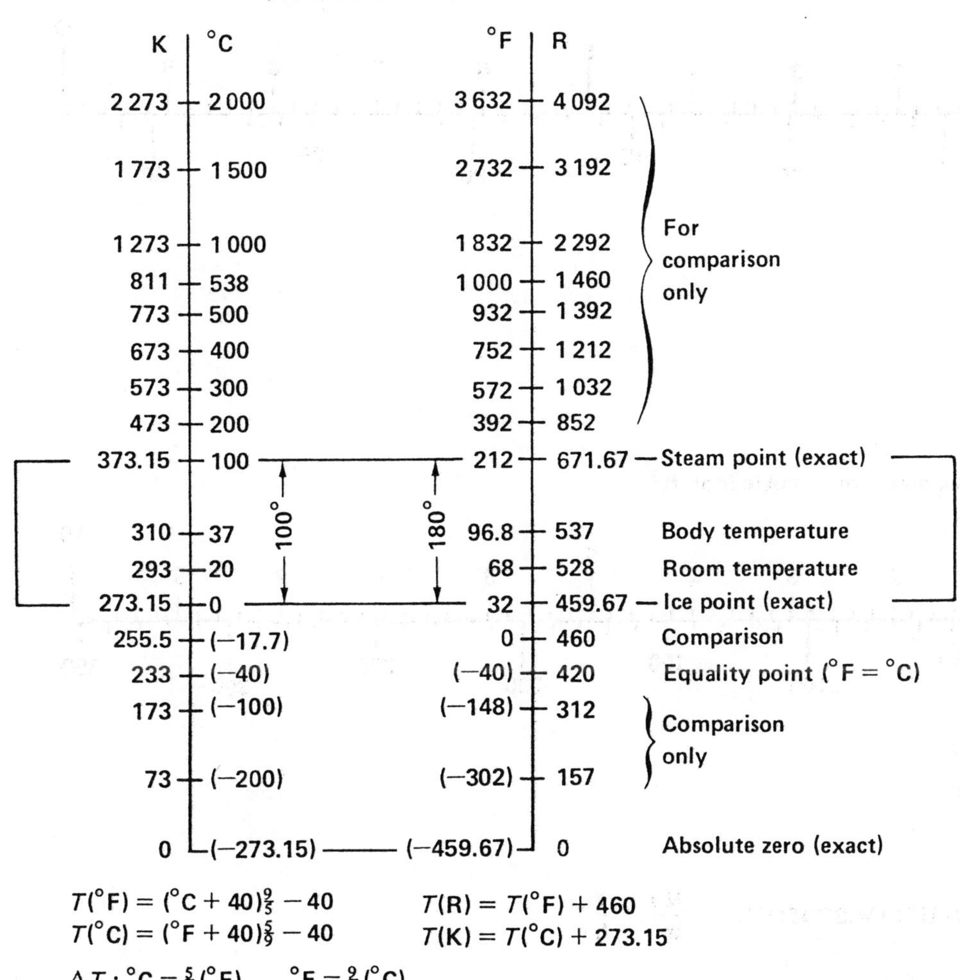

THERMAL CONDUCTIVITY: $\frac{W}{m\,K}$; $\frac{Btu}{h\,ft\,°F}$

$\frac{W}{m\,K}$ scale — liquids: 0, 2, 4, 6, 8, 0.1, 2, 4, 6, 8, 0.2, 2, 4, 6, 8, 0.3, 2, 4, 6, 8, 0.4, 2, 4, 6, 8, 0.5; gases: 0.001, 0.002, 0.003, 0.004, 0.005

$\frac{Btu}{h\,ft\,°F}$ scale — liquids: 0, 2, 4, 6, 8, 0.1, 2, 4, 6, 8, 0.2, 2, 4, 6, 8; gases: 0.001, 0.002

liquids

gases

$$\frac{W}{m\,K} \times 0.578 = \frac{Btu}{h\,ft\,°F}$$

$$\frac{Btu}{h\,ft\,°F} \times 1.731 = \frac{W}{m\,K}$$

Table 5 Unit conversion charts (*Continued*)

VELOCITY: $\frac{m}{s}$; $\frac{ft}{s}$

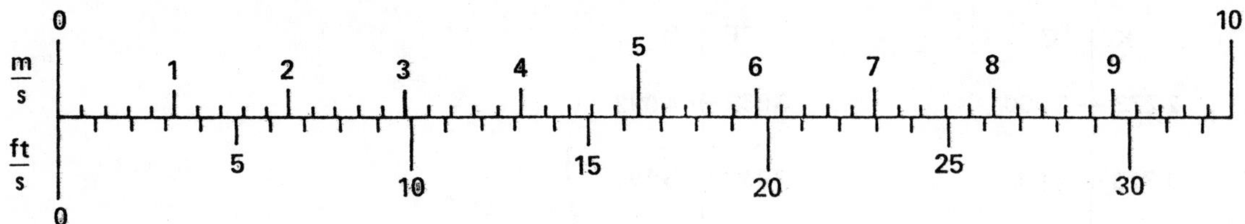

$\frac{m}{s} \times 3.28 = \frac{ft}{s}$

$\frac{ft}{s} \times 0.305 = \frac{m}{s}$

VOLUME: cubic meter, m^3; cubic foot, ft^3

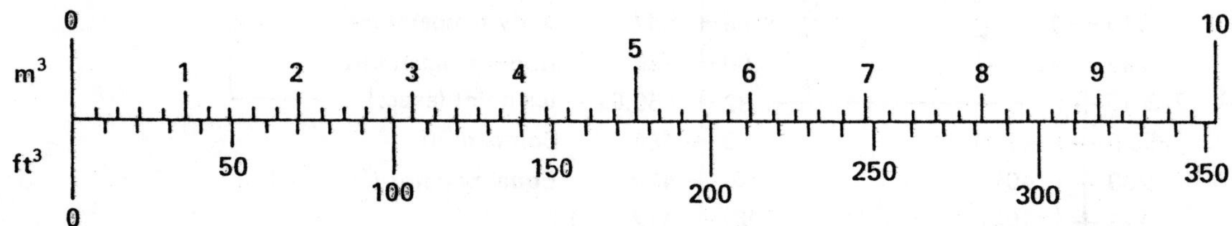

$m^3 \times 35.314 = ft^3$
$ft^3 \times 0.083 = m^3$

DYNAMIC (ABSOLUTE) VISCOSITY: $\frac{N\,s}{m^2}$; $\frac{lb_m}{h\,ft}$

centipoise $cP = 10^{-3}\,\frac{N\,s}{m^2} = mPa\,s$

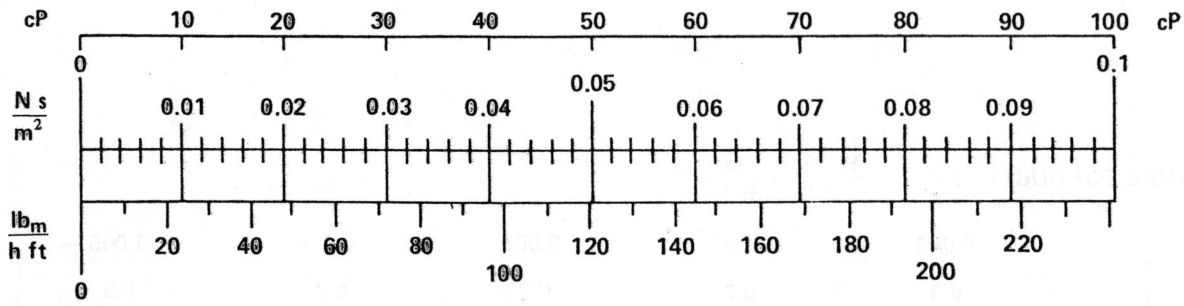

$cP \times 2.42 = \frac{lb_m}{h\,ft}$

$\frac{lb_m}{h\,ft} \times 0.413 = cP$

CHAPTER 1
Properties of Pure Fluids

Physical property data for various fluids are listed in Chapter 2, but it sometimes occurs that for a certain fluid no data are available or the range of given data is not sufficient. In order to cover these cases many methods for estimating properties of gases and liquids have been developed. Some of them are simple and easy to employ. However, due to the real nature of the phenomena, which is generally complicated, their results are not too reliable. More elaborate methods can give better results, but they are often restricted to limited groups of compounds (mostly organic). Some of these methods can be evaluated economically only with the aid of a computer. Very good collections of estimating methods have been published, especially Reid, et al. (1) or Hecht, et al. (2).

In this chapter an attempt is made to give some of the simpler rules and methods which may be sufficient in many cases. At least, they may help to decide if more elaborate methods or the services of a data center are required where not only the available data but also the most modern estimating methods are computerized. This step is recommended in all cases where the accuracy of the data is essential.

1.1. Critical Data

A. Correspondence principle

Sometimes, the critical data of a fluid are of interest in themselves. This is especially the case where processes are carried out in the neighborhood of the critical state which employ liquefaction or vaporization or where the existence of two phases is required or prohibited.

On the other hand, knowledge of the critical data may often be essential since by using critical temperature, critical pressure, and critical volume, correspondences in the thermodynamic behavior of fluids can be found. This leads to the widely employed principle of correspondence which uses "reduced" state parameters, i.e., dimensionless parameters of state obtained by dividing the real parameters by the corresponding critical parameters:

$$T_r = \frac{T}{T_c} \qquad p_r = \frac{p}{p_c} \qquad \tilde{V}_r = \frac{\tilde{V}}{\tilde{V}_c}$$

Ideally, it might be possible to link these reduced parameters together by a universal equation of state, valid for all fluids.

Numerous investigations, however, have shown that this is only possible for relatively small groups of similar fluids. Common equations of state even for "normal" fluids can be established only by using additional parameters. These, in many cases, are correlated again to the critical data. Riedel [3a] defines the critical parameter α_c:

$$\alpha_c = \left(\frac{d \ln p_{rs}}{d \ln T_{rs}}\right)_c \qquad (1)^\dagger$$

where p_{rs} and T_{rs} represent pairs of reduced pressure and reduced temperature values according to the saturation curve. The parameter is used in correlations with its value at the critical point, α_c, only.

An equivalent of the critical parameter α_c is the "acentric factor" ω defined by Pitzer [4]:

$$\omega = -(\log p_{rs(0.7)} + 1) \qquad (2)$$

where $p_{rs(0.7)}$ is the reduced pressure on the saturation curve at a reduced temperature of $T_{rs} = 0.7$. Correspondence relations using α_c or ω are sometimes recommended in the following paragraphs.

B. Critical temperature

Guldberg's rule can give a rough estimate of critical temperature:

$$T_c \approx 1.5\ T_b \qquad \text{or} \qquad \frac{T_b}{T_c} \approx \frac{2}{3} \qquad (3)$$

where T_b is the boiling temperature at normal pressure.

†Within a three-digit section the equation, figure, and table numbers do not have the three-digit identifier; for cross-references between sections, the appropriate three-digit number will be given.

Table 1 Atomic and structural constants used in Lydersen's method for calculating critical data of organic compounds[a,b]

Group	Δ_T	Δ_p	Δ_V
Basic value	–	–	0.040
$-CH_3$ and $-CH_2-$	0.020	0.225	0.055
$-CH_2-$ member of ring	0.013	0.183	0.044 5
$-CH$	0.012	0.209	0.051
$-CH$ member of ring	0.012	0.191	0.046
$=CH$ and $=CH_2$	0.018	0.197	0.045
$=CH$ member of ring	0.011	0.153	0.037
C	0.000	0.209	0.041
C member of ring	(–0.007)	(0.153)	(0.031)
$=C$ and $=C=$	0.000	0.197	0.036
Same, member of ring	0.011	0.153	0.036
$\equiv C-$ and $\equiv CH$	0.005	0.152	(0.036)
–F	0.018	0.222	0.018
–Cl	0.017	0.318	0.049
–Br	0.010	0.496	(0.070)
–I	(0.012)	(0.824)	(0.095)
–O–	0.021	0.159	0.020
–O– member of ring	(0.014)	(0.119)	(0.008)
–OH alcohols	0.082	0.060	0.018
–OH phenols	(0.035)	(–0.020)	(0.003)
CO	0.040	0.288	0.060
CO member of ring	(0.033)	(0.199)	(0.050)
–CHO	0.048	0.328	0.073
–COO–	0.047	0.467	0.080
–COOH	0.085	0.397	0.080
$-NH_2$	0.031	0.097	0.028
NH	0.031	0.137	(0.037)
NH member of ring	(0.024)	(0.089)	(0.027)
N	0.014	0.169	0.042
N– member of ring	(0.007)	(0.129)	(0.032)
–CN	(0.060)	(0.357)	(0.080)
–SH and –S–	0.015	0.268	0.055
–S– member of ring	(0.008)	(0.238)	(0.045)
=S	0.003	(0.238)	0.047
=O	(0.020)	(0.119)	(0.011)
$-NO_2$	(0.055)	(0.417)	(0.078)

[a] According to Lydersen [5].
[b] Uncertain values in parentheses.

(Real values of T_b/T_c vary from 0.55 to 0.72 with 0.364 for mercury.)

More reliable values for organic compounds are obtained by Lydersen's incremental method [5]:

$$\frac{T_b}{T_c} = 0.567 + \Sigma\,\Delta_T + (\Sigma\,\Delta_T)^2 \tag{4}$$

where $\Sigma\,\Delta_T$ is determined from atomic and structural constants as listed in Table 1. (The deviation is ±2% on average with a maximum of ±7%.)

C. Critical pressure

No simple rule exists for calculating critical pressure other than extrapolating the saturation curve (see Sec. 1.3B), if the critical temperature is known. Lydersen's incremental method [5] gives

$$p_c = \frac{\tilde{M}}{(\Sigma\,\Delta_p + 0.33)^2}\ \text{bar} \tag{5}$$

where 1 bar = 10^5 N/m^2, and again $\Sigma\,\Delta_p$ is composed of atomic and structural constants as listed in Table 1. (The deviation is ±2% on average with a maximum of ±20%.)

D. Critical volume

According to the rule of Cailletet and Mathias [6] [see Sec. 1.2A and Fig. 1.2(1)], the mean value of liquid and vapor density in the saturated state plotted against

Table 2 Numerical values of the functions $\phi(T_r)$ and $\psi(T_r)$ for the calculation of the critical parameter α_c according to Riedel [3a]

T_r	ϕ	ψ
0.35	6.003	1.321
0.36	5.719	1.248
0.37	5.452	1.179
0.38	5.200	1.115
0.39	4.963	1.055
0.40	4.739	0.999
0.41	4.527	0.946
0.42	4.326	0.896
0.43	4.135	0.850
0.44	3.955	0.805
0.45	3.783	0.764
0.46	3.619	0.724
0.47	3.463	0.687
0.48	3.315	0.652
0.49	3.174	0.619
0.50	3.039	0.588
0.51	2.909	0.558
0.52	2.786	0.530
0.53	2.668	0.503
0.54	2.555	0.477
0.55	2.447	0.453
0.56	2.343	0.430
0.57	2.243	0.409
0.58	2.147	0.388
0.59	2.055	0.368
0.60	1.967	0.349
0.61	1.882	0.331
0.62	1.800	0.314
0.63	1.721	0.298
0.64	1.645	0.283
0.65	1.572	0.268
0.66	1.501	0.254
0.67	1.433	0.240
0.68	1.367	0.227
0.69	1.303	0.215
0.70	1.242	0.203
0.71	1.182	0.192
0.72	1.125	0.181
0.73	1.069	0.171
0.74	1.015	0.161
0.75	0.962	0.152
0.76	0.911	0.143
0.77	0.862	0.134
0.78	0.814	0.126
0.79	0.767	0.118
0.80	0.722	0.110
0.81	0.677	0.103
0.82	0.635	0.096
0.83	0.593	0.089
0.84	0.552	0.082
0.85	0.512	0.076
0.86	0.473	0.070
0.87	0.435	0.064
0.88	0.398	0.058
0.89	0.361	0.053
0.90	0.325	0.047
0.91	0.290	0.042
0.92	0.256	0.037
0.93	0.222	0.032
0.94	0.189	0.027
0.95	0.157	0.023
0.96	0.124	0.018
0.97	0.093	0.013
0.98	0.061	0.009
0.99	0.031	0.004
1.00	0.000	0.000

temperature is nearly a straight line that includes the critical point. Thus, if some sets of saturated liquid and vapor densities are known together with the critical temperature, the critical specific volume can be estimated.

For organic fluids, Lydersen's incremental method [5] gives

$$\tilde{V}_c = \Sigma\, \Delta_V + 0.040 \quad \text{m}^3/\text{kmol or dm}^3/\text{mol} \tag{6}$$

where $\Sigma\, \Delta_V$ is composed of atomic and structural constants as listed in Table 1. (The deviation is ±3% on average with a maximum of ±10%.)

E. Critical parameter α_c and acentric factor ω

If α_c cannot be evaluated according to its original definition (see Sec. A), Riedel [3a] gives the equation

$$\alpha_c = \frac{\log (1/p_{rs}) - \phi(T_{rs})}{\psi(T_{rs})} + 7 \tag{7}$$

where p_{rs} and T_{rs} are the reduced vapor pressure and the reduced equilibrium temperature of one known pair of values of the saturation curve, and the functions

$$\phi(T_{rs}) = 0.118\,3\left(\frac{36}{T_{rs}} - 35 - T_{rs}^6\right) + 4.44 \log T_{rs} \tag{8}$$

and

$$\psi(T_{rs}) = 0.036\,4\left(\frac{36}{T_{rs}} - 35 - T_{rs}^6\right) + 2.52 \log T_{rs} \tag{9}$$

Values of both functions are tabulated in Table 2.

The evaluation of the acentric factor ω in Pitzer's definition [4] (see Sec. A) requires knowledge of the vapor equilibrium pressure at a reduced temperature of $T_{rs} = 0.7$. For cases where this value is not available, Riedel [3f] links ω to α_c by the equation

$$\omega = \frac{\alpha_c - 5.811}{4.919} \tag{10}$$

Edmister [7] gives the equation

$$\omega = 0.429 \frac{T_b/T_c}{1 - T_b/T_c} (\log p_c - 0.991\,5) \tag{11}$$

where T_b/T_c as well as p_c can be estimated, if necessary, by the methods described in Secs. B and C. The critical pressure p_c has to be given in bars.

1.2. Specific Volume, p-$\tilde{V}$-T Correlations

A. Specific volume of the liquid phase

According to the empirical rule of Cailletet and Mathias [6] the specific volumes $\tilde{V}_{ls}$ of the saturated liquid and $\tilde{V}_{vs}$ of the saturated vapor at the same temperature follow the relationship

$$\frac{\tilde{M}}{2}\left(\frac{1}{\tilde{V}_{ls}}+\frac{1}{\tilde{V}_{vs}}\right)=0.5(\rho_{ls}+\rho_{vs})$$
$$=a+bT+cT^2+\cdots \quad (1)$$

where the factors c, d, etc. are so small that for estimating purposes they may be neglected even in the neighborhood of the critical state. Thus, the rule states that in a plot of density versus temperature the mean values of saturated vapor and saturated liquid densities are represented by a straight line with a slope of b (see Fig. 1). If the specific volume of a liquid at low temperature is known together with its critical volume and the specific volume of the saturated vapor, then this rule gives the specific volume of the saturated liquid throughout the whole temperature range.

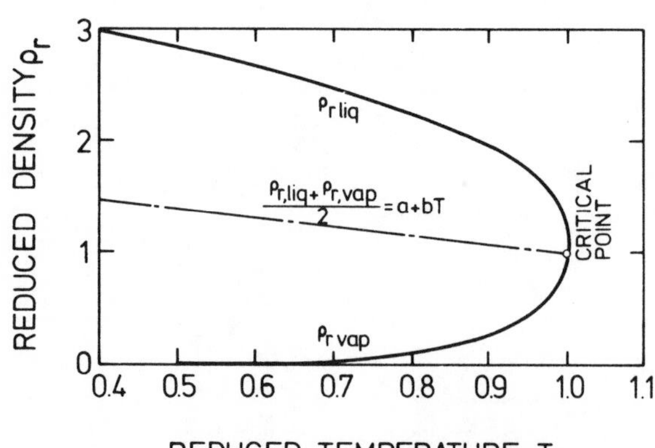

Figure 1 Rule of Cailletet and Mathias: Example, propane.

According to Watson [8], the specific volume of a liquid can be related to the reduced pressure p_r and the reduced temperature T_r by the *expansion factor* ω^* using the equation

$$V\omega^*=C$$

where ω^* is a function of p_r and T_r shown in Fig. 2 and C is a constant that can be determined if a reliable value of the specific volume of the liquid is known together with the critical pressure and temperature. This method also applies to temperatures below or pressures above the saturated state.

For normal fluids in the saturated state Riedel [3b] gives the equation

$$\tilde{V}_{lsr}=\{1+0.85(1-T_r)+[1.93+0.2(\alpha_c-7)(1-T_r)^3]\}^{-1} \quad (2)$$

As T_r approaches 0, this equation approaches

$$\frac{\tilde{V}_0}{\tilde{V}_c}=\frac{1}{3.78+0.2(\alpha_c-7)} \quad (3)$$

where $\tilde{V}_0$ is the (merely fictitious) specific volume at $T_r=0$. Timmermans [9] stated that for most liquids $\tilde{V}_0/\tilde{V}_c=Z_c$, where Z_c is the critical compressibility factor

$$Z_c=\frac{p_c\tilde{V}_c}{\tilde{R}T_c} \quad (4)$$

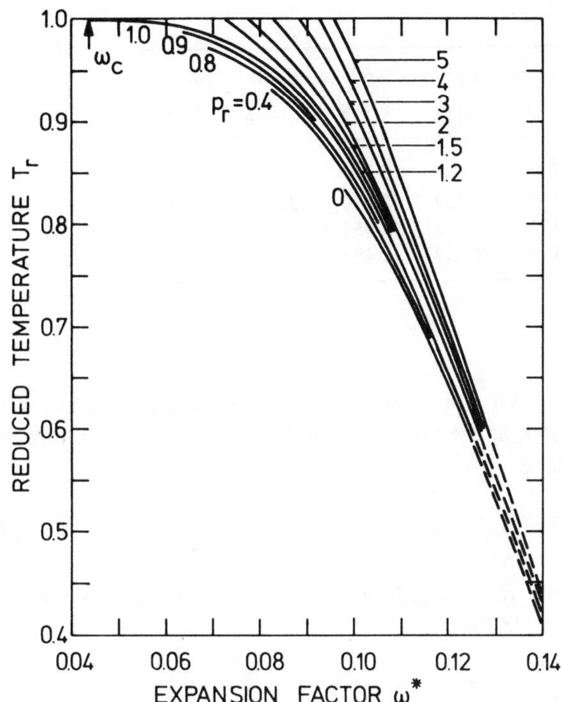

Figure 2 Expansion factor ω^* according to Watson [8].

with values commonly in the range of $Z_c = 0.265 \pm 0.015$. Riedel [3b] has tabulated the numerical values of $\tilde{V}_{lsr}/Z_c$ as a function of T_r. (See Table 1 which is reciprocal to Riedel's original table.) It should be noted that the influence of α_c is very small below $T_r = 0.8$.

Rackett [10] showed that the simpler equation

$$\tilde{V}_{lsr} = Z_c^{(1-T_r)^{0.285}} \tag{5}$$

is also a very good approximation. The critical compressibility factor Z_c in this equation can be estimated according to Riedel (see above) or according to Yamada and Gunn [11] using Pitzer's acentric factor ω:

$$Z_c = 0.291 - 0.087\,8\,\omega \tag{6}$$

B. Coefficient of volumetric thermal expansion of liquids

In the tables of Sec. 5.5, the densities of numerous liquids are given for various temperatures. From these or other known values, the coefficient of volumetric thermal expansion $\beta = [\partial\tilde{V}/(\tilde{V}\,\partial T)]_p$ can be calculated or graphically evaluated. If no values of density or specific volume are available, β can be estimated using the expansion factor ω^* (see Fig. 2).

From $\tilde{V} = C(\omega^*)^{-1}$ follows

$$\left(\frac{\partial\tilde{V}}{\tilde{V}\,\partial\omega^*}\right)_p = -\frac{1}{\omega^*} \tag{7}$$

and

$$\beta = \left(\frac{\partial\tilde{V}}{\tilde{V}\,\partial T}\right)_p = \left(\frac{\partial\tilde{V}}{\tilde{V}\,\partial\omega^*}\,\frac{\partial\omega^*}{\partial T}\right)_p = -\frac{1}{\omega^*}\left(\frac{\partial\omega^*}{\partial T}\right)_p \tag{8}$$

which can be graphically evaluated from Fig. 2.

Differentiation of the Rackett equation [9] (see Sec. A) gives

$$\beta = -\frac{0.285\ \ln Z_c}{T_c(1 - T_r)^{0.715}} \quad \mathrm{K}^{-1} \tag{9}$$

or with $Z_c \approx 0.265 \pm 0.015$

$$\beta = +\frac{0.380 \pm 0.015}{T_c(1 - T_r)^{0.715}} \quad \mathrm{K}^{-1} \tag{10}$$

Here, however, the fact is neglected that Rackett's equation is valid for the saturated liquid, i.e., for a pressure varying with temperature. This limits the validity of the above relationship to a temperature range not much above the normal boiling point.

C. Specific volume of gases

At low reduced pressures and high reduced temperatures, the ideal gas law applies quite correctly:

$$p\tilde{V} = \tilde{R}T \tag{11}$$

(If p is in N/m², $\tilde{V}$ in m³/kmol, and T in K, then $\tilde{R}$, the universal gas constant, has a value of 8 314.3 J/kmol.) At higher pressures or near the condensing temperature, the law of ideal gases no longer applies. However, the behavior of real gases and vapors can be related to that of ideal gases by introducing a factor of compressibility,

$$Z = \frac{p\tilde{V}}{\tilde{R}T} \tag{12}$$

The dimensionless factor Z is dependent upon temperature, pressure, and thermodynamical properties of the gas or vapor. It can be given as a function of reduced state parameters, thus permitting a fairly good prediction of the $p - \tilde{V} - T$ behavior of any gas.

In Fig. 3, Z is plotted in accordance with Nelson and Obert [12] as well as Riedel [3f] for $\alpha_c = 6$ in the range $0.8 < T_r < 15$. The deviations for different values of α_c are fairly small. Riedel [3f] gives more extensive tabulations of $\partial Z/\partial\alpha_c$.

The dotted lines in Fig. 3 apply to an "ideal reduced specific volume," $\tilde{V}_{r\,\mathrm{id}}$, which can be used without knowledge of the critical volume $\tilde{V}_c$ in the calculation of corresponding pressure-temperature values for a given specific volume:

$$\tilde{V}_{r\,\mathrm{id}} = \tilde{V}\,\frac{p_c}{\tilde{R}T_c} = \frac{Z\tilde{R}Tp_c}{p\tilde{R}T_c} = Z\,\frac{T_r}{p_r} \tag{13}$$

Table 1 Numerical values[a] of $\widetilde{V}_{lsr}/Z_c$

T_r	0.3	0.35	0.4	0.45	0.5	0.55	0.6	0.65	0.7
$\widetilde{V}_{lsr}/Z_c$	1.142	1.172	1.205	1.239	1.279	1.321	1.368	1.422	1.484
T_r	0.75	0.8	0.85	0.9	0.95	0.97	0.98	0.99	1.0
$\alpha_c = 6$	1.557	1.642	1.751	1.898	2.132	2.288	2.410	2.591	3.584
7	1.558	1.645	1.755	1.908	2.155	2.326	2.451	2.653	3.774
8	1.559	1.647	1.759	1.919	2.179	2.358	2.494	2.710	3.984

[a] Abbreviated from Riedel [3b].

D. Coefficient of volumetric thermal expansion of gases

For ideal gases, the volumetric thermal expansion is described by

$$\beta = \left(\frac{\partial \widetilde{V}}{\widetilde{V}\,\partial T}\right)_p = \frac{1}{T} \quad \mathrm{K}^{-1} \tag{14}$$

In any case where $Z \neq 1$ or $Z \neq$ const (e.g., at higher pressures), this equation is changed to

$$\beta = \left(\frac{\partial Z}{T\,\partial T}\right)_p \tag{15}$$

where $\partial Z/\partial T$ at constant (reduced) pressure can be graphically evaluated from Fig. 3.

Of course, it is also possible to calculate two values of $\widetilde{V}$ at the same pressure but different temperatures and take

$$\beta = \left(\frac{\Delta \widetilde{V}}{\widetilde{V}\,\Delta T}\right)_p \tag{16}$$

but this procedure also requires careful evaluation of $(\Delta Z/\Delta T)_p$.

E. Equations of state

Analytical formulations of the relationships among p, $\widetilde{V}$, and T are called equations of state. Many of them have been developed, the first one by van der Waals [13] in 1873:

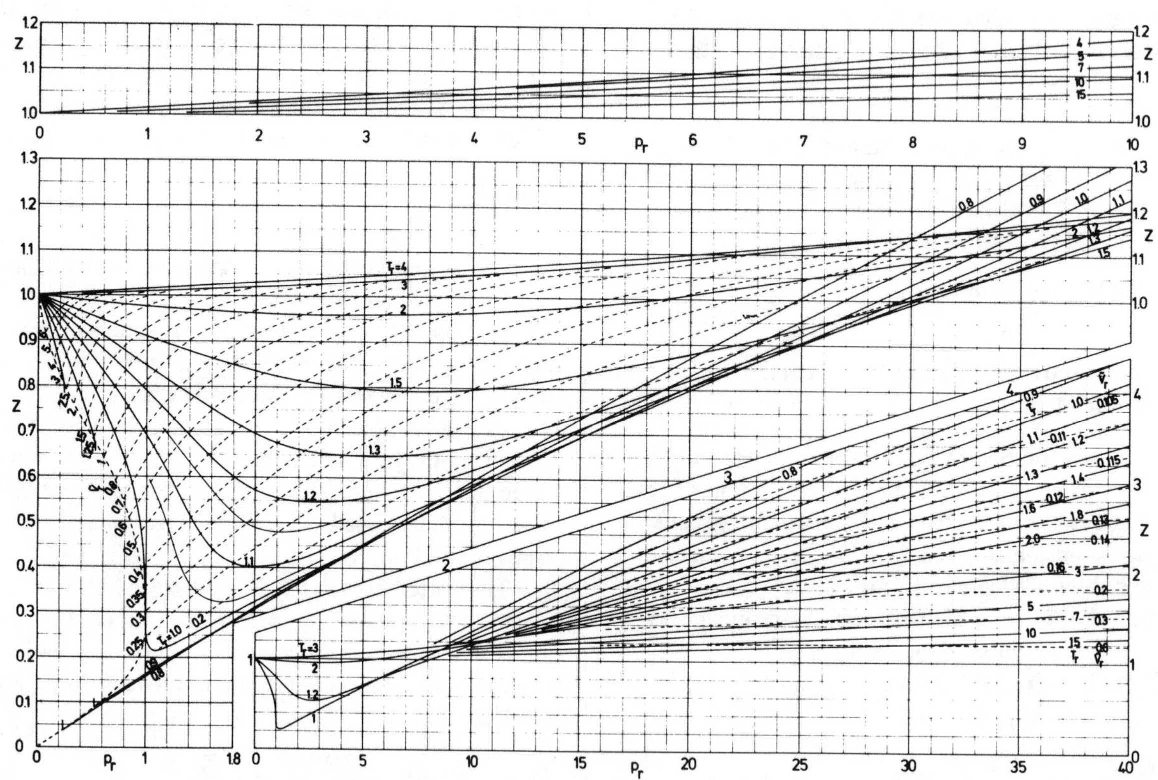

Figure 3 Compressibility factor $Z = p\widetilde{V}/\widetilde{R}T$ as a function of p_r and T_r for $\alpha_c = 6$ [3f, 12].

$$\left(p + \frac{a}{\tilde{V}^2}\right)(\tilde{V} - b) = \tilde{R}T \tag{17}$$

where a and b are specific constants approaching 0 for ideal gas conditions. This is one condition that should be fulfilled by any equation of state, but all of them should also satisfy the conditions at the critical point:

$$\left(\frac{dp}{d\tilde{V}}\right)_{T_c} = 0 \tag{18}$$

$$\left(\frac{d^2p}{d\tilde{V}^2}\right)_{T_c} = 0 \tag{19}$$

With this requirement, the values of a and b in the van der Waals equation become

$$a = \left(\frac{3}{4}\right)^3 \frac{(\tilde{R}T_c)^2}{p_c} \qquad b = \left(\frac{1}{2}\right)^3 \frac{\tilde{R}T_c}{p_c} \tag{20}$$

The best compromise between simplicity and accuracy seems to be provided by the Redlich-Kwong equation [14], which is also a two-constant equation:

$$\left[p + \frac{a}{T^{0.5}\tilde{V}(\tilde{V} + b)}\right](\tilde{V} - b) = \tilde{R}T \tag{21}$$

or

$$p = \frac{\tilde{R}T}{\tilde{V} - b} - \frac{aT^{-0.5}}{\tilde{V}(\tilde{V} + b)}$$

with the constants a and b derived by similar considerations as for the van der Waals equation:

$$a = 0.427\,5\,\frac{\tilde{R}^2 T_c^{2.5}}{p_c} \tag{22}$$

$$b = 0.086\,64\,\frac{\tilde{R}T_c}{p_c} \tag{23}$$

Soave [15] improved the accuracy of the Redlich-Kwong equation, replacing the term $aT^{-0.5}$ by a function of the reduced temperature and Pitzer's acentric factor ω—the whole equation then reading

$$p = \frac{\tilde{R}T}{\tilde{V} - b} - a_c[1 + (0.480 + 1.574\,\omega - 0.176\,\omega^2) \cdot (1 - T_r^{0.5})]^2 / \tilde{V}(\tilde{V} + b) \tag{24}$$

where

$$a_c = (aT^{-0.5})_c = 0.427\,5\,\frac{\tilde{R}^2 T_c^2}{p_c} \tag{25}$$

and

$$b = 0.086\,64\,\frac{\tilde{R}T_c}{p_c} \tag{26}$$

as above.

A further modification of the Redlich-Kwong equation, suggested by Peng and Robinson [16] is reported to be even more accurate for hydrocarbons and hydrates and for liquid densities than the Redlich-Kwong-Soave version:

$$p = \frac{\tilde{R}T}{\tilde{V} - b_1} - a_{c1}[1 + (0.375 + 1.542\,\omega - 0.207\,\omega^2) \cdot (1 - T_r^{0.5})]^2 / (\tilde{V} + b_1)^2 - 2\,b_1^2 \tag{27}$$

with a_{c1} and b_1 slightly different from a_c and b as a consequence of the different structure of the equation:

$$a_{c1} = 0.457\,2\,\frac{\tilde{R}^2 T_c^2}{p_c} \tag{28}$$

$$b_1 = 0.077\,8\,\frac{\tilde{R}T_c}{p_c} \tag{29}$$

For many compounds the virial coefficients for the virial equation of state,

$$Z = \frac{p\tilde{V}}{\tilde{R}T} = 1 + B(T) + C(T)\tilde{\rho}^2 + D(T)\tilde{\rho}^3 + \cdots \tag{30}$$

are given in the literature. B, the second virial coefficient, C, the third virial coefficient, etc., are functions of temperature only for pure fluids.

Other, more elaborate equations of state, e.g., Benedict-Webb-Rubin, or Lee-Kesler, are beyond the scope of this section. See Ref. [1].

1.3. Thermodynamic Properties

A. Normal boiling point

The normal boiling point of a pure fluid (at normal pressure, 760 mm Hg = 1. 013 3 bar) is rarely unknown since the experimental determination is very easy. The estimating methods presented below will therefore be more or less a last refuge.

For a large number of organic compounds the normal boiling point can be determined according to Ogata and Tsuchida [17] from structural constants listed in Table 1. With the constants p, y, and q taken from this table, the normal boiling temperature is

$$T_b = py + q \quad \text{K} \tag{1}$$

All other existing rules for estimating the normal boiling temperature of a fluid solely from the knowledge of its chemical composition are reported to be less accurate than simple considerations of analogy with similar compounds whose boiling points are known.

If some points of the vapor pressure-temperature curve are available, the normal boiling point can be found by extrapolating or interpolating. For this purpose it is either possible to draw a vapor pressure curve in a diagram like a Cox chart (see Sec. B) where it is nearly a straight line, or to use the Antoine equation. However, it has to be kept in mind that many liquid compounds will decompose at temperatures well below their normal boiling point.

B. Vapor pressure–temperature curves

A plot of log p versus T^{-1} for any compound gives a fairly straight curve that can be extrapolated or interpolated easily. Even if only two points of the whole function are known, e.g., the boiling point and the critical point, a line through these two points in the log p versus T^{-1} diagram is a good estimation for the relationship between saturation pressure and saturation temperature. It can be improved if it is drawn similar to a known curve or if more than two pairs of parameters are available for its determination.

A further improvement was first suggested by Cox [18]: A plot of log p over a function of T is graphically generated by drawing a straight line with a positive or negative slope into a coordinate system. The ordinate is then marked up in log p and the abscissa in T according to the known equilibrium values of a reference fluid, e.g., water. In such a diagram, since named Cox chart, the vapor pressure-temperature curves of other fluids are found to plot as almost straight lines too.

Hoffmann and Florin [19] gave an analytical function for the abscissa of a Cox chart, which is linear in τ^{-1} instead of T^{-1}:

$$\tau^{-1} = T^{-1} - (7.915\,1 - 2.672\,6 \log T) \times 10^{-3} - 0.862\,5\, T \times 10^{-6} \tag{2}$$

A tabulation of corresponding values of T^{-1} and τ^{-1} is given in Table 2. This tabulation is especially useful when values of a Cox chart are to be interpolated. Two Cox charts, for several fluids, prepared according to Hoffmann and Florin are presented in Figs. 1 and 2. Cox charts like these show an interesting and sometimes useful peculiarity: the vapor pressure-temperature curves of homologous series of compounds often (nearly) meet in one point. In many cases this "infinite

Table 1 Structural constants *p, q,* and *y* for the estimation of normal boiling points of organic liquids[a]

RX	p	q	(Ex)[b]
RH	1.615	63.8	Me,[c] *t*-Bu
RCl	1.348	197.7	
RBr	1.260	213.6	
RI	1.198	253.4	
ROH	0.896	277.6	Me, *t*-Bu
MeOR	1.217	191.2	Me
EtOR	1.137	221.8	
ROR	2.158	143.2	Me, Hep
PhOR	0.894	377.4	
$RONO_2$	1.016	280.5	
RSH	1.191	221.0	
RSMe	1.146	249.2	Me
RSEt	1.080	280.0	
RSR	1.937	214.4	Me, Hep
RNH_2	1.194	201.4	
RNHMe	1.180	215.2	
RNHEt	1.081	247.9	
RNHPr	0.991	282.8	
RNMe	1.193	218.7	Me
RNO_2	0.923	308.8	
HCOR	1.140	233.8	
MeCOR	1.022	270.6	
EtCOR	0.918	302.2	
RCN	0.960	292.2	
RCOCl	1.040	267.9	
HCOOR	1.073	244.6	
MeCOOR	1.000	273.2	(Standard series)
EtCOOR	0.963	297.5	
PhCOOR	0.766	425.9	
RCOOH	0.903	342.4	
RCOOMe	1.000	273.2	(Standard series)
RCOOEt	0.963	297.5	
RCOOPr	0.911	323.4	
RCOOPh	0.766	425.9	
$(RCO)_2O$	1.286	337.7	Hep
$ClCH_2COOR$	0.721	359.6	
$Cl_2CHCOOR$	0.745	372.3	
$BrCH_2COOR$	0.745	374.4	
$NCCH_2COOR$	0.565	433.5	
$CH_2{=}CHCOOR$	0.918	302.2	

R	*y*	*R*	*y*
Me	55.5	*t*-Am	122.0
Et	77.1	Neopent	125.0
n-Pr	102.0	*n*-Hex	171.0
iso-Pr	92.0	iso-Hex	168.0
n-Bu	124.0	*n*-Hep	191.5
sec-Bu	113.0	*n*-Oct	210.0
iso-Bu	116.5	Vinyl	71.0
t-Bu	96.0	Allyl	104.0
n-Am	149.0	2-Butenyl	127.0
iso-Am	140.5	Phenyl	197.0

[a] According to Ogata and Tsuchida [17].
[b] (Ex) stands for configurations for which deviations of more than ±5% are to be expected.
[c] Me = methyl; Ph = phenyl, etc.

point" is close to $\tau^{-1} = 0$ or $T = 1\,400$ K, and $p = 2\,250$ bars or $\log p = 3.352$.

Antoine [20] proposed a simple equation for the vapor pressure-temperature equilibrium which also results in a distortion of the T^{-1} scale in a $\log p$ versus $f(T)$ correlation:

$$\log p = A - \frac{B}{T - C} \tag{3}$$

The constants A, B, and C for Antoine's equation are often given in the literature (see also Sec. 2.2). Antoine suggested that $C = 13$ K, but better results are claimed for a rule given by Thomson [21]:

$$C = -0.3 + 0.034\, T_b \quad \text{K} \tag{4}$$

for fluids with $T_b < 125$ K and for monatomic elements, while

$$C = -18 + 0.19\, T_b \quad \text{K} \tag{5}$$

for all other fluids. Once a value for C is accepted, it is easy to calculate A and B from two known pairs of vapor pressure-temperature values, p_1, T_1 and p_2, T_2:

$$B = \frac{\log p_2 - \log p_1}{(T_1 - C)^{-1} - (T_2 - C)^{-1}} \tag{6}$$

$$A = \log p_1 + \frac{B}{T_1 - C} = \log p_2 + \frac{B}{T_2 - C} \tag{7}$$

The Antoine equation is not dimensionless. If tabulated values of A, B, and C are to be used, one has to make sure if $\ln p$ or $\log p$ must be taken, and in which of the possible units p and even T (K or R) will be given.

C. Heat of vaporization

According to Trouton's rule [22], the increase of the molar entropy during vaporization at the normal boiling point is very similar for many fluids, being close to

$$\frac{\Delta \tilde{H}_v}{T_b} = 88 \text{ kJ/kmol K} \tag{8}$$

where $\Delta \tilde{H}_v$ is the molar heat of vaporization in J/mol or kJ/kmol.

Trouton's rule is a good approach for nonassociating fluids with normal boiling temperatures over 170 K. For other fluids, the deviations are sometimes considerable; instead of 88 kJ/kmol, the real values are, for instance, hydrogen, 46 kJ/kmol K; nitrogen, 71 kJ/kmol K; and water, 108 kJ/kmol K.

Better results are achieved by employing a "Trouton parameter" which is a function of the reduced pressure and the reduced temperature at the boiling point, p_{rb} and T_{rb}. Thus Trouton's rule changes to

Table 2 Numerical values of T^{-1} and τ^{-1} as functions of temperature for the distortion of the T^{-1} scale[a]

ϑ °C	T K	$1/T$	$1/\tau$	ϑ °C	T K	$1/T$	$1/\tau$
−270	3.15	0.317 460	0.310 874	+ 60	333.15	0.003 002	0.001 541
−265	8.15	0.122 699	0.117 213	+ 70	343.15	0.002 914	0.001 479
−260	13.15	0.076 046	0.071 110	+ 80	353.15	0.002 832	0.001 422
−255	18.15	0.055 096	0.050 530	+ 90	363.15	0.002 754	0.001 368
−250	23.15	0.043 197	0.038 908	+100	373.15	0.002 680	0.001 317
−245	28.15	0.035 524	0.031 459	+110	383.15	0.002 610	0.001 269
−240	33.15	0.030 166	0.026 286	+120	393.15	0.002 544	0.001 224
−235	38.15	0.026 212	0.022 491	+130	403.15	0.002 481	0.001 181
−230	43.15	0.023 175	0.019 592	+140	413.15	0.002 420	0.001 141
−225	48.15	0.020 768	0.017 309	+150	423.15	0.002 363	0.001 103
−220	53.15	0.018 815	0.015 465	+160	433.15	0.002 309	0.001 067
−215	58.15	0.017 197	0.013 948	+170	443.15	0.002 257	0.001 033
−210	63.15	0.015 835	0.012 677	+180	453.15	0.002 207	0.001 000
−205	68.15	0.014 674	0.011 600	+190	463.15	0.002 159	0.000 969
−200	73.15	0.013 671	0.010 675	+200	473.15	0.002 114	0.000 940
−195	78.15	0.012 796	0.009 872	+210	483.15	0.002 070	0.000 912
−190	83.15	0.012 027	0.009 171	+220	493.15	0.002 028	0.000 885
−185	88.15	0.011 344	0.008 552	+230	503.15	0.001 988	0.000 859
−180	93.15	0.010 735	0.008 003	+240	513.15	0.001 949	0.000 834
−175	98.15	0.010 189	0.007 512	+250	523.15	0.001 912	0.000 811
−170	103.15	0.009 695	0.007 072	+260	533.15	0.001 876	0.000 789
−165	108.15	0.009 246	0.006 674	+270	543.15	0.001 841	0.000 767
−160	113.15	0.008 838	0.006 314	+280	553.15	0.001 808	0.000 746
−155	118.15	0.008 464	0.006 003	+290	563.15	0.001 776	0.000 726
−150	123.15	0.008 120	0.005 659	+300	573.15	0.001 745	0.000 707
−140	133.15	0.007 510	0.005 158	+310	583.15	0.001 715	0.000 689
−130	143.15	0.006 986	0.004 709	+320	593.15	0.001 686	0.000 671
−120	153.15	0.006 530	0.004 322	+330	603.15	0.001 658	0.000 654
−110	163.15	0.006 129	0.003 987	+340	613.15	0.001 631	0.000 647
−100	173.15	0.005 775	0.003 693	+350	623.15	0.001 605	0.000 621
− 90	183.15	0.005 460	0.003 435	+360	633.15	0.001 579	0.000 606
− 80	193.15	0.005 177	0.003 205	+370	643.15	0.001 555	0.000 591
− 70	203.15	0.004 923	0.003 000	+380	653.15	0.001 531	0.000 577
− 60	213.15	0.004 692	0.002 816	+390	663.15	0.001 508	0.000 562
− 50	223.15	0.004 481	0.002 651	+400	673.15	0.001 486	0.000 548
− 40	233.15	0.004 289	0.002 501	+420	693.15	0.001 443	0.000 522
− 30	243.15	0.004 113	0.002 364	+440	713.15	0.001 402	0.000 497
− 20	253.15	0.003 950	0.002 240	+460	733.15	0.001 364	0.000 474
− 10	263.15	0.003 800	0.002 126	+480	753.15	0.001 328	0.000 452
0	273.15	0.003 661	0.002 022	+500	773.15	0.001 293	0.000 431
+ 10	283.15	0.003 532	0.001 926	+520	793.15	0.001 261	0.000 411
+ 20	293.15	0.003 411	0.001 837	+540	813.15	0.001 230	0.000 391
+ 30	303.15	0.003 299	0.001 755	+560	833.15	0.001 200	0.000 373
+ 40	313.15	0.003 193	0.001 678	+580	853.15	0.001 172	0.000 355
+ 50	323.15	0.003 095	0.001 607	+600	873.15	0.001 145	0.000 338

[a] According to Hoffmann and Florin [19].

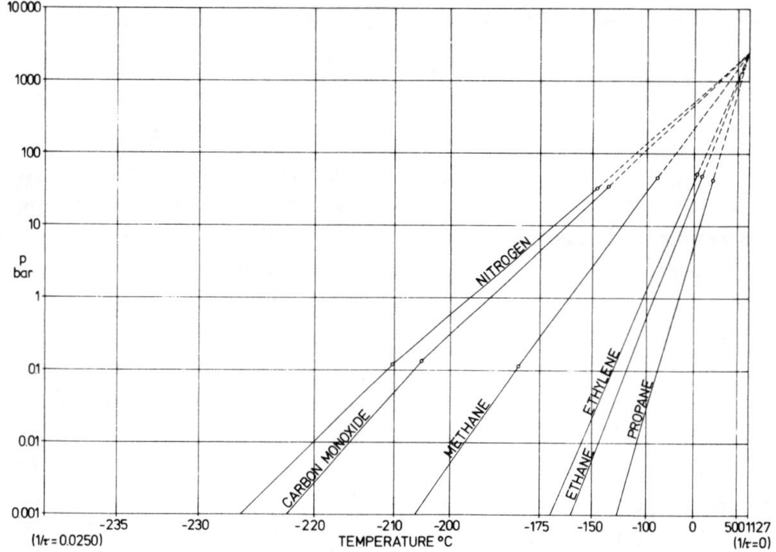

Figure 1 Cox chart [18] prepared according to Hoffmann and Florin [19] for the low-temperature range.

$$\frac{\Delta\tilde{H}_v}{T_b} = \frac{\tilde{R}(-\ln p_{rb})}{1 - T_{rb}} \tag{9}$$

An exact calculation of the heat of vaporization of pure (and mixed) fluids at any temperature is possible by evaluating the Clausius-Clapeyron equation:

$$\Delta\tilde{H}_v = T(\tilde{V}_{vap} - \tilde{V}_{liq})\left(\frac{dp}{dT}\right) \tag{10}$$

where dp/dT is the slope of the vapor pressure-temperature curve at the temperature of vaporization. At low pressures, $\tilde{V}_{liq}$ (the molar volume of the liquid phase) can be neglected in comparison to $\tilde{V}_{vap}$ (the molar volume of the saturated vapor). If, in addition, $\tilde{V}_{vap}$ is taken as the specific volume of the ideal gas (i.e., $Z = 1$), $\tilde{V}_{vap} = \tilde{R}T/p$, the equation is simplified to

$$\Delta\tilde{H}_v = T\frac{\tilde{R}T}{p}\frac{dp}{dT} = \tilde{R}\frac{T^2}{p}\frac{dp}{dT} = \tilde{R}\frac{d\ln p}{d(1/T)} \tag{11}$$

which is proportional to the slope of the vapor pressure-temperature curve in the log p versus $1/T$ diagram.

If, in such a plot, the slope of the saturation curve at the boiling point is estimated by drawing a straight line from the boiling point to the critical point, this last expression converts into the Trouton parameter given above.

Since the conditions $Z = 1$ and $\tilde{V}_{vap} \gg \tilde{V}_{liq}$ are valid only in the low-pressure range, other methods have been

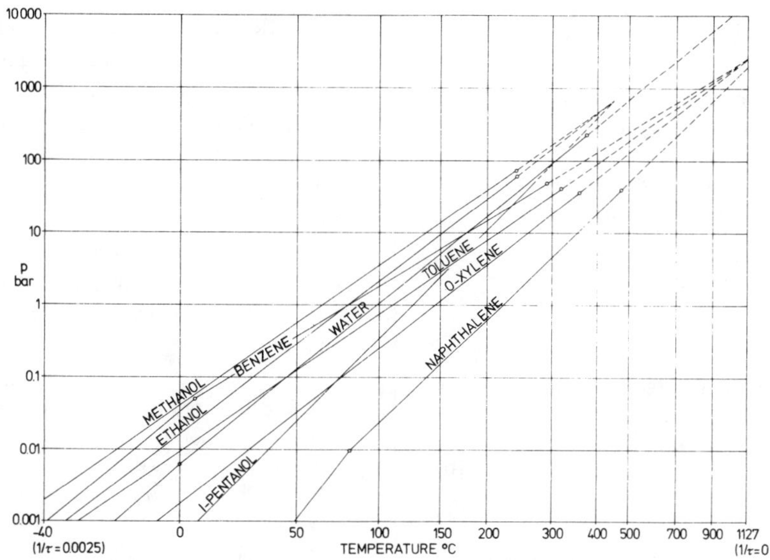

Figure 2 Cox chart [18] prepared according to Hoffmann and Florin [19] for the high-temperature range.

proposed for estimating the heat of vaporization at any temperature. If one value of $\Delta\tilde{H}_v$ is known (e.g., that at the normal boiling point), good results are obtained with the correlation

$$\frac{\Delta\tilde{H}_{v1}}{\Delta\tilde{H}_{v2}} = \left(\frac{1 - T_{r1}}{1 - T_{r2}}\right)^n \tag{12}$$

where $n = 0.38$ for nonassociating liquids, according to a recommendation given by Watson [8].

D. Specific heat capacity

(a) Ideal gases

The specific heat capacity (specific heat) of ideal gases applies to real gases in the range of low (reduced) pressures and high (reduced) temperatures. Roughly this range is limited by $T_r > 2$ for $p_r < 0.4$ and $T_r > 4$ for $p_r < 4$ [see Sec. D(b)]. The order of magnitude for the specific heat at constant pressure, $\tilde{C}_p$, is about 30 kJ/kmol K, increasing with increasing temperature.

According to the kinetic theory of gases, the internal energy of an (ideal) gas is identical with the kinetic energy of its molecules. The possibility of internal vibrations of the atoms within the molecules is neglected as well as the energy of interaction between molecules as a function of their separation distance.

In monatomic gases, the kinetic energy consists merely of translational energy while in polyatomic gases it is evenly distributed between translational energy and energy of rotation. Each of the possible vectors will pick up the same amount of energy: $\tilde{R}/2 = 4\ 157$ J/kmol K.

In ideal gases, the difference between the specific heat capacity at constant pressure, $\tilde{C}_p$, and at constant volume, $\tilde{C}_v$, amounts to

$$\tilde{C}_p - \tilde{C}_v = p\,\frac{d\tilde{V}}{dT} = \tilde{R} = 8\ 314.3 \text{ J/kmol K} \tag{13}$$

In consequence, the relationships for ideal gases, according to the kinetic theory, are as follows:

1. Monatomic (three degrees of freedom for translational movement in the three directions of space):

$$\tilde{C}_p = \tfrac{5}{2}\tilde{R} = 20.8 \text{ kJ/kmol K} \tag{14}$$

$$\tilde{C}_v = \tfrac{3}{2}\tilde{R} = 12.5 \text{ kJ/kmol K} \tag{15}$$

2. Diatomic (two additional degrees of freedom for vectors of rotation perpendicular to the connecting line between the two atoms):

$$\tilde{C}_p = \tfrac{7}{2}\tilde{R} = 29.1 \text{ kJ/kmol K} \tag{16}$$

$$\tilde{C}_v = \tfrac{5}{2}\tilde{R} = 20.8 \text{ kJ/kmol K} \tag{17}$$

3. Triatomic (one additional degree of freedom for a vector of rotation in the third possible direction):

$$\tilde{C}_p = \tfrac{8}{2}\tilde{R} = 33.3 \text{ kJ/kmol K} \tag{18}$$

$$\tilde{C}_v = \tfrac{6}{2}\tilde{R} = 24.9 \text{ kJ/kmol K} \tag{19}$$

The real specific heat capacities of monatomic gases sufficiently above the saturation temperature indeed have the values predicted by the kinetic theory of gases. The diatomic and polyatomic gases, however, have higher specific heat capacities because of the elastic vibrations of the molecules neglected by the theory. Such vibrations can be stimulated only by collisions that transmit the minimum quantum of energy, $h\nu$, where h is the Planck quantum ($h = 6.625\ 3 \times 10^{-34}$ J s) and ν is the frequency of the molecule as an elastic vibrator in s^{-1}. With increasing temperature the number of collisions fulfilling this requirement will also increase, thus increasing the proportion of vibrational energy within the total energy of a polyatomic (but still ideal) gas.

In real gases, the forces interacting between the molecules let the difference $\tilde{C}_p - \tilde{C}_v$ exceed $\tilde{R}$ if the distance between the molecules becomes small, i.e., at low temperatures or high pressures.

The specific heat capacity of a polyatomic ideal gas can be calculated from the vibration patterns of its molecules as determined by spectroscopic methods. However, an exact prediction of these patterns from the molecular structure alone is only feasible for very simple molecules. The vibration frequency of a single atom is strongly influenced by the kind of bonds connecting it with its immediate neighbors, but only to a lesser degree by the rest of the structure of the molecule. Thus, typical atomic groups give typical contributions to the specific heat of a polyatomic gas, no matter what other groups are present in the molecule. Benson [23] has developed a very accurate group contribution method also described in [1].

A different method requiring less tabulation has been reported by Sakiadis and Coates [24] and Gambill [25]. Here, a certain frequency is ascribed to the bending vibration and another frequency to the longitudinal vibration of many types of bonds. The real influence of any atom in the molecule on the vibration frequencies of all its atoms is thereby neglected. Nevertheless, the mean deviation is better than ±5% within a temperature range from 200 K to over 1 000 K (or the temperature of decomposition of nearly all organic compounds).

The molar specific heat of an ideal polyatomic gas, $\tilde{C}_{p\,\text{id}}$, is calculated by the equation

$$\frac{\tilde{C}_{p\,\text{id}}}{\tilde{R}} = 4 + \frac{n_r}{2} + \Sigma\, n_i\, \Delta_{\nu i} + \frac{3z - 6 - n_r - n}{n}\, \Sigma\, n_i\, \Delta_{\delta i} \tag{20}$$

where $\Delta_{\nu i} = f(\omega_{\nu i}/T)$ is the contribution of the longitudinal vibration of a certain type of bond
$\Delta_{\delta i} = f(\omega_{\delta i}/T)$ is the contribution of the bending vibration of a certain type of bond
$\omega_{\nu i}$ is the frequency, longitudinal
$\omega_{\delta i}$ is the frequency, bending
n_i = number of bonds of type i
n_r = number of bonds allowing a rotation of one part of a molecule relative to another part, i.e., C–C, O–O bonds, and C–O bonds in ethers and esters
n = total number of bonds in the molecule
z = number of atoms in the molecule

The vibration frequencies ω_i ascribed to various types of organic bonds are listed in Table 3, where the indices ν and δ again denote longitudinal and bending vibrations. $\Delta_{\nu i}$ and $\Delta_{\delta i}$ are functions of $\omega_{\nu i}/T$ and $\omega_{\delta i}/T$, respectively. They are plotted in Fig. 3.

The method is not applicable for strongly associating vapors since they deviate too much from the ideal gas conditions.

Table 3 Bond frequencies and bond increments for estimating specific heat capacities of gases, vapors, and liquids[a]

Type of bond	ω_ν	ω_δ	ΔG	ΔL
C–H (aliphatic)	2 914	1 274	5.10	4.16
C–C (aliphatic)	989	390	–1.10	1.07
C=C (aliphatic symmetric)	1 618	599	5.68	6.36
C=C (aliphatic asymmetric)	1 664	421	5.68	6.36
C≡C (aliphatic)	2 215	333		
C–H (aromatic)	3 045	1 318	5.10	4.16
C–C (aromatic)	989	390	–1.10	1.07
C=C (aromatic)	1 618	844	5.68	6.36
C–I	500	260		
C–Br	560	280	15.54	15.33
C–Cl	650	330	12.91	12.55
C–F	1 050	530		
C–S	650	330		
C=S	1 050	530		
S–S	500	260		
S–H	2 570	1 050		
C–N	990	390	0.40	0.24
C=N	1 620	845		
N–N	990	390		
N–H	2 920	1 320	5.57	5.00
N–O	1 030	205		
N=O	1 700	390	8.17	8.28
C–O	1 030	205	2.05	2.78
C=O	1 700	390	9.93	9.08
O–H	3 420	1 150	4.64	5.07
O–O	850	(300)		
Cycle			4.80	–0.43

[a]See Refs. [25, 26].

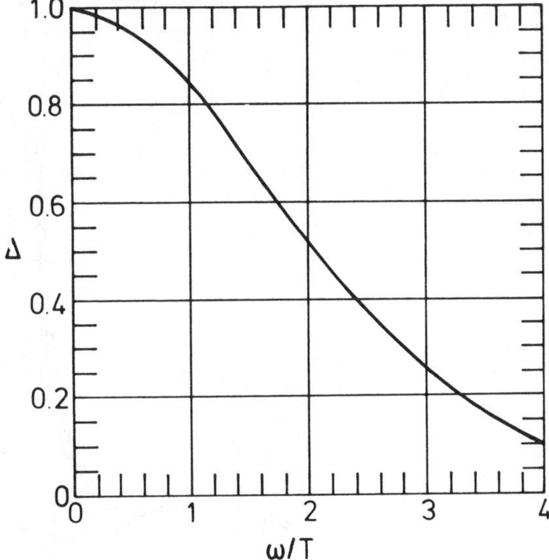

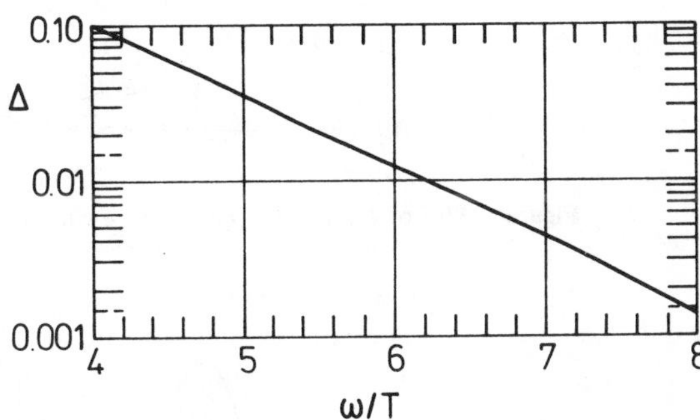

Figure 3 Plot of $\Delta = f(\omega/T)$ for the evaluation of Eqs. (20) and (23). Redrawn from [24].

(b) General fluids

For a real fluid in the whole range of reduced pressures and temperatures, the molar specific heat at constant pressure can be calculated by the equation

$$\tilde{C}_p = \tilde{C}_{p\,\text{id}} + \Delta\tilde{C}_p \tag{21}$$

The residual specific heat $\Delta\tilde{C}_p$ can be given as a function of the reduced pressure p_r, the reduced temperature T_r, and Pitzer's acentric factor ω:

$$\Delta\tilde{C}_p = \tilde{R}(\Delta_0 + \omega\,\Delta_1) \tag{22}$$

where Δ_0 and Δ_1 are functions of p_r and T_r given in Figs. 4 and 5 [4, 43].

The method is valid in the whole range of p_r and T_r but gives no reliable values in the surroundings of the critical state. Below the critical state, the curves are split up into a liquid range and a vapor range.

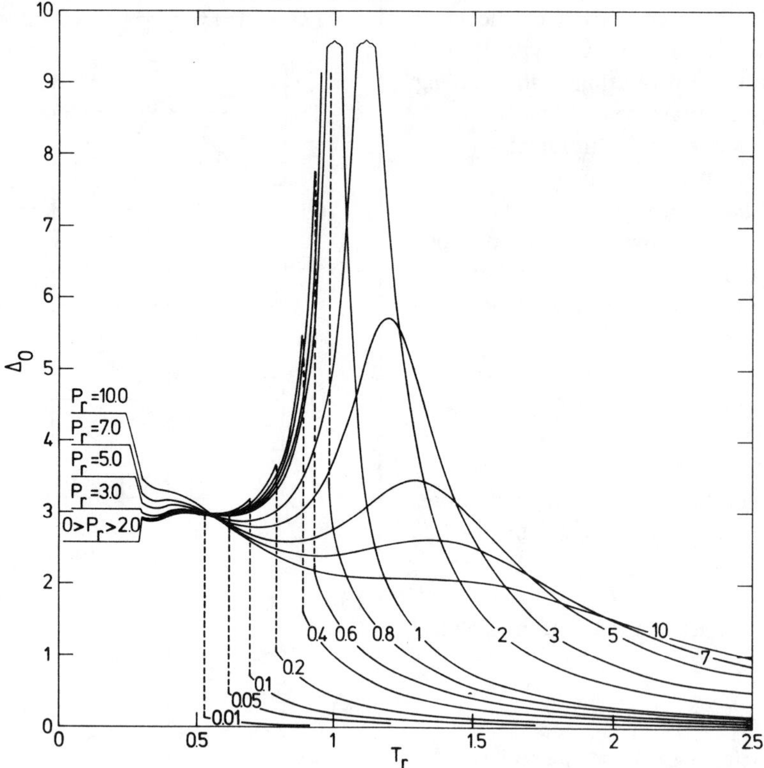

Figure 4 Plot of $\Delta_0 = f(T_r, p_r)$ for the evaluation of Eq. (22). Drawn according to Lee and Kesler [43].

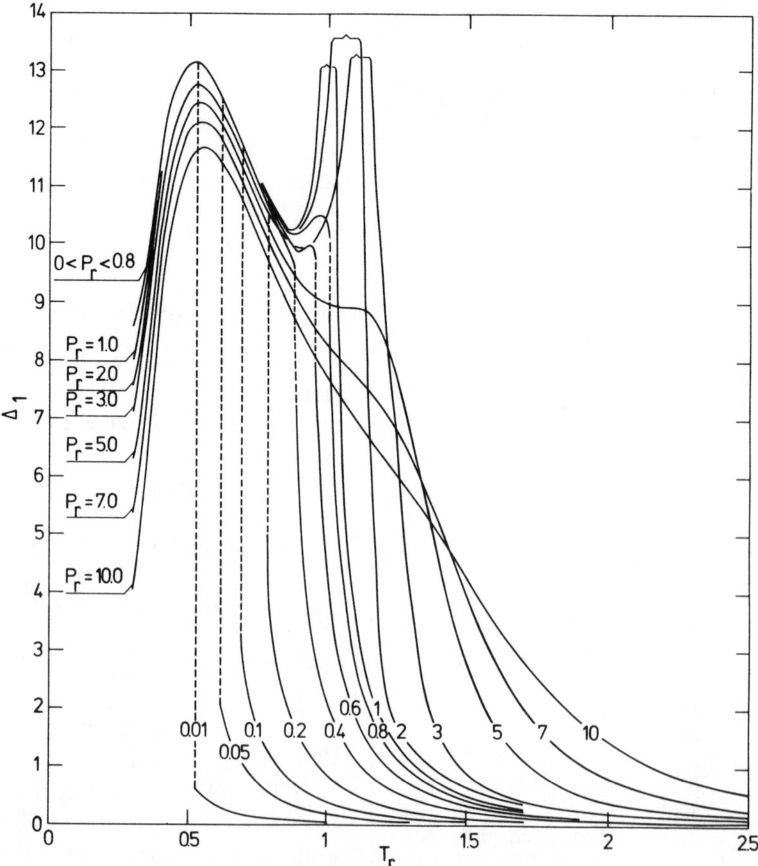

Figure 5 Plot of $\Delta_1 = f(T_r, p_r)$ for the evaluation of Eq. (22). Drawn according to Lee and Kesler [43].

(c) Liquids

The specific heat capacities of liquids below their normal boiling point are usually in the range of

$$1.6 \text{ kJ/kg K} < c_p = \left(\frac{\tilde{C}_p}{\tilde{M}}\right) < 2.1 \text{ kJ/kg K}$$

with exceptions to higher values (water, ammonia up to 4.6 kJ/kg K) and lower values (halogen compounds, down to 0.42 kJ/kg K). In the low-pressure range, the specific heat capacities of liquids rise with temperature. For the estimation a method may be used, according to Sakiadis and Coates [24], which is very similar to the one given in Sec. D(a). The corresponding equation is

$$\frac{\tilde{C}_{p\,\mathrm{id}}}{\tilde{R}\kappa} = 6 + n_r + \Sigma\, n_i\, \Delta_{vi} + F\,\frac{3z - 6 - n_r - n}{n}\,\Sigma\, n_i\, \Delta_{\delta i} \tag{23}$$

with the same meaning of the symbols as in Sec. D(a), and also employing Table 3 and Fig. 3. Of the additional factors F and κ, F is a correction factor to be taken from Fig. 6, and

$$\kappa = \left(\frac{\Sigma\, \Delta G}{\Sigma\, \Delta L}\right)^7$$

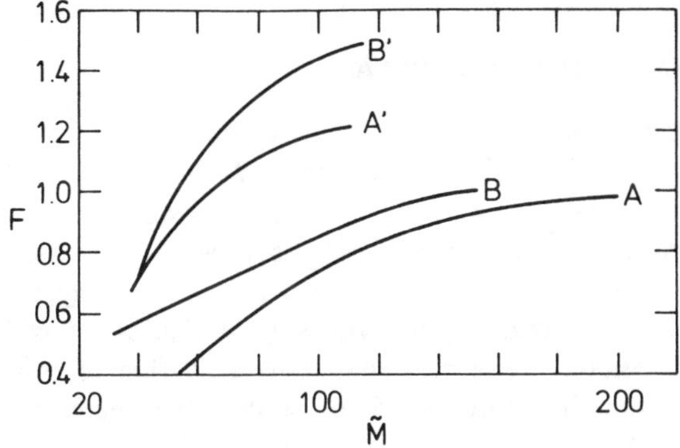

Figure 6 Correction factor F as a function of the molecular mass $\tilde{M}$ to be used in Eq. (23). Redrawn from [24].

Table 4 Values of the constants C and n for the estimation of specific heat capacities of liquids[a]

Row	C	n
Alcohols	3 560	−0.10
Acids (organic)	3 800	−0.152
Ketones	2 460	−0.013 5
Esters	2 510	−0.057 3
Amines (aliphatic)	6 820	−0.22
Nitriles (aliphatic)	3 030	−0.073 3
Hydrocarbons (aliphatic)	3 660	−0.113
Hydrocarbons (aromatic)	1 370	+0.048 5
Monochloro- (aliphatic)	640	+0.229
Polychloro- (aliphatic)	19 800	−0.63

[a] According to Pachaiappan et al. [26].

where ΔG and ΔL are added up from the increments given in Table 3.

A relative rotation of one part of a molecule in liquids is only possible with C–C bonds; therefore n_r denotes the number of aliphatic C–C bonds in a molecule. If there are types of bonds in a molecule for which no values of ΔG and ΔL are listed in Table 3, then κ is estimated on the basis of the remaining bond increments. Deviations of ±10% are to be expected, especially for the first members of homologous rows or for temperatures below 220 K (−50°C).

Pachaiappan et al. [26] recommend an equation by which the specific heat capacities of members of the same homologous row can be calculated alone from their molecular masses:

$$C_p = C\tilde{M}^n \tag{24}$$

The constants C and n can be evaluated if the specific heat capacity for at least two members of the row are known. For a number of homologous rows Table 4 gives values of C and n. The deviations are mostly within a range of ±5%.

1.4. Transport Properties

A. Viscosity

The term viscosity in this chapter stands for dynamic (absolute) viscosity η in N s/m^2, identical to kg/m s, while the kinematic viscosity ν in m^2/s is simply obtained as the quotient of dynamic (absolute) viscosity and density.

(a) Viscosity of gases

At ambient temperature, the viscosity of gases is usually in the range of $10 \times 10^6 < \eta < 20 \times 10^6$ N s/m^2, increasing with temperature. In the region of low pressures (0.1-10 bars), the dynamic (absolute) viscosity of gases and vapors is nearly independent of pressure. At higher pressures, however, the behavior of real gases deviates more and more from that of an ideal gas, the viscosity increasing considerably with pressure, especially in the neighborhood of the critical pressure.

For a great number of simple gases the viscosity can be determined from Fig. 1 developed by Lucas [27]. The critical temperature and pressure as well as the molecular mass of the gas must be known. The mean deviations to be expected are less than ±6% except for gases with large quantum effects, such as helium and hydrogen, with long molecular chains, paraffins above heptane, or with strong dipolar momentum, e.g., water or alcohols. Figure 1 gives a generalized viscosity $\xi\eta$ as a function of reduced pressure and temperature where

$$\xi = \frac{T_c^{1/6}\tilde{R}^{1/6}\tilde{L}^{1/3}}{\tilde{M}^{1/2}p_c^{2/3}} = 3.8 \times 10^9 \, \frac{T_c^{1/6}}{\tilde{M}^{1/2}p_c^{2/3}} \tag{1}$$

The equation is not dimensionless; the values must be given in the following units: T_c in K, $\tilde{R}$ in J/kmol K, $\tilde{M}$ in kg/kmol, p_c in N/m^2, and $\tilde{L}$ (the Loschmidt number) = $6.025\,2 \times 10^{26}$ kmol^{-1}. The value of ξ is then obtained in m^2/N s. Values of ξ are given in Table 1 for a number of gases.

(b) Viscosity of liquids

The viscosities of liquids vary within a range from about 1×10^{-4} N s/m^2 to infinity. Water at 20°C has a viscosity of 1×10^{-3} N s/m^2, decreasing rapidly with rising temperature. While in gases the viscosity is caused by the exchange of impulses between "layers" of different velocity, the viscosity of liquids stems from intermolecular forces that resist a dilation of adjacent layers. Therefore, the viscosity of liquids decreases with

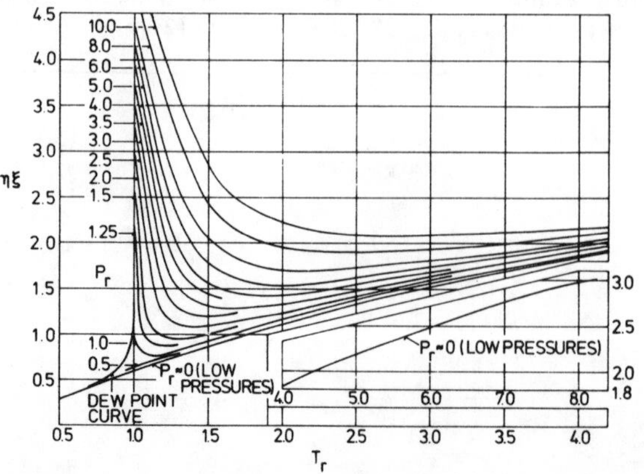

Figure 1 Plot of generalized viscosity $\eta\xi = f(T_r, p_r)$ to be used with Eq. (1) from Lucas [27].

Table 1 Numerical values of ξ for estimating gas viscosities using Fig. 1[a]

Compound	ξ, 10^6 m^2/N s	Compound	ξ, 10^6 m^2/N s
Noble gases		Ethanol	0.046 08
Helium	0.670 04	*i*-Propanol	0.046 76
Neon	0.081 61	*n*-Propanol	0.048 93
Argon	0.048 34	*n*-Butanol	0.047 11
Krypton	0.032 48	Ethers	
Xenon	0.026 20	Dimethylether	0.049 81
Elements		Diethylether	0.056 76
Hydrogen	0.403 06	Diphenylether	0.041 18
Deuterium	0.248 77	Ketones	
Nitrogen	0.071 20	Acetone	0.049 78
Oxygen	0.052 70	Esters	
(Air)	0.065 60	Methylacetate	0.044 49
Chlorine	0.031 61	Ethylformate	0.044 51
Paraffins		Ethylacetate	0.046 95
Methane	0.081 83	Ethylpropionate	0.048 08
Ethane	0.062 48	Propylacetate	0.048 22
Propane	0.058 39	*i*-Butylformate	0.043 60
n-Butane	0.056 14	*i*-Butylacetate	0.046 82
i-Butane	0.057 29	*n*-Butylacetate	0.047 77
n-Pentane	0.055 45	Organic halogen compounds	
i-Pentane	0.055 71	Methylchloride	0.041 21
Hexane	0.055 22	Methylbromide	0.026 19
2,2,3-Trimethylbutane	0.051 72	Dichlormethane	0.034 00
n-Heptane	0.055 39	Chloroform	0.031 93
n-Octane	0.055 67	R 22 (CHF_2Cl)	0.037 87
n-Nonane	0.056 18	R 21 ($CHFCl_2$)	0.037 87
Olefins, Diolefins		Carbon tetrafluoride	0.038 49
Ethylene	0.061 91	Carbon tetrachloride	0.031 96
Propylene	0.056 62	Ethylchloride	0.043 42
1-Butene	0.054 89	Organic sulfur compounds	
2-Butene	0.054 48	Thiophene	0.037 52
i-Butene	0.053 48	Anorganic compounds	
2-Pentene	0.049 54	Water	0.033 41
3-Methyl-1-butene	0.049 77	Hydrogen chloride	0.040 17
Hexene	0.054 37	Hydrogen bromide	0.026 96
1,3-Butadiene	0.053 36	Hydrogen iodide	0.022 47
1,5-Hexadiene	0.055 39	Hydrogen sulfide	0.040 35
Cyclic paraffins		Hydrogen cyanide	0.069 79
Cyclopropane	0.051 00	Sulfur dioxide	0.032 97
Cyclopentane	0.046 95	Ammonia	0.049 47
Cyclohexane	0.046 69	Nitrous oxide	0.039 72
Aromatic compounds		Nitric oxide	0.047 40
Benzene	0.042 68	Nitrosyl chloride	0.029 91
Toluene	0.043 97	Carbon monoxide	0.023 73
1,2,3-Trimethyl benzene	0.046 42	Carbon dioxide	0.039 17
Diphenylmethane	0.044 05	Carbonyl sulfide	0.039 41
Acetylenes		Carbon disulfide	0.031 44
Acetylene	0.057 13	Boron trifluoride	0.039 97
Methylacetylene	0.059 04	Phosphine	0.048 88
Alcohols		Silane	0.05957
Methanol	0.047 67		

[a]Both ξ and Fig. 1 are from Lucas [27].

rising temperature as the molecular movement tends to overcome the restraining forces.

The relationships are so complicated that there is no general theory. None of the many attempts have been successful in relating the viscosity of a liquid to any of its other properties. The method of Souders [28] described below employs atomic and structural constants that have been empirically determined. From these increments, listed in Table 2, the value of J in the following equation is added up:

$$\log(\log \eta + 4) = \frac{J}{\tilde{M}} \rho \times 10^{-3} - 2.9 \qquad (2)$$

Table 2 Atomic and structural increments for calculating *J* in Souders' equation for estimating liquid viscosities[a]

Atomic and group values					
C	50.2	OH	57.1	CH_2	55.6
H	2.7	Cl	60.0	COO	90.0
O	29.7	Br	79.0	COOH	104.4
N	37.0	I	110.0	NO_2	80.0

Structural values			
Double bond	−15.5	Ortho and para	+ 3.0
5-C ring	−24.0	Meta	+ 1.0
6-C ring	−21.0	$-CH{=}CHCH_2X$	+ 4.0
Side group on 6-C ring		$R_2{>}CHX$	+ 6.0
With $M < 17$	− 9.0	(X = negative group)	
With $M > 16$	−17.0		
$R_2{>}CHCH{<}R_2$	+ 8.0	$R_2{>}C{<}R_2$	+13.0
$H-C(=O)-R$	+10.0	$CH_3-C(=O)-R$	+ 5.0

[a] See Ref. [28].

The viscosity η is obtained in N s/m^2, if the molecular mass $\tilde{M}$ is given in kg/kmol and the density ρ is in kg/m^3. The method is adequate for a rough estimation of the viscosity at room temperature for liquids with freezing points below and boiling points above 20°C. It fails completely if sulfur is present in the compound. For other nonassociating liquids the deviation should be in the range of ±30%. For acids, the results tend to be too low; for multihalogenated hydrocarbons they are often too high.

For liquids below their normal boiling point, Thomas [29] suggested the equation

$$\log\left(8\,600\,\frac{\eta}{\rho^{1/2}}\right) = \Theta\left(\frac{1}{T_r} - 1\right) \tag{3}$$

The viscosity η is obtained in N s/m^2, if the density ρ is given in kg/m^3 and Θ is added up from increments taken from Table 3.

Table 3 Structural increments for calculating Θ in Thomas' equation for estimating liquid viscosities

Double bond	0.478
C_6H_5	0.385
CO in ketones and esters	0.105
CN in nitriles	0.381
S	0.043
C	−0.462
H	0.249
O	0.054
Cl	0.340
Br	0.326
I	0.335

The method is reported to be more accurate than Souders' method described above for aromatics above benzene, unsaturated compounds, and paraffins. It works also with compounds containing sulfur, but gives deviations up to ±80% for multihalogenated compounds, acids, alcohols, aldehydes, amines, naphthenes, and heterocyclic compounds. In judging the accuracy of estimating methods for liquid viscosities one should, however, keep in mind the vast range of possible values which lets a deviation of ±50% still be a good approach.

(c) Temperature dependency of the viscosity of liquids

As already mentioned, the viscosity of liquids decreases rapidly with rising temperature. It may serve as a rule of thumb that within some kelvins above the melting point the viscosity of many liquids is close to 2×10^{-3} N s/m^2, decreasing to about 0.3×10^{-3} N s/m^2 at the normal boiling temperature. Within this range of temperatures, the logarithm of the viscosity is an almost linear function of the reciprocal temperature. In approaching the freezing point, however, there is often a sharp rise in viscosity. On the contrary, above the temperature of the normal boiling point, the decrease of the viscosity is usually more rapid than suggested by the logarithmic function valid in the intermediate range.

Following theoretical considerations, Andrade [30] introduced the liquid density into the logarithmic function and thus obtained the equation

$$\eta = 10^{-6}\, A\rho^{1/3} e^{c\rho/T} \tag{4}$$

The constants A and c of Andrade's equation can be determined if two or more viscosity values, η_1, η_2, etc., are known at different temperatures, T_1, T_2, etc., together with the respective liquid densities, ρ_1, ρ_2, etc.:

$$c = \frac{\ln(\eta_1/\eta_2) + \frac{1}{3}\ln(\rho_2/\rho_1)}{\rho_1/T_1 - \rho_2/T_2} \tag{5}$$

$$A = 10^6\, \eta\rho^{-1/3} e^{-c\rho/T} \tag{6}$$

The equation is reported to be a very good approach for inorganic and organic liquids and even for molten metal. It is valid not only between the experimental values but also in extrapolation until close to the critical state. However, the uncertainty of all experimental values must be kept in mind; thus, A and c should not be determined from measured viscosity values too close together. If the liquid is highly associating, the values of c are extraordinarily high at lower temperatures but normalize as the temperature rises—that is, c is no longer a constant for such liquids.

Numerical values of A and c in Andrade's equation

are given in Table 4. They are valid for obtaining η in N s/m^2 if ρ is in kg/m^3. The tabulated values are based on experimental viscosity values between 273 and 373 K (0 and 100°C). Within this range, the deviations are mostly below ±1%; for liquids marked with +, deviations from ±5% to even ±20% must be expected, however.

Lewis and Squires [31] have developed a chart illustrated in Fig. 2 which shows the approximate change

Table 4 Numerical values for the constants A and c in Andrade's relationships, Eqs. (4) and (7)[a]

Compound	A	c	Compound	A	c
Elements			Fluorobenzene	5.14	0.695
Water	0.588	1.534	Chlorobenzene	5.48	0.703
Mercury	24.67	0.021	Iodobenzene	5.47	0.516
Chlorine	11.53	0.197	Sulfur compounds		
Bromine	7.08	0.213	Thiophene	4.40	0.739
Iodine	15.35	0.220	Methyl mercaptan	5.30	0.610
Paraffins			Ethyl mercaptan	5.08	0.779
Pentane	4.34	0.855	Carbon disulfide	7.29	0.356
Hexane	4.55	0.929	Alcohols		
Heptane	4.53	0.990	Methanol	2.69	1.171
Octane	4.37	1.098	Ethanol	2.28	1.491
i-Pentane	4.36	0.856	Propanol	1.05	1.986
i-Hexane	4.54	0.900	Butanol	0.783	2.174
i-Heptane	4.49	0.974	Allyl alcohol	1.33	1.609
Aromatic compounds			*i*-Propanol	0.352	2.466
Benzene	3.38	1.000	*i*-Butanol[b]	0.34	0.262
Ethyl benzene	4.58	0.922	Dimethyl ethylcarbinol[b]	0.097	3.111
o-Xylene	4.17	1.007	Trimethyl carbinol[b]	0.039	3.574
m-Xylene	4.65	0.893	Amyl alcohol, active	0.298	2.688
p-Xylene	4.36	0.931	Inactive	0.515	2.640
Toluene	4.39	0.912	Acids and anhydrides		
Olefins			Formic acid	2.25	1.036
Isoprene	4.49	0.731	Acetic acid	4.28	0.927
1,5-Hexadiene	4.16	0.854	Acetic anhydride	4.57	0.809
Trimethyl ethylene	4.70	0.721	Propionic acid	5.13	0.904
Halogen compounds			Propionic anhydride	4.18	0.952
Propyl chloride	5.02	0.655	Butyric acid	4.19	1.107
i-Propyl chloride	4.59	0.683	*i*-Butyric acid	4.55	1.048
i-Butyl chloride	4.43	0.797	Esters		
Allyl chloride	4.78	0.611	Methyl formate	5.24	0.571
Methylene chloride	5.77	0.422	Ethyl formate	4.92	0.675
Ethylene chloride	4.44	0.668	Propyl formate	4.51	0.799
Ethylidene chloride	4.98	0.557	Methyl acetate	4.63	0.668
Chloroform	6.07	0.412	Ethyl acetate	4.40	0.767
Carbon tetrachloride	3.97	0.560	Propyl acetate	4.06	0.891
Ethylene tetrachloride	6.80	0.436	Methyl propionate	4.98	0.719
Ethyl bromide	5.46	0.378	Ethyl propionate	4.33	0.838
Propyl bromide	5.27	0.473	Methyl butyrate	4.25	0.862
i-Propyl bromide	4.91	0.492	Methyl *i*-butyrate	4.37	0.827
i-Butyl bromide	4.80	0.605	Ethers		
Allyl bromide	5.08	0.444	Diethyl ether	4.44	0.571
Acetylene bromide	6.52	0.315	Methylpropyl ether	4.46	0.675
Propylene bromide	4.73	0.505	Ethylpropyl ether	4.25	0.843
i-Butylene bromide	3.76	0.656	Dipropyl ether	4.14	0.951
Ethylene bromide	4.79	0.446	Methyl *i*-butyl ether	4.34	0.825
Methyl iodide	5.40	0.247	Ethyl *i*-butyl ether	4.12	0.925
Ethyl iodide	5.68	0.319	Carbonyl compounds		
Propyl iodide	5.43	0.406	Acetaldehyde	4.80	0.610
i-Propyl iodide	5.30	0.410	Acetone	4.91	0.720
i-Butyl iodide	4.90	0.499	Methylethyl ketone	4.53	0.834
Allyl iodide	5.23	0.389	Diethyl ketone	4.76	0.848
			Methylpropyl ketone	4.72	0.884

[a] See Ref. [30].
[b] For these compounds deviations greater than ±5% are to be expected.

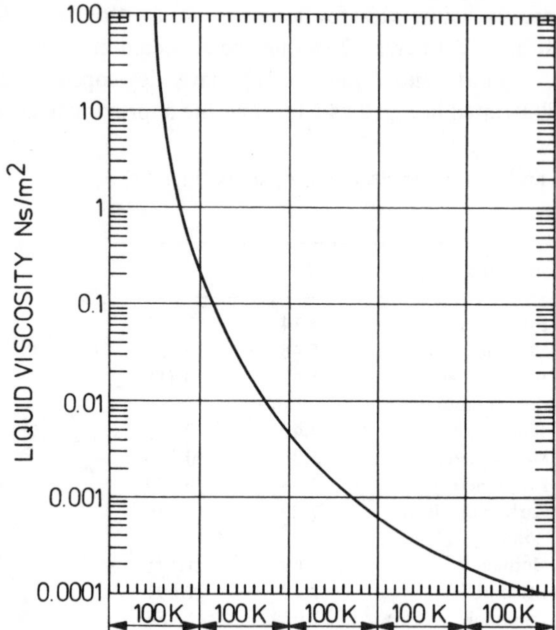

Figure 2 Change of liquid viscosity with temperature. Adapted from Lewis and Squires [31].

of viscosity with temperature for many fluids, except mercury but including water. To use this chart, one enters with a known viscosity value and follows the curve for a distance according to the temperature change in question.

For petroleum oils, the use of the ASTM Viscosity-Temperature Chart for Liquid Petroleum Products (ASTM D 341-39) is recommended.

(d) Pressure dependency of the viscosity of liquids

The viscosity of all liquids increases with pressure. Most data suggest that the logarithm of η is an almost linear function of pressure in a range of 1 000 to 4 000 bars. For water and mercury, the increase of viscosity is very little, especially at lower temperatures. For oils, Bondi [32] observed that, as a rule of thumb, an increase in pressure of 100 bars has about the same effect on viscosity as a decrease in temperature of 3 K.

The pressure dependency of viscosity is higher for liquids with high initial viscosity and for liquids with more branched molecules. Other influences are the liquid density and its change with pressure, i.e., the liquid compressibility. Andrade [30] gave the relationship

$$\frac{\eta_p}{\eta_0} = \left(\frac{\rho_p}{\rho_0}\right)^{1/6} \left(\frac{\kappa_0'}{\kappa_p'}\right)^{1/2} e^{c(\rho_p - \rho_0)/T} \tag{7}$$

where the subscripts p, 0 denote high pressure and low pressure conditions; κ' is the isothermal compressibility $\kappa' = 1/\tilde{V}(\partial \tilde{V}/\partial p)_T$ which can be estimated [e.g., from Fig. 5.1.2(3)] if no measured values are available. [Since Fig. 5.1.2(3) gives Z as a function of p_r, one has to calculate

$$\frac{1}{\tilde{V}}\left(\frac{\partial \tilde{V}}{\partial p}\right)_T = -\frac{1}{p}\left(\frac{\partial Z}{\partial p}\right)_T = -\frac{1}{p_c^2}\frac{1}{p_r}\left(\frac{\partial Z}{\partial p_r}\right)_T \tag{8}$$

where $(\partial Z/\partial p_r)_T$ is the slope of the curve and T_r = const at p_r and T_r of interest.] For c see Sec. A(c).

B. Thermal conductivity

(a) Thermal conductivity of gases

With the exceptions of hydrogen ($\lambda = 0.18$ W/m K at 300 K) and helium ($\lambda = 0.15$ W/m K at 300 K) the thermal conductivities of gases and vapors are in the range $0.01 < \lambda < 0.025$ W/m K. The thermal conductivity of gases is explained by the kinetic theory as the exchange of energy in collisions between the molecules of a gas. Therefore, λ is related to the heat capacity per unit volume of the gas, $\tilde{C}_v/\tilde{V}$, the average particle velocity v, and the mean free path l:

$$\lambda = \frac{1}{3}\frac{\tilde{C}_v}{\tilde{V}} vl \qquad \text{W/m K} \tag{9}$$

Usually, v and l are not available, but there is a close relationship between thermal conductivity λ and viscosity η. According to the kinetic theory of gases, the dynamic (absolute) viscosity is

$$\eta = \frac{1}{3}\rho vl = \frac{1}{3}\frac{\tilde{M}}{\tilde{V}} vl \qquad \text{N s/m}^2 \tag{10}$$

Therefore, the following relationship should be valid:

$$\lambda = \eta \frac{\tilde{C}_v}{\tilde{M}} \qquad \text{W/m K} \tag{11}$$

In reality, however, a correction of this relationship is required in order to get agreement with measured values. Based on theoretical considerations, but with simplifying assumptions, Eucken [33] proposed the following correction:

$$\lambda = \frac{\tilde{C}_v}{\tilde{M}}\frac{9\kappa - 5}{4} \tag{12}$$

where $\kappa = \tilde{C}_p/\tilde{C}_v$. Thus, Eucken's correction factor ideally should have the following values:

Monatomic gases: $\kappa = 1.67$; $(9\kappa - 5)/4 = 2.50$
Diatomic gases: $\kappa = 1.40$; $(9\kappa - 5)/4 = 1.90$
Triatomic gases: $\kappa = 1.33$; $(9\kappa - 5)/4 = 1.74$
Polyatomic gases: $\kappa \to 1.0$; $(9\kappa - 5)/4 \to 1.0$

Eucken's relationship gives a good approach in the pressure range from about 0.1 to 10 bars and in the temperature range from about 80 to 1 000 K. The calculated values seem to give a lower limit for actual values, but the deviations are normally less than −20%.

(b) Temperature dependency of the thermal conductivity of gases

In the range of low pressures, the thermal conductivity of gases increases with temperature. For monatomic and diatomic gases, the rise in conductivity is almost proportional:

$$\frac{\lambda_2}{\lambda_1} = \frac{T_2}{T_1} \tag{13}$$

For organic vapors, Owens and Thodos [34] recommend the relationship

$$\frac{\lambda_2}{\lambda_1} = \left(\frac{T_2}{T_1}\right)^m \tag{14}$$

with $m = 1.786$.

(c) Pressure dependency of the thermal conductivity of gases

With increasing pressure, the thermal conductivity of gases will also increase. At very low pressures (i.e., below about 0.001 bar), the mean free path of the molecules is partly or wholly limited by the walls confining the gas volume in question. In that range, the thermal conductivity increases linearly with pressure. Above 0.001 bar, the increase of λ is in the order of magnitude of 1% per 1 bar of increase in pressure.

For diatomic gases with the exception of hydrogen, Schaefer and Thodos [35] developed a corresponding-states correlation, shown in Fig. 3. Average errors are reported to be less than 2% within a wide range of gases and temperature/pressure conditions.

Stiel and Thodos [44] showed by dimensional analysis that the increase of thermal conductivity with pressure is a function of the increase of density. They succeeded in relating $(\lambda - \lambda_0)$ to ρ_r by a graphical correlation (Fig. 4) which they also evaluated in tabular form (see Table 5).

Within three ranges of reduced density ρ_r, the correlation can be approximated by the equations

$$(\lambda - \lambda_0)\frac{M^{1/2}T_c^{1/6}}{p_c^{2/3}} Z_c^5 = \begin{cases} 1.139 \cdot 10^{-8}(e^{1.124}\rho_r - 0.995) & (15) \\ 2.327 \cdot 10^{-8}(e^{0.690}\rho_r - 1.057) & (16) \\ 0.054\,7 \cdot 10^{-8}(e^{1.956}\rho_r - 79.70) & (17) \end{cases}$$

$0.04 < \rho_r < 0.06$ Eq. (15)

$0.6 < \rho_r < 2.0$ Eq. (16)

$2.0 < \rho_r < 2.8$ Eq. (17)

These equations are not identical to those given in the original publication, but they have been revised for closer approximation. As in the original, they are not dimensionless; the values of T_c and p_c must be given in K and N/m^2, respectively, and the resulting increase in thermal conductivity $(\lambda - \lambda_0)$, is in W/m K. λ is the thermal conductivity at low pressure but at the same temperature as the λ of interest.

The use of this correlation is not recommended for H_2, He, and polar compounds, but otherwise average

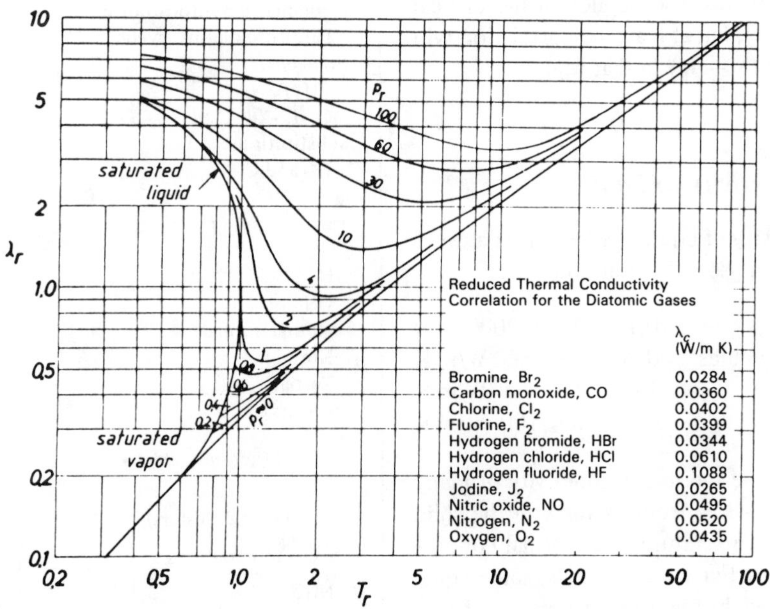

Figure 3 Reduced thermal conductivity $\lambda_r = \lambda/\lambda_c$ for diatomic gases as a function of T_r and ρ_r from Schaefer and Thodos [35].

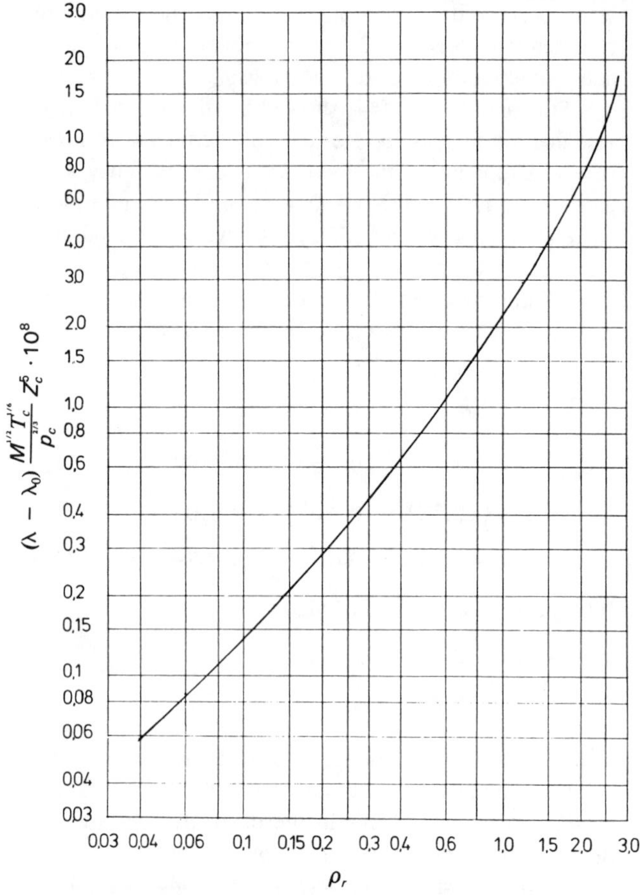

Figure 4 Residual thermal conductivity $(\lambda - \lambda_0)$ correlated to the reduced density ρ_r, according to Stiel and Thodos [44].

deviations of less than $\pm 10\%$ may be expected. There is some doubt whether this also applies in the critical region $(0.9 < p < 1.1)$, but at least the correlation will give a lower limit for $(\lambda - \lambda_0)$ in that region.

(d) Thermal conductivity of liquids

The thermal conductivity of liquids varies within a much wider range than that of gases. Typical values are:

Organic liquids:	$0.12 < \lambda < 0.20$ W/m K
Water and other polar liquids:	$0.20 < \lambda < 0.60$ W/m K
Salt molten:	$1.0 < \lambda < 4.0$ W/m K
Metal molten:	$10 < \lambda < 100$ W/m K

At the melting point, the thermal conductivity of many fluids will drop by 15%-30% from its value in the solid state. With increasing temperature, the thermal conductivity decreases further. For organic fluids, some equations are recommended by various authors, all based upon a fundamental equation given by Weber [36]:

Table 5 Tabulation of $(\lambda - \lambda_0)\,(M^{1/2}T_c^{1/6}/p_c^{1/3})\,Z_c^5 \cdot 10^8$ as a function of ρ_r

ρ_r	$f(\rho_r)$	ρ_r	$f(\rho_r)$	ρ_r	$f(\rho_r)$	ρ_r	$f(\rho_r)$
0.040	0.059	0.175	0.252	0.80	1.65	2.00	7.09
0.060	0.087	0.200	0.291	1.00	2.28	2.20	8.62
0.080	0.112	0.250	0.374	1.20	2.99	2.40	10.43
0.100	0.142	0.300	0.467	1.40	3.84	2.60	12.99
0.125	0.177	0.400	0.654	1.60	4.73	2.80	17.52
0.150	0.213	0.600	1.099	1.80	5.81		

$$\lambda = a\widetilde{C}_p \widetilde{V}^{-4/3} \tag{15}$$

[Dimensions for Eq. (15) are given below.] If accuracy is not required (deviations range from $\pm 15\%$ up to $\pm 50\%$), Weber's original vaue of $a = 3.59 \times 10^{-5}$ may be used. For a better estimate of the constant a, Robbins and Kingrea [37] employ a structural constant $\Sigma\,\Delta\lambda$ which is added up from structural increments listed in Table 6, and use the following relationship:

$$a = 10^{-4}\,\frac{(36.8 - 2\Sigma\,\Delta\lambda)(0.55/T_r)}{(\Delta\widetilde{H}_v/T_b) + \widetilde{R}\,\ln\,(273/T_b)} \tag{16}$$

The factor $0.55/T_r$ is to be replaced by 1.0 for liquids

Table 6 Structural increments $\Delta\lambda$ for estimating the thermal conductivity of liquids[a]

Atomic group		$\Delta\lambda$
Unbranched hydrocarbons:		
Paraffins, olefins, cyclic		0
Branches:		
CH_3	Each	1
C_2H_5-, $i-C_3H_7-$, C_4H_9-	Each	2
Substitutions:		
F– and Cl–	Each	1
	Maximum	3
Br–	One	4
	Two	6
I	One	5
OH–iso	Each	1
Norma normal	One	−1
Normal	Two	0
Tertiary	One	5
Oxygen:		
$-\overset{\vert}{C}{=}O$ (ketones, aldehydes)		0
$-\overset{\Vert}{C}{-}O$ (acids, esters)		0
–O– (ethers)		2
NH_2-	Each	1

[a]From Robbins and Kingrea [37].

with densities $\rho = 1\,000$ kg/m^3 and higher. The method fails for compounds containing sulfur.

Weber's equations are not dimensionless; i.e., the given constants apply for λ in W/m K, $\tilde{C}_p$ in kJ/kmol K, $\Delta\tilde{H}_v$ in kJ/kmol, $\tilde{V}$ in m^3/kmol, and $\tilde{R} = 8.314\,3$ kJ/kmol K. The index b stands for the normal boiling point.

(e) Influence of temperature and pressure on the thermal conductivity of liquids

The thermal conductivity of liquids generally decreases with rising temperature up to temperatures well above T_c, i.e., in the range of hypercritical fluids. (Compare also the liquid and hypercritical range in Fig. 3.) In a range sufficiently below the critical temperature, the order of magnitude of this decrease is about 1% per 10 K of increase in temperature. With rising pressure, there is a slight increase of thermal conductivity. Up to pressures of about 50 bars, this effect is negligible, especially at low temperatures. However, the influence of pressure increases in approaching the critical temperature.

C. Prandtl number

(a) Prandtl number of gases

In the range of low pressures, and at temperatures between 200 and 1 000 K (−80–700°C), the Prandtl number is, according to Eucken [33],

$$\mathrm{Pr} \equiv \frac{\tilde{C}_p \eta}{\tilde{M}\lambda} \approx \frac{4}{9 - 5/\kappa} \quad \text{with } \kappa = \frac{\tilde{C}_p}{\tilde{C}_v} \tag{17}$$

In the given range, it follows that

For monatomic gases: $\kappa = 1.67, \mathrm{Pr} = 0.67$
For diatomic gases: $\kappa = 1.40, \mathrm{Pr} = 0.73$
For triatomic gases: $\kappa = 1.33, \mathrm{Pr} = 0.76$
For polyatomic gases: $\kappa \to 1.0, \quad \mathrm{Pr} \to 1.0$

For estimating Prandtl numbers at higher pressures and temperatures, one should try to determine C_p, η, and λ separately.

(b) Prandtl numbers of liquids

If for a liquid, no exact values of C_p, η, and λ are available, the calculation of Pr from estimated values can sometimes be very inaccurate. In such cases, the relationship

$$\log \mathrm{Pr} = 0.027\,0 \frac{\Delta\tilde{H}_v}{T} - 1.80 \tag{18}$$

given by Gambill [38] could be used for a comparison. This relationship is valid for all practical purposes; deviations above ±50% are to be expected however. For water, Gambill reports slightly different values to give a better correlation:

$$\log \mathrm{Pr} = 0.024\,4 \frac{\Delta\tilde{H}_v}{T} - 2.20 \tag{19}$$

The equations are valid for T in K, ΔH_v (at T) in kJ/kmol, while log Pr denotes the decimal logarithm of the Prandtl number.

1.5. Surface Tension

A. Methods for estimating the surface tension from physical properties of the fluid

The molecules in the boundary layer between a liquid and its vapor are subject to different attractive forces due to the different densities in the liquid and in the gaseous phase. Obviously, the attraction toward the bulk of the liquid phase is normally much higher than toward the vapor. Thus, the surface resists any increase in area, or in other words, it requires mechanical work to bring a molecule from the bulk of the liquid into the boundary layer.

The surface tension σ is normally defined as the force per unit length in the plane of the surface, measured in N/m. The same result is obtained if σ is defined as the mechanical work (N m = J) necessary to increase the surface by the unit of area (m^2). The order of magnitude of the surface tension for various substances are:

Organic and inorganic liquids:	$0.01 < \sigma < 0.06$ N/m
Molten salt	$0.10 < \sigma < 0.20$ N/m
Metal melts	$0.30 < \sigma < 1.00$ N/m

where 1 N/m = 10^3 dynes/cm.

For the surface tension of nonassociating liquids at their normal boiling point, Walden [39] gave the simple equation

$$\sigma_b = 0.065\,6 \times 10^{-6} \left(\frac{\Delta\widetilde{H}_{v,b}}{\widetilde{V}_b}\right) \tag{1}$$

in N/m, if $\Delta\widetilde{H}_{v,b}$ (the heat of vaporization at the normal boiling point) is given in kJ/kmol and $\widetilde{V}_b$ (the specific volume of the liquid phase at the normal boiling point) in m^3/kmol.

Also, for nonassociating liquids, Riedel [3d] gives

$$\sigma = 10^{-3}(0.133\,\alpha_c - 0.281)(1 - T_r)^{11/9}\,T_c^{1/3}p_c^{2/3} \tag{2}$$

in N/m, if T_c is in K and p_c in bars. The critical parameter,

$$\alpha_c \equiv \left(\frac{\partial \ln p_r}{\partial \ln T_r}\right)_c$$

can be estimated according to Sec. 1.1E (p. 3).

For associating liquids, the values obtained by this equation tend to be too high so Hakim et al. [40] recommend at $T_r = 0.6$ (which is usually not far from the normal boiling point) the equation

$$\sigma_{0.6} = 10^{-3}\,p_c^{2/3}\,T_c^{1/3}\,(0.157\,4 + 0.359\,\omega - 1.769\,x - 0.51\,\omega^2 - 13.69\,x^2 + 1.298\,\omega x)$$

where ω is the acentric factor (see Sec. 1.1E) and x is the Stiel polar factor

$$x = \log p_{r(0.6)} + 1.70\,\omega + 1.552$$

where $\log p_{r(0.6)}$ is the decimal logarithm of the reduced pressure at a reduced temperature of $T_r = 0.6$.

If the surface tension at a certain temperature is known, the transition to a different temperature follows from Riedel's equation above:

$$\frac{\sigma_1}{\sigma_2} = \left(\frac{1 - T_{r,1}}{1 - T_{r,2}}\right)^m$$

with $m = 11/9 = 1.22 \cdots$ for nonassociating liquids.

Table 1 Structural increments for the calculation of the Parachor P^a

Atomic group	$\Delta P \times 10^3$	Atomic group	$\Delta P \times 10^3$
Elements		Special groups	
C	1.60	–COO–	11.35
H	2.76	–COOH	13.13
O	3.56	–OH	5.30
N	3.11	–O–	3.56
S	8.73	=O (ketone)	
P	7.20	3 C atoms	3.97
F	4.64	4 C atoms	3.56
Cl	9.82	5 C atoms	3.29
Br	12.09	6 C atoms	3.08
I	16.06	–CHO	11.74
Hydrocarbons		$-NH_2$	7.56
CH_3-	9.87	$-NO_2$ (nitrite)	13.16
$CH_3-CH(CH_3)-$	23.71	$-NO_3$ (nitrate)	16.54
$-CH_2-$	7.11	$-CO(NH_2)$	16.31
More than 12 $-CH_2-$	7.17	Additional increments	
$CH_3-CH_2-CH(CH_3)-$	30.57	Ethylene linkage	
$CH_3-CH_2-CH_2-CH(CH_3)-$	37.65	terminal	3.40
$CH_3-CH(CH_3)-CH_2-$	30.82	2,3 Position	3.15
$CH_3-CH_2-CH(C_2H_5)-$	37.26	3,4 Position	2.90
$CH_3-C(CH_3)_2$	30.31	Triple bond	7.22
$CH_3-CH_2-C(CH_3)_2$	36.91	Ring closure	
$CH_3-CH(CH_3)-CH(CH_3)-$	36.98	3-Membered	2.22
$CH_3-CH(CH_3)-C(CH_3)_2-$	43.31	4-Membered	1.07
C_6H_5-	33.72	5-Membered	0.53
		6-Membered	0.14

[a]According to Sugden [41] and Quale [42].

According to Hakim et al. [40], the exponent m for associating liquids has the value

$$m = 1.21 + 0.538\,5\,\omega - 14.61\,x - 1.65\,\omega^2 - 32.07\,x^2 + 22.03\,\omega x$$

with ω and x as given above.

B. Method for estimating the surface tension from the chemical structure

Walden [39] describes an additive method of estimating the surface tension based on numbers and types of bonds in a compound. The most successful additive method has been established by Sugden [41] and further developed by Quale [42].

In the equation

$$\sigma^{0.25} = P\left(\frac{1}{\tilde{V}_L} - \frac{1}{\tilde{V}_G}\right)$$

the "Parachor" P is added up from structural increments given in Table 1. In this table, the original values from Quale [42] have been multiplied by $1\,000^{-1.25} = 0.177\,9 \cdot 10^{-3}$ in order to give σ in N/m instead of $dm^3/kmol$. L and G denote the liquid and gas phase. The average deviations to be expected are ±3% with maximum deviations up to ±20%.

REFERENCES

1. Reid, R. C., Prausnitz, J. M., and Sherwood, T. K., *The Properties of Gases and Liquids,* McGraw-Hill, New York, 1977.
2. Hecht, G., Lehmann, H., Lehmann, H., Thielemann, I., Ruschitzky, E., Jacobi, B., and Holste, C., *Berechnung thermodynamischer Stoffwerte von Gasen und Flüssigkeiten,* VEB Deutscher Verlag für Grundstoff-Industrie, Leipzig, 1966.

3a. Riedel, L., Untersuchungen über eine Erweiterung des Theorems der übereinstimmenden Zustände; Teil I. Eine neue universelle Dampfdruckformel, *Chem. Ing. Tech.*, vol. 26, pp. 83–89, 1954;

b. Teil II. Die Flüssigkeitsdichte im Sättigungszustand, *ibid.*, vol. 26, pp. 259–264, 1954;

c. Teil III. Kritischer Koeffizient, Dichte des gesättigten Dampfes und Verdampfungswärme, *ibid.*, vol. 26, pp. 679–683, 1954;

d. Teil IV. Kompressibilität, Oberflächenspannung und Wärmeleitfähigkeit im flüssigen Zustand, *ibid.*, vol. 27, pp. 209–213, 1955;

e. Teil V. Bestimmung unbekannter kritischer Daten von nicht assoziierenden Stoffen, *ibid.*, vol. 27, pp. 475–480, 1955;

f. Teil VI. Die Zustandsfunktion des realen Gases, *ibid.*, vol. 28, pp. 557–562, 1956.

4. Pitzer, K. S., Lippmann, D. Z., Curl, R. F., Huggins, C. M., and Petersen, D. E., The Volumetric and Thermodynamic Properties of

Fluids. II. Compressibility Factor, Vapor Pressure, and Entropy of Vaporization, *J. Am. Chem. Soc.*, vol. 77, pp. 3433–3440, 1955.

5. Lydersen, A. L., Estimation of Critical Properties of Organic Compounds, College of Engineering, University of Wisconsin, Madison, WI., Eng. Expt. Sta. Rept. 3, 1955.
6. Cailletet, L., and Mathias, E., Recherches sur les Densités des Gaz Liquéfiés et de leurs Vapeurs Saturées, *C.R. Acad. Sci.* (Paris), vol. 102, pp. 1202–1207, 1886.
7. Edmister, W. C., Applied Hydrocarbon Thermodynamics. Part 4. Compressibility Factors and Equations of State, *Petrol. Refiner*, vol. 37, no. 4, pp. 173–179 April, 1958.
8. Watson, K. M., Thermodynamics of the Liquid State–Generalized Prediction of Properties, *Ind. Eng. Chem.*, vol. 35, pp. 398–406, 1943.
9. Timmermans, J., La Densité des Liquides Sous 0°, *Bull. Soc. Chim. Belg.*, vol. 26, pp. 205–215, 1912.
10. Rackett, H. G., Equation of State for Saturated Liquids, *J. Chem. Eng. Data*, vol. 15, no. 4, pp. 514–517, 1970.
11. Yamada, T., and Gunn, R. D., Saturated Liquid Molar Volumes, The Rackett Equation, *J. Chem. Eng. Data*, vol. 18, pp. 234–236, 1973.
12. Nelson, L. C., and Obert, E. F., Generalized *p-V-T* Properties of Gases, *Trans. ASME*, vol. 76, pp. 1057–1066, 1954.
13. van der Waals, J. D., Over de Continuiteit van den Gas–en Vloeisstoftoestand, thesis, University of Leiden, 1873.
14. Redlich, O., and Kwong, J. N. S., On the Thermodynamics of Solutions. V. An Equation of State, Fugacities of Gaseous Solutions, *Chem. Rev.*, vol. 44, pp. 233–244, 1949.
15. Soave, G., Equilibrium Constants from a Modified Redlich-Kwong Equation of State, *Chem. Eng. Sci.*, vol. 27, pp. 1197–1203, 1972.
16. Peng, D.-Y., and Robinson, D. B., A New Two Constant Equation of State, *Ind. Eng. Chem. Fund.*, vol. 15, no. 1, pp. 59–64, 1976.
17. Ogata, Y., and Tsuchida, M., Linear Boiling Point Relationships, *Ind. Eng. Chem.*, vol. 49, pp. 415–417, 1957.
18. Cox, E. R., Pressure–Temperature Chart for Hydrocarbon Vapors, *Ind. Eng. Chem.*, vol. 15, pp. 592–593, 1923.
19. Hoffmann, W., and Florin, F., Zweckmäßige Darstellung von Dampfdruckkurven, *Verfahrenstechnik, Z. VDI-Beiheft*, no. 2, pp. 47–51, 1943.
20. Antoine, C., Tensions de Diverses Vapeurs, *C.R. Acad. Sci.* (Paris), vol. 107, pp. 836–837, 1888.
21. Thomson, G. W., Determination of Vapor Pressure, Chapter 9; Vapor Pressure Temperature Relations, Sec. IV, *Technique of Organic Chemistry*, 3rd ed., ed. A. Weissberger, vol. I, part I, pp. 471–493, Interscience Publishers, New York, 1959.
22. Trouton, F., On Molecular Latent Heat, *Philos. Mag.*, vol. 18, no. 5, pp. 54–57, 1884.
23. Benson, S. W., Cruickshank, F. R., Golden, D. M., Haugen, G. R., O'Neal, H. E., Rodgers, A. S., Shaw, R., and Walsh, R., Additivity Rules for the Estimation of Thermochemical Properties, *Chem. Rev.*, vol. 69, pp. 279–324, 1969.
24. Sakiadis, B. C., and Coates, J., Prediction of Specific Heat of Organic Liquids, *AIChE J.*, vol. 2, no. 1, pp. 88–93, 1956.
25. Gambill, W. R., Predict Heat Capacities of Gases, *Chem. Eng.*, vol. 64, no. 9, pp. 267–270, 1957.
26. Pachaiappan, V., Ibrahim, S. H., and Kuloor, N. R., Simple Correlation for Determining A Liquid's Heat Capacity, *Chem. Eng.*, vol. 74, pp. 241–243, October 9, 1967.
27. Lucas, K., Ein einfaches Verfahren zur Berechnung der Viskosität von Gasen und Gasgemischen, *Chem. Ing. Tech.*, vol. 46, no. 4, p. 157, 1974.
28. Souders, M., Viscosity and Chemical Constitution, *J. Am. Chem. Soc.*, vol. 60, pp. 154–158, 1938.
29. Thomas, L. H., The Dependence of the Viscosities of Liquids on Reduced Temperature, and a Relation between Viscosity, Density, and Chemical Constitution, *J. Chem. Soc.*, vol. 1946, pp. 573–579, 1946.
30. Andrade, E. N. da C., A Theory of the Viscosity of Liquids. Part I, *Philos. Mag. J. Sci.*, vol. XVII, pp. 497–511, 1934; Part II, *ibid.*, vol. XVII, pp. 698–732, 1934.
31. Lewis, W. K., and Squires, L., The Mechanism of Oil Viscosity as Related to the Structure of Liquids, *Oil Gas J.*, pp. 92–96, November 15, 1934.
32. Bondi, A., Physical Chemistry of Lubricating Oils. Part I., *Petrol. Refiner*, vol. 25, no. 6, pp. 122–126, 1946; Part II., *ibid.*, vol. 25, no. 7, pp. 119–132, 1946; Part III., *ibid.*, vol. 25, no. 8, pp. 119–129, 1946.
33. Eucken, A., Über das Wärmeleitvermögen, die spezifische Wärme und die innere Reibung der Gase, *Phys. Z.*, vol. 14, pp. 324–336, 1913.
34. Owens, E. J., and Thodos, G., Thermal Conductivity: Correlation for Ethylene and Its Application to Gaseous Aliphatic Hydrocarbons and Their Derivatives at Moderate Pressures, *AIChE J.*, vol. 6, pp. 676–681, 1960.
35. Schaefer, C. A., and Thodos, G., Thermal Conductivity of Diatomic Gases: Liquid and Gaseous States, *AIChE J.*, vol. 5, no. 3, pp. 367–372, 1959.
36. Weber, H. F., Untersuchungen über die Wärmeleitung in Flüssigkeiten, *Ann. Phys. Chem.*, vol. X, pp. 103–129, 1880.
37. Robbins, L. A., and Kingrea, C. L., Estimate Thermal Conductivity, *Hydroc. Proc. Petrol. Refiner*, vol. 41, no. 5, pp. 133–136, 1962.
38. Gambill, W. R., Best Methods for Prandtl Number, *Chem. Eng.*, vol. 65, no. 17, pp. 121–124, 1958.
39. Walden, P., Über den Zusammenhang der Kapillaritätskonstanten mit der latenten Verdampfungswärme der Lösungsmittel, *Z. Phys. Chem.*, vol. 65, pp. 267–288, 1909.
40. Hakim, D. J., Steinberg, D., and Steil, L. I., Generalized Relationship for the Surface Tension of Polar Fluids, *Ind. Eng. Chem. Fund.*, vol. 10, no. 1, pp. 174–175, 1971.
41. Sugden, S., A Relation between Surface Tension, Density, and Chemical Composition, *J. Chem. Soc.*, vol. 125, pp. 1177–1189, 1924.
42. Quale, O. R., The Parachors of Organic Compounds, *Chem. Rev.*, vol. 53, pp. 439–585, 1953.
43. Lee, B. I., and Kesler, M. G., A Generalized Thermodynamic Correlation Based on Three-Parameter Corresponding States, *AIChE J.*, vol. 21, no. 3, pp. 510–527.
44. Stiel, L. I., and Thodos, G. The Thermal Conductivity of Nonpolar Substances in the Dense Gaseous Liquid Regions, *AIChE J.*, vol. 10, no. 1, pp. 26–30, 1964.

CHAPTER 2
Physical Property Data Tables for Pure Fluids

2.1. Introduction

This chapter contains tabular information for pure component thermophysical properties.

Four types of tables are used:

1. Properties of the liquid at temperatures below the normal boiling point (291 fluids).
2. Properties of the saturated liquid and vapor (64 fluids).
3. Transport properties of the superheated vapor (249 vapors).
4. Thermodynamic properties of the superheated vapor (11 vapors).

At low vapor pressures the saturated vapor and/or liquid value will be essentially the same as those for the superheated gas or subcooled liquid. However, as the critical pressure is approached there can be substantial deviation of the saturated properties from those one would normally expect to encounter. For this reason, values at saturation pressures greater than approximately 70% of the critical value should be used with extreme caution.

Figure 1 shows a typical pressure-enthalpy diagram for a single pure component. The critical point is indicated, as are the saturated vapor and liquid lines. As the temperature on the material is increased, the pressure must also increase to maintain the material in the saturated state. For this reason most of the properties for the saturated liquid and vapor are different from those frequently measured in the laboratory, or predicted by correlations. At low reduced pressures and temperatures, these differences are normally small and can be neglected. As the temperature and pressure approach the critical, however, the differences become larger and must be taken into account.

As an example of the behavior of properties for the saturated phase, consider the enthalpies. The liquid enthalpy behaves "normally" until a point approximately at A is reached. In the region of the critical, the enthalpy increases more rapidly with temperature than at lower temperature and pressure. At the critical, presumably the liquid enthalpy increases with no change in temperature. If this seems strange, consider the vapor enthalpy. At approximately point B, the vapor enthalpy no longer increases with temperature, but actually begins to decrease as the temperature is increased.

The behavior of the enthalpy is reflected in the specific heat capacity. For the saturated liquid, the heat capacity in the region of the critical becomes very large, reflecting the enthalpy increase with little or no change

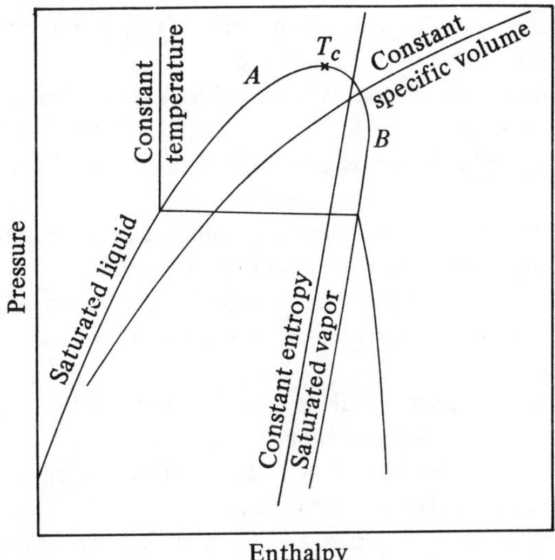

Figure 1 Typical pressure-enthalpy diagram for pure component.

in temperature. The saturated vapor enthalpy actually becomes negative to accommodate the decrease in enthalpy as the material approaches its critical temperature.

Values from the tables may be interpolated. However, there should be no extrapolation to higher temperatures and/or pressures.

Guidelines for interpolation of data are as follows:

Saturation (vapor) pressure values at intermediate temperatures can be interpolated by use of the linearity of log vapor pressure with reciprocal absolute temperature. For most of the compounds, particularly at temperatures removed from the critical temperature, this line should be straight.

Liquid density. Over restricted ranges of temperature liquid density, except in the region of the critical, should be plotted as a straight line function of temperature. In the region of the critical there will be some curvature.

Vapor density values should plot as a reasonably straight line with temperature.

Liquid enthalpy should plot as a smooth curve function of temperature.

Vapor enthalpy values should plot as a smooth curve function of temperature.

Heat of vaporization values can be satisfactorily interpolated by plotting heat of vaporization as a function of temperature.

Liquid heat capacities plot satisfactorily as smooth curves as a function of temperature.

Vapor heat capacities plot satisfactorily as smooth curves as a function of temperature.

Liquid viscosity can be interpolated by plotting log viscosity as a function of reciprocal absolute temperature. Contrary to most conceptions this will not, in general, be a straight line. However, curvature will be slight and will allow satisfactory interpolation of values.

Vapor viscosity values plot satisfactorily as smooth curves as a function of temperature.

Liquid thermal conductivity values can be plotted satisfactorily as a function of temperature. The line will *not* be straight. However curvature is slight and should allow for satisfactory interpolation.

Vapor thermal conductivity values can be interpolated by plotting as a function of temperature.

Liquid Prandtl number. Values for liquid Prandtl number can be interpolated by plotting as a function of temperature.

Vapor Prandtl numbers can be interpolated by plotting as a function of temperature.

Interfacial tension values can be interpolated by plotting as a function of temperature.

Coefficient of thermal expansion values can be interpolated by plotting as a function of temperature. Coefficient of thermal expansion is based on the density at the normal boiling point.

Naturally, mathematical interpolation is possible for any of the values, but the appropriate equation for interpolation of values must be used. This may be a source of trouble in some cases (liquid viscosity, for example) where the equation used does not truly linearize the property value. Analysis of the error between calculated and tabulated values will show if the errors are too large.

2.2. Sources

Table types (1) for liquids and (3) for vapors are derived principally from the tables of physical properties given in the Wärmeatlas (56) and are supplemented from other data collections.

An extensive search has been made to discover reported values for each property, but when no values were found, an estimated value is given based on well-tried prediction procedures such as those reported in Chapter 1. Such predicted values are indicated in parentheses.

Table type (2) for saturated fluids were prepared by R. N. Maddox and are based mainly on data published by Vargaftik (54), Touloukian (52), or on the Technical Data Book (48) and correlations developed by Thinh (50), Yaws et al. (2, 18, 34, 36, 45), and Gallant (14). The data for the saturated vapor is considerably less accurate than that for the liquid because there is little data reported in the literature.

Table type (4) for superheated vapors were prepared by Reynolds (41) from Equations of State.

Data for the constants were those preferred by Maddox or the compilers of the Wärmeatlas. Critical densities have been added when available and the normal vapor density calculated.

2.3. List of Fluids by Class

Data are included in the present tables for the following substances:

NORMAL PARAFFINS

- Methane
- Ethane
- Propane
- Butane
- Pentane
- Hexane
- Heptane
- Octane
- Nonane
- Decane
- Undecane
- Dodecane
- Tridecane
- Tetradecane
- Pentadecane
- Hexadecane

Heptadecane
Octadecane
Nonadecane
Eicosane
ISOPARAFFINS
2-Methylpropane
2-Methylbutane
2-Methylpentane
3-Methylpentane
2,2-Dimethylpropane
2,2-Dimethylbutane
2,3-Dimethylbutane
OLEFINS
Ethylene
Propylene
1-Butene
1-Pentene
1-Hexene
1-Heptene
1-Octene
Propadiene
1,2-Butadiene
1,3-Butadiene
1,2-Pentadiene
1,trans 3 Pentadiene
1,4-Pentadiene
2,3-Pentadiene
ACETYLENES
Acetylene
Methyl acetylene
Ethyl acetylene
Dimethyl acetylene
NAPHTHENES
Cyclopropane
Cyclobutane
Cyclopentane
Methylcyclopentane
Ethylcyclopentane
Propylcyclopentane
Butylcyclopentane
Pentylcyclohexane
Hexylcyclohexane
Cyclohexane
Methylcyclohexane
Ethylcyclohexane
Isopropylcyclohexane
Butylcyclohexane
Pentylcyclohexane
Hexycyclohexane
Cyclopentene
Cyclohexene
AROMATICS
Benzene
Toluene
Ethylbenzene
Vinylbenzene
Propylbenzene
Isopropylbenzene
Butylbenzene
Pentylbenzene
Hexylbenzene
o-Xylene
m-Xylene
p-Xylene
1,2,3-Trimethylbenzene
1,2,4-Trimethylbenzene
1,3,5-Trimethylbenzene
1,2,3,4-Tetramethylbenzene
1,2,3,5-Tetramethylbenzene
1,2,4,5-Tetramethylbenzene
Pentamethylbenzene
Hexamethylbenzene
Diphenyl
Diphenylmethane
Triphenylmethane
Naphthalene
1-Methylnaphthalene
2-Methylnaphthalene
1-Ethylnaphthalene
2-Ethylnaphthalene
ALCOHOLS
Methanol
Ethanol
1-Propanol
1-Butanol
1-Pentanol
1-Hexanol
1-Heptanol
1-Octanol
Isopropanol
Isobutanol
t-Butyl alcohol
Isopentanol
Ethylene glycol
1,3-Propylene glycol
Glycerol
Cyclohexanol
Benzyl alcohol
PHENOLS
Phenol
o-Cresol
m-Cresol
p-Cresol
ESTERS
Methyl formate
Ethyl formate
Propyl formate
Methyl acetate
Ethyl acetate
Propyl acetate

Methyl propionate
Ethyl propionate
Propyl propionate
Methyl butanoate
Ethyl butanoate
Methyl benzoate
Ethyl benzoate
Methyl salicylate

ALDEHYDES
Formaldehyde
Acetaldehyde
Paraldehyde
Furfural
Benzaldehyde
Salicylaldehyde

KETONES
Ketene
Acetone
Methyl ethyl ketone
Diethyl ketone
Dipropyl ketone
Acetophenone
Benzophenone

CARBOXYLIC ACIDS
Formic acid
Acetic acid
Propionic acid
Butyric acid
Pentanoic acid (Valeric acid)
Hexanoic acid (Caproic acid
Acetic anhydride
Propionic anhydride

ETHERS
Dimethyl ether
Diethyl ether
Methyl propyl ether
Ethyl propyl ether
Methyl *t*-butyl ether
Dipropyl ether
Ethylene oxide
Propylene oxide
Furan
1,4-Dioxan

HALOGENATED HYDROCARBONS
Fluoromethane
Difluoromethane
Trifluoromethane
Tetrafluoromethane
Chloromethane
Dichloromethane
Trichloromethane (Chloroform)
Tetrachloromethane (Carbon tetrachloride)
Dibromomethane
Bromomethane
Tribromomethane
Tetrabromomethane (carbon tetrabromide) (Bromoform)
Trichlorofluoromethane (Refrigerant 11)
Dichlorodifluoromethane (Refrigerant 12)
Chlorotrifluoromethane (Refrigerant 13)
Dichlorofluoromethane (Refrigerant 21)
Chlorodifluoromethane (Refrigerant 22)
1,1,1-Trifluoroethane
Chloroethane
1,1-Dichloroethane
1,2-Dichloroethane
1,1,1-Trichloroethane
1,1,2,2-Tetrachloroethane
Pentachloroethane
Bromoethane
1,2-Dibromoethane
Tetrachlorodifluoroethane (Refrigerant 112)
Trichlorotrifluoroethane (Refrigerant 113)
Dichlorotetrafluoroethane (Refrigerant 114)
Chloropropane
Chlorobutane
Chloropentane
Vinyl chloride
1,1-Dichloroethylene
Trichloroethylene
Tetrachloroethylene
Fluorobenzene
Chlorobenzene
Bromobenzene
Iodobenzene
Benzyl chloride
m-Chlorotoluene

AMINES
Methylamine
Dimethylamine
Trimethylamine
Ethylamine
Diethylamine
Triethylamine
Isopropylamine
Butylamine
Piperidine
Pyridine
Aniline
Methylaniline
Dimethylaniline
Diethylaniline

NITRILES
Acetonitrile
Propionitrile
Butyronitrile

AMIDES
Formamide

NITRO DERIVATIVES
Nitromethane
Nitrobenzene
o-Nitrotoluene
m-Nitrotoluene
p-Nitrotoluene

ORGANIC SULFUR COMPOUNDS
- Methyl mercaptan
- Ethyl mercaptan
- Dimethyl sulfide
- Diethyl sulfide
- Thiophene

INORGANIC SUBSTANCES
- Hydrogen fluoride
- Hydrogen chloride
- Hydrogen bromide
- Hydrogen iodide
- Hydrogen cyanide
- Hydrogen sulfide
- Air
- Ammonia
- Nitric oxide
- Nitrogen dioxide
- Nitrogen peroxide
- Nitrous oxide
- Carbon monoxide
- Carbon dioxide
- Cyanogen
- Cyanogen fluoride
- Carbonyl sulfide
- Carbon disulfide
- Sulfur dioxide
- Sulfur hexafluoride
- Phosgene
- Cyanogen chloride

ELEMENTS

- Helium
- Argon
- Neon
- Krypton
- Xenon
- Hydrogen
- Nitrogen
- Oxygen
- Fluorine
- Chlorine
- Bromine
- Iodine
- Sulfur
- Mercury

HEAT TRANSFER MEDIA
- Dowtherm A
- Dowtherm J

NOTATION

Symbols

Where *symbols* are used in the following tables, they conform with those listed on Table 3 with the addition of: $B_{e,1}$ Coefficient of thermal expansion for liquid phase

Subscripts

l indicates liquid
g indicates gas or vapor

Phases are represented by S (solid), L (liquid), V (vapor). Parentheses ( ) indicate that the value is estimated.

DATA TABLES

METHANE

Chemical formula: CH_4
Molecular weight: 16.042
Melting point: 90.66 K
Normal boiling point: 111.42 K

Critical temperature: 190.55 K
Critical pressure: 4.641 MPa
Critical density: 162 kg/m^3
Normal vapor density: 0.72 kg/m^3
(@ 0°C, 101.3 kPa)

Properties of the Liquid at Temperatures Below the Normal Boiling Point

	Temperature, °C												
	−150	−100	−75	−50	−25	0	20	50	100	150	200	250	300
	Temperature, K												
Property	123.15	173.15	198.15	223.15	248.15	273.15	293.15	323.15	373.15	423.15	473.15	523.15	573.15
Density, ρ_l (kg/m^3)	409	V	V	V	V	V	V	V	V	V	V	V	V
Specific heat capacity, $c_{p,l}$ (kJ/kg K)	3.517	V	V	V	V	V	V	V	V	V	V	V	V
Thermal conductivity, λ_l [(W/m^2)/(K/m)]	0.175	V	V	V	V	V	V	V	V	V	V	V	V
Dynamic viscosity, η_l (10^{-5} Ns/m^2)	8.8	V	V	V	V	V	V	V	V	V	V	V	V

Properties of the Saturated Liquid and Vapor

T_{sat}, K	111.42	120	130	140	150	160	170	180	185	190
p_{sat}, kPa	101	192	367	638	1 033	1 588	2 338	3 288	3 854	4 552
ρ_ℓ, kg/m^3	424.3	412.0	396.7	379.8	361.0	339.3	312.3	271.9	240.0	182.0
ρ_g, kg/m^3	1.79	3.26	5.95	10.03	16.08	25.03	38.57	59.14	76.28	120.9
h_ℓ, kJ/kg	716.3	747.0	784.1	821.9	860.0	901.4	948.4	1 011.1	1 057.0	1 133.4
h_g, kJ/kg	1 228.1	1 241.8	1 255.7	1 267.2	1 274.9	1 277.6	1 273.3	1 258.9	1 245.0	1 203.2
$\Delta h_{g,\ell}$, kJ/kg	511.8	494.8	471.8	445.3	414.9	376.2	324.9	247.8	188	69.8
$c_{p,\ell}$, kJ/(kg K)	3.43	3.53	3.63	3.77	3.94	4.12	5.16	7.45	11.3	70.5
$c_{p,g}$, kJ/(kg K)	2.07	2.11	2.19	2.33	2.53	2.90	3.62	5.95	6.33	277.5
η_ℓ, μNs/m^2	106.5	86.05	71.65	61.26	52.24	44.54	37.69	30.98	26.92	19.34
η_g, μNs/m^2	4.49	4.84	5.28	5.74	6.27	6.89	7.69	8.89	9.84	12.96
λ_ℓ, (mW/m^2)/(K/m)	193	178	163	148	133	118	103	88	80	73
λ_g, (mW/m^2)/(K/m)	12.1	12.9	16.4	19.6	23.0	27.6	33.7	39.9	45.3	62.0
Pr_ℓ	1.88	1.70	1.60	1.56	11.55	1.56	1.89	2.62	3.80	18.8
Pr_g	0.77	0.79	0.71	0.68	0.69	0.72	0.83	1.33	1.38	58.01
σ, mN/m	13.5	11.5	9.28	7.22	5.31	3.58	2.06	0.81	0.33	0.01
$\beta_{e,\ell}$, kK^{-1}	2.27	3.46	4.37	5.18	6.58	8.87	14.2	32.8	58.6	385

Transport Properties of the Superheated Vapor

	Temperature, °C												
	−150	−100	−50	0	25	100	200	300	400	600	800	1 000	1 200
	Temperature, K												
Property (at low pressure)	123.15	173.15	223.15	273.15	298.15	373.15	473.15	573.15	673.15	873.15	1 073.15	1 273.15	1 473.15
Specific heat capacity, $c_{p,g}$ (kJ/kg K)	1.876	1.968	2.068	2.165	2.227	2.449	2.805	3.178	3.534	4.157	4.664	5.062	5.372
Thermal conductivity, λ_g[(W/m^2)/(K/m)]	0.013	0.019	0.024	0.030	0.034	0.044	0.061	0.079	0.099	0.139	0.183	0.223	0.260
Dynamic viscosity, η_g(10^{-5} Ns/m^2)	0.480	0.670	0.845	1.02	1.10	1.33	1.61	1.86	2.07	2.49	2.88	3.22	3.54

ETHANE

Chemical formula: CH_3CH_3
Molecular weight: 30.068
Melting point: 89.88 K
Normal boiling point: 184.52 K

Critical temperature: 305.5 K
Critical pressure: 4.913 MPa
Critical density: 212 kg/m^3
Normal vapor density: 1.34 kg/m^3
(@ 0°C, 101.3 kPa)

Properties of the Liquid at Temperatures Below the Normal Boiling Point

	Temperature, °C												
	−150	−100	−75	−50	−25	0	20	50	100	150	200	250	300
	Temperature, K												
Property	123.15	173.15	198.15	223.15	248.15	273.15	293.15	323.15	373.15	423.15	473.15	523.15	573.15
Density, ρ_l(kg/m^3)	666	561	V	V	V	V	V	V	V	V	V	V	V
Specific heat capacity, $c_{p,l}$ (kJ/kg K)	2.319	2.403	V	V	V	V	V	V	V	V	V	V	V
Thermal conductivity, λ_l [$(W/m^2)/(K/m)$]	(0.199)	(0.155)	V	V	V	V	V	V	V	V	V	V	V
Dynamic viscosity, η_l (10^{-5} Ns/m^2)	42.2	22.2	V	V	V	V	V	V	V	V	V	V	V

Properties of the Saturated Liquid and Vapor

T_{sat}, K	184.52	200	210	230	240	260	270	280	290	300
p_{sat}, kPa	101	217	334	700	968	1 712	2 208	2 801	3 510	4 365
ρ_ℓ, kg/m^3	546.45	529.10	516.79	489.71	474.60	440.14	419.81	396.35	364.56	316.25
ρ_g, kg/m^3	2.04	4.09	6.21	12.75	17.56	31.65	42.03	55.96	77.10	119.18
h_ℓ, kJ/kg	399.52	437.50	462.53	515.34	541.41	598.79	629.79	663.10	700.28	753.08
h_g, kJ/kg	889.19	903.74	912.82	929.18	935.72	943.27	943.23	941.14	930.10	892.31
$\Delta h_{g,\ell}$, kJ/kg	489.67	466.24	450.29	414.84	394.31	344.48	313.44	278.04	229.82	139.23
$c_{p,\ell}$, kJ/(kg K)	2.42	2.48	2.54	2.66	2.70	3.00	3.18	3.42	3.80	
$c_{p,g}$, kJ/(kg K)	1.40	1.48	1.54	1.70	1.79	2.13	2.42	2.94	3.31	9.51
η_ℓ, $\mu Ns/m^2$	168	139	124	99.4	88.8	70.8	61.6	54.0	46.0	36.1
η_g, $\mu Ns/m^2$	6.00	6.59	7.03	7.89	8.42	9.85	10.9	12.1	14.2	19.0
λ_ℓ, $(mW/m^2)/(K/m)$	157	146	140	126	117	99.2	92.1	83.9	76.0	67.4
λ_g, $(mW/m^2)/(K/m)$	8.6	10.3	11.5	14.1	15.5	18.6	20.7	22.8	26.1	32.0
Pr_ℓ	2.59	2.37	2.26	2.10	2.05	2.14	2.12	2.2	2.23	
Pr_g	0.98	0.95	0.94	0.95	0.97	1.13	1.27	1.56	1.80	5.64
σ, mN/m	15.86	13.28	11.71	8.51	6.85	4.28	3.14	2.00	1.14	0.43
$\beta_{e,\ell}$, kK^{-1}	2.01	2.31	2.58	3.32	3.80	5.46	6.67	9.40	15.9	96.2

ETHANE

Transport Properties of the Superheated Vapor

Property (at low pressure)	Temperature, °C: −150	−100	−50	0	25	100	200	300	400	600	800	1 000	1 200
	Temperature, K: 123.15	173.15	223.15	273.15	298.15	373.15	473.15	573.15	673.15	873.15	1 073.15	1 273.15	1 473.15
Specific heat capacity, $c_{p,g}$ (kJ/kg K)	L	L	1.478	1.650	1.754	2.068	2.491	2.876	3.220	3.789	4.233	4.572	4.832
Thermal conductivity, λ_g[(W/m^2)/(K/m)]	L	L	0.013	0.018	0.021	0.034	0.047	0.064	0.081	0.118	0.154	0.186	0.217
Dynamic viscosity, η_g(10^{-5} Ns/m^2)	L	L	0.715	0.860	0.935	1.15	1.42	1.66	1.90	2.34	2.66	2.96	3.25

Thermodynamic Properties of the Superheated Vapor

Pressure, MPa	Property	300 K	340 K	380 K	420 K	460 K	500 K	540 K	580 K
0.070	v, m^3/kg	1.180	1.340	1.499	1.658	1.816	1.975	2.134	2.292
(177.8)	h, kJ/kg	764.36	838.62	919.15	1 006.11	1 099.55	1 199.46	1 305.74	1 418.27
	s, kJ/(kg K)	4.068 6	4.300 7	4.524 5	4.741 9	4.954 3	5.162 5	5.366 9	5.567 9
0.101 325	v, m^3/kg	0.813 2	0.924 0	1.034	1.144	1.254	1.364	1.474	1.583
(184.3)	h, kJ/kg	763.73	838.14	918.77	1 005.80	1 099.29	1 199.24	1 305.56	1 418.11
	s, kJ/(kg K)	3.964 8	4.197 4	4.421 4	4.639 0	4.851 6	5.059 8	5.264 3	5.465 3
0.40	v, m^3/kg	0.201 2	0.230 5	0.259 2	0.287 7	0.316 0	0.344 1	0.372 1	0.400 1
(214.3)	h, kJ/kg	757.62	833.52	915.14	1 002.86	1 096.86	1 197.19	1 303.80	1 416.60
	s, kJ/(kg K)	3.570 6	3.807 9	4.034 7	5.254 1	4.467 7	4.676 8	4.881 8	5.083 3
0.70	v, m^3/kg	0.112 2	0.129 6	0.146 5	0.163 2	0.179 6	0.195 8	0.212 0	0.228 1
(229.8)	h, kJ/kg	751.24	828.77	911.44	999.88	1 094.40	1 195.12	1 302.04	1 415.07
	s, kJ/(kg K)	3.400 8	3.643 3	3.873 0	4.094 2	4.309 0	4.518 9	4.724 5	4.926 4
1.0	v, m^3/kg	0.076 48	0.089 26	0.101 5	0.113 3	0.125 0	0.136 5	0.148 0	0.159 3
(240.9)	h, kJ/kg	744.60	823.91	907.68	996.87	1 091.93	1 193.05	1 300.27	1 413.55
	s, kJ/(kg K)	3.286 5	3.534 5	3.767 3	3.990 4	4.206 4	4.417 1	4.623 4	4.825 7
2.0	v, m^3/kg	0.034 51	0.042 05	0.048 82	0.055 20	0.061 35	0.067 36	0.073 26	0.079 10
(265.8)	h, kJ/kg	720.03	806.77	894.73	986.63	1 083.58	1 186.08	1 294.36	1 408.47
	s, kJ/(kg K)	3.035 3	3.306 7	3.551 2	3.781 1	4.001 5	4.215 1	4.423 3	4.627 1
4.0	v, m^3/kg	0.011 83	0.018 13	0.022 43	0.026 14	0.029 57	0.032 83	0.035 97	0.039 04
(295.5)	h, kJ/kg	644.02	766.56	866.58	965.16	1 066.43	1 171.98	1 282.52	1 398.37
	s, kJ/(kg K)	2.641 8	3.027 1	3.305 4	3.552 0	3.782 3	4.002 2	4.214 8	4.421 8
7.0	v, m^3/kg		0.007 27	0.011 06	0.013 74	0.016 04	0.018 12	0.020 09	0.021 97
	h, kJ/kg		678.32	817.80	930.66	1 039.90	1 150.64	1 264.88	1 383.51
	s, kJ/(kg K)		2.663 7	3.053 1	3.335 7	3.584 1	3.814 9	4.034 7	4.246 6
10	v, m^3/kg		0.003 97	0.006 72	0.008 92	0.010 74	0.012 35	0.013 83	0.015 23
	h, kJ/kg		589.90	764.34	894.94	1 013.29	1 129.67	1 247.78	1 369.25
	s, kJ/(kg K)		2.357 9	2.844 6	3.171 9	3.441 1	3.683 7	3.910 9	4.127 9
20	v, m^3/kg		0.002 77	0.003 42	0.004 29	0.005 21	0.006 09	0.006 91	0.007 69
	h, kJ/kg		532.36	668.78	808.21	941.97	1 071.47	1 199.79	1 329.20
	s, kJ/(kg K)		2.096 6	2.475 6	2.824 6	3.128 9	3.398 9	3.645 8	3.876 9
30	v, m^3/kg		0.002 52	0.002 87	0.003 32	0.003 84	0.004 38	0.004 92	0.005 44
	h, kJ/kg		526.36	646.92	774.99	905.97	1 037.43	1 169.40	1 302.75
	s, kJ/(kg K)		2.001 7	2.336 7	2.657 0	2.954 9	3.229 0	3.482 9	3.721 1

PROPANE

Chemical formula: $CH_3CH_2CH_3$
Molecular weight: 44.094
Melting point: 85.47 K
Normal boiling point: 231.10 K

Critical temperature: 370.00 K
Critical pressure: 4.264 MPa
Critical density: 225 kg/m^3
Normal vapor density: 1.97 kg/m^3 (@ 0°C, 101.3 kPa)

Properties of the Liquid at Temperatures Below the Normal Boiling Point

	Temperature, °C												
	−150	−100	−75	−50	−25	0	20	50	100	150	200	250	300
	Temperature, K												
Property	123.15	173.15	198.15	223.15	248.15	273.15	293.15	323.15	373.15	423.15	473.15	523.15	573.15
Density, ρ_l (kg/m^3)	696	646	619	590	V	V	V	V	V	V	V	V	V
Specific heat capacity, $c_{p,l}$ (kJ/kg K)	1.959	2.056	2.119	2.202	V	V	V	V	V	V	V	V	V
Thermal conductivity, λ_l [(W/m^2)/(K/m)]	–	–	–	(0.139)	V	V	V	V	V	V	V	V	V
Dynamic viscosity, η_l (10^{-5} Ns/m^2)	134.1	43.3	30.7	20.0	V	V	V	V	V	V	V	V	V

Properties of the Saturated Liquid and Vapor

T_{sat}, K	231.1	248.06	259.83	275.24	291.83	317.42	330.70	351.23	359.61	367.18
p_{sat}, kPa	101	203	304	507	810	1 520	2 026	3 039	3 545	4 052
ρ_ℓ, kg/m^3	582	562	549	528	504	460	434	381	347	300
ρ_g, kg/m^3	2.42	4.63	6.77	11.0	17.5	33.9	47.1	80.4	104	150
h_ℓ, kJ/kg	421.2	459.7	485.2	522.5	563.1	631.8	672.8	745.7	781.7	829.0
h_g, kJ/kg	847.4	866.7	879.2	895.6	911.0	929.5	937.4	942.4	934.9	915.6
$\Delta h_{g,\ell}$, kJ/kg	426.2	407.0	394.0	373.1	347.9	297.7	264.6	196.7	153.2	86.6
$c_{p,\ell}$, kJ/(kg K)	2.24	2.32	2.38	2.47	2.58	2.78	3.27	4.27	6.62	
$c_{p,g}$, kJ/(kg K)	1.37	1.51	1.65	1.88	2.27	3.37	4.14	6.16	7.01	11.42
η_ℓ, μNs/m^2	208.7	177.4	154.9	134.5	114.7	85.7	72.4	51.2	41.0	25.4
η_g, μNs/m^2	6.0	7.0	7.3	7.5	8.5	9.6	10.3	12.1	14.9	20.1
λ_ℓ, (mW/m^2)/(K/m)	134	124	117	108	99	84	76	64	59	
λ_g, (mW/m^2)/(K/m)	10.7	12.8	14.1	15.9	17.8	21.7	23.9	28.1	31.0	36.1
Pr_ℓ	3.49	3.26	3.15	3.08	2.99	2.84	3.07	3.41	4.60	
Pr_g	0.77	0.83	0.85	0.89	1.08	1.49	1.78	2.65	3.37	6.36
σ, mN/m	15.5	14.25	13.2	9.5	7.5	4.6	3.0	1.4	1.0	0.4
$\beta_{e,\ell}$, kK^{-1}	2.01	2.21	2.45	2.90	3.60	5.19	6.52	15.2	23.7	

PROPANE

Transport Properties of the Superheated Vapor

Property (at low pressure)	−150	−100	−50	0	25	100	200	300	400	600	800	1 000	1 200
	Temperature, °C												
	Temperature, K												
	123.15	173.15	223.15	273.15	298.15	373.15	473.15	573.15	673.15	873.15	1 073.15	1 273.15	1 473.15
Specific heat capacity, $c_{p,g}$ (kJ/kg K)	L	L	L	1.549	1.671	2.018	2.458	2.839	3.169	3.701	4.107	4.417	4.652
Thermal conductivity, λ_g[(W/m^2)/(K/m)]	L	L	L	0.015	0.018	0.027	0.042	0.059	0.079	0.133	0.142	0.166	0.188
Dynamic viscosity, η_g(10^{-5} Ns/m^2)	L	L	L	0.750	0.815	1.01	1.26	1.44	1.76	2.06	2.36	2.66	2.93

Thermodynamic Properties of the Superheated Vapor

Pressure, MPa	Property	250	300	350	400	450	500	550	600
		Value at different temperatures, K							
0.050	v, m^3/kg	0.929 1	1.123	1.315	1.505	1.695	1.885	2.074	2.263
(216.4)	h, kJ/kg	523.67	603.16	693.61	795.32	908.02	1 031.20	1 164.24	1 306.49
	s, kJ/(kg K)	2.410 6	2.699 9	2.978 3	3.249 6	3.514 8	3.774 2	4.027 7	4.275 1
0.101 325	v, m^3/kg	0.450 9	0.549 4	0.645 5	0.740 4	0.834 7	0.928 6	1.022	1.116
(231.3)	h, kJ/kg	521.02	601.52	692.49	794.51	907.40	1 030.72	1 163.85	1 306.17
	s, kJ/(kg K)	2.269 7	2.562 7	2.842 7	3.114 9	3.380 6	3.640 2	3.893 8	4.141 4
0.20	v, m^3/kg	0.220 8	0.273 6	0.323 8	0.372 8	0.421 1	0.469 1	0.516 9	0.564 5
(247.9)	h, kJ/kg	515.70	598.29	690.32	792.93	906.21	1 029.78	1 163.09	1 305.54
	s, kJ/(kg K)	2.126 0	2.426 7	2.710 0	2.983 7	3.250 3	3.510 5	3.764 5	4.012 3
0.40	v, m^3/kg		0.131 8	0.158 6	0.184 0	0.208 8	0.233 2	0.257 4	0.281 4
(267.9)	h, kJ/kg		591.44	685.79	789.70	903.77	1 027.87	1 161.55	1 304.27
	s, kJ/(kg K)		2.279 4	2.570 0	2.847 2	3.115 7	3.377 0	3.631 7	3.879 9
1.0	v, m^3/kg			0.059 15	0.070 65	0.081 35	0.091 63	0.101 7	0.111 6
(300.3)	h, kJ/kg			671.19	779.59	896.28	1 022.05	1 156.89	1 300.45
	s, kJ/(kg K)			2.366 9	2.656 2	2.930 8	3.195 7	3.452 6	3.702 3
2.0	v, m^3/kg			0.025 34	0.032 68	0.038 82	0.044 44	0.049 79	0.054 97
(330.4)	h, kJ/kg			641.60	761.13	883.12	1 012.06	1 148.99	1 294.01
	s, kJ/(kg K)			2.172 3	2.491 6	2.778 8	3.050 4	3.311 3	3.563 5
4.0	v, m^3/kg				0.013 16	0.017 45	0.020 85	0.023 89	0.026 73
(366.5)	h, kJ/kg				714.17	853.84	990.91	1 132.70	1 280.98
	s, kJ/(kg K)				2.270 2	2.599 4	2.888 2	3.158 4	3.416 4
7.0	v, m^3/kg				0.004 18	0.008 22	0.010 80	0.012 88	0.014 73
	h, kJ/kg				591.69	800.86	956.60	1 107.49	1 261.33
	s, kJ/(kg K)				1.906 4	2.401 9	2.730 3	3.017 9	3.285 6
10	v, m^3/kg				0.002 98	0.004 95	0.006 97	0.008 62	0.010 05
	h, kJ/kg				544.14	746.96	921.77	1 082.57	1 242.27
	s, kJ/(kg K)				1.761 9	2.240 0	2.608 8	2.915 4	3.193 3
20	v, m^3/kg				0.002 38	0.002 87	0.003 55	0.004 31	0.005 06
	h, kJ/kg				517.85	678.83	849.71	1 021.44	1 192.46
	s, kJ/(kg K)				1.631 4	2.009 8	2.370 2	2.697 6	2.995 2

BUTANE

Chemical formula: $CH_3CH_2CH_2CH_3$
Molecular weight: 58.12
Melting point: 134.82 K
Normal boiling point: 272.66 K

Critical temperature: 425.16 K
Critical pressure: 3.796 MPa
Critical density: 225.3 kg/m^3
Normal vapor density: 2.59 kg/m^3
(@ 0°C, 101.3 kPa)

Properties of the Liquid at Temperatures Below the Normal Boiling Point

	Temperature, °C												
	−150	−100	−75	−50	−25	0	20	50	100	150	200	250	300
	Temperature, K												
Property	123.15	173.15	198.15	223.15	248.15	273.15	293.15	323.15	373.15	423.15	473.15	523.15	573.15
Density, ρ_l (kg/m^3)	S	698	676	652	627	V	V	V	V	V	V	V	V
Specific heat capacity, $c_{p,l}$ (kJ/kg K)	S	2.001	2.052	2.119	2.194	V	V	V	V	V	V	V	V
Thermal conductivity, λ_l [(W/m^2)/(K/m)]	S	0.166	0.153	(0.129)	(0.122)	V	V	V	V	V	V	V	V
Dynamic viscosity, η_l (10^{-5} Ns/m^2)	S	74.7	49.7	35.5	26.7	V	V	V	V	V	V	V	V

Properties of the Saturated Liquid and Vapor

T_{sat}, K	273.15	289	305	321	337	353	369	385	405	425.16
p_{sat}, kPa	103	184	304	469	706	1 023	1 526	1 925	2 739	3 797
ρ_ℓ, kg/m^3	603	587	571	551	529	504	475	441	388	225.3
ρ_g, kg/m^3	2.81	4.81	7.53	11.6	17.4	25.1	35.6	51.3	80.7	225.3
h_ℓ, kJ/kg	−1 194	−1 158	−1 121	−1 081	−1 040	−997	−945	−896	−821	−665
h_g, kJ/kg	−809	−789	−769	−747	−725	−706	−681	−663	−648	−665
$\Delta h_{g,\ell}$, kJ/kg	385	369	352	334	315	291	264	233	173	
$c_{p,\ell}$, kJ/(kg K)	2.34	2.47	2.59	2.68	2.80	2.95	3.11	3.36	3.80	
$c_{p,g}$, kJ/(kg K)	1.67	1.76	1.88	2.00	2.15	2.33	2.62	3.03	4.76	
η_ℓ, μNs/m^2	206	179	154	131	112	95	80	65	51	
η_g, μNs/m^2	7.35	7.81	8.32	8.87	9.44	10.20	11.25	12.77	16.30	
λ_ℓ, (mW/m^2)/(K/m)	114.6	109.8	104.9	100.1	95.1	90.4	85.5	80.7	74.6	48.7
λ_g, (mW/m^2)/(K/m)	13.69	15.19	16.82	18.57	20.47	22.49	24.69	27.24	31.2	48.7
Pr_ℓ	4.20	4.02	3.80	3.51	3.30	3.11	2.89	2.72	2.59	
Pr_g	0.90	0.90	0.93	0.96	1.00	1.06	1.19	1.42	2.48	
σ, mN/m	14.8	12.8	11.0	9.10	7.29	5.54	4.03	2.75	1.34	
$\beta_{e,\ell}$, kK^{-1}	1.73	2.01	2.37	2.80	3.45	4.31	7.31	9.87	10.0	

Transport Properties of the Superheated Vapor

	Temperature, °C												
	−150	−100	−50	0	25	100	200	300	400	600	800	1 000	1 200
Property	Temperature, K												
(at low pressure)	123.15	173.15	223.15	273.15	298.15	373.15	473.15	573.15	673.15	873.15	1 073.15	1 273.15	1 473.15
Specific heat capacity, $c_{p,g}$ (kJ/kg K)	S	L	L	1.599	1.700	2.031	2.453	2.822	3.136	3.647	4.032	4.329	4.551
Thermal conductivity, λ_g[(W/m^2)/(K/m)]	S	L	L	0.014	0.016	0.024	0.037	0.052	0.066	0.097	0.126	0.152	(0.176)
Dynamic viscosity, η_g(10^{-5} Ns/m^2)	S	L	L	0.689	0.744	0.950	1.185	1.42	1.65	2.10	2.59	2.99	2.36

PENTANE

Chemical formula: $CH_3(CH_2)_3CH_3$
Molecular weight: 72.151
Melting point: 143.4 K
Normal boiling point: 309.2 K

Critical temperature: 469.6 K
Critical pressure: 3.369 MPa
Critical density: 273.3 kg/m^3
Normal vapor density: 3.22 kg/m^3
(@ 0 °C, 101.3 kPa)

Properties of the Liquid at Temperatures Below the Normal Boiling Point

	Temperature, °C												
	−150	−100	−75	−50	−25	0	20	50	100	150	200	250	300
	Temperature, K												
Property	123.15	173.15	198.15	223.15	248.15	273.15	293.15	323.15	373.15	423.15	473.15	523.15	573.15
Density, ρ_l(kg/m^3)	S	737	715	693	670	646	626	V	V	V	V	V	V
Specific heat capacity, $c_{p,l}$ (kJ/kg K)	S	1.972	2.001	2.060	2.123	2.206	2.273	V	V	V	V	V	V
Thermal conductivity, λ_l [(W/m^2)/(K/m)]	S	0.155	0.151	0.148	0.144	0.140	0.136	V	V	V	V	V	V
Dynamic viscosity, η_l (10^{-5} Ns/m^2)	S	125.0	66.0	48.4	36.4	27.7	22.7	V	V	V	V	V	V

Properties of the Saturated Liquid and Vapor

T_{sat}, K	309.2	335	350	365	380	395	410	425	440	469.6
p_{sat}, kPa	101.3	227	341	492	688	935	1 249	1 634	2 103	3 370
ρ_ℓ, kg/m^3	610.2	582.9	566.0	548.0	528.9	507.9	484.1	456.5	423.5	280.9
ρ_g, kg/m^3	3.00	6.36	9.41	13.51	18.99	26.11	36.21	49.73	68.96	184.1
h_ℓ, kJ/kg	319.8	383.8	423.3	458.2	504.7	546.6	588.5	637.3	686.2	846.7
h_g, kJ/kg	678.0	721.1	744.3	767.6	790.8	814.1	837.4	855.9	876.9	846.7
$\Delta h_{g,\ell}$, kJ/kg	358.2	337.3	321.0	309.4	286.1	267.5	248.9	218.6	190.7	
$c_{p,\ell}$, kJ/(kg K)	2.34	2.52	2.62	2.72	2.82	2.94	3.06	3.20	3.44	
$c_{p,g}$, kJ/(kg K)	1.79	1.96	2.05	2.16	2.28	2.48	2.66	2.96	3.37	
η_ℓ, μNs/m^2	196	159	140	123	108	95	83	72	60	
η_g, μNs/m^2	6.9	7.6	8.1	8.5	9.0	9.5	10.2	11.1	12.4	
λ_ℓ, (mW/m^2)/(K/m)	107	98	93	88	83	79	75	71	69	47
λ_g, (mW/m^2)/(K/m)	16.7	19.3	21.0	22.8	24.8	26.7	29.0	31.7	34.9	47
Pr_ℓ	4.29	4.09	3.94	3.80	3.67	3.53	3.36	3.16	2.84	
Pr_g	0.74	0.77	0.79	0.81	0.83	0.88	0.94	1.03	1.19	
σ, mN/m	14.3	11.3	9.7	8.1	6.7	5.2	3.8	2.8	1.6	
$\beta_{e,\ell}$, kK^{-1}	1.40	1.93	2.21	2.52	2.92	3.54	4.48	5.95	15.5	

Transport Properties of the Superheated Vapor

	Temperature, °C												
	−150	−100	−50	0	25	100	200	300	400	600	800	1 000	1 200
	Temperature, K												
Property (at low pressure)	123.15	173.15	223.15	273.15	298.15	373.15	473.15	573.15	673.15	873.15	1 073.15	1 273.15	1 473.15
Specific heat capacity, $c_{p,g}$ (kJ/kg K)	S	L	L	L	L	2.026	2.445	2.809	3.115	3.613	3.990	4.275	4.488
Thermal conductivity, λ_g[(W/m^2)/(K/m)]	S	L	L	L	L	0.021	0.034	0.047	0.061	0.090	0.117	0.142	(0.162)
Dynamic viscosity, η_g(10^{-5} Ns/m^2)	S	L	L	L	L	0.860	1.09	1.29	1.49	1.85	2.17	2.46	2.74

HEXANE

Chemical formula: $CH_3(CH_2)_4CH_3$
Molecular weight: 86.178
Melting point: 177.83 K
Normal boiling point: 341.88 K

Critical temperature: 507.44 K
Critical pressure: 3.031 MPa
Critical density: 232.8 kg/m^3
Normal vapor density: 3.85 kg/m^3 (@ 0°C, 101.3 kPa)

Properties of the Liquid at Temperatures Below the Normal Boiling Point

	Temperature, °C												
	−150	−100	−75	−50	−25	0	20	50	100	150	200	250	300
	Temperature, K												
Property	123.15	173.15	198.15	223.15	248.15	273.15	293.15	323.15	373.15	423.15	473.15	523.15	573.15
Density, ρ_l (kg/m^3)	S	S	742	721	700	678	659	631	V	V	V	V	V
Specific heat capacity, $c_{p,l}$ (kJ/kg K)	S	S	1.993	2.035	2.093	2.165	2.227	(2.37)	V	V	V	V	V
Thermal conductivity, λ_l [(W/m^2)/(K/m)]	S	S	0.156	0.146	0.137	0.127	0.120	0.110	V	V	V	V	V
Dynamic viscosity, η_l (10^{-5} Ns/m^2)	S	S	92.0	68.5	51.5	38.3	30.8	22.9	V	V	V	V	V

Properties of the Saturated Liquid and Vapor

T_{sat}, K	341.88	370	385	400	415	430	445	460	475	507.44
p_{sat}, kPa	101.3	228	331	465	638	854	1 124	1 457	1 859	3 031
ρ_ℓ, kg/m^3	613	585	568	551	532	511	488	463	432	233
ρ_g, kg/m^3	3.3	7.0	10.0	14.1	19.5	26.7	34.7	48.6	66.2	233
h_ℓ, kJ/kg	395.4	465.2	507.1	546.6	586.2	632.7	676.9	725.7	774.6	930.4
h_g, kJ/kg	728.0	776.9	802.5	830.4	856.0	879.2	907.1	930.4	953.7	930.4
$\Delta h_{g,\ell}$, kJ/kg	332.6	311.7	295.4	283.8	269.8	246.5	230.2	204.7	179.1	
$c_{p,\ell}$, kJ/(kg K)	2.39	2.58	2.68	2.78	2.89	3.00	3.12	3.26	3.46	
$c_{p,g}$, kJ/(kg K)	1.91	2.07	2.18	2.28	2.39	2.51	2.66	2.92	3.32	
η_ℓ, μNs/m^2	202.2	158.9	145.7	128.5	113.6	99.9	87.4	75.6	64.3	
η_g, μNs/m^2	7.3	7.9	8.4	8.8	9.4	10.1	10.9	11.9	13.3	
λ_ℓ, (mW/m^2)/(K/m)	100.4	91.0	87	82	78	74.5	67.8	63.0	56.0	49.8
λ_g, (mW/m^2)/(K/m)	16.7	18.8	20.4	22.4	24.2	26.2	28.5	30.4	33.8	49.8
Pr_ℓ	4.81	4.51	4.49	4.36	4.19	4.01	3.99	3.83	3.97	
Pr_g	0.83	0.87	0.91	0.91	0.92	0.97	1.02	1.14	1.31	
σ, mN/m	13.33	10.57	9.13	7.71	6.35	5.02	3.78	2.66	1.63	
$\beta_{e,\ell}$, kK^{-1}	1.28	1.86	2.17	2.46	2.88	3.41	4.14	5.41	20.0	

Transport Properties of the Superheated Vapor

	Temperature, °C												
	−150	−100	−50	0	25	100	200	300	400	600	800	1 000	1 200
Property (at low pressure)	Temperature, K												
	123.15	173.15	223.15	273.15	298.15	373.15	473.15	573.15	673.15	873.15	1 073.15	1 273.15	1 473.15
Specific heat capacity, $c_{p,g}$ (kJ/kg K)	S	S	L	L	L	2.026	2.441	2.801	3.120	3.583	3.957	4.237	4.446
Thermal conductivity, λ_g[(W/m^2)/(K/m)]	S	S	L	L	L	0.019	0.030	0.043	0.056	0.084	0.109	0.132	(0.152)
Dynamic viscosity, η_g(10^{-5} Ns/m^2)	S	S	L	L	L	0.822	1.04	1.23	1.48	1.90	2.12	2.40	2.66

HEPTANE

Chemical formula: $CH_3(CH_2)_5CH_3$
Molecular weight: 100.198
Melting point: 182.6 K
Normal boiling point: 371.6 K

Critical temperature: 540.61 K
Critical pressure: 2.736 MPa
Critical density: 234.1 kg/m^3
Normal vapor density: 4.47 kg/m^3
(@ 0°C, 101.3 kPa)

Properties of the Liquid at Temperatures Below the Normal Boiling Point

	Temperature, °C												
	−150	−100	−75	−50	−25	0	20	50	100	150	200	250	300
	Temperature, K												
Property	123.15	173.15	198.15	223.15	248.15	273.15	293.15	323.15	373.15	423.15	473.15	523.15	573.15
Density, ρ_l (kg/m^3)	S	S	761	741	721	701	684	658	V	V	V	V	V
Specific heat capacity, $c_{p,l}$ (kJ/kg K)	S	S	2.104	2.035	2.081	2.144	2.198	2.307	V	V	V	V	V
Thermal conductivity, λ_l [(W/m^2)/(K/m)]	S	S	0.156	0.148	0.139	0.131	0.124	0.114	V	V	V	V	V
Dynamic viscosity, η_l (10^{-5} Ns/m^2)	S	S	129.0	96.6	72.5	52.6	41.3	30.2	V	V	V	V	V

Properties of the Saturated Liquid and Vapor

T_{sat}, K	371.6	380	400	420	440	460	480	500	520	540.6
p_{sat}, kPa	101.3	130	219	349	529	721	1 094	1 513	2 046	2 736
ρ_ℓ, kg/m^3	614	606	585	563	540	512	484	448	397	234
ρ_g, kg/m^3	3.46	4.36	7.23	11.5	17.4	25.6	37.8	56.5	88.3	234
h_ℓ, kJ/kg	453.6	474.5	530.3	586.2	639.7	702.5	760.6	825.7	895.5	1 004.8
h_g, kJ/kg	765.3	786.2	823.4	860.6	897.8	937.4	972.3	1 009.5	1 035.1	1 004.8
$\Delta h_{g,\ell}$, kJ/kg	319.7	311.7	293.1	274.4	258.1	234.9	211.7	183.8	139.6	
$c_{p,\ell}$, kJ/(kg K)	2.57	2.61	2.72	2.82	2.93	3.03	3.19	3.39	3.85	
$c_{p,g}$, kJ/(kg K)	1.98	2.01	2.15	2.26	2.41	2.51	2.72	3.05	3.60	
η_ℓ, μNs/m^2	201	186	159	135	115	97	82	67	54	41
η_g, μNs/m^2	7.3	7.7	8.3	9.0	9.7	10.7	12.1	14.0	17.6	41
λ_ℓ, (mW/m^2)/(K/m)	98.0	95.5	88.8	82.9	76.1	69.9	61.1	52.0	41.8	
λ_g, (mW/m^2)/(K/m)	18.0	18.4	20.1	22.6	24.9	27.7	29.9	33.0	36.0	
Pr_ℓ	5.27	5.08	4.87	4.59	4.43	4.27	4.28	4.37	4.97	
Pr_g	0.80	0.84	0.89	0.90	0.94	0.97	1.10	1.29	1.76	
σ, mN/m	12.6	11.8	10.0	8.3	6.6	5.1	3.6	2.2	0.8	
$\beta_{e,\ell}$, kK^{-1}	1.6	1.7	1.90	2.2	2.7	3.2	4.2	6.8	24.5	

Transport Properties of the Superheated Vapor

	Temperature, °C												
	−150	−100	−50	0	25	100	200	300	400	600	800	1 000	1 200
Property (at low pressure)	Temperature, K												
	123.15	173.15	223.15	273.15	298.15	373.15	473.15	573.15	673.15	873.15	1 073.15	1 273.15	1 473.15
Specific heat capacity, $c_{p,g}$ (kJ/kg K)	S	S	L	L	L	2.026	2.437	2.793	3.070	3.571	3.936	4.212	4.417
Thermal conductivity, λ_g[(W/m^2)/(K/m)]	S	S	L	L	L	0.017	0.029	0.041	0.054	0.080	0.104	0.124	(0.142)
Dynamic viscosity, η_g(10^{-5} Ns/m^2)	S	S	L	L	L	0.76	0.95	1.14	1.32	1.65	1.97	2.26	(2.55)

OCTANE

Chemical formula: $CH_3(CH_2)_6CH_3$
Molecular weight: 114.224
Melting point: 216.35 K
Normal boiling point: 398.8 K

Critical temperature: 568.8 K
Critical pressure: 2.486 MPa
Critical density: 232 kg/m^3
Normal vapor density: 5.10 kg/m^3
(@ 0 °C, 101.3 kPa)

Properties of the Liquid at Temperatures Below the Normal Boiling Point

Property	Temperature, °C −150	−100	−75	−50	−25	0	20	50	100	150	200	250	300
	Temperature, K 123.15	173.15	198.15	223.15	248.15	273.15	293.15	323.15	373.15	423.15	473.15	523.15	573.15
Density, ρ_l (kg/m^3)	S	S	S	757	738	719	703	678	635	V	V	V	V
Specific heat capacity, $c_{p,l}$ (kJ/kg K)	S	S	S	2.043	2.064	2.131	2.186	2.303	(2.51)	V	V	V	V
Thermal conductivity, λ_l [(W/m^2)/(K/m)]	S	S	S	0.152	0.144	0.137	0.131	0.122	0.107	V	V	V	V
Dynamic viscosity, η_l (10^{-5} Ns/m^2)	S	S	S	137.0	102.0	71.4	54.6	38.9	24.4	V	V	V	V

Properties of the Saturated Liquid and Vapor

T_{sat}, K	398.8	415	435	455	475	495	515	535	555	568.8
p_{sat}, kPa	101.3	156	252	386	569	809	1 127	1 535	2 052	2 486
ρ_ℓ, kg/m^3	611	595	575	553	529	502	470	432	373	232
ρ_g, kg/m^3	3.80	5.67	8.98	13.7	20.4	29.9	44.0	65.0	105	232
h_ℓ, kJ/kg	514.1	558.2	609.4	667.6	725.7	786.2	851.3	921.1	990.9	1 088.6
h_g, kJ/kg	814.1	844.3	883.9	923.4	965.3	1 004.8	1 046.7	1 081.6	1 109.5	1 088.6
$\Delta h_{g,\ell}$, kJ/kg	300.0	286.1	274.5	255.8	239.6	218.6	195.4	160.5	118.6	
$c_{p,\ell}$, kJ/(kg K)	2.50	2.61	2.74	2.89	3.03	3.18	3.36	3.60	4.23	
$c_{p,g}$, kJ/(kg K)	2.11	2.19	2.30	2.42	2.56	2.73	2.96	3.30	4.80	
η_ℓ, μNs/m^2	203	174	149	126	107	91	75	61	48	
η_g, μNs/m^2	7.4	7.9	8.5	9.2	10.0	11.1	12.5	14.7	19.4	
λ_ℓ, (mW/m^2)/(K/m)	98	93	88	83	78	73	68	62	56	47.7
λ_g, (mW/m^2)/(K/m)	18.4	20.3	22.5	24.3	26.7	29.1	31.6	34.4	38.6	47.7
Pr_ℓ	5.18	4.88	4.64	4.39	4.16	3.96	3.71	3.54	3.63	
Pr_g	0.85	0.85	0.87	0.92	0.96	1.04	1.21	1.41	2.41	
σ, mN/m	11.9	10.5	8.9	7.3	5.8	4.4	2.9	1.8	0.55	
$\beta_{e,\ell}$, kK^{-1}	1.62	1.79	2.05	2.39	2.88	3.60	4.82	6.96	15.9	

Transport Properties of the Superheated Vapor

Property (at low pressure)	Temperature, °C −150	−100	−50	0	25	100	200	300	400	600	800	1 000	1 200
	Temperature, K 123.15	173.15	223.15	273.15	298.15	373.15	473.15	573.15	673.15	873.15	1 073.15	1 273.15	1 473.15
Specific heat capacity, $c_{p,g}$ (kJ/kg K)	S	S	L	L	L	L	2.437	2.788	3.081	3.559	3.919	4.191	4.396
Thermal conductivity, λ_g[(W/m^2)/(K/m)]	S	S	L	L	L	L	0.026	0.038	0.050	0.076	0.099	0.119	(0.136)
Dynamic viscosity, η_g(10^{-5} Ns/m^2)	S	S	L	L	L	L	0.845	(1.04)	(1.21)	(1.52)	(1.82)	(2.09)	(2.36)

NONANE

Chemical formula: $CH_3(CH_2)_7CH_3$
Molecular weight: 128.26
Melting point: 219.7 K
Normal boiling point: 423.97 K

Critical temperature: 594.63 K
Critical pressure: 2.289 MPa
Critical density: 234 kg/m^3
Normal vapor density: 5.72 kg/m
(@ 0 °C, 101.3 kPa)

Properties of the Liquid at Temperatures Below the Normal Boiling Point

	Temperature, °C												
	−150	−100	−75	−50	−25	0	20	50	100	150	200	250	300
	Temperature, K												
Property	123.15	173.15	198.15	223.15	248.15	273.15	293.15	323.15	373.15	423.15	473.15	523.15	573.15
Density, ρ_l (kg/m^3)	S	S	S	769	751	733	718	694	653	609	V	V	V
Specific heat capacity, $c_{p,l}$ (kJ/kg K)	S	S	S	2.047	2.060	2.127	2.177	(2.29)	(2.48)	(2.74)	V	V	V
Thermal conductivity, λ_l [(W/m^2)/(K/m)]	S	S	S	0.150	0.142	0.136	0.132	0.124	0.111	0.096	V	V	V
Dynamic viscosity, η_l (10^{-5} Ns/m^2)	S	S	S	193.0	143.1	96.4	71.3	49.2	30.1	19.9	V	V	V

Properties of the Saturated Liquid and Vapor

T_{sat}, K	423.97	435	455	475	495	515	535	555	575	594.63
p_{sat}, kPa	101.3	134	214	338	496	717	965	1 320	1 750	2 289
ρ_ℓ, kg/m^3	614	602	581	560	535	510	479	444	394	234
ρ_g, kg/m^3	3.94	5.18	8.22	12.6	18.7	27.3	39.6	58.0	89.8	234
h_ℓ, kJ/kg	195	226	282	340	402	464	529	594	664	754
h_g, kJ/kg	490	513	555	598	640	683	724	762	792	754
$\Delta h_{g,\ell}$, kJ/kg	295	287	273	258	238	219	195	168	128	
$c_{p,\ell}$, kJ/(kg K)	2.72	2.77	2.87	2.97	3.10	3.23	3.38	3.59	4.00	
$c_{p,g}$, kJ/(kg K)	2.24	2.30	2.40	2.51	2.63	2.77	2.95	3.25	4.05	
η_ℓ, μNs/m^2	213	177	153	132	112	94	78	63	50	
η_g, μNs/m^2	7.4	7.6	7.9	8.3	8.7	9.1	9.7	10.6	12.2	
λ_ℓ, (mW/m^2)/(K/m)	95	92	87	82	77	72	66	60	55	49
λ_g, (mW/m^2)/(K/m)	20.8	22.0	24.1	26.4	28.7	31.2	33.8	36.6	40.0	49
Pr_ℓ	6.10	5.33	5.04	4.78	4.51	4.22	3.99	3.77	3.64	
Pr_g	0.80	0.79	0.79	0.79	0.80	0.81	0.85	0.94	1.24	
σ, mN/m	11.3	10.4	8.86	7.33	5.86	4.46	3.13	1.90	1.81	
$\beta_{e,\ell}$, kK^{-1}	1.60	1.71	1.94	2.27	2.70	3.33	4.36	6.31	11.7	

Transport Properties of the Superheated Vapor

	Temperature, °C												
	− 150	− 100	− 50	0	25	100	200	300	400	600	800	1 000	1 200
Property (at low pressure)	Temperature, K												
	123.15	173.15	223.15	273.15	298.15	373.15	473.15	573.15	673.15	873.15	1 073.15	1 273.15	1 473.15
Specific heat capacity, $c_{p,g}$ (kJ/kg K)	S	S	L	L	L	L	2.433	2.784	3.077	3.550	3.906	4.178	4.379
Thermal conductivity, λ_g[(W/m^2)/(K/m)]	S	S	L	L	L	L	0.026	0.037	0.048	0.071	0.094	0.115	—
Dynamic viscosity, η_g(10^{-5} Ns/m^2)	S	S	L	L	L	L	0.84	(1.00)	(1.16)	(1.47)	(1.76)	(2.04)	—

DECANE

Chemical formula: $CH_3(CH_2)_8CH_3$
Molecular weight: 142.3
Melting point: 243.51 K
Normal boiling point: 447.4 K

Critical temperature: 617.6 K
Critical pressure: 2.096 MPa
Critical density: 235.9 kg/m^3
Normal vapor density: 6.35 kg/m^3
(@ 0°C, 101.3 kPa)

Properties of the Liquid at Temperatures Below the Normal Boiling Point

	Temperature, °C												
	−150	−100	−75	−50	−25	0	20	50	100	150	200	250	300
	Temperature, K												
Property	123.15	173.15	198.15	223.15	248.15	273.15	293.15	323.15	373.15	423.15	473.15	523.15	573.15
Density, ρ_l (kg/m^3)	S	S	S	S	762	745	730	707	667	625	V	V	V
Specific heat capacity, $c_{p,l}$ (kJ/kg K)	S	S	S	S	2.085	2.127	2.173	2.265	2.474	(2.71)	V	V	V
Thermal conductivity, λ_l [(W/m^2)/(K/m)]	S	S	S	S	0.143	0.132	0.126	0.117	0.104	0.093	V	V	V
Dynamic viscosity, η_l (10^{-5} Ns/m^2)	S	S	S	S	200.0	129.2	92.1	61.5	36.4	23.7	V	V	V

Properties of the Saturated Liquid and Vapor

T_{sat}, K	447.31	460	480	500	520	540	560	580	600	617.6
p_{sat}, kPa	101.3	141	227	329	479	675	927	1 250	1 650	2 096
ρ_ℓ, kg/m^3	621	608	588	564	538	513	479	445	392	235.9
ρ_g, kg/m^3	4.13	5.60	8.73	13.2	19.4	28.1	40.7	59.6	93.1	235.9
h_ℓ, kJ/kg	265	299	357	416	476	538	603	673	743	824
h_g, kJ/kg	542	569	613	658	702	746	789	829	859	824
$\Delta h_{g,\ell}$, kJ/kg	277	270	256	242	226	208	186	156	116	
$c_{p,\ell}$, kJ/(kg K)	2.79	2.85	2.94	3.06	3.16	3.29	3.44	3.65	4.06	
$c_{p,g}$, kJ/(kg K)	2.33	2.39	2.49	2.59	2.71	2.85	3.03	3.33	4.20	
η_ℓ, μNs/m^2	205	180	149	127	108	91	75	61	48	
η_g, μNs/m^2	8.1	8.3	8.7	9.1	9.6	10.1	10.7	11.6	13.3	
λ_ℓ, (mW/m^2)/(K/m)	91	88	83	78	73	68	63	58	54	49
λ_g, (mW/m^2)/(K/m)	21.8	23.1	25.3	27.5	29.8	32.1	34.7	37.5	40.9	49
Pr_ℓ	6.29	5.83	5.28	4.98	4.68	4.40	4.10	3.84	3.61	
Pr_g	0.87	0.86	0.86	0.86	0.87	0.90	0.93	1.03	1.37	
σ, mN/m	10.64	9.68	8.20	6.77	5.39	4.07	2.83	1.68	0.67	
$\beta_{e,\ell}$, kK^{-1}	1.61	1.70	1.99	2.30	2.71	3.41	4.58	6.60	12.8	

Transport Properties of the Superheated Vapor

	Temperature °C												
	− 150	− 100	− 50	0	25	100	200	300	400	600	800	1 000	1 200
	Temperature, K												
Property (at low pressure)	123.15	173.15	223.15	273.15	298.15	373.15	473.15	573.15	673.15	873.15	1 073.15	1 273.15	1 473.15
Specific heat capacity, $c_{p,g}$ (kJ/kg K)	S	S	S	L	L	L	2.433	2.780	3.073	3.542	3.898	4.166	4.362
Thermal conductivity, λ_g[(W/m^2)/(K/m)]	S	S	S	L	L	L	0.023	0.034	0.046	0.070	0.091	0.104	—
Dynamic viscosity, η_g(10^{-5} Ns/m^2)	S	S	S	L	L	L	0.82	(0.98)	(1.14)	(1.44)	(1.73)	(2.00)	—

UNDECANE

Chemical formula: $C_{11}H_{24}$
Molecular weight: 156.31
Melting point: 247.6 K
Boiling point: 469.1 K

Critical temperature: 638.8 K
Critical pressure: 1.97 MPa
Critical density: 237 kg/m^3
Normal vapor density: 6.97 kg/m^3
(@ 0°C, 101.3 kPa)

Properties of the Liquid at Temperatures Below the Normal Boiling Point

	Temperature, °C												
	−150	−100	−75	−50	−25	0	20	50	100	150	200	250	300
	Temperature, K												
Property	123.15	173.15	198.15	223.15	248.15	273.15	293.15	323.15	373.15	423.15	473.15	523.15	573.15
Density, ρ_l (kg/m^3)	S	S	S	S	—	756	751	727	689	648	605	V	V
Specific heat capacity, $c_{p,l}$ (kJ/kg K)	S	S	S	S	S	2.148	2.173	(2.27)	(2.43)	(2.66)	(2.94)	V	V
Thermal conductivity, λ_l [(W/m^2)/(K/m)]	S	S	S	S	S	0.147	0.143	0.137	0.128	0.118	0.109	V	V
Dynamic viscosity, η_l (10^{-5} Ns/m^2)	S	S	S	S	S	226.8	148.9	02.2	51.1	32.3	21.7	V	V

Transport Properties of the Superheated Vapor

	Temperature, °C												
	− 150	− 100	− 50	0	25	100	200	300	400	600	800	1 000	1 200
Property (at low pressure)	Temperature, K												
	123.15	173.15	223.15	273.15	298.15	373.15	473.15	573.15	673.15	873.15	1 073.15	1 273.15	1 473.15
Specific heat capacity, $c_{p,g}$ (kJ/kg K)	S	S	S	L	L	L	2.433	2.776	3.069	3.538	3.900	4.153	4.354
Thermal conductivity, λ_g[(W/m^2)/(K/m)]	S	S	S	L	L	L	(0.022)	(0.032)	(0.044)	(0.067)	(0.087)	(0.100)	—
Dynamic viscosity, η_g(10^{-5} Ns/m^2)	S	S	S	L	L	L	(0.79)	(0.94)	(1.09)	(1.38)	(1.66)	(1.93)	—

DODECANE

Chemical formula: $C_{12}H_{26}$
Molecular weight: 170.34
Melting point: 263.0 K
Boiling point: 489.5 K

Critical temperature: 658.3 K
Critical pressure: 1.82 MPa
Critical density: 237 kg/m^3
Normal vapor density: 7.60 kg/m^3
(@ 0°C, 101.3 kPa)

Properties of the Liquid at Temperatures Below the Normal Boiling Point

	Temperature, °C												
	−150	−100	−75	−50	−25	0	20	50	100	150	200	250	300
	Temperature, K												
Property	123.15	173.15	198.15	223.15	248.15	273.15	293.15	323.15	373.15	423.15	473.15	523.15	573.15
Density, ρ_l (kg/m^3)	S	S	S	S	772	754	740	718	679	638	V	V	V
Specific heat capacity, $c_{p,l}$ (kJ/kg K)	S	S	S	S	2.114	2.123	2.173	(2.29)	(2.45)	(2.67)	V	V	V
Thermal conductivity, λ_l [(W/m^2)/(K/m)]	S	S	S	S	(0.154)	0.146	0.141	0.134	0.124	0.113	V	V	V
Dynamic viscosity, η_l (10^{-5} Ns/m^2)	S	S	S	S	280.0	171.7	117.6	75.7	43.3	27.9	V	V	V

Transport Properties of the Superheated Vapor

	Temperature, °C												
	−150	−100	−50	0	25	100	200	300	400	600	800	1 000	1 200
Property (at low pressure)	Temperature, K												
	123.15	173.15	223.15	273.15	298.15	373.15	473.15	573.15	673.15	873.15	1 073.15	1 273.15	1 473.15
Specific heat capacity, $c_{p,g}$ (kJ/kg K)	S	S	S	L	L	L	L	2.776	3.065	3.529	3.881	4.145	4.342
Thermal conductivity, λ_g[(W/m^2)/(K/m)]	S	S	S	L	L	L	L	(0.031)	(0.042)	(0.064)	(0.084)	(0.096)	—
Dynamic viscosity, η_g(10^{-5} Ns/m^2)	S	S	S	L	L	L	L	(0.91)	(1.05)	(1.34)	(1.61)	(1.87)	—

TRIDECANE

Chemical formula: $C_{13}H_{28}$
Molecular weight: 184.37
Melting point: 267.8 K
Boiling point: 508.6 K

Critical temperature: 675.8 K
Critical pressure: 1.72 MPa
Critical density: 240 kg/m^3
Normal vapor density: 8.23 kg/m^3
(@ 0°C, 101.3 kPa)

Properties of the Liquid at Temperatures Below the Normal Boiling Point

	Temperature, °C												
	−150	−100	−75	−50	−25	0	20	50	100	150	200	250	300
	Temperature, K												
Property	123.15	173.15	198.15	223.15	248.15	273.15	293.15	323.15	373.15	423.15	473.15	523.15	573.15
Density, ρ_l (kg/m^3)	S	S	S	S	S	770	756	735	698	658	615	V	V
Specific heat capacity, $c_{p,l}$ (kJ/kg K)	S	S	S	S	S	1.971	2.160	(2.27)	(2.43)	(2.64)	(2.91)	V	V
Thermal conductivity, λ_l [(W/m^2)/(K/m)]	S	S	S	S	S	0.148	0.146	0.141	0.132	0.123	0.115	V	V
Dynamic viscosity, η_l (10^{-5} Ns/m^2)	S	S	S	S	S	297.6	186.5	111.0	59.5	37.2	24.8	V	V

Transport Properties of the Superheated Vapor

	Temperature, °C												
	− 150	− 100	− 50	0	25	100	200	300	400	600	800	1 000	1 200
Property	Temperature, K												
(at low pressure)	123.15	173.15	223.15	273.15	298.15	373.15	473.15	573.15	673.15	873.15	1 073.15	1 273.15	1 473.15
Specific heat capacity, $c_{p,g}$ (kJ/kg K)	S	S	S	L	L	L	L	2.772	3.061	3.525	3.873	4.141	4.333
Thermal conductivity, λ_g[(W/m^2)/(K/m)]	S	S	S	L	L	L	L	(0.030)	(0.040)	(0.063)	(0.082)	(0.094)	—
Dynamic viscosity, η_g(10^{-5} Ns/m^2)	S	S	S	L	L	L	L	(0.88)	(1.03)	(1.31)	(1.57)	(1.83)	—

TETRADECANE

Chemical formula: $C_{14}H_{30}$
Molecular weight: 198.39
Melting point: 279.0 K
Boiling point: 526.7 K

Critical temperature: 694 K
Critical pressure: 1.62 MPa
Critical density: 240 kg/m^3
Normal vapor density: 8.85 kg/m^3
(@ 0 °C, 101.3 kPa)

Properties of the Liquid at Temperatures Below the Normal Boiling Point

	Temperature, °C												
	−150	−100	−75	−50	−25	0	20	50	100	150	200	250	300
	Temperature, K												
Property	123.15	173.15	198.15	223.15	248.15	273.15	293.15	323.15	373.15	423.15	473.15	523.15	573.15
Density, ρ_l (kg/m^3)	S	S	S	S	S	S	763	742	705	666	624	579	V
Specific heat capacity, $c_{p,l}$ (kJ/kg K)	S	S	S	S	S	S	2.186	(2.27)	(2.43)	(2.63)	(2.89)	(3.26)	
Thermal conductivity, λ_l [(W/m^2)/(K/m)]	S	S	S	S	S	S	0.145	0.140	0.130	0.120	0.102	0.093	
Dynamic viscosity, η_l (10^{-5} Ns/m^2)	S	S	S	S	S	S	231.9	132.4	70.1	43.1	28.7	20.8	V

Transport Properties of the Superheated Vapor

	Temperature, °C												
	−150	−100	−50	0	25	100	200	300	400	600	800	1 000	1 200
Property (at low pressure)	Temperature, K												
	123.15	173.15	223.15	273.15	298.15	373.15	473.15	573.15	673.15	873.15	1 073.15	1 273.15	1 473.15
Specific heat capacity, $c_{p,g}$ (kJ/kg K)	S	S	S	S	L	L	L	2.772	3.056	3.521	3.869	4.132	4.329
Thermal conductivity, λ_g[(W/m^2)/(K/m)]	S	S	S	S	L	L	L	(0.029)	(0.039)	(0.061)	(0.080)	(0.092)	—
Dynamic viscosity, η_g(10^{-5} Ns/m^2)	S	S	S	S	L	L	L	(0.86)	(1.00)	(1.27)	(1.53)	(1.78)	—

PENTADECANE

Chemical formula: $C_{15}H_{32}$
Molecular weight: 212.42
Melting point: 283.0 K
Boiling point: 480.8 K

Critical temperature: 707 K
Critical pressure: 1.52 MPa
Critical density: 240 kg/m^3
Normal vapor density: 9.48 kg/m^3
(@ 0 °C, 101.3 kPa)

Properties of the Liquid at Temperatures Below the Normal Boiling Point

Property	Temperature, °C −150	−100	−75	−50	−25	0	20	50	100	150	200	250	300
	Temperature, K 123.15	173.15	198.15	223.15	248.15	273.15	293.15	323.15	373.15	423.15	473.15	523.15	573.15
Density, ρ_l (kg/m^3)	S	S	S	S	S	S	768	748	712	673	631	V	V
Specific heat capacity, $c_{p,l}$ (kJ/kg K)	S	S	S	S	S	S	2.151	(2.27)	(2.43)	(2.62)	(2.86)	V	V
Thermal conductivity, λ_l [(W/m^2)/(K/m)]	S	S	S	S	S	S	0.141	0.135	0.125	0.115	0.105	V	V
Dynamic viscosity, η_l (10^{-5} Ns/m^2)	S	S	S	S	S	S	286.3	156.7	78.7	47.9	31.6	V	V

Transport Properties of the Superheated Vapor

Property (at low pressure)	Temperature °C − 150	− 100	− 50	0	25	100	200	300	400	600	800	1 000	1 200
	Temperature, K 123.15	173.15	223.15	273.15	298.15	373.15	473.15	573.15	673.15	873.15	1 073.15	1 273.15	1 473.15
Specific heat capacity, $c_{p,g}$ (kJ/kg K)	S	S	S	S	L	L	L	2.767	3.056	3.517	3.864	4.128	4.321
Thermal conductivity, λ_g[(W/m^2)/(K/m)]	S	S	S	S	L	L	L	(0.028)	(0.038)	(0.058)	(0.077)	(0.089)	—
Dynamic viscosity, η_g(10^{-5} Ns/m^2)	S	S	S	S	L	L	L	(0.83)	(0.97)	(1.23)	(1.36)	(1.73)	—

HEXADECANE

Chemical formula: $C_{16}H_{34}$
Molecular weight: 226.45
Melting point: 291.0 K
Boiling point: 560.0 K

Critical temperature: 717 K
Critical pressure: 1.42 MPa
Critical density: 240 kg/m^3
Normal vapor density: 10.1 kg/m^3
(@ 0°C, 101.3 kPa)

Properties of the Liquid at Temperatures Below the Normal Boiling Point

	Temperature, °C												
	−150	−100	−75	−50	−25	0	20	50	100	150	200	250	300
Property	**Temperature, K**												
	123.15	173.15	198.15	223.15	248.15	273.15	293.15	323.15	373.15	423.15	473.15	523.15	573.15
Density, ρ_l (kg/m^3)	S	S	S	S	S	S	775	753	717	679	638	594	V
Specific heat capacity, $c_{p,l}$ (kJ/kg K)	S	S	S	S	S	S	2.144	2.272	(2.43)	(2.62)	(2.86)	(3.18)	V
Thermal conductivity, λ_l [(W/m^2)/(K/m)]	S	S	S	S	S	S	0.150	0.144	0.134	0.124	0.108	0.099	V
Dynamic viscosity, η_l (10^{-5} Ns/m^2)	S	S	S	S	S	S	350.6	184.1	89.5	53.6	35.2	(25.3)	V

Transport Properties of the Superheated Vapor

	Temperature °C												
	− 150	− 100	− 50	0	25	100	200	300	400	600	800	1 000	1 200
Property (at low pressure)	**Temperature, K**												
	123.15	173.15	223.15	273.15	298.15	373.15	473.15	573.15	673.15	873.15	1 073.15	1 273.15	1 473.15
Specific heat capacity, $c_{p,g}$ (kJ/kg K)	S	S	S	S	L	L	L	2.767	3.056	3.517	3.860	4.124	4.317
Thermal conductivity, λ_g[(W/m^2)/(K/m)]	S	S	S	S	L	L	L	(0.026)	(0.036)	(0.056)	(0.075)	(0.086)	—
Dynamic viscosity, η_g(10^{-5} Ns/m^2)	S	S	S	S	L	L	L	(0.80)	(0.93)	(1.19)	(1.44)	(1.67)	—

HEPTADECANE

Chemical formula: $C_{17}H_{36}$
Molecular weight: 240.47
Melting point: 295.0 K
Boiling point: 575.65 K

Critical temperature: 733 K
Critical pressure: 1.32 MPa
Critical density: 240 kg/m^3
Normal vapor density: 10.7 kg/m^3
(@ 0 °C, 101.3 kPa)

Properties of the Liquid at Temperatures Below the Normal Boiling Point

	Temperature, °C												
	−150	−100	−75	−50	−25	0	20	50	100	150	200	250	300
	Temperature, K												
Property	123.15	173.15	198.15	223.15	248.15	273.15	293.15	323.15	373.15	423.15	473.15	523.15	573.15
Density, ρ_l (kg/m^3)	S	S	S	S	S	S	S	758	722	684	644	600	555
Specific heat capacity, $c_{p,l}$ (kJ/kg K)	S	S	S	S	S	S	S	2.274	2.427	(2.61)	(2.85)	(3.16)	(3.60)
Thermal conductivity, λ_l [(W/m^2)/(K/m)]	S	S	S	S	S	S	S	0.139	0.129	0.120	0.111	0.102	0.094
Dynamic viscosity, η_l (10^{-5} Ns/m^2)	S	S	S	S	S	S	S	214.4	101.1	59.7	39.0	27.2	19.9

Transport Properties of the Superheated Vapor

	Temperature, °C												
	−150	−100	−50	0	25	100	200	300	400	600	800	1 000	1 200
Property (at low pressure)	Temperature, K												
	123.15	173.15	223.15	273.15	298.15	373.15	473.15	573.15	673.15	873.15	1 073.15	1 273.15	1 473.15
Specific heat capacity, $c_{p,g}$ (kJ/kg K)	S	S	S	S	L	L	L	L	3.052	3.513	3.856	4.120	4.312
Thermal conductivity, λ_g[(W/m^2)/(K/m)]	S	S	S	S	L	L	L	L	(0.035)	(0.054)	(0.072)	(0.083)	—
Dynamic viscosity, η_g(10^{-5} Ns/m^2)	S	S	S	S	L	L	L	L	(0.90)	(1.14)	(1.38)	(1.61)	—

OCTADECANE

Chemical formula: $C_{18}H_{38}$
Molecular weight: 254.50
Melting point: 301.3 K
Boiling point: 589.4 K

Critical temperature: 745 K
Critical pressure: 1.21 MPa
Critical density: 240 kg/m^3
Normal vapor density: 11.4 kg/m^3
(@ 0°C, 101.3 kPa)

Properties of the Liquid at Temperatures Below the Normal Boiling Point

	Temperature, °C												
	−150	−100	−75	−50	−25	0	20	50	100	150	200	250	300
	Temperature, K												
Property	123.15	173.15	198.15	223.15	248.15	273.15	293.15	323.15	373.15	423.15	473.15	523.15	573.15
Density, ρ_l (kg/m^3)	S	S	S	S	S	S	S	762	727	689	649	606	561
Specific heat capacity, $c_{p,l}$ (kJ/kg K)	S	S	S	S	S	S	S	2.276	2.428	(2.61)	(2.84)	(3.14)	(3.56)
Thermal conductivity, λ_l [(W/m^2)/(K/m)]	S	S	S	S	S	S	S	0.141	0.132	0.122	0.113	0.105	0.096
Dynamic viscosity, η_l (10^{-5} Ns/m^2)	S	S	S	S	S	S	S	248.5	113.5	66.2	43.1	30.1	21.9

Transport Properties of the Superheated Vapor

	Temperature °C												
	− 150	− 100	− 50	0	25	100	200	300	400	600	800	1 000	1 200
Property	Temperature, K												
(at low pressure)	123.15	173.15	223.15	273.15	298.15	373.15	473.15	573.15	673.15	873.15	1 073.15	1 273.15	1 473.15
Specific heat capacity, $c_{p,g}$ (kJ/kg K)	S	S	S	S	L	L	L	L	3.052	3.513	3.856	4.116	4.308
Thermal conductivity, λ_g[(W/m^2)/(K/m)]	S	S	S	S	S	L	L	L	(0.033)	(0.052)	(0.069)	(0.080)	—
Dynamic viscosity, η_g(10^{-5} Ns/m^2)	S	S	S	S	S	L	L	L	(0.86)	(1.10)	(1.33)	(1.55)	—

NONADECANE

Chemical formula: $C_{19}H_{40}$
Molecular weight: 268.52
Melting point: 305.0 K
Boiling point: 603.1 K

Critical temperature: 756 K
Critical pressure: 1.12 MPa
Critical density: 240 kg/m^3
Normal vapor density: 12.0 kg/m^3
(@ 0 °C, 101.3 kPa)

Properties of the Liquid at Temperatures Below the Normal Boiling Point

	Temperature, °C												
	−150	−100	−75	−50	−25	0	20	50	100	150	200	250	300
	Temperature, K												
Property	123.15	173.15	198.15	223.15	248.15	273.15	293.15	323.15	373.15	423.15	473.15	523.15	573.15
Density, ρ_l (kg/m^3)	S	S	S	S	S	S	S	766	731	694	654	612	567
Specific heat capacity, $c_{p,l}$ (kJ/kg K)	S	S	S	S	S	S	S	(2.28)	(2.43)	(2.61)	(2.83)	(3.12)	(3.50)
Thermal conductivity, λ_l [(W/m^2)/(K/m)]	S	S	S	S	S	S	S	0.143	0.134	0.125	0.116	0.107	0.099
Dynamic viscosity, η_l (10^{-5} Ns/m^2)	S	S	S	S	S	S	S	286.1	126.7	73.1	47.2	32.2	23.5

Transport Properties of the Superheated Vapor

	Temperature, °C												
	− 150	− 100	− 50	0	25	100	200	300	400	600	800	1 000	1 200
Property (at low pressure)	Temperature, K												
	123.15	173.15	223.15	273.15	298.15	373.15	473.15	573.15	673.15	873.15	1 073.15	1 273.15	1 473.15
Specific heat capacity, $c_{p,g}$ (kJ/kg K)	S	S	S	S	S	L	L	L	3.052	3.509	3.852	4.111	4.304
Thermal conductivity, λ_g[(W/m^2)/(K/m)]	S	S	S	S	S	L	L	L	(0.031)	(0.049)	(0.065)	(0.077)	—
Dynamic viscosity, η_g(10^{-5} Ns/m^2)	S	S	S	S	S	L	L	L	(0.82)	(1.05)	(1.27)	(1.48)	—

EICOSANE

Chemical formula: $C_{20}H_{42}$
Molecular weight: 282.56
Melting point: 310.0 K
Boiling point: 617.0 K

Critical temperature: 767 K
Critical pressure: 1.11 MPa
Critical density: 240 kg/m^3
Normal vapor density: 12.6 kg/m^3
(@ 0°C, 101.3 kPa)

Properties of the Liquid at Temperatures Below the Normal Boiling Point

	Temperature, °C												
	−150	−100	−75	−50	−25	0	20	50	100	150	200	250	300
	Temperature, K												
Property	123.15	173.15	198.15	223.15	248.15	273.15	293.15	323.15	373.15	423.15	473.15	523.15	573.15
Density, ρ_l (kg/m^3)	S	S	S	S	S	S	S	769	742	698	659	617	572
Specific heat capacity, $c_{p,l}$ (kJ/kg K)	S	S	S	S	S	S	S	(2.28)	(2.43)	(2.61)	(2.83)	(3.11)	(3.47)
Thermal conductivity, λ_l [(W/m^2)/(K/m)]	S	S	S	S	S	S	S	0.145	0.136	0.127	0.118	0.109	0.101
Dynamic viscosity, η_l (10^{-5} Ns/m^2)	S	S	S	S	S	S	S	327.6	140.8	80.1	51.5	34.6	25.2

Transport Properties of the Superheated Vapor

	Temperature °C												
	− 150	− 100	− 50	0	25	100	200	300	400	600	800	1 000	1 200
Property (at low pressure)	Temperature, K												
	123.15	173.15	223.15	273.15	298.15	373.15	473.15	573.15	673.15	873.15	1 073.15	1 273.15	1 473.15
Specific heat capacity, $c_{p,g}$ (kJ/kg K)	S	S	S	S	S	L	L	L	3.048	3.509	3.848	4.107	4.300
Thermal conductivity, λ_g[(W/m^2)/(K/m)]	S	S	S	S	S	L	L	L	(0.031)	(0.049)	(0.065)	(0.076)	—
Dynamic viscosity, η_g(10^{-5} Ns/m^2)	S	S	S	S	S	L	L	L	(0.76)	(0.93)	(1.22)	(1.43)	—

2-METHYLPROPANE
ISOBUTANE

Chemical formula: C_4H_{10}
Molecular weight: 58.12
Melting point: 261.4 K
Boiling point: 113.55 K

Critical temperature: 408.1 K
Critical pressure: 3.647 MPa
Critical density: 221 kg/m^3
Normal vapor density: 2.59 kg/m^3
(@ 0 °C, 101.3 kPa)

Properties of the Liquid at Temperatures Below the Normal Boiling Point

	Temperature, °C												
	−150	−100	−75	−50	−25	0	20	50	100	150	200	250	300
	Temperature, K												
Property	123.15	173.15	198.15	223.15	248.15	273.15	293.15	323.15	373.15	423.15	473.15	523.15	573.15
Density, ρ_l (kg/m^3)	724	680	659	634	609	V	V	V	V	V	V	V	V
Specific heat capacity, $c_{p,l}$ (kJ/kg K)	1.729	1.884	1.968	2.060	2.177	V	V	V	V	V	V	V	V
Thermal conductivity, λ_l [(W/m^2)/(K/m)]	0.182	0.155	0.143	0.130	0.118	V	V	V	V	V	V	V	V
Dynamic viscosity, η_l (10^{-5} Ns/m^2)	1100	112	61.7	39.1	28.3	V	V	V	V	V	V	V	V

Thermodynamic Properties of the Superheated Vapor

T_{sat}, K	261.4	285	300	315	330	345	360	375	390	408.1
p_{sat}, kPa	101.3	233	355	553	805	1 132	1 540	2 066	2 697	3 647
ρ_ℓ, kg/m^3	594	567	550	529	508	485	455	422	377	221
ρ_g, kg/m^3	2.87	6.33	9.81	14.6	21.2	30.0	42.2	59.7	87.3	221
h_ℓ, kJ/kg	232.5	286.1	323.3	360.5	397.7	437.3	481.5	528.0	574.5	697.8
h_g, kJ/kg	597.8	623.4	646.6	667.6	686.2	704.8	721.1	737.3	744.3	697.8
$\Delta h_{g,\ell}$, kJ/kg	365.2	337.3	323.3	307.1	288.5	267.5	239.6	209.3	169.8	
$c_{p,\ell}$, kJ/(kg K)	2.12	2.34	2.45	2.56	2.68	2.79	2.95	3.16	3.59	
$c_{p,g}$, kJ/(kg K)	1.53	1.69	1.81	1.94	2.09	2.28	2.55	3.00	4.18	
η_ℓ, μNs/m^2	240	190	145	116	101	86	74	61	44	34
η_g, μNs/m^2	6.7	7.4	8.0	8.6	9.1	9.7	10.4	11.3	12.9	34
λ_ℓ, (mW/m^2)/(K/m)	100	92	87	83	80	77	73	68	61	42.4
λ_g, (mW/m^2)/(K/m)	11.6	14.4	16.3	18.3	20.4	22.7	25.2	27.9	31.2	42.4
Pr_ℓ	5.09	4.83	4.08	3.58	3.38	3.12	2.99	2.83	2.59	
Pr_g	0.88	0.88	0.89	0.91	0.93	0.97	1.05	1.22	1.73	
σ, mN/m	14.1	11.4	9.7	8.1	6.5	5.0	3.6	2.3	1.1	
$\beta_{e,\ell}$, kK^{-1}	1.87	2.22	2.52	2.91	3.44	4.20	5.40	7.56	12.8	

Transport Properties of the Superheated Vapor

	Temperature, °C												
	−150	−100	−50	0	25	100	200	300	400	600	800	1 000	1 200
Property (at low pressure)	Temperature, K												
	123.15	173.15	223.15	273.15	298.15	373.15	473.15	573.15	673.15	873.15	1 073.15	1 273.15	1 473.15
Specific heat capacity, $c_{p,g}$ (kJ/kg K)	L	L	L	1.540	1.666	2.022	2.458	2.834	3.153	3.663	4.049	4.342	4.559
Thermal conductivity, λ_g[(W/m^2)/(K/m)]	L	L	L	0.014	0.016	0.025	0.038	(0.053)	(0.068)	(0.099)	(0.128)	(0.154)	(0.178)
Dynamic viscosity, η_g(10^{-5} Ns/m^2)	L	L	L	0.689	0.758	0.947	1.202	1.450	(1.63)	(2.01)	(2.36)	(2.67)	(2.97)

2-METHYLBUTANE (ISOPENTANE)

Chemical formula: $(CH_3)_2C(CH_3)_2$
Molecular weight: 72.146
Melting point: 113.25 K
Normal boiling point: 301.0 K

Critical temperature: 460.4 K
Critical pressure: 3.380 MPa
Critical density: 236 kg/m^3
Normal vapor density: 3.22 kg/m^3
(@ 0°C, 101.3 kPa)

Properties of the Liquid at Temperatures Below the Normal Boiling Point

	Temperature, °C												
	−150	−100	−75	−50	−25	0	20	50	100	150	200	250	300
	Temperature, K												
Property	123.15	173.15	198.15	223.15	248.15	273.15	293.15	323.15	373.15	423.15	473.15	523.15	573.15
Density, ρ_l (kg/m^3)	(760)	727	706	687	663	639	620	V	V	V	V	V	V
Specific heat capacity, $c_{p,l}$ (kJ/kg K)	1.725	1.838	1.905	1.980	2.068	2.169	2.265	V	V	V	V	V	V
Thermal conductivity, λ_l [(W/m^2)/(K/m)]	0.182	0.159	0.148	0.137	0.127	0.117	0.110	V	V	V	V	V	V
Dynamic viscosity, η_l (10^{-5} Ns/m^2)	1720	154	84.4	55.0	37.8	27.9	22.5	V	V	V	V	V	V

Properties of the Saturated Liquid and Vapor

T_{sat}, K	301.0	325	340	355	370	385	400	415	430	460.4
p_{sat}, kPa	101.3	217	328	476	667	913	1 222	1 603	2 070	3 380
ρ_ℓ, kg/m^3	613	586	569	552	532	511	488	461	428	236
ρ_g, kg/m^3	3.07	6.30	9.41	13.5	18.9	26.2	36.0	49.3	68.3	236
h_ℓ, kJ/kg	290.8	348.9	383.8	423.3	465.2	504.7	546.6	588.5	639.7	790.8
h_g, kJ/kg	632.7	669.9	693.1	716.4	739.7	765.3	783.9	807.1	825.7	790.8
$\Delta h_{g,\ell}$, kJ/kg	341.9	321.0	309.3	293.1	274.5	260.6	237.3	218.6	186.0	
$c_{p,\ell}$, kJ/(kg K)	2.29	2.43	2.53	2.62	2.71	2.80	2.91	3.05	3.24	
$c_{p,g}$, kJ/(kg K)	1.72	1.86	1.96	2.08	2.20	2.34	2.54	2.81	3.28	
η_ℓ, μNs/m^2	201	163	146	126	111	97	84	71	60	
η_g, μNs/m^2	7.4	8.0	8.5	9.0	9.5	9.9	10.5	11.2	12.2	
λ_ℓ, (mW/m^2)/(K/m)	103	95	91	87	84	80	75	70	63	48.3
λ_g, (mW/m^2)/(K/m)	14.7	17.4	19.3	21.5	23.4	25.0	27.6	29.4	33.4	48.3
Pr_ℓ	4.47	4.17	4.06	3.79	3.58	3.40	3.22	3.05	2.99	
Pr_g	0.87	0.86	0.86	0.87	0.89	0.93	0.97	1.07	1.20	
σ, mN/m	14.0	11.5	9.9	8.4	7.0	5.6	4.3	3.0	1.9	
$\beta_{e,\ell}$, kK^{-1}	1.73	2.03	2.27	2.58	2.99	3.54	4.36	5.66	8.10	

Transport Properties of the Superheated Vapor

	Temperature, °C												
	−150	−100	−50	0	25	100	200	300	400	600	800	1 000	1 200
Property (at low pressure)	Temperature, K												
	123.15	173.15	223.15	273.15	298.15	373.15	473.15	573.15	673.15	873.15	1 073.15	1 273.15	1 473.15
Specific heat capacity, $c_{p,g}$ (kJ/kg K)	L	L	L	L	L	2.031	2.462	2.830	3.140	3.634	4.007	4.291	4.505
Thermal conductivity, λ_g[(W/m^2)/(K/m)]	L	L	L	L	L	0.022	0.034	(0.049)	(0.063)	(0.092)	(0.120)	(0.144)	(0.164)
Dynamic viscosity, η_g(10^{-5} Ns/m^2)	L	L	L	L	L	0.89	(1.11)	(1.32)	(1.52)	(1.90)	(2.23)	(2.53)	(2.80)

2-METHYLPENTANE

Chemical formula: C_6H_{14}
Molecular weight: 86.18
Melting point: 119.5 K
Boiling point: 333.4 K

Critical temperature: 497.5 K
Critical pressure: 3.01 MPa
Critical density: 235 kg/m^3
Normal vapor density: 3.85 kg/m^3 (@ 0 °C, 101.3 kPa)

Properties of the Liquid at Temperatures Below the Normal Boiling Point

	Temperature, °C												
	−150	−100	−75	−50	−25	0	20	50	100	150	200	250	300
	Temperature, K												
Property	123.15	173.15	198.15	223.15	248.15	273.15	293.15	323.15	373.15	423.15	473.15	523.15	573.15
Density, ρ_l (kg/m^3)	(779)	(752)	(734)	(714)	(693)	672	653	625	V	V	V	V	V
Specific heat capacity, $c_{p,l}$ (kJ/kg K)	1.712	1.813	1.884	1.955	2.043	2.139	2.227	(2.39)	V	V	V	V	V
Thermal conductivity, λ_l [(W/m^2)/(K/m)]	(0.180)	0.160	0.151	0.140	0.132	0.122	0.115	0.103	V	V	V	V	V
Dynamic viscosity, η_l (10^{-5} Ns/m^2)	2820	204	110	68.1	48.0	37.0	30.0	22.5	V	V	V	V	V

Transport Properties of the Superheated Vapor

	Temperature °C												
	− 150	− 100	− 50	0	25	100	200	300	400	600	800	1 000	1 200
Property (at low pressure)	Temperature, K												
	123.15	173.15	223.15	273.15	298.15	373.15	473.15	573.15	673.15	873.15	1 073.15	1 273.15	1 473.15
Specific heat capacity, $c_{p,g}$ (kJ/kg K)	L	L	L	L	L	2.018	2.445	2.809	3.119	3.605	—	—	—
Thermal conductivity, λ_g[(W/m^2)/(K/m)]	L	L	L	L	L	(0.020)	(0.032)	(0.045)	(0.059)	(0.086)	—	—	—
Dynamic viscosity, η_g(10^{-5} Ns/m^2)	L	L	L	L	L	(0.83)	(1.03)	(1.23)	(1.42)	(1.78)	—	—	—

3-METHYLPENTANE

Chemical formula: C_6H_{14}
Molecular weight: 86.18
Melting point: 155.0 K
Boiling point: 336.4 K

Critical temperature: 504.4 K
Critical pressure: 3.12 MPa
Critical density: 235 kg/m^3
Normal vapor density: 3.85 kg/m^3
(@ 0°C, 101.3 kPa)

Properties of the Liquid at Temperatures Below the Normal Boiling Point

	Temperature, °C												
	−150	−100	−75	−50	−25	0	20	50	100	150	200	250	300
	Temperature, K												
Property	123.15	173.15	198.15	223.15	248.15	273.15	293.15	323.15	373.15	423.15	473.15	523.15	573.15
Density, ρ_l (kg/m^3)	S	(763)	(751)	725	705	683	665	637	V	V	V	V	V
Specific heat capacity, $c_{p,l}$ (kJ/kg K)	S	1.800	1.859	1.926	2.010	2.131	2.190	(2.34)	V	V	V	V	V
Thermal conductivity, λ_l [(W/m^2)/(K/m)]	S	0.159	0.149	0.140	0.131	0.122	0.115	0.105	V	V	V	V	V
Dynamic viscosity, η_l (10^{-5} Ns/m^2)	S	(95)	(76)	(61)	(46)	39.4	32.4	25.3	V	V	V	V	V

Transport Properties of the Superheated Vapor

	Temperature, °C												
	−150	−100	−50	0	25	100	200	300	400	600	800	1 000	1 200
Property (at low pressure)	Temperature, K												
	123.15	173.15	223.15	273.15	298.15	373.15	473.15	573.15	673.15	873.15	1 073.15	1 273.15	1 473.15
Specific heat capacity, $c_{p,g}$ (kJ/kg K)	S	L	L	L	L	2.052	2.470	2.822	3.128	3.609	—	—	—
Thermal conductivity, λ_g[(W/m^2)/(K/m)]	S	L	L	L	L	(0.020)	(0.032)	(0.045)	(0.059)	(0.087)	—	—	—
Dynamic viscosity, η_g(10^{-5} Ns/m^2)	S	L	L	L	L	(0.83)	(1.04)	(1.24)	(1.43)	(1.80)	—	—	—

2,2-DIMETHYL PROPANE (NEOPENTANE)

Chemical formula: $C(CH_3)_4$
Molecular weight: 72.15
Melting point: 256.58 K
Normal boiling point: 282.65 K

Critical temperature: 433.78 K
Critical pressure: 3.196 MPa
Critical density: 238 kg/m^3
Normal vapor density: 3.22 kg/m^3
(@ 0°C, 101.3 kPa)

Properties of the Liquid at Temperatures Below the Normal Boiling Point

Property	Temperature, °C / Temperature, K												
	−150	−100	−75	−50	−25	0	20	50	100	150	200	250	300
	123.15	173.15	198.15	223.15	248.15	273.15	293.15	323.15	373.15	423.15	473.15	523.15	573.15
Density, ρ_l (kg/m^3)	S	S	S	S	S	615	V	V	V	V	V	V	V
Specific heat capacity, $c_{p,l}$ (kJ/kg K)	S	S	S	S	S	2.19	V	V	V	V	V	V	V
Thermal conductivity, λ_l [(W/m^2)/(K/m)]	S	S	S	S	S	0.105	V	V	V	V	V	V	V
Dynamic viscosity, η_l (10^{-5} Ns/m^2)	S	S	S	S	S	32.9	V	V	V	V	V	V	V

Properties of the Saturated Liquid and Vapor

T_{sat}, K	282.65	305	320	335	350	365	380	395	410	433.78
p_{sat}, kPa	101.3	215	362	482	676	945	1 280	1 700	2 200	3 196
ρ_ℓ, kg/m^3	603	575	558	543	519	497	472	443	407	238
ρ_g, kg/m^3	3.28	6.38	9.91	14.5	20.7	29.0	40.2	55.8	79.1	238
h_ℓ, kJ/kg	−111	−50	−25	13	51	91	128	175	221	329
h_g, kJ/kg	204	237	260	282	305	327	347	366	379	329
$\Delta h_{g,\ell}$, kJ/kg	315	297	284	269	254	236	219	191	158	
$c_{p,\ell}$, kJ/(kg K)	2.14	2.27	2.37	2.48	2.63	2.77	2.96	3.20	3.59	
$c_{p,g}$, kJ/(kg K)	1.65	1.80	1.91	2.03	2.16	2.32	2.53	2.85	3.51	
η_ℓ, μNs/m^2	280	211	169	140	112	94	82	69	61	36
η_g, μNs/m^2	6.89	7.44	7.88	8.27	8.73	9.25	9.87	10.3	11.8	36
λ_ℓ, (mW/m^2)/(K/m)	90	84	80	76	72	67	63	58	54	42
λ_g, (mW/m^2)/(K/m)	13.2	15.0	16.3	18.0	19.8	21.6	24.2	26.6	30.8	42
Pr_ℓ	6.66	5.70	5.01	4.57	4.09	3.89	3.85	3.81	4.06	
Pr_g	0.86	0.89	0.92	0.93	0.95	0.99	1.03	1.10	1.34	
σ, mN/m	12.9	10.6	9.11	7.67	6.27	4.93	3.65	2.45	1.34	
$\beta_{e,\ell}$, kK^{-1}	1.80	2.11	2.38	2.72	3.18	3.83	4.81	6.47	9.99	

Transport Properties of the Superheated Vapor

Property (at low pressure)	Temperature, °C / Temperature, K												
	−150	−100	−50	0	25	100	200	300	400	600	800	1 000	1 200
	123.15	173.15	223.15	273.15	298.15	373.15	473.15	573.15	673.15	873.15	1 073.15	1 273.15	1 473.15
Specific heat capacity, $c_{p,g}$ (kJ/kg K)	S	S	S	L	1.687	2.052	2.504	2.881	3.199	3.689	4.057	4.329	4.538
Thermal conductivity, λ_g[(W/m^2)/(K/m)]	S	S	S	L	(0.015)	(0.024)	(0.037)	(0.051)	(0.066)	(0.095)	(0.122)	(0.141)	(0.160)
Dynamic viscosity, η_g(10^{-5} Ns/m^2)	S	S	S	L	(0.71)	(0.86)	(1.07)	(1.25)	(1.41)	(1.71)	(1.96)	(2.20)	(2.43)

2,2-DIMETHYLBUTANE

Chemical formula: C_6H_{14}
Molecular weight: 86.18
Melting point: 173.3 K
Boiling point: 322.9 K

Critical temperature: 488.7 K
Critical pressure: 3.08 MPa
Critical density: 240 kg/m^3
Normal vapor density: 3.85 kg/m^3
(@ 0°C, 101.3 kPa)

Properties of the Liquid at Temperatures Below the Normal Boiling Point

	Temperature, °C												
	−150	−100	−75	−50	−25	0	20	50	100	150	200	250	300
	Temperature, K												
Property	123.15	173.15	198.15	223.15	248.15	273.15	293.15	323.15	373.15	423.15	473.15	523.15	573.15
Density, ρ_l (kg/m^3)	S	S	(737)	(714)	(690)	665	647	(639)	V	V	V	V	V
Specific heat capacity, $c_{p,l}$ (kJ/kg K)	S	S	1.809	1.892	1.976	2.081	2.169	(2.31)	V	V	V	V	V
Thermal conductivity, λ_l [(W/m^2)/(K/m)]	S	S	0.138	0.129	0.120	0.112	0.106	0.095	V	V	V	V	V
Dynamic viscosity, η_l (10^{-5} Ns/m^2)	S	S	—	(100)	(67)	47.8	37.6	(31)	V	V	V	V	V

Transport Properties of the Superheated Vapor

	Temperature °C												
	− 150	− 100	− 50	0	25	100	200	300	400	600	800	1 000	1 200
Property (at low pressure)	Temperature, K												
	123.15	173.15	223.15	273.15	298.15	373.15	473.15	573.15	673.15	873.15	1 073.15	1 273.15	1 473.15
Specific heat capacity, $c_{p,g}$ (kJ/kg K)	S	S	L	L	L	2.026	2.570	2.843	3.157	3.634	(3.99)	(4.25)	—
Thermal conductivity, λ_g[(W/m^2)/(K/m)]	S	S	L	L	L	(0.021)	(0.033)	(0.047)	(0.061)	(0.090)	(0.115)	(0.133)	—
Dynamic viscosity, η_g(10^{-5} Ns/m^2)	S	S	L	L	L	0.836	(1.07)	(1.27)	(1.47)	(1.84)	(2.17)	(2.47)	—

2,3-DIMETHYLBUTANE

Chemical formula: C_6H_{14}
Molecular weight: 86.18
Melting point: 144.6 K
Boiling point: 331.2 K

Critical temperature: 499 K
Critical pressure: 3.13 MPa
Critical density: 241 kg/m^3
Normal vapor density: 3.85 kg/m^3
(@ 0°C, 101.3 kPa)

Properties of the Liquid at Temperatures Below the Normal Boiling Point

	Temperature, °C												
	−150	−100	−75	−50	−25	0	20	50	100	150	200	250	300
	Temperature, K												
Property	123.15	173.15	198.15	223.15	248.15	273.15	293.15	323.15	373.15	423.15	473.15	523.15	573.15
Density, ρ_l (kg/m^3)	S	(759)	(741)	(722)	(701)	681	663	635	V	V	V	V	V
Specific heat capacity, $c_{p,l}$ (kJ/kg K)	S	1.754	1.821	1.901	1.985	2.077	2.165	(2.35)	V	V	V	V	V
Thermal conductivity, λ_l [(W/m^2)/(K/m)]	S	0.148	0.139	0.130	0.121	0.113	0.107	0.099	V	V	V	V	V
Dynamic viscosity, η_l (10^{-5} Ns/m^2)	S	(152)	(114)	(86)	(65)	49.5	38.5	27.6	V	V	V	V	V

Transport Properties of the Superheated Vapor

	Temperature, °C												
	− 150	− 100	− 50	0	25	100	200	300	400	600	800	1 000	1 200
Property (at low pressure)	Temperature, K												
	123.15	173.15	223.15	273.15	298.15	373.15	473.15	573.15	673.15	873.15	1 073.15	1 273.15	1 473.15
Specific heat capacity, $c_{p,g}$ (kJ/kg K)	S	L	L	L	L	2.031	2.458	2.826	3.128	3.613	(3.96)	(4.24)	—
Thermal conductivity, λ_g[(W/m^2)/(K/m)]	S	L	L	L	L	(0.020)	(0.033)	(0.046)	(0.060)	(0.089)	(0.115)	(0.135)	—
Dynamic viscosity, η_g(10^{-5} Ns/m^2)	S	L	L	L	L	0.85	(1.05)	(1.25)	(1.45)	(1.81)	(2.15)	(2.45)	—

ETHYLENE

Chemical formula: $H_2C{:}CH_2$
Molecular weight: 28.052
Normal boiling point: 169.43 K
Melting point: 104 K

Critical temperature: 282.65 K
Critical pressure: 5.060 MPa
Critical density: 220 kg/m^3
Normal vapor density: 1.26 kg/m^3
(@ 0°C, 101.3 kPa)

Properties of the Liquid at Temperatures Below the Normal Boiling Point

	Temperature, °C												
	−150	−100	−75	−50	−25	0	20	50	100	150	200	250	300
	Temperature, K												
Property	123.15	173.15	198.15	223.15	248.15	273.15	293.15	323.15	373.15	423.15	473.15	523.15	573.15
Density, ρ_l (kg/m^3)	630	V	V	V	V	V	V	V	V	V	V	V	V
Specific heat capacity, $c_{p,l}$ (kJ/kg K)	2.433	V	V	V	V	V	V	V	V	V	V	V	V
Thermal conductivity, λ_l [(W/m^2)/(K/m)]	0.242	V	V	V	V	V	V	V	V	V	V	V	V
Dynamic viscosity, η_l (10^{-5} Ns/m^2)	41.0	V	V	V	V	V	V	V	V	V	V	V	V

Properties of the Saturated Liquid and Vapor

T_{sat}, K	169.43	183	193	203	213	223	233	243	263	281
p_{sat}, kPa	101.3	213	341	518	755	1 063	1 453	1 938	3 240	4 899
ρ_ℓ, kg/m^3	567.92	547.95	532.88	517.17	500.61	482.84	463.41	441.61	385.64	287.43
ρ_g, kg/m^3	2.09	4.24	6.60	9.81	14.01	19.47	26.58	36.12	69.58	152.70
h_ℓ, kJ/kg	−662.49	−624.50	−600.49	−578.48	−552.50	−526.51	−498.48	−468.50	−396.51	−301.15
h_g, kJ/kg	−179.97	−163.61	−155.64	−151.12	−145.06	−141.54	−140.04	−140.71	−152.62	−213.38
$\Delta h_{g,\ell}$, kJ/kg	482.52	460.89	444.85	427.36	407.44	384.97	358.44	327.79	243.89	94.56
$c_{p,\ell}$, kJ/(kg K)	2.32	2.46	2.54	2.61	2.67	2.73	2.80	2.93	3.89	
$c_{p,g}$, kJ/(kg K)	1.31	1.35	1.40	1.47	1.56	1.67	1.82	2.02	2.91	
η_ℓ, μNs/m^2	162.0	138.5	124.1	112.1	102.6	94.4	86.4	77.6	55.9	28.7
η_g, μNs/m^2	6.04	6.56	6.96	7.37	7.81	8.29	9.82	9.44	11.30	16.25
λ_ℓ, (mW/m^2)/(K/m)	192	178	168	158	147	137	126	116	94.7	77.0
λ_g, (mW/m^2)/(K/m)	6.44	7.62	8.62	9.71	11.0	12.4	14.0	15.9	21.9	
Pr_ℓ	1.96	1.91	1.88	1.85	1.86	1.88	1.92	1.96	2.30	
Pr_g	1.23	1.16	1.13	1.11	1.10	1.11	1.14	1.20	1.50	
σ, mN/m	16.46	13.99	12.23	10.52	8.88	7.29	5.78	4.35	1.80	0.16
$\beta_{e,\ell}$, kK^{-1}	2.52	2.83	3.12	3.47	3.88	4.50	5.40	6.70	10.4	

Transport Properties of the Superheated Vapor

Property (at low pressure)	Temperature, °C: −150	−100	−50	0	25	100	200	300	400	600	800	1 000	1 200
	Temperature, K: 123.15	173.15	223.15	273.15	298.15	373.15	473.15	573.15	673.15	873.15	1 073.15	1 273.15	1 473.15
Specific heat capacity, $c_{p,g}$ (kJ/kg K)	L	1.654	1.319	1.461	1.553	1.830	2.177	2.479	2.738	3.157	3.475	3.722	3.910
Thermal conductivity, λ_g[(W/m^2)/(K/m)]	L	0.009	0.013	0.017	0.021	0.031	0.044	0.060	0.075	0.106	0.136	0.162	0.188
Dynamic viscosity, η_g(10^{-5} Ns/m^2)	L	0.592	0.770	0.939	1.02	1.27	1.55	1.78	2.01	2.44	2.79	3.07	3.45

Thermodynamic Properties of the Superheated Vapor

Pressure, MPa	Property	250	275	300	325	350	375	400	425
		Value at different temperatures, K							
0.101 325	v, m^3/kg	0.723 9	0.798 3	0.872 5	0.946 4	1.020	1.094	1.167	1.241
(169.4)	h, kJ/kg	636.60	672.44	710.12	749.81	791.63	835.64	881.87	930.30
	s, kJ/(kg K)	3.659 2	3.795 8	3.926 9	4.053 9	4.177 8	4.299 3	4.418 6	4.536 0
0.20	v, m^3/kg	0.363 0	0.401 5	0.439 5	0.477 4	0.515 1	0.552 6	0.590 1	0.627 5
(181.9)	h, kJ/kg	634.30	670.58	708.57	748.50	790.50	834.66	881.00	929.53
	s, kJ/(kg K)	3.451 4	3.589 6	3.721 8	3.849 6	3.974 1	4.095 9	4.215 5	4.333 2
0.50	v, m^3/kg	0.140 6	0.156 9	0.172 8	0.188 4	0.203 9	0.219 2	0.234 4	0.249 6
(202.3)	h, kJ/kg	627.07	664.76	703.78	744.46	787.04	831.65	878.36	927.19
	s, kJ/(kg K)	3.159 9	3.303 6	3.439 3	3.569 5	3.695 7	3.818 8	3.939 4	4.057 7
1.0	v, m^3/kg	0.066 16	0.075 20	0.083 77	0.092 04	0.100 1	0.108 1	0.115 9	0.123 7
(221.3)	h, kJ/kg	614.02	654.54	695.48	737.54	781.16	826.57	873.91	923.25
	s, kJ/(kg K)	2.917 8	3.072 3	3.214 7	3.349 4	3.478 7	3.603 9	3.726 1	3.845 8
2.0	v, m^3/kg	0.028 09	0.034 02	0.039 12	0.043 79	0.048 22	0.052 49	0.056 65	0.060 72
(244.3)	h, kJ/kg	581.93	631.43	677.50	722.94	768.96	816.15	864.87	915.31
	s, kJ/(kg K)	2.617 6	2.806 5	2.966 8	3.112 3	3.248 7	3.378 9	3.504 7	3.627 0
5.0	v, m^3/kg			0.011 37	0.014 54	0.017 00	0.019 17	0.021 16	0.023 04
(282.0)	h, kJ/kg			602.17	670.35	727.99	782.62	836.57	890.92
	s, kJ/(kg K)			2.502 1	2.720 8	2.891 8	3.042 5	3.181 8	3.313 6

PROPYLENE

Chemical formula: $CH_3CH{:}CH_2$
Molecular weight: 42.078
Melting point: 87.9 K
Normal boiling point: 225.45 K

Critical temperature: 364.8 K
Critical pressure: 4.610 MPa
Critical density: 233 kg/m^3
Normal vapor density: 1.90 kg/m^3
(@ 0 °C, 101.3 kPa)

Properties of the Liquid at Temperatures Below the Normal Boiling Point

	Temperature, °C												
	−150	−100	−75	−50	−25	0	20	50	100	150	200	250	300
	Temperature, K												
Property	123.15	173.15	198.15	223.15	248.15	273.15	293.15	323.15	373.15	423.15	473.15	523.15	573.15
Density, ρ_l (kg/m^3)	729	671	641	612	V	V	V	V	V	V	V	V	V
Specific heat capacity, $c_{p,l}$ (kJ/kg K)	2.098	2.085	2.123	2.177	V	V	V	V	V	V	V	V	V
Thermal conductivity, λ_l [(W/m^2)/(K/m)]	0.217	0.179	0.160	0.145	V	V	V	V	V	V	V	V	V
Dynamic viscosity, η_l (10^{-5} Ns/m^2)	129.1	37.0	26.5	19.2	V	V	V	V	V	V	V	V	V

Properties of the Saturated Liquid and Vapor

T_{sat}, K	225.45	240	255	270	285	300	315	330	345	365
p_{sat}, kPa	101.3	187	333	530	820	1 210	1 710	2 410	3 190	4 610
ρ_ℓ, kg/m^3	611	587	575	556	535	509	481	443	390	233
ρ_g, kg/m^3	2.15	3.93	6.68	10.70	16.4	24.4	35.6	51.7	77.2	233
h_ℓ, kJ/kg	−309.0	−273.7	−237.3	−199.8	−161.2	−121.3	−79.8	−36.2	10.4	119.5
h_g, kJ/kg	130.2	150.0	164.8	182.3	196.3	212.0	229.7	220.2	210.5	119.5
$\Delta h_{g,\ell}$, kJ/kg	439.2	423.7	402.1	382.1	357.5	333.3	309.5	256.4	220.9	
$c_{p,\ell}$, kJ/(kg K)	2.39	2.45	2.55	2.64	2.72	2.85	3.10	3.40	3.77	
$c_{p,g}$, kJ/(kg K)	1.31	1.40	1.49	1.62	1.78	1.96	2.23	2.62	3.71	
η_ℓ, μNs/m^2	151	132	108	101	99.2	90.3	80.9	78.7	61.1	32
η_g, μNs/m^2	6.62	7.14	7.53	8.04	8.74	9.26	10.1	11.4	12.7	32
λ_ℓ, (mW/m^2)/(K/m)	119	111	104	98.6	93.6	90.9	88.0	83.3	76.1	49.3
λ_g, (mW/m^2)/(K/m)	9.52	11.2	13.0	14.9	17.1	19.4	22.2	25.4	29.6	49.3
$\Pr_\ell$	3.03	2.91	2.65	2.70	2.88	2.83	2.85	3.21	3.03	
$\Pr_g$	0.91	0.89	0.86	0.87	0.91	0.94	1.01	1.18	1.69	
σ, mN/m	16.5	14.7	12.6	10.5	8.7	6.5	5.1	3.4	2.0	
$\beta_{e,\ell}$, kK^{-1}	1.99	2.22	2.52	2.90	3.42	4.17	5.32	7.38	12.2	

Transport Properties of the Superheated Vapor

Property (at low pressure)	Temperature °C −150	−100	−50	0	25	100	200	300	400	600	800	1 000	1 200
	Temperature, K 123.15	173.15	223.15	273.15	298.15	373.15	473.15	573.15	673.15	873.15	1 073.15	1 273.15	1 473.15
Specific heat capacity, $c_{p,g}$ (kJ/kg K)	L	L	L	1.424	1.520	1.800	2.160	2.479	2.755	3.203	3.542	3.802	3.998
Thermal conductivity, λ_g[(W/m^2)/(K/m)]	L	L	L	0.014	0.017	0.026	0.039	0.054	0.069	0.099	0.127	0.155	0.180
Dynamic viscosity, η_g(10^{-5} Ns/m^2)	L	L	L	0.780	0.860	1.07	1.34	1.59	1.82	2.23	2.62	2.97	3.29

Thermodynamic Properties of the Superheated Vapor

Pressure, MPa	Property	Value at different temperatures, K 260	300	340	380	420	460	500	540
0.070	v, m^3/kg	0.722 9	0.838 5	0.953 3	1.068	1.181	1.295	1.409	1.522
(217.6)	h, kJ/kg	545.62	604.29	668.72	738.90	815.18	897.33	985.06	1 078.00
	s, kJ/(kg K)	2.492 4	2.702 1	2.903 2	3.098 4	3.289 2	3.475 9	3.658 7	3.837 5
0.101 325	v, m^3/kg	0.496 0	0.576 7	0.656 6	0.735 9	0.815 0	0.893 7	0.972 4	1.051
(225.4)	h, kJ/kg	544.60	603.51	668.01	738.40	814.77	896.98	984.75	1 077.73
	s, kJ/(kg K)	2.416 7	2.627 3	2.828 9	3.024 5	3.215 4	3.402 3	3.585 2	3.764 0
0.20	v, m^3/kg	0.245 6	0.288 0	0.329 5	0.370 3	0.410 9	0.451 2	0.491 3	0.531 4
(241.6)	h, kJ/kg	541.28	601.00	666.04	736.81	813.45	895.87	983.80	1 076.91
	s, kJ/(kg K)	2.273 8	2.487 3	2.690 6	2.887 2	3.078 9	3.266 2	3.449 5	3.628 5
0.40	v, m^3/kg		0.139 6	0.161 4	0.182 6	0.203 4	0.224 0	0.244 4	0.264 6
(261.0)	h, kJ/kg		595.69	661.96	733.55	810.77	893.61	981.87	1 075.23
	s, kJ/(kg K)		2.338 3	2.545 4	2.744 4	2.937 5	3.125 8	3.309 7	3.489 3
0.70	v, m^3/kg		0.075 76	0.089 33	0.102 1	0.114 5	0.126 6	0.138 5	0.150 3
(279.3)	h, kJ/kg		587.03	655.54	728.51	806.67	890.19	978.96	1 072.72
	s, kJ/(kg K)		2.207 5	2.421 8	2.624 6	2.820 0	3.009 9	3.194 9	3.375 2
1.0	v, m^3/kg		0.049 91	0.060 39	0.069 89	0.078 91	0.087 64	0.096 19	0.104 6
(292.5)	h, kJ/kg		577.25	648.73	723.31	802.50	886.73	976.03	1 070.20
	s, kJ/(kg K)		2.113 6	2.337 2	2.544 5	2.742 5	2.934 0	3.120 1	3.301 2
2.0	v, m^3/kg			0.026 08	0.032 12	0.037 34	0.042 19	0.046 81	0.051 29
(321.9)	h, kJ/kg			621.71	704.43	787.93	874.92	966.15	1 061.76
	s, kJ/(kg K)			2.141 7	2.371 8	2.580 7	2.778 4	2.968 6	3.152 5
4.0	v, m^3/kg				0.012 56	0.016 41	0.019 44	0.022 15	0.024 67
(357.1)	h, kJ/kg				654.01	754.88	849.80	945.84	1 044.76
	s, kJ/(kg K)				2.132 5	2.385 2	2.601 1	2.801 3	2.991 5
7.0	v, m^3/kg				0.003 14	0.007 16	0.009 70	0.011 64	0.013 34
	h, kJ/kg				490.13	689.17	807.70	914.06	1 019.08
	s, kJ/(kg K)				1.649 8	2.150 4	2.420 5	2.642 2	2.844 3
10	v, m^3/kg				1.002 63	0.003 99	0.005 96	0.007 56	0.008 91
	h, kJ/kg				463.63	618.60	762.41	881.75	993.83
	s, kJ/(kg K)				1.557 8	1.944 9	2.272 6	2.521 6	2.737 2

1-BUTENE

Chemical formula: C_4H_8
Molecular weight: 56.11
Melting point: 87.8 K
Boiling point: 266.9 K

Critical temperature: 419.6 K
Critical pressure: 4.02 MPa
Critical density: 233 kg/m^3
Normal vapor density: 2.50 kg/m^3
(@ 0 °C, 101.3 kPa)

Properties of the Liquid at Temperatures Below the Normal Boiling Point

	Temperature, °C												
	−150	−100	−75	−50	−25	0	20	50	100	150	200	250	300
	Temperature, K												
Property	123.15	173.15	198.15	223.15	248.15	273.15	293.15	323.15	373.15	423.15	473.15	523.15	573.15
Density, ρ_l (kg/m^3)	(755)	712	691	668	645	V	V	V	V	V	V	V	V
Specific heat capacity, $c_{p,l}$ (kJ/kg K)	1.888	1.909	1.959	2.022	2.102	V	V	V	V	V	V	V	V
Thermal conductivity, λ_l [(W/m^2)/(K/m)]	0.204	0.175	0.161	0.147	0.135	V	V	V	V	V	V	V	V
Dynamic viscosity, η_l (10^{-5} Ns/m^2)	242	64.0	42.0	30.0	22.5	V	V	V	V	V	V	V	V

Transport Properties of the Superheated Vapor

	Temperature, °C												
	−150	−100	−50	0	25	100	200	300	400	600	800	1 000	1 200
Property (at low pressure)	Temperature, K												
	123.15	173.15	223.15	273.15	298.15	373.15	473.15	573.15	673.15	873.15	1 073.15	1 273.15	1 473.15
Specific heat capacity, $c_{p,g}$ (kJ/kg K)	L	L	L	1.482	1.595	1.905	2.278	2.596	2.868	3.303	3.634	3.885	4.078
Thermal conductivity, λ_g[(W/m^2)/(K/m)]	L	L	L	0.012	0.015	0.023	0.035	(0.048)	(0.062)	(0.090)	(0.117)	(0.140)	(0.162)
Dynamic viscosity, η_g(10^{-5} Ns/m^2)	L	L	L	(0.72)	0.79	0.98	1.21	1.44	(1.66)	(2.06)	(2.40)	(2.71)	(3.00)

1-PENTENE

Chemical formula: C_5H_{10}
Molecular weight: 70.13
Melting point: 107.9 K
Boiling point: 303.1 K

Critical temperature: 464.7 K
Critical pressure: 4.05 MPa
Critical density: 230 kg/m^3
Normal vapor density: 3.13 kg/m^3
(@ 0°C, 101.3 kPa)

Properties of the Liquid at Temperatures Below the Normal Boiling Point

	Temperature, °C												
	−150	−100	−75	−50	−25	0	20	50	100	150	200	250	300
	Temperature, K												
Property	123.15	173.15	198.15	223.15	248.15	273.15	293.15	323.15	373.15	423.15	473.15	523.15	573.15
Density, ρ_l (kg/m^3)	(800)	755	734	711	687	662	642	V	V	V	V	V	V
Specific heat capacity, $c_{p,l}$ (kJ/kg K)	(1.84)	1.859	1.901	1.955	2.026	2.119	2.219	V	V	V	V	V	V
Thermal conductivity, λ_l [(W/m^2)/(K/m)]	0.196	0.172	0.160	0.149	0.138	0.127	0.119	V	V	V	V	V	V
Dynamic viscosity, η_l (10^{-5} Ns/m^2)	56.9	99.1	64.0	43.1	31.0	24.0	20.9	V	V	V	V	V	V

Transport Properties of the Superheated Vapor

	Temperature, °C												
	−150	−100	−50	0	25	100	200	300	400	600	800	1 000	1 200
Property	Temperature, K												
(at low pressure)	123.15	173.15	223.15	273.15	298.15	373.15	473.15	573.15	673.15	873.15	1 073.15	1 273.15	1 473.15
Specific heat capacity, $c_{p,g}$ (kJ/kg K)	L	L	L	L	L	1.938	2.311	2.629	2.901	3.337	3.668	3.919	4.111
Thermal conductivity, λ_g[(W/m^2)/(K/m)]	L	L	L	L	L	(0.021)	(0.033)	(0.045)	(0.058)	(0.086)	(0.112)	(0.135)	(0.157)
Dynamic viscosity, η_g(10^{-5} Ns/m^2)	L	L	L	L	L	0.858	(1.12)	(1.36)	(1.57)	(1.95)	(2.30)	(2.62)	(2.92)

1-HEXENE

Chemical formula: C_6H_{12}
Molecular weight: 84.16
Melting point: 133.3 K
Boiling point: 336.6 K

Critical temperature: 504 K
Critical pressure: 3.17 MPa
Critical density: 240 kg/m^3
Normal vapor density: 3.75 kg/m^3 (@ 0 °C, 101.3 kPa)

Properties of the Liquid at Temperatures Below the Normal Boiling Point

	Temperature, °C												
	−150	−100	−75	−50	−25	0	20	50	100	150	200	250	300
	Temperature, K												
Property	123.15	173.15	198.15	223.15	248.15	273.15	293.15	323.15	373.15	423.15	473.15	523.15	573.15
Density, ρ_l (kg/m^3)	S	770	750	732	713	693	673	645	V	V	V	V	V
Specific heat capacity, $c_{p,l}$ (kJ/kg K)	S	1.857	1.882	1.941	2.004	2.090	2.161	(2.28)	V	V	V	V	V
Thermal conductivity, λ_l [(W/m^2)/(K/m)]	S	0.169	0.159	0.149	0.139	0.130	0.122	0.112	V	V	V	V	V
Dynamic viscosity, η_l (10^{-5} Ns/m^2)	S	166	94	63.1	44.0	33.0	28.0	23.0	V	V	V	V	V

Transport Properties of the Superheated Vapor

	Temperature, °C												
	−150	−100	−50	0	25	100	200	300	400	600	800	1 000	1 200
Property (at low pressure)	Temperature, K												
	123.15	173.15	223.15	273.15	298.15	373.15	473.15	573.15	673.15	873.15	1 073.15	1 273.15	1 473.15
Specific heat capacity, $c_{p,g}$ (kJ/kg K)	S	L	L	L	L	1.951	2.328	2.650	2.922	3.362	3.689	3.944	4.128
Thermal conductivity, λ_g[(W/m^2)/(K/m)]	S	L	L	L	L	0.019	0.031	0.043	(0.056)	(0.082)	(0.108)	(0.130)	(0.150)
Dynamic viscosity, η_g(10^{-5} Ns/m^2)	S	L	L	L	L	0.869	1.08	1.30	1.50	1.87	2.18	2.46	(2.66)

1-HEPTENE

Chemical formula: C_7H_{14}
Molecular weight: 98.195
Melting point: 154.3 K
Boiling point: 366.8 K

Critical temperature: 537.2 K
Critical pressure: 2.84 MPa
Critical density: 241 kg/m^3
Normal vapor density: 4.38 kg/m^3 (@ 0 °C, 101.3 kPa)

Properties of the Liquid at Temperatures Below the Normal Boiling Point

	Temperature, °C												
	−150	−100	−75	−50	−25	0	20	50	100	150	200	250	300
	Temperature, K												
Property	123.15	173.15	198.15	223.15	248.15	273.15	293.15	323.15	373.15	423.15	473.15	523.15	573.15
Density, ρ_l (kg/m^3)	S	(788)	(770)	(752)	(733)	713	697	671	V	V	V	V	V
Specific heat capacity, $c_{p,l}$ (kJ/kg K)	S	1.842	1.880	1.922	1.989	2.072	2.148	(2.26)	V	V	V	V	V
Thermal conductivity, λ_l [(W/m^2)/(K/m)]	S	0.167	0.158	0.149	0.140	0.132	0.125	0.115	V	V	V	V	V
Dynamic viscosity, η_l (10^{-5} Ns/m^2)	S	296	150	89.4	61	44.0	35.0	27.0	V	V	V	V	V

Transport Properties of the Superheated Vapor

	Temperature, °C												
	− 150	− 100	− 50	0	25	100	200	300	400	600	800	1 000	1 200
Property (at low pressure)	Temperature, K												
	123.15	173.15	223.15	273.15	298.15	373.15	473.15	573.15	673.15	873.15	1 073.15	1 273.15	1 473.15
Specific heat capacity, $c_{p,g}$ (kJ/kg K)	S	L	L	L	L	1.959	2.340	2.663	2.939	3.379	3.710	3.957	4.145
Thermal conductivity, λ_g[(W/m^2)/(K/m)]	S	L	L	L	L	0.017	0.028	0.039	(0.052)	(0.077)	(0.100)	(0.120)	(0.136)
Dynamic viscosity, η_g(10^{-5} Ns/m^2)	S	L	L	L	L	(0.77)	(0.97)	(1.16)	(1.35)	(1.70)	(2.01)	(2.27)	(2.47)

1-OCTENE

Chemical formula: C_8H_{16}
Molecular weight: 112.21
Melting point: 171.45 K
Boiling point: 394.45 K

Critical temperature: 566.7 K
Critical pressure: 2.53 MPa
Critical density: 242 kg/m^3
Normal vapor density: 5.01 kg/m^3
(@ 0°C, 101.3 kPa)

Properties of the Liquid at Temperatures Below the Normal Boiling Point

	Temperature, °C												
	−150	−100	−75	−50	−25	0	20	50	100	150	200	250	300
	Temperature, K												
Property	123.15	173.15	198.15	223.15	248.15	273.15	293.15	323.15	373.15	423.15	473.15	523.15	573.15
Density, ρ_l (kg/m^3)	S	S	S	(773)	(744)	732	715	689	(646)	V	V	V	V
Specific heat capacity, $c_{p,l}$ (kJ/kg K)	S	S	S	1.943	1.992	2.067	2.129	(2.24)	(2.45)	V	V	V	V
Thermal conductivity, λ_l [(W/m^2)/(K/m)]	S	S	S	0.150	0.142	0.134	0.128	0.119	0.105	V	V	V	V
Dynamic viscosity, η_l (10^{-5} Ns/m^2)	S	S	S	133	86	61.3	47.0	34.7	23.4	V	V	V	V

Transport Properties of the Superheated Vapor

	Temperature, °C												
	−150	−100	−50	0	25	100	200	300	400	600	800	1 000	1 200
Property (at low pressure)	Temperature, K												
	123.15	173.15	223.15	273.15	298.15	373.15	473.15	573.15	673.15	873.15	1 073.15	1 273.15	1 473.15
Specific heat capacity, $c_{p,g}$ (kJ/kg K)	S	L	L	L	L	L	2.349	2.675	2.952	3.391	3.718	3.969	4.157
Thermal conductivity, λ_g[(W/m^2)/(K/m)]	S	L	L	L	L	L	0.028	(0.039)	(0.052)	(0.077)	(0.100)	(0.116)	(0.130)
Dynamic viscosity, η_g(10^{-5} Ns/m^2)	S	L	L	L	L	L	(0.85)	(1.04)	(1.21)	(1.52)	(1.82)	(2.09)	(2.36)

PROPADIENE (ALLENE)

Chemical formula: C_3H_4
Molecular weight: 40.06
Melting point: 136.9 K
Boiling point: 238.7 K

Critical temperature: 393 K
Critical pressure: 5.47 MPa
Critical density: 247 kg/m^3
Normal vapor density: 1.79 kg/m^3 (@ 0 °C, 101.3 kPa)

Properties of the Liquid at Temperatures Below the Normal Boiling Point

	Temperature, °C												
	−150	−100	−75	−50	−25	0	20	50	100	150	200	250	300
	Temperature, K												
Property	123.15	173.15	198.15	223.15	248.15	273.15	293.15	323.15	373.15	423.15	473.15	523.15	573.15
Density, ρ_l (kg/m^3)	S	(791)	(712)	682	V	V	V	V	V	V	V	V	V
Specific heat capacity, $c_{p,l}$ (kJ/kg K)	S	(1.87)	(1.94)	(2.02)	V	V	V	V	V	V	V	V	V
Thermal conductivity, λ_l [(W/m^2)/(K/m)]	S	(0.206)	(0.187)	(0.170)	V	V	V	V	V	V	V	V	V
Dynamic viscosity, η_l (10^{-5} Ns/m^2)	S	(46)	(32)	(23)	V	V	V	V	V	V	V	V	V

Transport Properties of the Superheated Vapor

	Temperature, °C												
	−150	−100	−50	0	25	100	200	300	400	600	800	1 000	1 200
Property (at low pressure)	Temperature, K												
	123.15	173.15	223.15	273.15	298.15	373.15	473.15	573.15	673.15	873.15	1 073.15	1 273.15	1 473.15
Specific heat capacity, $c_{p,g}$ (kJ/kg K)	S	L	L	1.386	1.474	1.717	2.001	2.244	2.445	2.763	3.006	3.195	3.333
Thermal conductivity, λ_g[(W/m^2)/(K/m)]	S	L	L	(0.013)	(0.016)	(0.024)	(0.035)	(0.046)	(0.058)	(0.081)	(0.109)	(0.136)	(0.161)
Dynamic viscosity, η_g(10^{-5} Ns/m^2)	S	L	L	(0.79)	(0.85)	(1.06)	(1.31)	(1.56)	(1.79)	(2.21)	(2.60)	(2.95)	(3.37)

1,2-BUTADIENE

Chemical formula: $CH_3CH{:}C{:}CH_2$
Molecular weight: 54.09
Melting point: 137.0 K
Normal boiling point: 284.0 K

Critical temperature: 443.7 K
Critical pressure: 4.500 MPa
Critical density: 246.8 kg/m^3
Normal vapor density: 2.41 kg/m^3
(@ 0°C, 101.3 kPa)

Properties of the Liquid at Temperatures Below the Normal Boiling Point

Property	Temperature, °C												
	−150	−100	−75	−50	−25	0	20	50	100	150	200	250	300
	Temperature, K												
	123.15	173.15	198.15	223.15	248.15	273.15	293.15	323.15	373.15	423.15	473.15	523.15	573.15
Density, ρ_l (kg/m^3)	S	(785)	(760)	(730)	(704)	(676)	V	V	V	V	V	V	V
Specific heat capacity, $c_{p,l}$ (kJ/kg K)	S	2.035	2.041	2.086	2.136	2.208	V	V	V	V	V	V	V
Thermal conductivity, λ_l [(W/m^2)/(K/m)]	S	(0.196)	(0.181)	(0.167)	(0.153)	(0.140)	V	V	V	V	V	V	V
Dynamic viscosity, η_l (10^{-5} Ns/m^2)	S	(75)	(50)	(36)	(27)	(22)	V	V	V	V	V	V	V

Properties of the Saturated Liquid and Vapor

T_{sat}, K	284.0	300	315	330	345	360	375	390	400	443.7
p_{sat}, kPa	101.3	189	265	445	661	945	1 310	1 770	2 140	4 500
ρ_ℓ, kg/m^3	651	643	625	605	585	563	537	507	485	246.8
ρ_g, kg/m^3	2.32	4.04	6.43	9.80	14.4	20.7	29.2	40.7	50.8	246.8
h_ℓ, kJ/kg	−197	−166	−131	−94	−57	−19	19	61	88	255
h_g, kJ/kg	237	257	275	293	311	327	341	354	359	255
$\Delta h_{g,\ell}$, kJ/kg	434	423	406	387	368	346	322	293	271	
$c_{p,\ell}$, kJ/(kg K)	2.20	2.24	2.30	2.41	2.49	2.60	2.72	2.87	3.01	
$c_{p,g}$, kJ/(kg K)	1.48	1.56	1.65	1.75	1.87	2.01	2.18	2.43	2.68	
η_ℓ, μNs/m^2	200	185	170	150	134	116	100	85	76	43
η_g, μNs/m^2	7.40	7.78	8.27	8.76	9.26	9.77	10.4	11.0	11.5	43
λ_ℓ, (mW/m^2)/(K/m)	126	119	113	107	102	98	93	88	82	49
λ_g, (mW/m^2)/(K/m)	12.5	14.1	15.8	17.5	19.3	21.2	23.3	25.6	27.3	49
Pr_ℓ	3.62	3.48	3.46	3.38	3.27	3.08	2.92	2.77	2.79	
Pr_g	0.88	0.86	0.86	0.88	0.90	0.93	0.97	1.04	1.13	
σ, mN/m	18.0	15.7	13.9	12.1	10.4	8.65	7.00	5.30	4.10	
$\beta_{e,\ell}$, kK^{-1}	1.71	1.89	2.10	2.35	2.66	3.77	4.11	4.85	5.74	

Transport Properties of the Superheated Vapor

Property (at low pressure)	Temperature, °C												
	−150	−100	−50	0	25	100	200	300	400	600	800	1 000	1 200
	Temperature, K												
	123.15	173.15	223.15	273.15	298.15	373.15	473.15	573.15	673.15	873.15	1 073.15	1 273.15	1 473.15
Specific heat capacity, $c_{p,g}$ (kJ/kg K)	S	L	L	L	1.482	1.738	2.043	2.311	2.525	2.910	3.190	3.400	3.559
Thermal conductivity, λ_g[(W/m^2)/(K/m)]	S	L	L	L	(0.14)	(0.022)	(0.033)	(0.045)	(0.057)	(0.081)	(0.105)	(0.126)	(0.145)
Dynamic viscosity, η_g(10^{-5} Ns/m^2)	S	L	L	L	(0.80)	(0.99)	(1.23)	(1.47)	(1.69)	(2.11)	(2.50)	(2.85)	(3.15)

1,3-BUTADIENE

Chemical formula: $CH_2{:}CHCH{:}CH_2$
Molecular weight: 54.088

Melting point: 164.24 K
Normal boiling point: 268.69 K

Critical temperature: 425.15 K
Critical pressure: 4.330 MPa
Critical density: 245 kg/m^3
Normal vapor density: 2.41 kg/m^3
(@ 0 °C, 101.3 kPa)

Properties of the Liquid at Temperatures Below the Normal Boiling Point

	Temperature, °C												
	−150	−100	−75	−50	−25	0	20	50	100	150	200	250	300
	Temperature, K												
Property	123.15	173.15	198.15	223.15	248.15	273.15	293.15	323.15	373.15	423.15	473.15	523.15	573.15
Density, ρ_l (kg/m^3)	S	755	728	701	673	V	V	V	V	V	V	V	V
Specific heat capacity, $c_{p,l}$ (kJ/kg K)	S	2.002	2.068	2.171	2.277	V	V	V	V	V	V	V	V
Thermal conductivity, λ_l [(W/m^2)/(K/m)]	S	0.190	0.175	0.161	0.147	V	V	V	V	V	V	V	V
Dynamic viscosity, η_l (10^{-5} Ns/m^2)	S	(72)	(48)	(34)	(26)	V	V	V	V	V	V	V	V

Properties of the Saturated Liquid and Vapor

T_{sat}, K	268.69	285	300	315	330	350	370	390	410	425.15
p_{sat}, kPa	101.3	184	298	458	676	1 080	1 630	2 370	3 350	4 330
ρ_ℓ, kg/m^3	650	631	612	593	572	541	507	464	405	245
ρ_g, kg/m^3	2.53	4.41	6.97	10.5	15.5	26.7	38.7	53.5	101	245
h_ℓ, kJ/kg	493.3	529.0	563.0	598.4	635.0	686.5	741.5	800.1	867.5	
h_g, kJ/kg	908.7	928.4	946.4	964.4	981.7	1 003.8	1 023.2	1 038.3	1 041.4	
$\Delta h_{g,\ell}$, kJ/kg	415.4	399.4	383.4	366.0	346.7	317.3	281.7	238.2	173.9	
$c_{p,\ell}$, kJ/(kg K)	2.14	2.22	2.29	2.37	2.47	2.63	2.81	3.03	3.28	
$c_{p,g}$, kJ/(kg K)	1.42	1.53	1.63	1.74	1.87	2.07	2.35	2.92	4.54	
η_ℓ, μNs/m^2	200	164	138	117	99.5	84.3	71.3	61.7	53.4	
η_g, μNs/m^2	7.57	8.43	8.91	9.41	9.94	10.7	11.6	12.6	15.0	
λ_ℓ, (mW/m^2)/(K/m)	126	117	110	103	95.6	86.6	78.0	69.8	62.0	53.7
λ_g, (mW/m^2)/(K/m)	9.54	12.3	14.7	17.3	19.9	23.7	27.5	31.6	37.8	53.7
Pr_ℓ	3.40	3.11	2.87	2.69	2.57	2.56	2.57	2.68	2.83	
Pr_g	1.13	1.05	0.99	0.95	0.93	0.93	1.02	1.16	1.30	
σ, mN/m	16.6	14.5	12.7	10.9	9.12	6.86	4.73	2.75	1.00	
$\beta_{e,\ell}$, kK^{-1}	1.79	1.98	2.20	2.45	2.84	3.57	4.88	7.64	20.0	

Transport Properties of the Superheated Vapor

	Temperature, °C												
	−150	−100	−50	0	25	100	200	300	400	600	800	1 000	1 200
Property (at low pressure)	Temperature, K												
	123.15	173.15	223.15	273.15	298.15	373.15	473.15	573.15	673.15	873.15	1 073.15	1 273.15	1 473.15
Specific heat capacity, $c_{p,g}$ (kJ/kg K)	S	L	L	L	1.474	1.779	2.127	2.403	2.621	2.964	3.224	3.421	3.571
Thermal conductivity, λ_g[(W/m^2)/(K/m)]	S	L	L	L	0.016	0.022	0.034	(0.046)	(0.059)	(0.083)	(0.106)	(0.127)	(0.146)
Dynamic viscosity, η_g(10^{-5} Ns/m^2)	S	L	L	L	(0.74)	0.998	(1.23)	(1.47)	(1.69)	(2.09)	(2.45)	(2.76)	(3.06)

1,2-PENTADIENE

Chemical formula: C_5H_8
Molecular weight: 68.12
Melting point: 135.9 K
Boiling point: 318 K

Critical temperature: 503 K
Critical pressure: 4.07 MPa
Critical density: 247 kg/m^3
Normal vapor density: 3.04 kg/m^3 (@ 0°C, 101.3 kPa)

Properties of the Liquid at Temperatures Below the Normal Boiling Point

Property	Temperature, °C: −150	−100	−75	−50	−25	0	20	50	100	150	200	250	300
	Temperature, K: 123.15	173.15	198.15	223.15	248.15	273.15	293.15	323.15	373.15	423.15	473.15	523.15	573.15
Density, ρ_l (kg/m^3)	S	(801)	(778)	(755)	(733)	(710)	621	V	V	V	V	V	V
Specific heat capacity, $c_{p,l}$ (kJ/kg K)	S	1.934	1.960	2.002	2.059	2.132	2.20	V	V	V	V	V	V
Thermal conductivity, λ_l [(W/m^2)/(K/m)]	S	(0.174)	(0.163)	(0.152)	(0.141)	(0.131)	(0.123)	V	V	V	V	V	V
Dynamic viscosity, η_l (10^{-5} Ns/m^2)	S	(115)	(72)	(48)	(36)	(27)	(22)	V	V	V	V	V	V

1,trans 3-PENTADIENE

Chemical formula: C_5H_8
Molecular weight: 68.12
Melting point: 185.7 K
Boiling point: 315.2 K

Critical temperature: 496 K
Critical pressure: 3.99 MPa
Critical density: 247 kg/m^3
Normal vapor density: 3.04 kg/m^3 (@ 0°C, 101.3 kPa)

Properties of the Liquid at Temperatures Below the Normal Boiling Point

Property	Temperature, °C: −150	−100	−75	−50	−25	0	20	50	100	150	200	250	300
	Temperature, K: 123.15	173.15	198.15	223.15	248.15	273.15	293.15	323.15	373.15	423.15	473.15	523.15	573.15
Density, ρ_l (kg/m^3)	S	S	(763)	(740)	(718)	(695)	676	V	V	V	V	V	V
Specific heat capacity, $c_{p,l}$ (kJ/kg K)	S	S	1.919	1.953	2.009	2.089	2.169	V	V	V	V	V	V
Thermal conductivity, λ_l [(W/m^2)/(K/m)]	S	S	(0.163)	(0.152)	(0.141)	(0.131)	(0.123)	V	V	V	V	V	V
Dynamic viscosity, η_l (10^{-5} Ns/m^2)	S	S	(72)	(48)	(36)	(27)	(22)	V	V	V	V	V	V

Transport Properties of the Superheated Vapor

Property (at low pressure)	Temperature, °C: −150	−100	−50	0	25	100	200	300	400	600	800	1 000	1 200
	Temperature, K: 123.15	173.15	223.15	273.15	298.15	373.15	473.15	573.15	673.15	873.15	1 073.15	1 273.15	1 473.15
Specific heat capacity, $c_{p,g}$ (kJ/kg K)	S	S	L	L	L	1.817	2.169	2.449	2.684	3.052	3.328	3.542	3.697
Thermal conductivity, λ_g[(W/m^2)/(K/m)]	S	S	L	L	L	(0.021)	(0.031)	(0.042)	(0.054)	(0.080)	(0.106)	(0.129)	(0.151)
Dynamic viscosity, η_g(10^{-5} Ns/m^2)	S	S	L	L	L	(0.91)	(1.15)	(1.37)	(1.59)	(1.98)	(2.33)	(2.65)	(2.95)

1,4-PENTADIENE

Chemical formula: C_5H_8
Molecular weight: 68.12
Melting point: 124.65 K
Boiling point: 299.1 K

Critical temperature: 460.2 K
Critical pressure: 3.62 MPa
Critical density: 247 kg/m^3
Normal vapor density: 3.04 kg/m^3
(@ 0 °C, 101.3 kPa)

Properties of the Liquid at Temperatures Below the Normal Boiling Point

	Temperature, °C												
	−150	−100	−75	−50	−25	0	20	50	100	150	200	250	300
	Temperature, K												
Property	123.15	173.15	198.15	223.15	248.15	273.15	293.15	323.15	373.15	423.15	473.15	523.15	573.15
Density, ρ_l (kg/m^3)	S	(783)	(758)	(733)	(708)	(683)	660	V	V	V	V	V	V
Specific heat capacity, $c_{p,l}$ (kJ/kg K)	S	1.863	1.880	1.926	1.993	2.085	2.165	V	V	V	V	V	V
Thermal conductivity, λ_l [(W/m^2)/(K/m)]	S	(0.174)	(0.163)	(0.152)	(0.141)	(0.131)	(0.123)	V	V	V	V	V	V
Dynamic viscosity, η_l (10^{-5} Ns/m^2)	S	(115)	(72)	(48)	(36)	(27)	(22)	V	V	V	V	V	V

2,3-PENTADIENE

Chemical formula: C_5H_8
Molecular weight: 68.12
Melting point: 147.45 K
Boiling point: 321.35 K

Critical temperature: 499.2 K
Critical pressure: 4.197 MPa
Critical density: 247 kg/m^3
Normal vapor density: 3.04 kg/m^3
(@ 0 °C, 101.3 kPa)

Properties of the Liquid at Temperatures Below the Normal Boiling Point

	Temperature, °C												
	−150	−100	−75	−50	−25	0	20	50	100	150	200	250	300
	Temperature, K												
Property	123.15	173.15	198.15	223.15	248.15	273.15	293.15	323.15	373.15	423.15	473.15	523.15	573.15
Density, ρ_l (kg/m^3)	S	(807)	(783)	(759)	(736)	(713)	695	V	V	V	V	V	V
Specific heat capacity, $c_{p,l}$ (kJ/kg K)	S	1.999	2.019	2.052	2.100	2.162	2.221	V	V	V	V	V	V
Thermal conductivity, λ_l [(W/m^2)/(K/m)]	S	(0.174)	(0.163)	(0.152)	(0.141)	(0.131)	(0.112)	V	V	V	V	V	V
Dynamic viscosity, η_l (10^{-5} Ns/m^2)	S	(115)	(72)	(48)	(36)	(27)	(22)	V	V	V	V	V	V

ACETYLENE

Chemical formula: $HC\vdots CH$
Molecular weight: 26.036
Melting point: 192.2 K
Normal boiling point: 189.2 K

Critical temperature: 308.7 K
Critical pressure: 6.240 MPa
Critical density: 230 kg/m^3
Normal vapor density: 1.17 kg/m^3 (@ 0°C, 101.3 kPa)

Properties of the Liquid at Temperatures Below the Normal Boiling Point

	Temperature, °C												
	−150	−100	−75	−50	−25	0	20	50	100	150	200	250	300
	Temperature, K												
Property	123.15	173.15	198.15	223.15	248.15	273.15	293.15	323.15	373.15	423.15	473.15	523.15	573.15
Density, ρ_l (kg/m^3)	S	S	612	V	V	V	V	V	V	V	V	V	V
Specific heat capacity, $c_{p,l}$ (kJ/kg K)	S	S	(3.1)	V	V	V	V	V	V	V	V	V	V
Thermal conductivity, λ_l [(W/m^2)/(K/m)]	S	S	(0.54)	V	V	V	V	V	V	V	V	V	V
Dynamic viscosity, η_l (10^{-5} Ns/m^2)	S	S	(16)	V	V	V	V	V	V	V	V	V	V

Properties of the Saturated Liquid and Vapor

T_{sat}, K	192.2	200	210	230	240	250	270	280	290	308.7
p_{sat}, kPa	128	189	304	689	986	1 370	2 450	3 190	4 080	6 240
ρ_ℓ, kg/m^3	617	606	590	556	538	519	473	445	411	231
ρ_g, kg/m^3	2.16	3.11	4.86	10.8	15.6	22.1	41.0	54.0	73.2	231
h_ℓ, kJ/kg	−369.5	−351.8	−331.6	−271.6	−236.1	−202.0	−136.2	−100.5	−66.2	104.7
h_g, kJ/kg	214.8	222.9	230.9	248.6	254.4	257.6	257.0	252.8	238.8	104.7
$\Delta h_{g,\ell}$, kJ/kg	584.3	574.7	562.5	520.2	490.5	459.6	393.2	363.3	305.0	
$c_{p,\ell}$, kJ/(kg K)	3.09	3.12	3.15	3.27	3.35	3.46	3.87	4.25	5.14	
$c_{p,g}$, kJ/(kg K)	1.47	1.51	1.59	1.80	1.93	2.14	2.64	2.93	3.39	
η_ℓ, μNs/m^2	169	156	146	127	116	99.6	79.9	66.9	56.9	
η_g, μNs/m^2	7.35	7.67	8.10	8.68	9.03	9.44	10.5	11.2	12.1	
λ_ℓ, (mW/m^2)/(K/m)	55.9	54.3	51.9	48.0	48.3	44.9	43.0	42.2	41.5	
λ_g, (mW/m^2)/(K/m)	11.2	12.0	13.1	15.4	16.9	18.5	22.6	25.0	28.3	
Pr_ℓ	9.34	8.96	8.86	8.65	8.39	7.68	7.19	6.75	7.04	
Pr_g	0.96	0.97	0.98	1.01	1.03	1.09	1.23	1.31	1.45	
σ, mN/m	19.1	17.6	15.6	11.8	9.92	8.12	4.64	3.11	1.73	
$\beta_{e,\ell}$, kK^{-1}	2.21	2.45	2.79	3.56	4.02	4.73	7.28	8.97	15.6	

Transport Properties of the Superheated Vapor

	Temperature, °C												
	−150	−100	−50	0	25	100	200	300	400	600	800	1 000	1 200
Property (at low pressure)	Temperature, K												
	123.15	173.15	223.15	273.15	298.15	373.15	473.15	573.15	673.15	873.15	1 073.15	1 273.15	1 473.15
Specific heat capacity, $c_{p,g}$ (kJ/kg K)	S	S	1.503	1.616	1.687	1.871	2.047	2.177	2.286	2.462	2.613	2.734	2.834
Thermal conductivity, λ_g[(W/m^2)/(K/m)]	S	S	0.013	0.018	0.021	0.030	0.042	0.053	0.066	0.087	0.107	0.125	0.143
Dynamic viscosity, η_g(10^{-5} Ns/m^2)	S	S	0.785	0.960	1.04	1.28	1.55	1.83	2.08	2.53	2.93	3.30	3.65

METHYLACETYLENE (PROPYNE)

Chemical formula: C_3H_4
Molecular weight: 40.06
Melting point: 170.5 K
Boiling point: 250.0 K

Critical temperature: 402.4 K
Critical pressure: 5.62 MPa
Critical density: 244 kg/m^3
Normal vapor density: 1.79 kg/m^3
(@ 0 °C, 101.3 kPa)

Properties of the Liquid at Temperatures Below the Normal Boiling Point

	Temperature, °C												
	−150	−100	−75	−50	−25	0	20	50	100	150	200	250	300
	Temperature, K												
Property	123.15	173.15	198.15	223.15	248.15	273.15	293.15	323.15	373.15	423.15	473.15	523.15	573.15
Density, ρ_l (kg/m^3)	S	(770)	(740)	710	673	V	V	V	V	V	V	V	V
Specific heat capacity, $c_{p,l}$ (kJ/kg K)	S	(2.17)	(2.24)	(2.32)	(2.42)	V	V	V	V	V	V	V	V
Thermal conductivity, λ_l [(W/m^2)/(K/m)]	S	(0.19)	(0.175)	(0.16)	(0.145)	V	V	V	V	V	V	V	V
Dynamic viscosity, η_l (10^{-5} Ns/m^2)	S	(44)	(31)	(24)	(19)	V	V	V	V	V	V	V	V

Transport Properties of the Superheated Vapor

	Temperature, °C												
	− 150	− 100	− 50	0	25	100	200	300	400	600	800	1 000	1 200
Property	Temperature, K												
(at low pressure)	123.15	173.15	223.15	273.15	298.15	373.15	473.15	573.15	673.15	873.15	1 073.15	1 273.15	1 473.1
Specific heat capacity, $c_{p,g}$ (kJ/kg K)	S	L	L	1.436	1.516	1.738	1.997	2.233	2.416	2.734	2.977	3.165	3.312
Thermal conductivity, λ_g[(W/m^2)/(K/m)]	S	L	L	0.016	0.019	0.026	0.037	0.048	0.059	0.082	0.103	0.124	0.143
Dynamic viscosity, η_g(10^{-5} Ns/m^2)	S	L	L	0.806	0.886	1.088	1.348	1.586	1.807	2.206	2.564	2.891	3.196

ETHYLACETYLENE (1-BUTYNE)

Chemical formula: C_4H_6
Molecular weight: 54.09
Melting point: 143.15 K
Boiling point: 281.85 K

Critical temperature: 463.7 K
Critical pressure: 4.71 MPa
Critical density: 245 kg/m^3
Normal vapor density: 2.41 kg/m^3 (@ 0°C, 101.3 kPa)

Properties of the Liquid at Temperatures Below the Normal Boiling Point

Property	Temperature, °C: −150	−100	−75	−50	−25	0	20	50	100	150	200	250	300
	Temperature, K: 123.15	173.15	198.15	223.15	248.15	273.15	293.15	323.15	373.15	423.15	473.15	523.15	573.15
Density, ρ_l (kg/m^3)	S	(780)	(755)	(730)	705	678	V	V	V	V	V	V	V
Specific heat capacity, $c_{p,l}$ (kJ/kg K)	S	2.125	2.158	2.218	2.300	2.398	V	V	V	V	V	V	V
Thermal conductivity, λ_l [(W/m^2)/(K/m)]	S	(0.226)	(0.197)	(0.172)	(0.151)	(0.133)	V	V	V	V	V	V	V
Dynamic viscosity, η_l (10^{-5} Ns/m^2)	S	—	(53)	(40)	(35)	(24)	V	V	V	V	V	V	V

Transport Properties of the Superheated Vapor

Property (at low pressure)	Temperature, °C: −150	−100	−50	0	25	100	200	300	400	600	800	1 000	1 200
	Temperature, K: 123.15	173.15	223.15	273.15	298.15	373.15	473.15	573.15	673.15	873.15	1 073.15	1 273.15	1 473.15
Specific heat capacity, $c_{p,g}$ (kJ/kg K)	S	L	L	L	1.507	1.763	2.064	2.324	2.504	2.901	3.174	3.383	3.542
Thermal conductivity, λ_g[(W/m^2)/(K/m)]	S	L	L	L	(0.013)	(0.021)	(0.032)	(0.043)	(0.055)	(0.079)	(0.101)	(0.122)	(0.142)
Dynamic viscosity, η_g(10^{-5} Ns/m^2)	S	L	L	L	(0.77)	(0.95)	(1.19)	(1.42)	(1.63)	(2.03)	(2.37)	(2.70)	(3.00)

DIMETHYLACETYLENE (2-BUTYNE)

Chemical formula: C_4H_6
Molecular weight: 54.09
Melting point: 240.85 K
Boiling point: 300.35 K

Critical temperature: 485.2 K
Critical pressure: 5.13 MPa
Critical density: 245 kg/m^3
Normal vapor density: 2.41 kg/m^3
(@ 0 °C, 101.3 kPa)

Properties of the Liquid at Temperatures Below the Normal Boiling Point

Property	Temperature, °C												
	−150	−100	−75	−50	−25	0	20	50	100	150	200	250	300
	Temperature, K												
	123.15	173.15	198.15	223.15	248.15	273.15	293.15	323.15	373.15	423.15	473.15	523.15	573.15
Density, ρ_l (kg/m^3)	S	S	S	S	(742)	(715)	691	V	V	V	V	V	V
Specific heat capacity, $c_{p,l}$ (kJ/kg K)	S	S	S	S	2.190	2.257	2.311	V	V	V	V	V	V
Thermal conductivity, λ_l [(W/m^2)/(K/m)]	S	S	S	S	(0.167)	(0.146)	(0.133)	V	V	V	V	V	V
Dynamic viscosity, η_l (10^{-5} Ns/m^2)	S	S	S	S	(35)	(27)	(23)	V	V	V	V	V	V

Transport Properties of the Superheated Vapor

Property (at low pressure)	Temperature, °C												
	−150	−100	−50	0	25	100	200	300	400	600	800	1 000	1 200
	Temperature, K												
	123.15	173.15	223.15	273.15	298.15	373.15	473.15	573.15	673.15	873.15	1 073.15	1 273.15	1 473.15
Specific heat capacity, $c_{p,g}$ (kJ/kg K)	S	S	S	L	L	1.671	1.964	2.232	2.466	2.847	3.140	3.358	3.525
Thermal conductivity, λ_g[(W/m^2)/(K/m)]	S	S	S	L	L	(0.021)	(0.033)	(0.043)	(0.054)	(0.076)	(0.098)	(0.119)	(0.139)
Dynamic viscosity, η_g(10^{-5} Ns/m^2)	S	S	S	L	L	(0.93)	(1.20)	(1.41)	(1.63)	(2.03)	(2.37)	(2.70)	(3.00)

CYCLOPROPANE

Chemical formula: C_3H_6
Molecular weight: 42.08
Melting point: 145.7 K
Boiling point: 240.4 K

Critical temperature: 397.8 K
Critical pressure: 5.49 MPa
Critical density: 248 kg/m^3
Normal vapor density: 1.88 kg/m^3
(@ 0 °C, 101.3 kPa)

Properties of the Liquid at Temperatures Below the Normal Boiling Point

	Temperature, °C												
	−150	−100	−75	−50	−25	0	20	50	100	150	200	250	300
	Temperature, K												
Property	123.15	173.15	198.15	223.15	248.15	273.15	293.15	323.15	373.15	423.15	473.15	523.15	573.15
Density, ρ_l (kg/m^3)	S	740	716	693	V	V	V	V	V	V	V	V	V
Specific heat capacity, $c_{p,l}$ (kJ/kg K)	S	1.779	1.809	1.863	V	V	V	V	V	V	V	V	V
Thermal conductivity, λ_l [(W/m^2)/(K/m)]	S	S	(0.166)	(0.145)	V	V	V	V	V	V	V	V	V
Dynamic viscosity, η_l (10^{-5} Ns/m^2)	S	S	(40)	(30)	V	V	V	V	V	V	V	V	V

Transport Properties of the Superheated Vapor

	Temperature, °C												
	−150	−100	−50	0	25	100	200	300	400	600	800	1 000	1 200
Property (at low pressure)	Temperature, K												
	123.15	173.15	223.15	273.15	298.15	373.15	473.15	573.15	673.15	873.15	1 073.15	1 273.15	1 473.15
Specific heat capacity, $c_{p,g}$ (kJ/kg K)	S	L	L	1.210	1.331	1.696	2.144	2.512	2.814	3.282	3.627	3.887	4.085
Thermal conductivity, λ_g[(W/m^2)/(K/m)]	S	L	L	0.016	0.019	0.026	0.038	0.049	0.060	0.081	0.103	0.123	0.134
Dynamic viscosity, η_g(10^{-5} Ns/m^2)	S	L	L	0.80	0.87	1.08	1.34	1.60	1.81	2.22	2.56	2.89	3.20

CYCLOBUTANE

Chemical formula: C_4H_8
Molecular weight: 56.11
Melting point: 182.4 K
Boiling point: 285.7 K

Critical temperature: 459.9 K
Critical pressure: 4.99 MPa
Critical density: 267 kg/m^3
Normal vapor density: 2.50 kg/m^3
(@ 0°C, 101.3 kPa)

Properties of the Liquid at Temperatures Below the Normal Boiling Point

	Temperature, °C												
	–150	–100	–75	–50	–25	0	20	50	100	150	200	250	300
	Temperature, K												
Property	123.15	173.15	198.15	223.15	248.15	273.15	293.15	323.15	373.15	423.15	473.15	523.15	573.15
Density, ρ_l (kg/m^3)	S	S	791	769	747	724	V	V	V	V	V	V	V
Specific heat capacity, $c_{p,l}$ (kJ/kg K)	S	S	1.617	1.688	1.756	1.842	V	V	V	V	V	V	V
Thermal conductivity, λ_l [(W/m^2)/(K/m)]	S	S	(0.171)	(0.154)	(0.140)	(0.127)	V	V	V	V	V	V	V
Dynamic viscosity, η_l (10^{-5} Ns/m^2)	S	S	(60)	(45)	(34)	(26)	V	V	V	V	V	V	V

Transport Properties of the Superheated Vapor

	Temperature, °C												
	– 150	– 100	– 50	0	25	100	200	300	400	600	800	1 000	1 200
	Temperature, K												
Property (at low pressure)	123.15	173.15	223.15	273.15	298.15	373.15	473.15	573.15	673.15	873.15	1 073.15	1 273.15	1 473.15
Specific heat capacity, $c_{p,g}$ (kJ/kg K)	S	S	L	L	1.287	1.658	2.107	2.494	2.818	3.325	3.667	3.917	4.166
Thermal conductivity, λ_g[(W/m^2)/(K/m)]	S	S	L	L	(0.016)	(0.024)	(0.036)	(0.049)	(0.060)	(0.087)	(0.109)	(0.130)	(0.147)
Dynamic viscosity, η_g(10^{-5} Ns/m^2)	S	S	L	L	(0.82)	(1.01)	(1.27)	(1.53)	(1.76)	(2.14)	(2.51)	(2.86)	(3.14)

CYCLOPENTANE

Chemical formula: $CH_2\langle(CH_2CH_2)_2\rangle$
Molecular weight: 70.135
Melting point: 179.25 K
Normal boiling point: 322.45 K

Critical temperature: 511.8 K
Critical pressure: 4.508 MPa
Critical density: 272 kg/m^3
Normal vapor density: 3.13 kg/m^3
(@ 0°C, 101.3 kPa)

Properties of the Liquid at Temperatures Below the Normal Boiling Point

	Temperature, °C												
	−150	−100	−75	−50	−25	0	20	50	100	150	200	250	300
Property	Temperature, K												
	123.15	173.15	198.15	223.15	248.15	273.15	293.15	323.15	373.15	423.15	473.15	523.15	573.15
Density, ρ_l (kg/m^3)	S	S	(825)	(810)	(790)	765	745	V	V	V	V	V	V
Specific heat capacity, $c_{p,l}$ (kJ/kg K)	S	S	1.444	1.507	1.591	1.696	1.800	V	V	V	V	V	V
Thermal conductivity, λ_l [(W/m^2)/(K/m)]	S	S	0.165	0.156	0.148	0.139	0.133	V	V	V	V	V	V
Dynamic viscosity, η_l (10^{-5} Ns/m^2)	S	S	(135)	(105)	78.1	55.5	43.8	V	V	V	V	V	V

Properties of the Saturated Liquid and Vapor

T_{sat}, K	322.4	350	370	390	410	430	450	470	490	511.8
p_{sat}, kPa	101.3	239	406	642	948	1 309	1 861	2 742	3 562	4 508
ρ_ℓ, kg/m^3	706	680	656	635	607	577	547	505	455	272
ρ_g, kg/m^3	10.8	20.0	36.3	52.1	69.2	97.5	117	141	176	272
h_ℓ, kJ/kg	−288.7	−217.3	−179.4	−155.3	−92.8	−44.2	−3.8	62.2	127.4	227.1
h_g, kJ/kg	104.9	138.6	163.9	189.6	215.4	240.5	264.1	284.1	294.9	227.1
$\Delta h_{g,\ell}$, kJ/kg	393.6	355.9	343.3	334.9	318.2	284.7	267.9	221.9	167.5	
$c_{p,\ell}$, kJ/(kg K)	1.92	2.05	2.12	2.22	2.37	2.53	2.73	3.04	3.78	
$c_{p,g}$, kJ/(kg K)	1.34	1.47	1.61	1.79	1.93	2.11	2.40	2.81	3.91	
η_ℓ, μNs/m^2	320	255	200	160	128	109	92	76	58	43
η_g, μNs/m^2	7.9	8.9	9.7	10.5	11.2	11.9	12.8	13.9	15.9	43
λ_ℓ, (mW/m^2)/(K/m)	125	117	111	105	98	91	85	79	68	51.1
λ_g, (mW/m^2)/(K/m)	16.3	18.5	20.8	23.2	25.7	28.4	31.3	34.8	39.0	51.1
Pr_ℓ	4.92	4.47	3.82	3.38	3.10	3.03	2.95	2.92	3.22	
Pr_g	0.65	0.71	0.75	0.81	0.84	0.88	0.98	1.12	1.59	
σ, mN/m	18.6	15.1	13.0	11.1	8.7	6.6	4.7	2.9	1.3	
$\beta_{e,\ell}$, kK^{-1}	1.49	1.75	1.99	2.30	2.73	3.36	4.36	6.22	11.05	

Transport Properties of the Superheated Vapor

	Temperature, °C												
	−150	−100	−50	0	25	100	200	300	400	600	800	1 000	1 200
Property (at low pressure)	Temperature, K												
	123.15	173.15	223.15	273.15	298.15	373.15	473.15	573.15	673.15	873.15	1 073.15	1 273.15	1 473.15
Specific heat capacity, $c_{p,g}$ (kJ/kg K)	S	S	L	L	L	1.557	2.026	2.433	2.772	3.308	3.697	3.990	4.208
Thermal conductivity, λ_g[(W/m^2)/(K/m)]	S	S	L	L	L	0.015	(0.030)	(0.044)	(0.059)	(0.091)	(0.120)	(0.142)	(0.160)
Dynamic viscosity, η_g(10^{-5} Ns/m^2)	S	S	L	L	L	0.922	1.142	(1.41)	(1.63)	(2.04)	(2.42)	(2.76)	(3.70)

METHYLCYCLOPENTANE

Chemical formula: C_6H_{12}
Molecular weight: 84.16
Melting point: 130.7 K
Boiling point: 345.0 K

Critical temperature: 532.7 K
Critical pressure: 3.79 MPa
Critical density: 264 kg/m^3
Normal vapor density: 3.75 kg/m^3
(@ 0 °C, 101.3 kPa)

Properties of the Liquid at Temperatures Below the Normal Boiling Point

	Temperature, °C												
	−150	−100	−75	−50	−25	0	20	50	100	150	200	250	300
	Temperature, K												
Property	123.15	173.15	198.15	223.15	248.15	273.15	293.15	323.15	373.15	423.15	473.15	523.15	573.15
Density, ρ_l (kg/m^3)	S	(855)	(833)	(811)	(789)	767	749	720	V	V	V	V	V
Specific heat capacity, $c_{p,l}$ (kJ/kg K)	S	1.516	1.545	1.612	1.696	1.779	1.863	(2.01)	V	V	V	V	V
Thermal conductivity, λ_l [(W/m^2)/(K/m)]	S	0.157	0.149	0.142	0.135	0.129	0.123	0.111	V	V	V	V	V
Dynamic viscosity, η_l (10^{-5} Ns/m^2)	S	(300)	(190)	(130)	93.1	65.0	50.7	36.5	V	V	V	V	V

Transport Properties of the Superheated Vapor

	Temperature, °C												
	−150	−100	−50	0	25	100	200	300	400	600	800	1 000	1 200
Property (at low pressure)	Temperature, K												
	123.15	173.15	223.15	273.15	298.15	373.15	473.15	573.15	673.15	873.15	1 073.15	1 273.15	1 473.15
Specific heat capacity, $c_{p,g}$ (kJ/kg K)	S	L	L	L	L	1.677	2.123	2.516	2.843	3.354	3.730	4.011	4.220
Thermal conductivity, λ_g[(W/m^2)/(K/m)]	S	L	L	L	L	(0.019)	(0.030)	(0.043)	(0.057)	(0.086)	(0.113)	(0.135)	(0.153)
Dynamic viscosity, η_g(10^{-5} Ns/m^2)	S	L	L	L	L	0.883	(1.10)	(1.31)	(1.52)	(1.91)	(2.27)	(2.60)	(2.94)

ETHYLCYCLOPENTANE

Chemical formula: C_7H_{14}
Molecular weight: 98.19
Melting point: 134.7 K
Boiling point: 376.6 K

Critical temperature: 569.5 K
Critical pressure: 3.39 MPa
Critical density: 262 kg/m^3
Normal vapor density: 4.38 kg/m^3
(@ 0 °C, 101.3 kPa)

Properties of the Liquid at Temperatures Below the Normal Boiling Point

	Temperature, °C												
	−150	−100	−75	−50	−25	0	20	50	100	150	200	250	300
	Temperature, K												
Property	123.15	173.15	198.15	223.15	248.15	273.15	293.15	323.15	373.15	423.15	473.15	523.15	573.15
Density, ρ_l (kg/m^3)	S	(863)	(843)	(823)	(803)	(783)	763	739	691	V	V	V	V
Specific heat capacity, $c_{p,l}$ (kJ/kg K)	S	1.527	1.569	1.630	1.707	1.770	1.873	(2.01)	(2.22)	V	V	V	V
Thermal conductivity, λ_l [(W/m^2)/(K/m)]	S	(0.155)	(0.148)	(0.141)	(0.134)	(0.127)	(0.122)	(0.114)	(0.102)	V	V	V	V
Dynamic viscosity, η_l (10^{-5} Ns/m^2)	S	—	—	(150)	(101)	72.4	56.7	41.2	27.0	V	V	V	V

Transport Properties of the Superheated Vapor

	Temperature, °C												
	− 150	− 100	− 50	0	25	100	200	300	400	600	800	1 000	1 200
Property	Temperature, K												
(at low pressure)	123.15	173.15	223.15	273.15	298.15	373.15	473.15	573.15	673.15	873.15	1 073.15	1 273.15	1 473.15
Specific heat capacity, $c_{p,g}$ (kJ/kg K)	S	L	L	L	L	L	2.173	2.558	2.876	3.375	3.747	4.019	4.229
Thermal conductivity, λ_g[(W/m^2)/(K/m)]	S	L	L	L	L	L	(0.028)	(0.040)	(0.054)	(0.082)	(0.107)	(0.127)	(0.145)
Dynamic viscosity, η_g(10^{-5} Ns/m^2)	S	L	L	L	L	L	(1.03)	(1.23)	(1.42)	(1.79)	(2.14)	(2.46)	(2.78)

PROPYLCYCLOPENTANE

Chemical formula: C_8H_{16}
Molecular weight: 112.22
Melting point: 155.8 K
Boiling point: 404.1 K

Critical temperature: 603 K
Critical pressure: 3.00 MPa
Critical density: 264 kg/m^3
Normal vapor density: 5.01 kg/m^3
(@ 0 °C, 101.3 kPa)

Properties of the Liquid at Temperatures Below the Normal Boiling Point

	Temperature, °C												
	−150	−100	−75	−50	−25	0	20	50	100	150	200	250	300
	Temperature, K												
Property	123.15	173.15	198.15	223.15	248.15	273.15	293.15	323.15	373.15	423.15	473.15	523.15	573.15
Density, ρ_l (kg/m^3)	S	(864)	(846)	(828)	810	792	777	752	710	V	V	V	V
Specific heat capacity, $c_{p,l}$ (kJ/kg K)	S	(1.48)	(1.57)	(1.66)	1.749	1.840	1.915	2.031	2.331	V	V	V	V
Thermal conductivity, λ_l [(W/m^2)/(K/m)]	S	(0.155)	(0.149)	(0.143)	(0.137)	(0.132)	(0.127)	(0.121)	(0.109)	V	V	V	V
Dynamic viscosity, η_l (10^{-5} Ns/m^2)	S	—	—	(215)	140	89.8	68.3	48.8	33.0	V	V	V	V

Transport Properties of the Superheated Vapor

	Temperature, °C												
	−150	−100	−50	0	25	100	200	300	400	600	800	1 000	1 200
Property (at low pressure)	Temperature, K												
	123.15	173.15	223.15	273.15	298.15	373.15	473.15	573.15	673.15	873.15	1 073.15	1 273.15	1 473.15
Specific heat capacity, $c_{p,g}$ (kJ/kg K)	S	L	L	L	L	L	2.202	2.579	2.893	3.387	3.751	4.024	4.229
Thermal conductivity, λ_g[(W/m^2)/(K/m)]	S	L	L	L	L	L	(0.025)	(0.037)	(0.050)	(0.078)	(0.101)	(0.115)	(0.137)
Dynamic viscosity, η_g(10^{-5} Ns/m^2)	S	L	L	L	L	L	(0.95)	(1.13)	(1.31)	(1.66)	(1.99)	(2.30)	(2.62)

BUTYLCYCLOPENTANE

Chemical formula: C_9H_{18}
Molecular weight: 126.25
Melting point: 164.95 K
Boiling point: 429.95 K

Critical temperature: 616.7 K
Critical pressure: 2.58 MPa
Critical density: 262 kg/m^3
Normal vapor density: 5.64 kg/m^3
(@ 0 °C, 101.3 kPa)

Properties of the Liquid at Temperatures Below the Normal Boiling Point

	Temperature, °C												
	−150	−100	−75	−50	−25	0	20	50	100	150	200	250	300
	Temperature, K												
Property	123.15	173.15	198.15	223.15	248.15	273.15	293.15	323.15	373.15	423.15	473.15	523.15	573.15
Density, ρ_l (kg/m^3)	S	(867)	(850)	(833)	816	799	784	762	723	(681)	V	V	V
Specific heat capacity, $c_{p,l}$ (kJ/kg K)	S	1.446	1.480	1.528	1.592	1.664	1.732	1.840	2.026	(2.22)	V	V	V
Thermal conductivity. λ_l [(W/m^2)/(K/m)]	S	(0.152)	(0.144)	(0.139)	(0.134)	(0.128)	(0.125)	(0.118)	(0.108)	(0.098)	V	V	V
Dynamic viscosity, η_l (10^{-5} Ns/m^2)	S	—	—	(260)	(175)	120.7	89.0	60.8	38.0	(26)	V	V	V

Transport Properties of the Superheated Vapor

	Temperature, °C												
	−150	−100	−50	0	25	100	200	300	400	600	800	1 000	1 200
Property (at low pressure)	Temperature, K												
	123.15	173.15	223.15	273.15	298.15	373.15	473.15	573.15	673.15	873.15	1 073.15	1 273.15	1 473.15
Specific heat capacity, $c_{p,g}$ (kJ/kg K)	S	L	L	L	L	L	2.227	2.600	2.910	3.395	3.760	4.028	4.229
Thermal conductivity, λ_g[(W/m^2)/(K/m)]	S	L	L	L	L	L	(0.022)	(0.033)	(0.045)	(0.069)	(0.091)	(0.103)	(0.125)
Dynamic viscosity, η_g(10^{-5} Ns/m^2)	S	L	L	L	L	L	(0.85)	(1.01)	(1.17)	(1.49)	(1.79)	(2.07)	(2.38)

PENTYLCYCLOPENTANE

Chemical formula: $C_{10}H_{20}$
Molecular weight: 140.28
Melting point: 181 K
Boiling point: 453.15 K

Critical temperature: 639.2 K
Critical pressure: 2.37 MPa
Critical density: —
Normal vapor density: 6.25 kg/m^3
(@ 0 °C, 101.3 kPa)

Properties of the Liquid at Temperatures Below the Normal Boiling Point

	Temperature, °C												
	−150	−100	−75	−50	−25	0	20	50	100	150	200	250	300
	Temperature, K												
Property	123.15	173.15	198.15	223.15	248.15	273.15	293.15	323.15	373.15	423.15	473.15	523.15	573.15
Density, ρ_l (kg/m^3)	S	S	(862)	(844)	(825)	(807)	791	(770)	(733)	(695)	V	V	V
Specific heat capacity, $c_{p,l}$ (kJ/kg K)	S	S	—	—	(1.62)	(1.69)	(1.76)	(1.86)	(2.05)	(2.25)	V	V	V
Thermal conductivity, λ_l [(W/m^2)/(K/m)]	S	S	(0.146)	(0.141)	(0.136)	(0.131)	(0.127)	(0.121)	(0.111)	(0.101)	V	V	V
Dynamic viscosity, η_l (10^{-5} Ns/m^2)	S	S	—	(430)	255	163.1	115.5	75.3	45.0	(30)	V	V	V

HEXYLCYCLOPENTANE

Chemical formula: $C_{11}H_{22}$
Molecular weight: 154.30
Melting point: 194 K
Boiling point: 476.3 K

Critical temperature: 660.1 K
Critical pressure: 2.14 MPa
Critical density: —
Normal vapor density: 6.88 kg/m^3
(@ 0 °C, 101.3 kPa)

Properties of the Liquid at Temperatures Below the Normal Boiling Point

	Temperature, °C												
	−150	−100	−75	−50	−25	0	20	50	100	150	200	250	300
	Temperature, K												
Property	123.15	173.15	198.15	223.15	248.15	273.15	293.15	323.15	373.15	423.15	473.15	523.15	573.15
Density, ρ_l (kg/m^3)	S	S	(866)	(848)	(830)	(813)	797	(778)	(742)	(705)	(665)	V	V
Specific heat capacity, $c_{p,l}$ (kJ/kg K)	S	S	—	—	(1.66)	(1.72)	(1.78)	(1.88)	(2.06)	(2.25)	(2.45)	V	V
Thermal conductivity, λ_l [(W/m^2)/(K/m)]	S	S	(0.147)	(0.142)	(0.137)	(0.132)	(0.127)	(0.123)	(0.113)	(0.104)	(0.095)	V	V
Dynamic viscosity, η_l (10^{-5} Ns/m^2)	S	S	—	(560)	350	215.1	149.4	93.8	54.0	(33)	—	V	V

CYCLOHEXANE

Chemical formula: $CH_2\langle(CH_2CH_2)_2\rangle CH_2$
Molecular weight: 84.132
Melting point: 279.7 K
Normal boiling point: 353.87 K

Critical temperature: 554.15 K
Critical pressure: 4.075 MPa
Critical density: 273 kg/m^3
Normal vapor density: 3.75 kg/m^3
(@ 0 °C, 101.3 kPa)

Properties of the Liquid at Temperatures Below the Normal Boiling Point

	Temperature, °C												
	−150	−100	−75	−50	−25	0	20	50	100	150	200	250	300
	Temperature, K												
Property	123.15	173.15	198.15	223.15	248.15	273.15	293.15	323.15	373.15	423.15	473.15	523.15	573.15
Density, ρ_l (kg/m^3)	S	S	S	S	S	S	779	750	V	V	V	V	V
Specific heat capacity, $c_{p,l}$ (kJ/kg K)	S	S	S	S	S	S	2.081	2.119	V	V	V	V	V
Thermal conductivity, λ_l [(W/m^2)/(K/m)]	S	S	S	S	S	S	0.120	0.113	V	V	V	V	V
Dynamic viscosity, η_l (10^{-5} Ns/m^2)	S	S	S	S	S	S	98.0	60.6	V	V	V	V	V

Properties of the Saturated Liquid and Vapor

T_{sat}, K	353.87	360	385	410	435	460	485	510	535	554.15
p_{sat}, kPa	101.3	124	234	414	689	1 069	1 586	2 275	3 723	4 075
ρ_ℓ, kg/m^3	715	710	685	660	630	590	555	510	445	273
ρ_g, kg/m^3	3.02	3.75	6.76	11.6	18.9	29.0	42.9	64.6	144	273
h_ℓ, kJ/kg	314.0	325.6	379.1	434.9	500.1	558.2	628.0	707.1	790.8	895.5
h_g, kJ/kg	674.2	681.34	717.2	751.9	791.5	820.44	859.8	902.1	933.0	895.5
$\Delta h_{g,\ell}$, kJ/kg	360.2	355.7	338.1	316.9	291.4	262.2	231.8	195.0	142.2	
$c_{p,\ell}$, kJ/(kg K)	1.84	1.86	1.93	1.99	2.07	2.13	2.23	2.49	2.82	
$c_{p,g}$, kJ/(kg K)	1.60	1.64	1.79	1.95	2.15	2.36	2.57	3.11	7.05	
η_ℓ, μNs/m^2	400	370	280	230	190	160	135	115	95.0	54.0
η_g, μNs/m^2	8.60	8.71	9.44	10.18	10.72	11.58	12.33	13.56	17.73	54.0
λ_ℓ, (mW/m^2)/(K/m)	102	100	92.6	86.5	77.9	69.2	60.6	56.5	48.3	57.9
λ_g, (mW/m^2)/(K/m)	16.5	17.4	20.1	22.9	26.0	30.4	36.1	39.4	45.9	57.9
Pr_ℓ	7.21	6.88	5.83	5.29	5.05	4.92	4.96	5.07	5.55	
Pr_g	0.83	0.82	0.84	0.87	0.87	0.90	0.88	1.07	2.72	
σ, mN/m	18.0	17.2	13.0	11.0	8.1	5.2	2.0	1.5	0.5	
$\beta_{e,\ell}$, kK^{-1}	1.10	1.30	1.48	1.70	2.31	2.50	3.05	4.89	15.8	

Transport Properties of the Superheated Vapor

	Temperature, °C												
	−150	−100	−50	0	25	100	200	300	400	600	800	1 000	1 200
Property (at low pressure)	Temperature, K												
	123.15	173.15	223.15	273.15	298.15	373.15	473.15	573.15	673.15	873.15	1 073.15	1 273.15	1 473.15
Specific heat capacity, $c_{p,g}$ (kJ/kg K)	S	S	S	S	L	1.645	2.139	2.571	2.939	3.504	3.902	4.183	4.384
Thermal conductivity, λ_g[(W/m^2)/(K/m)]	S	S	S	S	L	0.018	(0.030)	(0.044)	(0.059)	(0.091)	(0.120)	(0.145)	(0.167)
Dynamic viscosity, η_g(10^{-5} Ns/m^2)	S	S	S	S	L	0.878	1.09	1.29	(1.49)	(1.89)	(2.29)	(2.64)	(2.99)

METHYLCYCLOHEXANE

Chemical formula: C_7H_{14}
Molecular weight: 98.19
Melting point: 146.6 K
Boiling point: 374.1 K

Critical temperature: 572.1 K
Critical pressure: 3.48 MPa
Critical density: 267 kg/m^3
Normal vapor density: 4.38 kg/m^3
(@ 0 °C, 101.3 kPa)

Properties of the Liquid at Temperatures Below the Normal Boiling Point

	Temperature, °C												
	−150	−100	−75	−50	−25	0	20	50	100	150	200	250	300
	Temperature, K												
Property	123.15	173.15	198.15	223.15	248.15	273.15	293.15	323.15	373.15	423.15	473.15	523.15	573.15
Density, ρ_l (kg/m^3)	S	870	850	828	808	787	769	742	697	V	V	V	V
Specific heat capacity, $c_{p,l}$ (kJ/kg K)	S	1.470	1.537	1.608	1.687	1.779	1.863	(1.97)	(2.23)	V	V	V	V
Thermal conductivity, λ_l [(W/m^2)/(K/m)]	S	0.135	0.130	0.125	0.121	0.116	0.112	0.107	0.098	V	V	V	V
Dynamic viscosity, η_l (10^{-5} Ns/m^2)	S	(600)	(360)	(235)	155.0	99.3	73.5	50.0	30.0	V	V	V	V

Transport Properties of the Superheated Vapor

	Temperature, °C												
	−150	−100	−50	0	25	100	200	300	400	600	800	1 000	1 200
Property (at low pressure)	Temperature, K												
	123.15	173.15	223.15	273.15	298.15	373.15	473.15	573.15	673.15	873.15	1 073.15	1 273.15	1 473.15
Specific heat capacity, $c_{p,g}$ (kJ/kg K)	S	L	L	L	L	L	2.236	2.650	2.998	3.534	3.910	4.183	4.379
Thermal conductivity, λ_g[(W/m^2)/(K/m)]	S	L	L	L	L	L	(0.028)	(0.042)	(0.057)	(0.088)	(0.114)	(0.130)	(0.152)
Dynamic viscosity, η_g(10^{-5} Ns/m^2)	S	L	L	L	L	L	(1.04)	(1.24)	(1.43)	(1.81)	(2.16)	(2.49)	(2.84)

ETHYLCYCLOHEXANE

Chemical formula: C_8H_{16}
Molecular weight: 112.22
Melting point: 161.8 K
Boiling point: 404.9 K

Critical temperature: 609 K
Critical pressure: 3.03 MPa
Critical density: 249 kg/m^3
Normal vapor density: 5.01 kg/m^3
(@ 0°C, 101.3 kPa)

Properties of the Liquid at Temperatures Below the Normal Boiling Point

Property	Temperature, °C: −150	−100	−75	−50	−25	0	20	50	100	150	200	250	300
	Temperature, K: 123.15	173.15	198.15	223.15	248.15	273.15	293.15	323.15	373.15	423.15	473.15	523.15	573.15
Density, ρ_l (kg/m^3)	S	(876)	(858)	(840)	(822)	(803)	787	(780)	(722)	V	V	V	V
Specific heat capacity, $c_{p,l}$ (kJ/kg K)	S	1.470	1.537	1.604	1.691	1.779	1.863	(1.98)	(2.22)	V	V	V	V
Thermal conductivity, λ_l [(W/m^2)/(K/m)]	S	(0.128)	(0.124)	(0.120)	(0.117)	(0.114)	(0.111)	(0.107)	(0.100)	V	V	V	V
Dynamic viscosity, η_l (10^{-5} Ns/m^2)	S	—	(450)	(290)	180.0	114.2	84.2	58.2	37.0	V	V	V	V

Transport Properties of the Superheated Vapor

Property (at low pressure)	Temperature, °C: −150	−100	−50	0	25	100	200	300	400	600	800	1 000	1 200
	Temperature, K: 123.15	173.15	223.15	273.15	298.15	373.15	473.15	573.15	673.15	873.15	1 073.15	1 273.15	1 473.15
Specific heat capacity, $c_{p,g}$ (kJ/kg K)	S	L	L	L	L	L	2.265	2.667	3.006	3.534	3.902	4.170	4.367
Thermal conductivity, λ_g[(W/m^2)/(K/m)]	S	L	L	L	L	L	(0.026)	(0.038)	(0.050)	(0.076)	(0.100)	(0.119)	(0.138)
Dynamic viscosity, η_g(10^{-5} Ns/m^2)	S	L	L	L	L	L	(0.92)	(1.10)	(1.27)	(1.61)	(1.92)	(2.22)	(2.54)

ISOPROPYLCYCLOHEXANE

Chemical formula: C_9H_{18}
Molecular weight: 126.24
Melting point: 183.4 K
Boiling point: 427.7 K

Critical temperature: 640 K
Critical pressure: 2.84 MPa
Critical density: 272 kg/m^3
Normal vapor density: 5.63 kg/m^3
(@ 0 °C, 101.3 kPa)

Properties of the Liquid at Temperatures Below the Normal Boiling Point

Property	Temperature, °C −150 Temperature, K 123.15	−100 173.15	−75 198.15	−50 223.15	−25 248.15	0 273.15	20 293.15	50 323.15	100 373.15	150 423.15	200 473.15	250 523.15	300 573.15
Density, ρ_l (kg/m^3)	S	S	(876)	(856)	(837)	817	803	(778)	(739)	(699)	V	V	V
Specific heat capacity, $c_{p,l}$ (kJ/kg K)	S	S	(1.56)	(1.63)	(1.70)	(1.79)	(1.86)	(1.99)	(2.21)	(2.49)	V	V	V
Thermal conductivity, λ_l [(W/m^2)/(K/m)]	S	S	(0.115)	(0.112)	(0.109)	(0.106)	(0.104)	(0.101)	(0.096)	(0.090)	V	V	V
Dynamic viscosity, η_l (10^{-5} Ns/m^2)	S	S	S	(400)	235.1	140.8	100.6	67.2	42.0	(29)	V	V	V

BUTYLCYCLOHEXANE

Chemical formula: $C_{10}H_{20}$
Molecular weight: 140.27
Melting point: 198.4 K
Boiling point: 454.1 K

Critical temperature: 688.3 K
Critical pressure: 2.41 MPa
Critical density: 266 kg/m^3
Normal vapor density: 6.26 kg/m^3
(@ 0 °C, 101.3 kPa)

Properties of the Liquid at Temperatures Below the Normal Boiling Point

Property	Temperature, °C −150 Temperature, K 123.15	−100 173.15	−75 198.15	−50 223.15	−25 248.15	0 273.15	20 293.15	50 323.15	100 373.15	150 423.15	200 473.15	250 523.15	300 573.15
Density, ρ_l (kg/m^3)	S	S	S	(860)	(843)	(825)	811	(790)	(754)	(718)	V	V	V
Specific heat capacity, $c_{p,l}$ (kJ/kg K)	S	S	S	1.630	1.695	1.778	1.897	1.960	2.160	2.37	V	V	V
Thermal conductivity, λ_l [(W/m^2)/(K/m)]	S	S	S	(0.117)	(0.114)	(0.112)	(0.109)	(0.107)	0.101	(0.096)	V	V	V
Dynamic viscosity, η_l (10^{-5} Ns/m^2)	S	S	S	(500)	(310)	191.0	131.4	83.1	50.1	(30)	V	V	V

PENTYLCYCLOHEXANE

Chemical formula: $C_{11}H_{22}$
Molecular weight: 154.30
Melting point: 215.65 K
Boiling point: 475.95 K

Critical temperature: 667.2 K
Critical pressure: 2.21 MPa
Critical density: —
Normal vapor density: 6.88 kg/m^3
(@ 0 °C, 101.3 kPa)

Properties of the Liquid at Temperatures Below the Normal Boiling Point

	Temperature, °C												
	−150	−100	−75	−50	−25	0	20	50	100	150	200	250	300
	Temperature, K												
Property	123.15	173.15	198.15	223.15	248.15	273.15	293.15	323.15	373.15	423.15	473.15	523.15	573.15
Density, ρ_l (kg/m^3)	S	S	S	(855)	(837)	(819)	805	(783)	(746)	(708)	(670)	V	V
Specific heat capacity, $c_{p,l}$ (kJ/kg K)	S	S	S	(1.68)	(1.75)	(1.84)	(1.92)	(2.04)	(2.25)	—	—	V	V
Thermal conductivity, λ_l [(W/m^2)/(K/m)]	S	S	S	(0.116)	(0.113)	(0.111)	(0.109)	(0.106)	(0.101)	(0.095)	(0.090)	V	V
Dynamic viscosity, η_l (10^{-5} Ns/m^2)	S	S	S	(850)	(470)	264.1	172.3	102.7	57.0	(31)	—	V	V

HEXYLCYCLOHEXANE

Chemical formula: $C_{12}H_{24}$
Molecular weight: 168.34
Melting point: 233 K
Boiling point: 501 K

Critical temperature: 685 K
Critical pressure: 2.04 MPa
Critical density: —
Normal vapor density: 7.51 kg/m^3
(@ 0 °C, 101.3 kPa)

Properties of the Liquid at Temperatures Below the Normal Boiling Point

	Temperature, °C												
	−150	−100	−75	−50	−25	0	20	50	100	150	200	250	300
	Temperature, K												
Property	123.15	173.15	198.15	223.15	248.15	273.15	293.15	323.15	373.15	423.15	473.15	523.15	573.15
Density, ρ_l (kg/m^3)	S	S	S	S	(838)	(825)	810	(790)	(755)	(718)	(681)	V	V
Specific heat capacity, $c_{p,l}$ (kJ/kg K)	S	S	S	S	(1.77)	(1.85)	(1.93)	(2.05)	(2.25)	—	—	V	V
Thermal conductivity, λ_l [(W/m^2)/(K/m)]	S	S	S	S	(0.115)	(0.113)	(0.110)	(0.107)	(0.101)	(0.096)	(0.091)	V	V
Dynamic viscosity, η_l (10^{-5} Ns/m^2)	S	S	S	S	(620)	352.1	222.0	126.8	66.0	(34)	—	V	V

CYCLOPENTENE

Chemical formula: C_5H_8
Molecular weight: 68.12
Melting point: 138.1 K
Boiling point: 317.4 K

Critical temperature: 506 K
Critical pressure: 4.97 MPa
Critical density: 278 kg/m^3
Normal vapor density: 3.04 kg/m^3
(@ 0 °C, 101.3 kPa)

Properties of the Liquid at Temperatures Below the Normal Boiling Point

	Temperature, °C												
	−150	−100	−75	−50	−25	0	20	50	100	150	200	250	300
	Temperature, K												
Property	123.15	173.15	198.15	223.15	248.15	273.15	293.15	323.15	373.15	423.15	473.15	523.15	573.15
Density, ρ_l (kg/m^3)	S	(892)	(867)	(842)	(817)	792	770	V	V	V	V	V	V
Specific heat capacity, $c_{p,l}$ (kJ/kg K)	S	1.453	1.470	1.532	1.608	1.696	1.779	V	V	V	V	V	V
Thermal conductivity, λ_l [(W/m^2)/(K/m)]	S	0.181	0.173	0.165	0.158	0.150	0.145	V	V	V	V	V	V
Dynamic viscosity, η_l (10^{-5} Ns/m^2)	S	—	—	(75)	(56)	43.6	34.7	V	V	V	V	V	V

Transport Properties of the Superheated Vapor

	Temperature °C												
	− 150	− 100	− 50	0	25	100	200	300	400	600	800	1 000	1 200
Property (at low pressure)	Temperature, K												
	123.15	173.15	223.15	273.15	298.15	373.15	473.15	573.15	673.15	873.15	1 073.15	1 273.15	1 473.15
Specific heat capacity, $c_{p,g}$ (kJ/kg K)	S	L	L	L	L	1.430	1.834	2.192	2.493	2.961	3.272	3.498	3.724
Thermal conductivity, λ_g[(W/m^2)/(K/m)]	S	L	L	L	L	(0.017)	(0.028)	(0.040)	(0.054)	(0.081)	(0.108)	(0.128)	(0.146)
Dynamic viscosity, η_g(10^{-5} Ns/m^2)	S	L	L	L	L	(0.95)	(1.18)	(1.41)	(1.63)	(2.04)	(2.42)	(2.76)	(3.10)

CYCLOHEXENE

Chemical formula: C_6H_{10}
Molecular weight: 82.15
Melting point: 169.7 K
Boiling point: 356.1 K

Critical temperature: 560.4 K
Critical pressure: 4.35 MPa
Critical density: 287 kg/m^3
Normal vapor density: 3.67 kg/m^3
(@ 0 °C, 101.3 kPa)

Properties of the Liquid at Temperatures Below the Normal Boiling Point

	Temperature, °C												
	−150	−100	−75	−50	−25	0	20	50	100	150	200	250	300
	Temperature, K												
Property	123.15	173.15	198.15	223.15	248.15	273.15	293.15	323.15	373.15	423.15	473.15	523.15	573.15
Density, ρ_l (kg/m^3)	S	(925)	(900)	(875)	852	830	811	780	V	V	V	V	V
Specific heat capacity, $c_{p,l}$ (kJ/kg K)	S	1.419	1.478	1.537	1.616	1.708	1.792	(1.91)	V	V	V	V	V
Thermal conductivity, λ_l [(W/m^2)/(K/m)]	S	0.158	0.154	0.149	0.145	0.140	0.134	0.125	V	V	V	V	V
Dynamic viscosity, η_l (10^{-5} Ns/m^2)	S	(500)	(300)	(200)	(128)	(84)	65.0	45.5	V	V	V	V	V

Transport Properties of the Superheated Vapor

	Temperature, °C												
	−150	−100	−50	0	25	100	200	300	400	600	800	1 000	1 200
Property (at low pressure)	Temperature, K												
	123.15	173.15	223.15	273.15	298.15	373.15	473.15	573.15	673.15	873.15	1 073.15	1 273.15	1 473.15
Specific heat capacity, $c_{p,g}$ (kJ/kg K)	S	L	L	L	L	1.719	2.095	2.427	2.722	3.173	3.471	3.688	3.905
Thermal conductivity, λ_g[(W/m^2)/(K/m)]	S	L	L	L	L	(0.018)	(0.029)	(0.042)	(0.055)	(0.083)	(0.108)	(0.127)	(0.150)
Dynamic viscosity, η_g(10^{-5} Ns/m^2)	S	L	L	L	L	(0.89)	(1.11)	(1.32)	(1.53)	(1.93)	(2.30)	(2.65)	(3.00)

BENZENE

Chemical formula: C_6H_6
Molecular weight: 78.108
Melting point: 278.7 K
Normal boiling point: 353.25 K

Critical temperature: 562.6 K
Critical pressure: 4.924 MPa
Critical density: 301.6 kg/m^3
Normal vapor density: 3.49 kg/m^3
(@ 0 °C, 101.3 kPa)

Properties of the Liquid at Temperatures Below the Normal Boiling Point

Property	Temperature, °C												
	−150	−100	−75	−50	−25	0	20	50	100	150	200	250	300
	Temperature, K												
	123.15	173.15	198.15	223.15	248.15	273.15	293.15	323.15	373.15	423.15	473.15	523.15	573.15
Density, ρ_l (kg/m^3)	S	S	S	S	S	S	879	847	V	V	V	V	V
Specific heat capacity, $c_{p,l}$ (kJ/kg K)	S	S	S	S	S	S	1.729	1.821	V	V	V	V	V
Thermal conductivity, λ_l [(W/m^2)/(K/m)]	S	S	S	S	S	S	0.144	0.134	V	V	V	V	V
Dynamic viscosity, η_l (10^{-5} Ns/m^2)	S	S	S	S	S	S	64.9	43.6	V	V	V	V	V

Properties of the Saturated Liquid and Vapor

T_{sat}, K	353.3	375	400	425	450	475	500	525	550	562.6
p_{sat}, kPa	101.3	191	354	607	975	1 484	2 166	3 060	4 218	4 924
ρ_ℓ, kg/m^3	823	798	767	735	699	660	615	559	475	304
ρ_g, kg/m^3	2.74	4.90	8.87	14.8	23.6	36.1	54.2	82.0	133	304
h_ℓ, kJ/kg	−154.3	−113.0	−62.1	−8.9	47.5	106.8	169.7	238.6	322.8	432.6
h_g, kJ/kg	243.4	270.3	302.1	334.8	367.9	400.7	432.3	460.4	478.5	432.6
$\Delta h_{g,\ell}$, kJ/kg	397.7	383.3	364.2	343.7	320.4	293.9	262.6	221.8	155.7	
$c_{p,\ell}$, kJ/(kg K)	1.88	1.98	2.08	2.20	2.32	2.45	2.60	2.83		
$c_{p,g}$, kJ/(kg K)	1.29	1.40	1.53	1.67	1.81	2.01	2.32	2.73		
η_ℓ, μNs/m^2	321	258	205	166	138	116	97.9	80.7	59.6	
η_g, μNs/m^2	9.26	9.87	10.7	11.5	12.5	13.7	15.0	16.8	19.1	
λ_ℓ, (mW/m^2)/(K/m)	131	126	119	112	106	100	93.5	87.1	77.4	62.6
λ_g, (mW/m^2)/(K/m)	14.8	17.1	19.8	23.0	26.7	31.0	35.7	41.1	50.2	62.6
Pr_ℓ	4.61	4.05	3.58	3.26	3.02	2.84	2.75	2.62		
Pr_g	0.81	0.82	0.83	0.84	0.85	0.89	0.97	1.12		
σ, mN/m	21.2	18.5	15.5	12.7	9.89	7.26	4.80	2.55	0.65	
$\beta_{e,\ell}$, kK^{-1}	1.15	1.37	1.67	2.02	2.49	3.13	4.32	6.98	16.0	

Transport Properties of the Superheated Vapor

Property (at low pressure)	Temperature, °C												
	−150	−100	−50	0	25	100	200	300	400	600	800	1 000	1 200
	Temperature, K												
	123.15	173.15	223.15	273.15	298.15	373.15	473.15	573.15	673.15	873.15	1 073.15	1 273.15	1 473.15
Specific heat capacity, $c_{p,g}$ (kJ/kg K)	S	S	S	L	L	1.336	1.679	1.959	2.186	2.525	2.767	2.943	3.077
Thermal conductivity, λ_g[(W/m^2)/(K/m)]	S	S	S	L	L	0.020	0.030	(0.036)	(0.047)	(0.070)	(0.092)	(0.112)	(0.130)
Dynamic viscosity, η_g(10^{-5} Ns/m^2)	S	S	S	L	L	0.951	1.20	1.45	(1.65)	(2.10)	(2.53)	(2.95)	(3.35)

TOLUENE

Chemical formula: $C_6H_5CH_3$
Molecular weight: 92.134
Melting point: 178.16 K
Normal boiling point: 383.78 K

Critical temperature: 594.0 K
Critical pressure: 4.050 MPa
Critical density: 290 kg/m^3
Normal vapor density: 4.11 kg/m^3
(@ 0°C, 101.3 kPa)

Properties of the Liquid at Temperatures Below the Normal Boiling Point

	Temperature, °C												
	−150	−100	−75	−50	−25	0	20	50	100	150	200	250	300
	Temperature, K												
Property	123.15	173.15	198.15	223.15	248.15	273.15	293.15	323.15	373.15	423.15	473.15	523.15	573.15
Density, ρ_l (kg/m^3)	S	S	955	932	908	885	867	839	793	V	V	V	V
Specific heat capacity, $c_{p,l}$ (kJ/kg K)	S	S	1.465	1.507	1.553	1.612	1.717	1.800	1.968	V	V	V	V
Thermal conductivity, λ_l [(W/m^2)/(K/m)]	S	S	0.156	0.152	0.148	0.144	0.141	0.136	0.128	V	V	V	V
Dynamic viscosity, η_l (10^{-5} Ns/m^2)	S	S	500	212	117.0	77.3	58.6	41.9	26.9	V	V	V	V

Properties of the Saturated Liquid and Vapor

T_{sat}, K	383.78	400	425	450	475	500	525	550	575	594.0
p_{sat}, kPa	101.3	158	285	487	776	1 230	1 820	2 570	3 450	4 050
ρ_ℓ, kg/m^3	778	760	733	702	670	632	590	541	469	290
ρ_g, kg/m^3	2.91	4.42	7.88	13.2	20.9	32.2	48.8	74.3	120.8	290
h_ℓ, kJ/kg	695	725	762	797	832	866	901	932	948	899
h_g, kJ/kg	336	381	432	488	551	604	667	734	808	899
$\Delta h_{g,\ell}$, kJ/kg	359	344	330	309	281	262	234	198	140	
$c_{p,\ell}$, kJ/(kg K)	1.81	1.88	1.96	2.07	2.21	2.38	2.59	2.86	3.34	
$c_{p,g}$, kJ/(kg K)	1.13	1.18	1.25	1.33	1.42	1.54	1.71	2.04	3.33	
η_ℓ, μNs/m^2	251	220	183	153	134	118	104	92	82	50
η_g, μNs/m^2	9.0	9.5	10.1	10.7	11.4	12.5	13.7	15.0	17.2	50
λ_ℓ, (mW/m^2)/(K/m)	113	108	103	98	92	86	79	72	65	37.8
λ_g, (mW/m^2)/(K/m)	11.1	12.3	14.2	16.0	17.8	19.8	21.9	24.4	27.8	37.8
Pr_ℓ	4.02	3.83	3.48	3.23	3.22	3.27	3.41	3.65	4.21	
Pr_g	0.92	0.91	0.89	0.89	0.91	0.97	1.07	1.25	2.06	
σ, mN/m	18.0	16.3	13.8	11.4	8.99	6.74	4.62	2.66	0.95	
$\beta_{e,\ell}$, kK^{-1}	1.38	1.43	1.67	2.07	2.52	3.11	3.98	6.15	14.3	

Transport Properties of the Superheated Vapor

	Temperature, °C												
	−150	−100	−50	0	25	100	200	300	400	600	800	1 000	1 200
Property (at low pressure)	Temperature, K												
	123.15	173.15	223.15	273.15	298.15	373.15	473.15	573.15	673.15	873.15	1 073.15	1 273.15	1 473.15
Specific heat capacity, $c_{p,g}$ (kJ/kg K)	S	S	L	L	L	L	1.758	2.047	2.286	2.650	2.914	3.102	3.245
Thermal conductivity, λ_g[(W/m^2)/(K/m)]	S	S	L	L	L	L	0.032	0.042	(0.052)	(0.072)	(0.092)	(0.112)	(0.130)
Dynamic viscosity, η_g(10^{-5} Ns/m^2)	S	S	L	L	L	L	1.12	1.33	1.545	1.95	(2.33)	(2.68)	(3.01)

ETHYL BENZENE

Chemical formula: $C_6H_5 \cdot C_2H_5$
Molecular weight: 106.2
Melting point: 178.1 K
Normal boiling point: 409.3 K

Critical temperature: 617.1 K
Critical pressure: 3.610 MPa
Critical density: 284 kg/m^3
Normal vapor density: 4.74 kg/m^3 (@ 0°C, 101.3 kPa)

Properties of the Liquid at Temperatures Below the Normal Boiling Point

	Temperature, °C												
	−150	−100	−75	−50	−25	0	20	50	100	150	200	250	300
	Temperature, K												
Property	123.15	173.15	198.15	223.15	248.15	273.15	293.15	323.15	373.15	423.15	473.15	523.15	573.15
Density, ρ_l (kg/m^3)	S	S	951	929	907	885	866	840	797	V	V	V	V
Specific heat capacity, $c_{p,l}$ (kJ/kg K)	S	S	1.503	1.549	1.612	1.675	1.738	1.880	2.219	V	V	V	V
Thermal conductivity, λ_l [(W/m^2)/(K/m)]	S	S	0.151	0.147	0.142	0.137	0.131	0.126	0.110	V	V	V	V
Dynamic viscosity, η_l (10^{-5} Ns/m^2)	S	S	560	246	135.0	89.5	67.9	48.2	30.8	V	V	V	V

Properties of the Saturated Liquid and Vapor

T_{sat}, K	409.3	433	453	473	493	513	553	573	593	613
p_{sat}, kPa	101.3	189	297	447	645	903	1 640	2 150	2 770	3 450
ρ_ℓ, kg/m^3	751	727	705	682	657	630	568	529	478	388
ρ_g, kg/m^3	3.26	5.76	8.88	13.2	19.0	26.9	52.1	72.9	106	189
h_ℓ, kJ/kg	−42.8	9.0	53.3	101.6	159.3	207.9	306.7	368.2	427.8	481.6
h_g, kJ/kg	298.8	335.6	367.3	399.3	431.4	463.3	524.4	550.8	570.2	557.0
$\Delta h_{g,\ell}$, kJ/kg	341.6	326.6	314.0	297.7	272.1	255.4	217.7	192.6	142.4	75.4
$c_{p,\ell}$, kJ/(kg K)	1.98	2.04	2.11	2.19	2.28	2.38	2.63	2.85	3.22	4.20
$c_{p,g}$, kJ/(kg K)	1.67	1.77	1.86	1.95	2.05	2.16	2.48	2.78	3.52	17.5
η_ℓ, μNs/m^2	230	197	176	158	142	129	108	99	91	84
η_g, μNs/m^2	9.0	9.6	10.3	11.0	11.7	12.4	13.7	14.8	16.6	22.6
λ_ℓ, (mW/m^2)/(K/m)	99	95	90	85	80	75	65	60	55	50
λ_g, (mW/m^2)/(K/m)	16.9	19.4	21.5	23.6	25.9	28.2	33.2	36.1	39.5	45.8
Pr_ℓ	4.60	4.31	4.13	4.07	4.05	4.09	4.37	4.70	5.32	7.02
Pr_g	0.87	0.88	0.89	0.91	0.93	0.95	1.02	1.14	1.48	8.64
σ, mN/m	16.6	14.5	12.5	10.5	8.8	7.1	3.9	2.5	1.2	0.23
$\beta_{e,\ell}$, kK^{-1}	1.35	1.52	1.70	1.93	2.23	2.64	4.14	5.80	9.88	

Transport Properties of the Superheated Vapor

	Temperature, °C												
	−150	−100	−50	0	25	100	200	300	400	600	800	1 000	1 200
Property (at low pressure)	Temperature, K												
	123.15	173.15	223.15	273.15	298.15	373.15	473.15	573.15	673.15	873.15	1 073.15	1 273.15	1 473.15
Specific heat capacity, $c_{p,g}$ (kJ/kg K)	S	S	L	L	L	L	2.056	2.156	2.399	2.767	3.055	3.232	3.383
Thermal conductivity, λ_g[(W/m^2)/(K/m)]	S	S	L	L	L	L	0.022	0.032	(0.043)	(0.066)	(0.086)	(0.106)	(0.124)
Dynamic viscosity, η_g(10^{-5} Ns/m^2)	S	S	L	L	L	L	(1.02)	(1.22)	(1.41)	(1.79)	(2.17)	(2.52)	(2.85)

VINYLBENZENE (STYRENE)

Chemical formula: C_8H_8
Molecular weight: 104.15
Melting point: 242.5 K
Boiling point: 418.3 K

Critical temperature: 647 K
Critical pressure: 3.99 MPa
Critical density: 294 kg/m^3
Normal vapor density: 4.65 kg/m^3
(@ 0°C, 101.3 kPa)

Properties of the Liquid at Temperatures Below the Normal Boiling Point

	Temperature, °C												
	−150	−100	−75	−50	−25	0	20	50	100	150	200	250	300
	Temperature, K												
Property	123.15	173.15	198.15	223.15	248.15	273.15	293.15	323.15	373.15	423.15	473.15	523.15	573.15
Density, ρ_l (kg/m^3)	S	S	S	S	(942)	922	907	881	838	V	V	V	V
Specific heat capacity, $c_{p,l}$ (kJ/kg K)	S	S	S	S	1.595	1.654	1.717	1.834	2.052	V	V	V	V
Thermal conductivity, λ_l [(W/m^2)/(K/m)]	S	S	S	S	0.148	0.142	0.137	0.131	0.119	V	V	V	V
Dynamic viscosity, η_l (10^{-5} Ns/m^2)	S	S	S	S	(160)	104.7	74.9	50.2	30.9	V	V	V	V

Transport Properties of the Superheated Vapor

	Temperature, °C												
	−150	−100	−50	0	25	100	200	300	400	600	800	1 000	1 200
Property (at low pressure)	Temperature, K												
	123.15	173.15	223.15	273.15	298.15	373.15	473.15	573.15	673.15	873.15	1 073.15	1 273.15	1 473.15
Specific heat capacity, $c_{p,g}$ (kJ/kg K)	S	S	S	L	L	L	1.771	2.035	2.248	2.575	2.805	2.977	3.102
Thermal conductivity, λ_g[(W/m^2)/(K/m)]	S	S	S	L	L	L	(0.023)	(0.033)	(0.043)	(0.065)	(0.087)	(0.107)	(0.125)
Dynamic viscosity, η_g(10^{-5} Ns/m^2)	S	S	S	L	L	L	(1.02)	(1.22)	(1.42)	(1.80)	(2.18)	(2.53)	(2.85)

PROPYLBENZENE

Chemical formula: C_9H_{12}
Molecular weight: 120.19
Melting point: 173.7 K
Boiling point: 432.4 K

Critical temperature: 638.3 K
Critical pressure: 3.20 MPa
Critical density: 273 kg/m^3
Normal vapor density: 5.36 kg/m^3
(@ 0 °C, 101.3 kPa)

Properties of the Liquid at Temperatures Below the Normal Boiling Point

Property	Temperature, °C												
	−150	−100	−75	−50	−25	0	20	50	100	150	200	250	300
	Temperature, K												
	123.15	173.15	198.15	223.15	248.15	273.15	293.15	323.15	373.15	423.15	473.15	523.15	573.15
Density, ρ_l (kg/m^3)	S	S	937	918	898	879	860	833	793	750	V	V	V
Specific heat capacity, $c_{p,l}$ (kJ/kg K)	S	S	1.540	1.617	1.652	1.715	1.769	1.858	(2.03)	(2.25)	V	V	V
Thermal conductivity, λ_l [(W/m^2)/(K/m)]	S	S	0.148	0.143	0.137	0.132	0.128	0.122	0.112	0.102	V	V	V
Dynamic viscosity, η_l (10^{-5} Ns/m^2)	S	S	—	(320)	191.0	118.3	85.7	58.4	36.0	25.0	V	V	V

Transport Properties of the Superheated Vapor

Property (at low pressure)	Temperature, °C												
	− 150	− 100	− 50	0	25	100	200	300	400	600	800	1 000	1 200
	Temperature, K												
	123.15	173.15	223.15	273.15	298.15	373.15	473.15	573.15	673.15	873.15	1 073.15	1 273.15	1 473.15
Specific heat capacity, $c_{p,g}$ (kJ/kg K)	S	S	L	L	L	L	1.930	2.227	2.474	2.851	3.128	3.328	3.479
Thermal conductivity, λ_g[(W/m^2)/(K/m)]	S	S	L	L	L	L	(0.023)	(0.033)	(0.043)	(0.064)	(0.085)	(0.104)	(0,122)
Dynamic viscosity, η_g(10^{-5} Ns/m^2)	S	S	L	L	L	L	(0.96)	(1.15)	(1.33)	(1.69)	(2.03)	(2.36)	(2.69)

ISOPROPYLBENZENE (CUMENE)

Chemical formula: C_9H_{12}
Molecular weight: 120.19
Melting point: 177.1 K
Boiling point: 425.6 K

Critical temperature: 631 K
Critical pressure: 3.21 MPa
Critical density: 280 kg/m^3
Normal vapor density: 5.36 kg/m^3
(@ 0°C, 101.3 kPa)

Properties of the Liquid at Temperatures Below the Normal Boiling Point

	Temperature, °C												
	−150	−100	−75	−50	−25	0	20	50	100	150	200	250	300
	Temperature, K												
Property	123.15	173.15	198.15	223.15	248.15	273.15	293.15	323.15	373.15	423.15	473.15	523.15	573.15
Density, ρ_l (kg/m^3)	S	S	(937)	(917)	(897)	878	862	837	(795)	(747)	V	V	V
Specific heat capacity, $c_{p,l}$ (kJ/kg K)	S	S	(1.53)	(1.58)	(1.64)	(1.71)	1.762	1.853	(2.03)	(2.25)	V	V	V
Thermal conductivity, λ_l [(W/m^2)/(K/m)]	S	S	0.141	0.136	0.131	0.126	0.122	0.116	0.106	0.097	V	V	V
Dynamic viscosity, η_l (10^{-5} Ns/m^2)	S	S	—	(340)	(180)	107.4	78.9	54.0	(33.9)	(23.5)	V	V	V

Transport Properties of the Superheated Vapor

	Temperature, °C												
	−150	−100	−50	0	25	100	200	300	400	600	800	1 000	1 200
Property (at low pressure)	Temperature, K												
	123.15	173.15	223.15	273.15	298.15	373.15	473.15	573.15	673.15	873.15	1 073.15	1 273.15	1 473.15
Specific heat capacity, $c_{p,g}$ (kJ/kg K)	S	S	L	L	L	L	1.930	2.236	2.483	2.860	3.132	3.337	3.488
Thermal conductivity, λ_g[(W/m^2)/(K/m)]	S	S	L	L	L	L	(0.024)	(0.034)	(0.044)	(0.066)	(0.087)	(0.106)	(0.124)
Dynamic viscosity, η_g(10^{-5} Ns/m^2)	S	S	L	L	L	L	(0.98)	(1.17)	(1.35)	(1.72)	(2.06)	(2.39)	(2.72)

BUTYLBENZENE

Chemical formula: $C_{10}H_{14}$
Molecular weight: 134.22
Melting point: 185.2 K
Boiling point: 456.4 K

Critical temperature: 660.5 K
Critical pressure: 2.89 MPa
Critical density: 269.7 kg/m^3
Normal vapor density: 5.99 kg/m^3
(@ 0°C, 101.3 kPa)

Properties of the Liquid at Temperatures Below the Normal Boiling Point

	Temperature, °C												
	−150	−100	−75	−50	−25	0	20	50	100	150	200	250	300
	Temperature, K												
Property	123.15	173.15	198.15	223.15	248.15	273.15	293.15	323.15	373.15	423.15	473.15	523.15	573.15
Density, ρ_l (kg/m^3)	S	S	934	914	895	877	869	838	798	756	V	V	V
Specific heat capacity, $c_{p,l}$ (kJ/kg K)	S	S	1.507	1.566	1.629	1.712	1.767	1.896	2.050	(2.21)	V	V	V
Thermal conductivity, λ_l [(W/m^2)/(K/m)]	S	S	0.147	0.142	0.137	0.132	0.128	0.123	0.113	0.104	V	V	V
Dynamic viscosity, η_l (10^{-5} Ns/m^2)	S	S	(2700)	601	250.0	146.6	103.6	68.5	41.1	28.0	V	V	V

Transport Properties of the Superheated Vapor

	Temperature, °C												
	−150	−100	−50	0	25	100	200	300	400	600	800	1 000	1 200
Property (at low pressure)	Temperature, K												
	123.15	173.15	223.15	273.15	298.15	373.15	473.15	573.15	673.15	873.15	1 073.15	1 273.15	1 473.15
Specific heat capacity, $c_{p,g}$ (kJ/kg K)	S	S	L	L	L	L	1.980	2.282	2.533	2.918	3.199	3.404	3.559
Thermal conductivity, λ_g[(W/m^2)/(K/m)]	S	S	L	L	L	L	(0.023)	(0.032)	(0.042)	(0.063)	(0.083)	(0.099)	(0.117)
Dynamic viscosity, η_g(10^{-5} Ns/m^2)	S	S	L	L	L	L	(0.91)	(1.09)	(1.27)	(1.61)	(1.94)	(2.25)	(2.56)

PENTYLBENZENE

Chemical formula: $C_{11}H_{16}$
Molecular weight: 148.24
Melting point: 194.85 K
Boiling point: 504.15 K

Critical temperature: 725.3 K
Critical pressure: 2.99 MPa
Critical density: 270 kg/m^3
Normal vapor density: 6.62 kg/m^3
(@ 0 °C, 101.3 kPa)

Properties of the Liquid at Temperatures Below the Normal Boiling Point

	Temperature, °C												
	−150	−100	−75	−50	−25	0	20	50	100	150	200	250	300
	Temperature, K												
Property	123.15	173.15	198.15	223.15	248.15	273.15	293.15	323.15	373.15	423.15	473.15	523.15	573.15
Density, ρ_l (kg/m^3)	S	S	(926)	(908)	(890)	873	859	837	799	(758)	(716)	V	V
Specific heat capacity, $c_{p,l}$ (kJ/kg K)	S	S	(1.59)	(1.64)	(1.70)	(1.76)	(1.82)	(1.90)	(2.07)	(2.29)	(2.55)	V	V
Thermal conductivity, λ_l [(W/m^2)/(K/m)]	S	S	0.147	0.142	0.138	0.134	0.130	0.125	0.117	0.108	0.100	V	V
Dynamic viscosity, η_l (10^{-5} Ns/m^2)	S	S	—	(550)	(333)	201.0	133.5	85.3	49.2	32.0	(21)	V	V

Transport Properties of the Superheated Vapor

	Temperature, °C												
	−150	−100	−50	0	25	100	200	300	400	600	800	1 000	1 200
Property (at low pressure)	Temperature, K												
	123.15	173.15	223.15	273.15	298.15	373.15	473.15	573.15	673.15	873.15	1 073.15	1 273.15	1 473.15
Specific heat capacity, $c_{p,g}$ (kJ/kg K)	S	S	L	L	L	L	L	2.328	2.579	2.964	3.253	3.467	3.622
Thermal conductivity, λ_g[(W/m^2)/(K/m)]	S	S	L	L	L	L	L	(0.031)	(0.040)	(0.061)	(0.080)	(0.096)	(0.110)
Dynamic viscosity, η_g(10^{-5} Ns/m^2)	S	S	L	L	L	L	L	(1.04)	(1.21)	(1.53)	(1.85)	(2.15)	(2.43)

HEXYLBENZENE

Chemical formula: $C_{12}H_{18}$
Molecular weight: 162.26
Melting point: 206.35 K
Boiling point: 536.15 K

Critical temperature: 696.8 K
Critical pressure: 2.554 MPa
Critical density: 261 kg/m^3
Normal vapor density: 7.24 kg/m^3
(@ 0 °C, 101.3 kPa)

Properties of the Liquid at Temperatures Below the Normal Boiling Point

	Temperature, °C												
	−150	−100	−75	−50	−25	0	20	50	100	150	200	250	300
	Temperature, K												
Property	123.15	173.15	198.15	223.15	248.15	273.15	293.15	323.15	373.15	423.15	473.15	523.15	573.15
Density, ρ_l (kg/m^3)	S	S	S	(905)	(890)	875	861	844	(811)	(780)	(748)	(716)	V
Specific heat capacity, $c_{p,l}$ (kJ/kg K)	S	S	S	(1.65)	(1.72)	(1.775)	(1.830)	(1.92)	(2.09)	(2.29)	(2.55)	(2.81)	V
Thermal conductivity, λ_l [(W/m^2)/(K/m)]	S	S	S	0.142	0.137	0.133	0.130	0.125	0.117	0.108	0.101	0.093	V
Dynamic viscosity, η_l (10^{-5} Ns/m^2)	S	S	S	(800)	(450)	260.1	167.6	103.6	57.4	36.0	(23)	(15)	V

o-XYLENE

Chemical formula: $C_6H_4)CH_3)_2$
Molecular weight: 106.168
Melting point: 248.0 K
Normal boiling point: 417.6 K

Critical temperature: 630.4 K
Critical pressure: 3.729 MPa
Critical density: 288 kg/m^3
Normal vapor density: 4.74 kg/m^3
(@ 0 °C, 101.3 kPa)

Properties of the Liquid at Temperatures Below the Normal Boiling Point

	Temperature, °C												
	−150	−100	−75	−50	−25	0	20	50	100	150	200	250	300
	Temperature, K												
Property	123.15	173.15	198.15	223.15	248.15	273.15	293.15	323.15	373.15	423.15	473.15	523.15	573.15
Density, ρ_l (kg/m^3)	S	S	S	S	(918)	905	881	854	808	V	V	V	V
Specific heat capacity, $c_{p,l}$ (kJ/kg K)	S	S	S	S	1.633	1.683	1.733	1.839	(2.00)	V	V	V	V
Thermal conductivity, λ_l [(W/m^2)/(K/m)]	S	S	S	S	(0.146)	0.138	0.134	0.129	0.122	V	V	V	V
Dynamic viscosity, η_l (10^{-5} Ns/m^2)	S	S	S	S	(180)	110.0	80.7	55.6	34.4	V	V	V	V

Properties of the Saturated Liquid and Vapor

T_{sat}, K	417.56	430	455	480	505	530	555	580	605	630.4
p_{sat}, kPa	101.3	142	249	422	655	995	1 420	2 040	2 750	3 729
ρ_ℓ, kg/m^3	764	751	725	696	666	632	594	549	489	288
ρ_g, kg/m^3	3.17	4.31	7.54	12.5	19.6	30.0	45.1	67.8	107	288
h_ℓ, kJ/kg	6.5	30.4	83.2	145.0	201.7	263.7	331.0	400.6	476.9	581.2
h_g, kJ/kg	349.9	369.5	409.8	450.5	491.4	531.7	570.5	605.8	631.8	581.2
$\Delta h_{g,\ell}$, kJ/kg	343.4	339.1	326.6	305.5	289.7	268.0	239.5	205.2	154.9	
$c_{p,\ell}$, kJ/(kg K)	2.16	2.20	2.30	2.40	2.52	2.65	2.81	3.02	3.44	
$c_{p,g}$, kJ/(kg K)	1.71	1.76	1.86	1.97	2.08	2.22	2.40	2.70	3.49	
η_ℓ, μNs/m^2	253	231	197	171	147	124	102	82	65	55
η_g, μNs/m^2	9.2	9.5	10.2	10.8	11.5	12.2	13.1	14.3	16.3	55
λ_ℓ, (mW/m^2)/(K/m)	99	96	91	85	80	74	69	63	58	52
λ_g, (mW/m^2)/(K/m)	18.4	19.3	21.9	24.4	27.0	29.9	33.0	36.5	40.6	52
Pr_ℓ	5.51	5.28	5.00	4.83	4.65	4.43	4.14	3.92	3.87	
Pr_g	0.86	0.87	0.87	0.88	0.89	0.91	0.95	1.06	1.40	
σ, mN/m	17.9	16.4	14.1	11.4	9.2	7.0	4.9	3.0	1.3	
$\beta_{e,\ell}$, kK^{-1}	1.31	1.39	1.59	1.85	2.20	2.71	3.54	5.11	9.36	

Transport Properties of the Superheated Vapor

	Temperature, °C												
	−150	−100	−50	0	25	100	200	300	400	600	800	1 000	1 200
Property	Temperature, K												
(at low pressure)	123.15	173.15	223.15	273.15	298.15	373.15	473.15	573.15	673.15	873.15	1 073.15	1 273.15	1 473.15
Specific heat capacity, $c_{p,g}$ (kJ/kg K)	S	S	S	L	L	L	1.855	2.139	2.378	2.751	3.023	3.224	3.370
Thermal conductivity, λ_g[(W/m^2)/(K/m)]	S	S	S	L	L	L	0.024	0.034	0.044	(0.065)	(0.085)	(0.105)	(0.123)
Dynamic viscosity, η_g(10^{-5} Ns/m^2)	S	S	S	L	L	L	(1.02)	(1.21)	(1.41)	(1.79)	(2.17)	(2.52)	(2.85)

m-XYLENE

Chemical formula: $C_6H_4(CH_3)_2$
Molecular weight: 106.168
Melting point: 225.5 K
Normal boiling point: 412.3 K

Critical temperature: 617.0 K
Critical pressure: 3.543 MPa
Critical density: 283 kg/m³
Normal vapor density: 4.74 kg/m³ (@ 0°C, 101.3 kPa)

Properties of the Liquid at Temperatures Below the Normal Boiling Point

Property	−150	−100	−75	−50	−25	0	20	50	100	150	200	250	300
Temperature, K	123.15	173.15	198.15	223.15	248.15	273.15	293.15	323.15	373.15	423.15	473.15	523.15	573.15
Density, ρ_l (kg/m³)	S	S	S	S	902	881	866	838	795	V	V	V	V
Specific heat capacity, $c_{p,l}$ (kJ/kg K)	S	S	S	S	1.587	1.650	1.717	1.806	(1.98)	V	V	V	V
Thermal conductivity, λ_l [(W/m²)/(K/m)]	S	S	S	S	0.144	0.138	0.132	0.126	0.114	V	V	V	V
Dynamic viscosity, η_l (10^{-5} Ns/m²)	S	S	S	S	125	80.2	61.5	44.5	29.0	V	V	V	V

(Column headings: Temperature, °C)

Properties of the Saturated Liquid and Vapor

T_{sat}, K	412	430	455	480	505	530	555	580	605	617
p_{sat}, kPa	101.3	162	281	472	741	1 121	1 601	2 264	3 052	3 543
ρ_ℓ, kg/m³	752	731	700	673	642	604	570	505	396	283
ρ_g, kg/m³	3.22	4.98	8.69	14.3	22.6	34.6	52.6	81.3	140	283
h_ℓ, kJ/kg	−34.6	−1.2	50.2	114.7	172.6	233.9	299.2	371.4	460.6	516.7
h_g, kJ/kg	310.8	337.9	376.8	416.1	455.2	493.5	529.5	559.8	571.1	516.7
$\Delta h_{g,\ell}$, kJ/kg	345.4	339.1	326.6	301.4	282.6	259.6	230.3	188.4	100.5	
$c_{p,\ell}$, kJ/(kg K)	2.13	2.19	2.29	2.40	2.54	2.69	2.87	3.16	3.86	
$c_{p,g}$, kJ/(kg K)	1.66	1.73	1.84	1.95	2.08	2.24	2.47	2.91	5.27	
η_ℓ, μNs/m²	232	207	182	158	135	112	90	71	55	54
η_g, μNs/m²	9.1	9.5	10.0	10.7	11.4	12.2	13.2	14.6	18.3	54
λ_ℓ, (mW/m²)/(K/m)	92	88	82	77	72	67	62	57	52	50
λ_g, (mW/m²)/(K/m)	16.9	19.4	22.1	24.6	27.2	29.9	33.1	37.1	42.2	50
Pr_ℓ	5.38	5.18	5.08	4.92	4.76	4.54	4.17	3.94	4.08	
Pr_g	0.89	0.85	0.83	0.85	0.87	0.91	0.99	1.15	2.29	
σ, mN/m	16.7	14.8	12.5	10.2	8.0	5.9	3.9	2.1	0.5	
$\beta_{e,\ell}$, kK⁻¹	1.38	1.50	1.73	2.04	2.47	3.14	4.29	6.83	18.0	

Transport Properties of the Superheated Vapor

Property (at low pressure)	−150	−100	−50	0	25	100	200	300	400	600	800	1 000	1 200
Temperature, K	123.15	173.15	223.15	273.15	298.15	373.15	473.15	573.15	673.15	873.15	1 073.15	1 273.15	1 473.15
Specific heat capacity, $c_{p,g}$ (kJ/kg K)	S	S	S	L	L	L	1.825	2.119	2.366	2.742	3.019	3.220	3.370
Thermal conductivity, λ_g[(W/m²)/(K/m)]	S	S	S	L	L	L	(0.023)	(0.033)	(0.043)	(0.064)	(0.084)	(0.104)	(0.122)
Dynamic viscosity, η_g(10^{-5} Ns/m²)	S	S	S	L	L	L	(1.00)	(1.20)	(1.39)	(1.76)	(2.14)	(2.49)	(2.82)

(Column headings: Temperature, °C)

p-XYLENE

Chemical formula: $C_6H_4(CH_3)_2$
Molecular weight: 106.168
Melting point: 286.4 K
Normal boiling point: 411.5 K

Critical temperature: 616.3 K
Critical pressure: 3.510 MPa
Critical density: 280 kg/m^3
Normal vapor density: 4.74 kg/m^3
(@ 0 °C, 101.3 kPa)

Properties of the Liquid at Temperatures Below the Normal Boiling Point

	Temperature, °C												
	−150	−100	−75	−50	−25	0	20	50	100	150	200	250	300
	Temperature, K												
Property	123.15	173.15	198.15	223.15	248.15	273.15	293.15	323.15	373.15	423.15	473.15	523.15	573.15
Density, ρ_l (kg/m^3)	S	S	S	S	S	S	861	833	790	V	V	V	V
Specific heat capacity, $c_{p,l}$ (kJ/kg K)	S	S	S	S	S	S	1.700	1.784	1.959	V	V	V	V
Thermal conductivity, λ_l [(W/m^2)/(K/m)]	S	S	S	S	S	S	0.133	0.126	0.114	V	V	V	V
Dynamic viscosity, η_l (10^{-5} Ns/m^2)	S	S	S	S	S	S	64.3	45.6	29.0	V	V	V	V

Properties of the Saturated Liquid and Vapor

T_{sat}, K	412	430	455	480	505	530	555	580	605	618
p_{sat}, kPa	101.3	161	287	476	752	1 120	1 590	2 270	3 100	3 510
ρ_ℓ, kg/m^3	753	734	706	676	643	607	565	513	432	280
ρ_g, kg/m^3	3.23	5.08	8.83	14.5	22.8	34.9	52.9	81.8	142	280
h_ℓ, kJ/kg	−35.7	−2.4	49.7	113.8	171.4	232.3	297.2	368.9	466.7	514.0
h_g, kJ/kg	309.7	337.7	376.3	415.2	454.0	491.9	527.5	557.3	567.2	514.0
$\Delta h_{g,\ell}$, kJ/kg	345.4	339.1	326.6	301.4	282.6	259.6	230.3	188.4	100.5	
$c_{p,\ell}$, kJ/(kg K)	2.11	2.18	2.27	2.39	2.52	2.67	2.86	3.15	3.88	
$c_{p,g}$, kJ/(kg K)	1.64	1.72	1.83	1.94	2.07	2.23	2.46	2.92	5.51	
η_ℓ, μNs/m^2	232	205	176	155	132	109	89	70	54	53
η_g, μNs/m^2	8.8	9.5	10.1	10.7	11.4	12.2	13.1	14.6	18.0	53
λ_ℓ, (mW/m^2)/(K/m)	98	93	87	81	75	69	63	57	52	49
λ_g, (mW/m^2)/(K/m)	17.1	19.7	22.1	24.6	27.2	30.0	33.0	37.1	42.3	49
Pr_ℓ	5.00	4.81	4.59	4.57	4.44	4.22	4.04	3.87	4.03	
Pr_g	0.84	1.83	0.84	0.84	0.87	0.91	0.98	1.15	2.34	
σ, mN/m	16.0	14.2	11.8	10.0	7.8	5.7	3.8	2.0	0.5	
$\beta_{e,\ell}$, kK^{-1}	1.57	1.74	1.83	1.97	2.62	3.03	4.87	12.3	42.7	

Transport Properties of the Superheated Vapor

	Temperature, °C												
	−150	−100	−50	0	25	100	200	300	400	600	800	1 000	1 200
Property (at low pressure)	Temperature, K												
	123.15	173.15	223.15	273.15	298.15	373.15	473.15	573.15	673.15	873.15	1 073.15	1 273.15	1 473.15
Specific heat capacity, $c_{p,g}$ (kJ/kg K)	S	S	S	S	L	L	1.813	2.106	2.353	2.734	3.010	3.215	3.366
Thermal conductivity, λ_g[(W/m^2)/(K/m)]	S	S	S	S	L	L	(0.023)	(0.033)	(0.043)	(0.064)	(0.084)	(0.104)	(0.122)
Dynamic viscosity, η_g(10^{-5} Ns/m^2)	S	S	S	S	L	L	(1.00)	(1.20)	(1.39)	(1.76)	(2.14)	(2.49)	(2.82)

1,2,3-TRIMETHYLBENZENE

Chemical formula: C_9H_{12}
Molecular weight: 120.19
Melting point: 247.7 K
Boiling point: 449.2 K

Critical temperature: 664.5 K
Critical pressure: 3.46 MPa
Critical density: 290 kg/m^3
Normal vapor density: 5.36 kg/m^3
(@ 0°C, 101.3 kPa)

Properties of the Liquid at Temperatures Below the Normal Boiling Point

	Temperature, °C												
	−150	−100	−75	−50	−25	0	20	50	100	150	200	250	300
	Temperature, K												
Property	123.15	173.15	198.15	223.15	248.15	273.15	293.15	323.15	373.15	423.15	473.15	523.15	573.15
Density, ρ_l (kg/m^3)	S	S	S	S	(929)	910	894	(871)	(830)	(787)	V	V	V
Specific heat capacity, $c_{p,l}$ (kJ/kg K)	S	S	S	S	1.656	1.733	1.787	(1.86)	(2.02)	(2.17)	V	V	V
Thermal conductivity, λ_l [(W/m^2)/(K/m)]	S	S	S	S	0.143	0.137	0.133	0.127	0.117	0.107	V	V	V
Dynamic viscosity, η_l (10^{-5} Ns/m^2)	S	S	S	S	(215)	(135)	(101)	(81)	(52)	(34)	V	V	V

Transport Properties of the Superheated Vapor

	Temperature, °C												
	−150	−100	−50	0	25	100	200	300	400	600	800	1 000	1 200
Property (at low pressure)	Temperature, K												
	123.15	173.15	223.15	273.15	298.15	373.15	473.15	573.15	673.15	873.15	1 073.15	1 273.15	1 473.15
Specific heat capacity, $c_{p,g}$ (kJ/kg K)	S	S	S	L	L	L	1.871	2.160	2.407	2.801	3.086	3.299	3.458
Thermal conductivity, λ_g[(W/m^2)/(K/m)]	S	S	S	L	L	L	(0.023)	(0.032)	(0.042)	(0.063)	(0.084)	(0.102)	(0.120)
Dynamic viscosity, η_g(10^{-5} Ns/m^2)	S	S	S	L	L	L	(0.97)	(1.16)	(1.34)	(1.71)	(2.06)	(2.39)	(2.70)

1,2,4-TRIMETHYLBENZENE

Chemical formula: C_9H_{12}
Molecular weight: 120.19
Melting point: 227.0 K
Boiling point: 442.5 K

Critical temperature: 649.1 K
Critical pressure: 3.23 MPa
Critical density: 280 kg/m^3
Normal vapor density: 5.36 kg/m^3
(@ 0 °C, 101.3 kPa)

Properties of the Liquid at Temperatures Below the Normal Boiling Point

	Temperature, °C												
	−150	−100	−75	−50	−25	0	20	50	100	150	200	250	300
	Temperature, K												
Property	123.15	173.15	198.15	223.15	248.15	273.15	293.15	323.15	373.15	423.15	473.15	523.15	573.15
Density, ρ_l (kg/m^3)	S	S	S	S	(913)	(892)	876	(852)	(811)	(770)	V	V	V
Specific heat capacity, $c_{p,l}$ (kJ/kg K)	S	S	S	S	1.641	1.704	1.758	1.837	2.006	(2.19)	V	V	V
Thermal conductivity, λ_l [(W/m^2)/(K/m)]	S	S	S	S	0.141	0.136	0.132	0.126	0.115	0.105	V	V	V
Dynamic viscosity, η_l (10^{-5} Ns/m^2)	S	S	S	S	(215)	(135)	101	81	(52)	(34)	V	V	V

Transport Properties of the Superheated Vapor

	Temperature, °C												
	− 150	− 100	− 50	0	25	100	200	300	400	600	800	1 000	1 200
Property (at low pressure)	Temperature, K												
	123.15	173.15	223.15	273.15	298.15	373.15	473.15	573.15	673.15	873.15	1 073.15	1 273.15	1 473.15
Specific heat capacity, $c_{p,g}$ (kJ/kg K)	S	S	S	L	L	L	1.876	2.165	2.412	2.805	3.090	3.303	3.458
Thermal conductivity, λ_g[(W/m^2)/(K/m)]	S	S	S	L	L	L	(0.023)	(0.032)	(0.042)	(0.062)	(0.084)	(0.100)	(0.118)
Dynamic viscosity, η_g(10^{-5} Ns/m^2)	S	S	S	L	L	L	(0.95)	(1.14)	(1.32)	(1.67)	(2.01)	(2.34)	(2.65)

1,3,5-TRIMETHYLBENZENE

Chemical formula: C_9H_{12}
Molecular weight: 120.19
Melting point: 228.4 K
Boiling point: 437.9 K

Critical temperature: 637.3 K
Critical pressure: 3.13 MPa
Critical density: 278 kg/m^3
Normal vapor density: 5.36 kg/m^3 (@ 0 °C, 101.3 kPa)

Properties of the Liquid at Temperatures Below the Normal Boiling Point

	Temperature, °C												
	−150	−100	−75	−50	−25	0	20	50	100	150	200	250	300
	Temperature, K												
Property	123.15	173.15	198.15	223.15	248.15	273.15	293.15	323.15	373.15	423.15	473.15	523.15	573.15
Density, ρ_l (kg/m^3)	S	S	S	S	(900)	(881)	865	841	(799)	(753)	V	V	V
Specific heat capacity, $c_{p,l}$ (kJ/kg K)	S	S	S	S	1.578	1.667	1.729	(1.82)	(1.97)	(2.14)	V	V	V
Thermal conductivity, λ_l [(W/m^2)/(K/m)]	S	S	S	S	0.144	0.141	0.136	0.129	0.116	0.108	V	V	V
Dynamic viscosity, η_l (10^{-5} Ns/m^2)	S	S	S	S	(140)	(98)	73.5	51.4	37.1	(20)	V	V	V

Transport Properties of the Superheated Vapor

	Temperature, °C												
	−150	−100	−50	0	25	100	200	300	400	600	800	1 000	1 200
Property	Temperature, K												
(at low pressure)	123.15	173.15	223.15	273.15	298.15	373.15	473.15	573.15	673.15	873.15	1 073.15	1 273.15	1 473.15
Specific heat capacity, $c_{p,g}$ (kJ/kg K)	S	S	S	L	L	L	1.863	2.160	2.412	2.805	3.094	3.303	3.462
Thermal conductivity, λ_g[(W/m^2)/(K/m)]	S	S	S	L	L	L	(0.023)	(0.032)	(0.042)	(0.062)	(0.084)	(0.100)	(0.118)
Dynamic viscosity, η_g(10^{-5} Ns/m^2)	S	S	S	L	L	L	0.95	(1.13)	(1.32)	(1.67)	(2.01)	(2.34)	(2.65)

1,2,3,4-TETRAMETHYLBENZENE

Chemical formula: $C_{10}H_{14}$
Molecular weight: 134.22
Melting point: 271.35 K
Boiling point: 477.55 K

Critical temperature: 700.1 K
Critical pressure: 3.22 MPa
Critical density: 280 kg/m^3
Normal vapor density: 5.99 kg/m^3 (@ 0 °C, 101.3 kPa)

Properties of the Liquid at Temperatures Below the Normal Boiling Point

	Temperature, °C												
	−150	−100	−75	−50	−25	0	20	50	100	150	200	250	300
	Temperature, K												
Property	123.15	173.15	198.15	223.15	248.15	273.15	293.15	323.15	373.15	423.15	473.15	523.15	573.15
Density, ρ_l (kg/m^3)	S	S	S	S	S	(921)	905	(882)	(843)	(805)	(767)	V	V
Specific heat capacity, $c_{p,l}$ (kJ/kg K)	S	S	S	S	S	1.739	1.790	(1.87)	(2.02)	(2.21)	(2.44)	V	V
Thermal conductivity, λ_l [(W/m^2)/(K/m)]	S	S	S	S	S	(0.143)	(0.139)	(0.133)	(0.127)	(0.114)	(0.105)	V	V
Dynamic viscosity, η_l (10^{-5} Ns/m^2)	S	S	S	S	S	(180)	(130)	(83)	(49)	(33)	(23)	V	V

1,2,3,5-TETRAMETHYLBENZENE

Chemical formula: $C_{10}H_{14}$
Molecular weight: 134.22
Melting point: 249.15 K
Boiling point: 471.05 K

Critical temperature: 686.8 K
Critical pressure: 3.27 MPa
Critical density: 280 kg/m^3
Normal vapor density: 5.99 kg/m^3
(@ 0°C, 101.3 kPa)

Properties of the Liquid at Temperatures Below the Normal Boiling Point

	Temperature, °C												
	−150	−100	−75	−50	−25	0	20	50	100	150	200	250	300
	Temperature, K												
Property	123.15	173.15	198.15	223.15	248.15	273.15	293.15	323.15	373.15	423.15	473.15	523.15	573.15
Density, ρ_l (kg/m^3)	S	S	S	S	S	(905)	890	(867)	(828)	(790)	V	V	V
Specific heat capacity, $c_{p,l}$ (kJ/kg K)	S	S	S	S	S	1.729	1.779	(.84)	(1.96)	(2.14)	V	V	V
Thermal conductivity, λ_l [(W/m^2)/(K/m)]	S	S	S	S	S	(0.143)	(0.139)	(0.133)	(0.127)	(0.114)	V	V	V
Dynamic viscosity, η_l (10^{-5} Ns/m^2)	S	S	S	S	S	(125)	(90)	(63)	(41)	(28)	V	V	V

1,2,4,5-TETRAMETHYLBENZENE

Chemical formula: $C_{10}H_{14}$
Molecular weight: 134.22
Melting point: 352.0 K
Boiling point: 470.0 K

Critical temperature: 675 K
Critical pressure: 2.94 MPa
Critical density: 280 kg/m^3
Normal vapor density: 5.99 kg/m^3
(@ 0°C, 101.3 kPa)

Properties of the Liquid at Temperatures Below the Normal Boiling Point

	Temperature, °C												
	−150	−100	−75	−50	−25	0	20	50	100	150	200	250	300
	Temperature, K												
Property	123.15	173.15	198.15	223.15	248.15	273.15	293.15	323.15	373.15	423.15	473.15	523.15	573.15
Density, ρ_l (kg/m^3)	S	S	S	S	S	S	S	S	(819)	(779)	V	V	V
Specific heat capacity, $c_{p,l}$ (kJ/kg K)	S	S	S	S	S	S	S	S	2.068	2.198	V	V	V
Thermal conductivity, λ_l [(W/m^2)/(K/m)]	S	S	S	S	S	S	S	S	0.123	0.114	V	V	V
Dynamic viscosity, η_l (10^{-5} Ns/m^2)	S	S	S	S	S	S	S	S	(37)	(26)	V	V	V

Transport Properties of the Superheated Vapor

	Temperature, °C												
	−150	−100	−50	0	25	100	200	300	400	600	800	1 000	1 200
Property	Temperature, K												
(at low pressure)	123.15	173.15	223.15	273.15	298.15	373.15	473.15	573.15	673.15	873.15	1 073.15	1 273.15	1 473.15
Specific heat capacity, $c_{p,g}$ (kJ/kg K)	S	S	S	S	S	L	1.962	2.246	2.492	2.882	—	—	—
Thermal conductivity, λ_g[(W/m^2)/(K/m)]	S	S	S	S	S	L	(0.023)	(0.032)	(0.042)	(0.062)	—	—	—
Dynamic viscosity, η_g(10^{-5} Ns/m^2)	S	S	S	S	S	L	(0.90)	(1.08)	(1.26)	(1.60)	—	—	—

PENTAMETHYLBENZENE

Chemical formula: $C_{11}H_{16}$
Molecular weight: 148.24
Melting point: 327.45 K
Boiling point: 504.15 K

Critical temperature: 725.3 K
Critical pressure: 2.99 MPa
Critical density: 274 kg/m^3
Normal vapor density: 6.61 kg/m^3
(@ 0°C, 101.3 kPa)

Properties of the Liquid at Temperatures Below the Normal Boiling Point

	Temperature, °C												
	−150	−100	−75	−50	−25	0	20	50	100	150	200	250	300
	Temperature, K												
Property	123.15	173.15	198.15	223.15	248.15	273.15	293.15	323.15	373.15	423.15	473.15	523.15	573.15
Density, ρ_l (kg/m^3)	S	S	S	S	S	S	S	S	859	(821)	(782)	V	V
Specific heat capacity, $c_{p,l}$ (kJ/kg K)	S	S	S	S	S	S	S	S	2.031	2.236	(2.55)	V	V
Thermal conductivity, λ_l [(W/m^2)/(K/m)]	S	S	S	S	S	S	S	S	0.122	0.113	0.105	V	V
Dynamic viscosity, η_l (10^{-5} Ns/m^2)	S	S	S	S	S	S	S	S	(66)	(40)	(27)	V	V

HEXAMETHYLBENZENE

Chemical formula: $C_{12}H_{18}$
Molecular weight: 162.26
Melting point: 438.65 K
Boiling point: 536.15 K

Critical temperature: 754.2 K
Critical pressure: 2.72 MPa
Critical density: 274 kg/m^3
Normal vapor density: 7.24 kg/m^3
(@ 0°C, 101.3 kPa)

Properties of the Liquid at Temperatures Below the Normal Boiling Point

	Temperature, °C												
	−150	−100	−75	−50	−25	0	20	50	100	150	200	250	300
	Temperature, K												
Property	123.15	173.15	198.15	223.15	248.15	273.15	293.15	323.15	373.15	423.15	473.15	523.15	573.15
Density, ρ_l (kg/m^3)	S	S	S	S	S	S	S	S	S	S	(800)	(760)	V
Specific heat capacity, $c_{p,l}$ (kJ/kg K)	S	S	S	S	S	S	S	S	S	S	2.458	(2.64)	V
Thermal conductivity, λ_l [(W/m^2)/(K/m)]	S	S	S	S	S	S	S	S	S	S	0.110	0.103	V
Dynamic viscosity, η_l (10^{-5} Ns/m^2)	S	S	S	S	S	S	S	S	S	S	(32)	(23)	V

DIPHENYL

Chemical formula: $C_{12}H_{10}$
Molecular weight: 154.21
Melting point: 342.4 K
Boiling point: 528.4 K

Critical temperature: 789 K
Critical pressure: 3.85 MPa
Critical density: 323 kg/m^3
Normal vapor density: 6.88 kg/m^3
(@ 0 °C, 101.3 kPa)

Properties of the Liquid at Temperatures Below the Normal Boiling Point

	Temperature, °C												
	−150	−100	−75	−50	−25	0	20	50	100	150	200	250	300
	Temperature, K												
Property	123.15	173.15	198.15	223.15	248.15	273.15	293.15	323.15	373.15	423.15	473.15	523.15	573.15
Density, ρ_l (kg/m^3)	S	S	S	S	S	S	S	S	970	930	888	848	V
Specific heat capacity, $c_{p,l}$ (kJ/kg K)	S	S	S	S	S	S	S	S	1.779	1.968	2.190	2.407	V
Thermal conductivity, λ_l [(W/m^2)/(K/m)]	S	S	S	S	S	S	S	S	0.134	0.127	0.119	0.112	V
Dynamic viscosity, η_l (10^{-5} Ns/m^2)	S	S	S	S	S	S	S	S	97.0	55.6	37.2	26.9	V

Transport Properties of the Superheated Vapor

	Temperature, °C												
	− 150	− 100	− 50	0	25	100	200	300	400	600	800	1 000	1 200
Property (at low pressure)	Temperature, K												
	123.15	173.15	223.15	273.15	298.15	373.15	473.15	573.15	673.15	873.15	1 073.15	1 273.15	1 473.15
Specific heat capacity, $c_{p,g}$ (kJ/kg K)	S	S	S	S	S	L	L	1.931	2.140	2.463	2.675	2.838	—
Thermal conductivity, λ_g[(W/m^2)/(K/m)]	S	S	S	S	S	L	L	0.027	(0.039)	(0.059)	(0.079)	(0.099)	—
Dynamic viscosity, η_g(10^{-5} Ns/m^2)	S	S	S	S	S	L	L	1.07	(1.28)	(1.68)	(2.08)	(2.48)	—

DIPHENYLMETHANE

Chemical formula: $C_{13}H_{12}$
Molecular weight: 168.24
Melting point: 300.0 K
Boiling point: 537.5 K

Critical temperature: 772.5 K
Critical pressure: 2.98 MPa
Critical density: 302 kg/m^3
Normal vapor density: 7.51 kg/m^3
(@ 0°C, 101.3 kPa)

Properties of the Liquid at Temperatures Below the Normal Boiling Point

	Temperature, °C												
	−150	−100	−75	−50	−25	0	20	50	100	150	200	250	300
	Temperature, K												
Property	123.15	173.15	198.15	223.15	248.15	273.15	293.15	323.15	373.15	423.15	473.15	523.15	573.15
Density, ρ_l (kg/m^3)	S	S	S	S	S	S	S	984	944	904	864	824	V
Specific heat capacity, $c_{p,l}$ (kJ/kg K)	S	S	S	S	S	S	S	1.631	1.760	(1.91)	(2.06)	(2.20)	V
Thermal conductivity, λ_l [(W/m^2)/(K/m)]	S	S	S	S	S	S	S	0.132	0.127	0.121	0.115	(0.109)	V
Dynamic viscosity, η_l (10^{-5} Ns/m^2)	S	S	S	S	S	S	S	180.1	93.1	56.0	37.5	27.0	V

TRIPHENYLMETHANE

Chemical formula: $C_{19}H_{16}$
Molecular weight: 244.32
Melting point: 366.25 K
Boiling point: 632.15 K

Critical temperature: (930) K
Critical pressure: (3.2) MPa
Critical density: —
Normal vapor density: 10.9 kg/m^3
(@ 0°C, 101.3 kPa)

Properties of the Liquid at Temperatures Below the Normal Boiling Point

	Temperature, °C												
	−150	−100	−50	0	25	100	200	300	400	600	800	1 000	1 200
	Temperature, K												
Property (at low pressure)	123.15	173.15	223.15	273.15	298.15	373.15	473.15	573.15	673.15	873.15	1 073.15	1 273.15	1 473.15
Density, ρ_l (kg/m^3)	S	S	S	S	S	S	S	S	1014	975	(937)	(899)	(860)
Specific heat capacity, $c_{p,l}$ (kJ/kg K)	S	S	S	S	S	S	S	S	1.851	1.913	(1.98)	(2.04)	(2.10)
Thermal conductivity, λ_l [(W/m^2)/(K/m)]	S	S	S	S	S	S	S	S	(0.136)	(0.132)	(0.129)	(0.125)	(0.121)
Dynamic viscosity, η_l (10^{-5} Ns/m^2)	S	S	S	S	S	S	S	S	322.1	(155)	(85)	(53)	(40)

NAPHTHALENE

Chemical formula: $C_{10}H_8$
Molecular weight: 128.17
Melting point: 353.5 K
Boiling point: 491.1 K

Critical temperature: 748.4 K
Critical pressure: 4.05 MPa
Critical density: 314 kg/m^3
Normal vapor density: 5.72 kg/m^3
(@ 0 °C, 101.3 kPa)

Properties of the Liquid at Temperatures Below the Normal Boiling Point

	Temperature, °C												
	–150	–100	–75	–50	–25	0	20	50	100	150	200	250	300
	Temperature, K												
Property	123.15	173.15	198.15	223.15	248.15	273.15	293.15	323.15	373.15	423.15	473.15	523.15	573.15
Density, ρ_l (kg/m^3)	S	S	S	S	S	S	S	S	963	922	(878)	V	V
Specific heat capacity, $c_{p,l}$ (kJ/kg K)	S	S	S	S	S	S	S	S	1.805	1.993	2.139	V	V
Thermal conductivity, λ_l [(W/m^2)/(K/m)]	S	S	S	S	S	S	S	S	0.137	0.130	0.123	V	V
Dynamic viscosity, η_l (10^{-5} Ns/m^2)	S	S	S	S	S	S	S	S	77.4	52.0	37.5	V	V

Transport Properties of the Superheated Vapor

	Temperature, °C												
	– 150	– 100	– 50	0	25	100	200	300	400	600	800	1 000	1 200
	Temperature, K												
Property (at low pressure)	123.15	173.15	223.15	273.15	298.15	373.15	473.15	573.15	673.15	873.15	1 073.15	1 273.15	1 473.15
Specific heat capacity, $c_{p,g}$ (kJ/kg K)	S	S	S	S	S	L	L	1.891	2.102	2.373	2.618	2.760	2.902
Thermal conductivity, λ_g[(W/m^2)/(K/m)]	S	S	S	S	S	L	L	(0.028)	(0.037)	(0.055)	(0.073)	(0.088)	(0.103)
Dynamic viscosity, η_g(10^{-5} Ns/m^2)	S	S	S	S	S	L	L	(1.16)	(1.35)	(1.73)	(2.11)	(2.59)	(2.97)

1-METHYLNAPHTHALENE

Chemical formula: $C_{11}H_{10}$
Molecular weight: 142.20
Melting point: 242.7 K
Boiling point: 517.8 K

Critical temperature: 772 K
Critical pressure: 3.57 MPa
Critical density: (320) kg/m^3
Normal vapor density: 6.35 kg/m^3 (@ 0°C, 101.3 kPa)

Properties of the Liquid at Temperatures Below the Normal Boiling Point

	Temperature, °C												
	−150	−100	−75	−50	−25	0	20	50	100	150	200	250	300
	Temperature, K												
Property	123.15	173.15	198.15	223.15	248.15	273.15	293.15	323.15	373.15	423.15	473.15	523.15	573.15
Density, ρ_l (kg/m^3)	S	S	S	S	(1053)	1035	1020	998	962	(924)	(884)	V	V
Specific heat capacity, $c_{p,l}$ (kJ/kg K)	S	S	S	S	1.424	1.509	1.564	1.661	(1.82)	(1.95)	(2.11)	V	V
Thermal conductivity, λ_l [(W/m^2)/(K/m)]	S	S	S	S	(0.136)	(0.133)	(0.130)	(0.127)	(0.120)	(0.114)	(0.107)	V	V
Dynamic viscosity, η_l (10^{-5} Ns/m^2)	S	S	S	S	(1400)	(571)	333	167	(85)	(50)	(32)	V	V

Transport Properties of the Superheated Vapor

	Temperature, °C												
	−150	−100	−50	0	25	100	200	300	400	600	800	1 000	1 200
Property (at low pressure)	Temperature, K												
	123.15	173.15	223.15	273.15	298.15	373.15	473.15	573.15	673.15	873.15	1 073.15	1 273.15	1 473.15
Specific heat capacity, $c_{p,g}$ (kJ/kg K)	S	S	S	L	L	L	L	1.987	2.205	2.529	2.740	2.890	3.039
Thermal conductivity, λ_g[(W/m^2)/(K/m)]	S	S	S	L	L	L	L	(0.027)	(0.036)	(0.053)	(0.071)	(0.086)	(0.101)
Dynamic viscosity, η_g(10^{-5} Ns/m^2)	S	S	S	L	L	L	L	(1.09)	(1.27)	(1.61)	(1.95)	(2.29)	(2.63)

2-METHYLNAPHTHALENE

Chemical formula: $C_{11}H_{10}$
Molecular weight: 142.20
Melting point: 307.7 K
Boiling point: 514.2 K

Critical temperature: 761 K
Critical pressure: 3.51 MPa
Critical density: (320) kg/m^3
Normal vapor density: 6.35 kg/m^3
(@ 0 °C, 101.3 kPa)

Properties of the Liquid at Temperatures Below the Normal Boiling Point

	Temperature, °C												
	−150	−100	−75	−50	−25	0	20	50	100	150	200	250	300
	Temperature, K												
Property	123.15	173.15	198.15	223.15	248.15	273.15	293.15	323.15	373.15	423.15	473.15	523.15	573.15
Density, ρ_l (kg/m^3)	S	S	S	S	S	S	S	984	947	908	(868)	V	V
Specific heat capacity, $c_{p,l}$ (kJ/kg K)	S	S	S	S	S	S	S	1.632	1.805	(1.98)	(2.17)	V	V
Thermal conductivity, λ_l [(W/m^2)/(K/m)]	S	S	S	S	S	S	S	0.127	0.120	(0.114)	(0.107)	V	V
Dynamic viscosity, η_l (10^{-5} Ns/m^2)	S	S	S	S	S	S	S	136.1	(72)	(43)	(30)	V	V

Transport Properties of the Superheated Vapor

	Temperature, °C												
	−150	−100	−50	0	25	100	200	300	400	600	800	1 000	1 200
Property (at low pressure)	Temperature, K												
	123.15	173.15	223.15	273.15	298.15	373.15	473.15	573.15	673.15	873.15	1 073.15	1 273.15	1 473.15
Specific heat capacity, $c_{p,g}$ (kJ/kg K)	S	S	S	S	S	L	L	1.974	2.191	2.517	2.730	2.881	3.033
Thermal conductivity, λ_g[(W/m^2)/(K/m)]	S	S	S	S	S	L	L	(0.027)	(0.036)	(0.053)	(0.070)	(0.086)	(0.101)
Dynamic viscosity, η_g(10^{-5} Ns/m^2)	S	S	S	S	S	L	L	(1.09)	(1.27)	(1.61)	(1.95)	(2.29)	(2.63)

1-ETHYLNAPHTHALENE

Chemical formula: $C_{12}H_{12}$
Molecular weight: 156.22
Melting point: 246.15 K
Boiling point: 531.45 K

Critical temperature: 775.6 K
Critical pressure: 3.99 MPa
Critical density: (325) kg/m^3
Normal vapor density: 6.97 kg/m^3 (@ 0 °C, 101.3 kPa)

Properties of the Liquid at Temperatures Below the Normal Boiling Point

	Temperature, °C												
	−150	−100	−75	−50	−25	0	20	50	100	150	200	250	300
	Temperature, K												
Property	123.15	173.15	198.15	223.15	248.15	273.15	293.15	323.15	373.15	423.15	473.15	523.15	573.15
Density, ρ_l (kg/m^3)	S	S	S	S	(1.039)	1.022	1008	987	950	(913)	(874)	(832)	V
Specific heat capacity, $c_{p,l}$ (kJ/kg K)	S	S	S	S	(1.52)	(1.58)	(1.63)	(1.71)	(1.85)	(1.98)	(2.10)	(2.22)	V
Thermal conductivity, λ_l [(W/m^2)/(K/m)]	S	S	S	S	(0.134)	(0.131)	(0.129)	(0.126)	(0.121)	(0.115)	(0.110)	(0.104)	V
Dynamic viscosity, η_l (10^{-5} Ns/m^2)	S	S	S	S	(1400)	(571)	(333)	(167)	(85)	(50)	(32)	(19)	V

Transport Properties of the Superheated Vapor

	Temperature, °C											
	−150	−100	−50	0	25	100	200	300	400	600	800	1 000	1 200
Property	Temperature, K												
(at low pressure)	123.15	173.15	223.15	273.15	298.15	373.15	473.15	573.15	673.15	873.15	1 073.15	1 273.15	1 473.15
Specific heat capacity, $c_{p,g}$ (kJ/kg K)	S	S	S	L	L	L	L	2.069	2.287	2.621	2.838	2.993	3.149
Thermal conductivity, λ_g[(W/m^2)/(K/m)]	S	S	S	L	L	L	L	(0.027)	(0.035)	(0.053)	(0.070)	(0.085)	(0.099)
Dynamic viscosity, η_g(10^{-5} Ns/m^2)	S	S	S	L	L	L	L	(1.03)	(1.20)	(1.54)	(1.86)	(2.18)	(2.50)

2-ETHYLNAPHTHALENE

Chemical formula: $C_{12}H_{12}$
Molecular weight: 156.22
Melting point: 265.75 K
Boiling point: 531.05 K

Critical temperature: 771.8 K
Critical pressure: 3.17 MPa
Critical density: (325) kg/m^3
Normal vapor density: 6.97 kg/m^3
(@ 0 °C, 101.3 kPa)

Properties of the Liquid at Temperatures Below the Normal Boiling Point

	Temperature, °C												
	−150	−100	−75	−50	−25	0	20	50	100	150	200	250	300
	Temperature, K												
Property	**123.15**	**173.15**	**198.15**	**223.15**	**248.15**	**273.15**	**293.15**	**323.15**	**373.15**	**423.15**	**473.15**	**523.15**	**573.15**
Density, ρ_l (kg/m^3)	S	S	S	S	S	1008	992	(971)	(935)	(896)	(855)	(810)	V
Specific heat capacity, $c_{p,l}$ (kJ/kg K)	S	S	S	S	S	(1.58)	(1.63)	(1.71)	(1.85)	(1.98)	(2.10)	(2.22)	V
Thermal conductivity, λ_l [(W/m^2)/(K/m)]	S	S	S	S	S	(0.131)	(0.129)	(0.126)	(0.121)	(0.115)	(0.110)	(0.104)	V
Dynamic viscosity, η_l (10^{-5} Ns/m^2)	S	S	S	S	S	(350)	(240)	(136)	(72)	(43)	(30)	(17)	V

Transport Properties of the Superheated Vapor

	Temperature, °C												
	−150	−100	−50	0	25	100	200	300	400	600	800	1 000	1 200
Property (at low pressure)	Temperature, K												
	123.15	173.15	223.15	273.15	298.15	373.15	473.15	573.15	673.15	873.15	1 073.15	1 273.15	1 473.15
Specific heat capacity, $c_{p,g}$ (kJ/kg K)	S	S	S	L	L	L	L	2.053	2.275	2.610	2.829	2.986	3.143
Thermal conductivity, λ_g[(W/m^2)/(K/m)]	S	S	S	L	L	L	L	(0.027)	(0.036)	(0.053)	(0.070)	(0.085)	(0.099)
Dynamic viscosity, η_g(10^{-5} Ns/m^2)	S	S	S	L	L	L	L	(1.09)	(1.27)	(1.61)	(1.96)	(2.29)	(2.62)

METHANOL (METHYL ALCOHOL)

Chemical formula: CH_3OH
Molecular weight: 32.00
Melting point: 175.15 K
Normal boiling point: 337.85 K

Critical temperature: 513.15 K
Critical pressure: 7.950 MPa
Critical density: 275 kg/m^3
Normal vapor density: 1.43 kg/m^3 (@ 0 °C, 101.3 kPa)

Properties of the Liquid at Temperatures Below the Normal Boiling Point

	Temperature, °C												
	−150	−100	−75	−50	−25	0	20	50	100	150	200	250	300
	Temperature, K												
Property	123.15	173.15	198.15	223.15	248.15	273.15	293.15	323.15	373.15	423.15	473.15	523.15	573.15
Density, ρ_l (kg/m^3)	S	S	905	900	858	810	792	765	V	V	V	V	V
Specific heat capacity, $c_{p,l}$ (kJ/kg K)	S	S	2.194	1.240	2.294	2.386	2.495	2.680	V	V	V	V	V
Thermal conductivity, λ_l [(W/m^2)/(K/m)]	S	S	0.235	0.225	0.215	0.208	0.202	0.193	V	V	V	V	V
Dynamic viscosity, η_l (10^{-5} Ns/m^2)	S	S	479	230.6	128.0	81.7	58.4	39.6	V	V	V	V	V

Properties of the Saturated Liquid and Vapor

T_{sat}, K	337.85	353.2	373.2	393.2	413.2	433.2	453.2	473.2	493.2	511.7
p_{sat}, kPa	101.3	178.4	349.4	633.3	1 076	1 736	2 678	3 970	5 675	7 775
ρ_ℓ, kg/m^3	751.0	735.5	714.0	690.0	664.0	634.0	598.0	553.0	490.0	363.5
ρ_g, kg/m^3	1.222	2.084	3.984	7.142	12.16	19.94	31.86	50.75	86.35	178.9
h_ℓ, kJ/kg	0.0	45	108	176	249	328	413	506		
h_g, kJ/kg	1 101	1 115	1 130	1 144	1 171	1 171	1 169	1 151		
$\Delta h_{g,\ell}$, kJ/kg	1 101	1 070	1 022	968	922	843	756	645	482	
$c_{p,\ell}$, kJ/(kg K)	2.88	3.03	3.26	3.52	3.80	4.11	4.45	4.81		
$c_{p,g}$, kJ/(kg K)	1.55	1.61	1.69	1.83	1.99	2.20	2.56	3.65	5.40	
η_ℓ, μNs/m^2	326	271	214	170	136	109	88.3	71.6	58.3	41.6
η_g, μNs/m^2	11.1	11.6	12.4	13.1	14.0	14.9	16.0	17.4	20.1	26.0
λ_ℓ, (mW/m^2)/(K/m)	191.4	187.0	181.3	178.5	170.0	164.0	158.7	153.0	147.3	142.0
λ_g, (mW/m^2)/(K/m)	18.3	20.6	23.2	26.2	29.7	33.8	39.4	46.9	60.0	98.7
Pr_ℓ	5.13	4.67	4.15	3.61	3.34	2.82	2.56	2.42		
Pr_g	0.94	0.91	0.90	0.92	0.94	0.97	1.04	1.35	1.81	
σ, mN/m	18.75	17.5	15.7	13.6	11.5	9.3	6.9	4.5	2.1	0.09
$\beta_{e,\ell}$, kK^{-1}	0.42	1.20	1.69	2.00	2.49	3.15	4.21	6.40	17.2	

Transport Properties of the Superheated Vapor

	Temperature, °C												
	−150	−100	−50	0	25	100	200	300	400	600	800	1 000	1 200
Property (at low pressure)	Temperature, K												
	123.15	173.15	223.15	273.15	298.15	373.15	473.15	573.15	673.15	873.15	1 073.15	1 273.15	1 473.15
Specific heat capacity, $c_{p,g}$ (kJ/kg K)	S	S	L	L	L	1.595	1.823	2.064	2.273	2.629	3.01	3.23	3.40
Thermal conductivity, λ_g[(W/m^2)/(K/m)]	S	S	L	L	L	0.026	0.045	0.055	0.071	0.104	0.136	0.167	0.197
Dynamic viscosity, η_g(10^{-5} Ns/m^2)	S	S	L	L	L	1.22	1.56	1.89	2.20	2.79	3.33	3.82	4.28

ETHANOL (ETHYL ALCOHOL)

Chemical formula: CH_3CH_2OH
Molecular weight: 46.1
Melting point: 158.65 K
Normal boiling point: 351.45 K

Critical temperature: 516.25 K
Critical pressure: 6.390 MPa
Critical density: 280 kg/m^3
Normal vapor density: 2.06 kg/m^3
(@ 0 °C, 101.3 kPa)

Properties of the Liquid at Temperatures Below the Normal Boiling Point

Property	Temperature, °C												
	−150	−100	−75	−50	−25	0	20	50	100	150	200	250	300
	Temperature, K												
	123.15	173.15	198.15	223.15	248.15	273.15	293.15	323.15	373.15	423.15	473.15	523.15	573.15
Density, ρ_l (kg/m^3)	S	892	870	850	825	806	789	763	V	V	V	V	V
Specific heat capacity, $c_{p,l}$ (kJ/kg K)	S	1.901	1.947	2.014	2.093	2.232	2.395	2.801	V	V	V	V	V
Thermal conductivity, λ_l [(W/m^2)/(K/m)]	S	0.197	0.193	0.188	0.183	0.177	0.173	0.165	V	V	V	V	V
Dynamic viscosity, η_l (10^{-5} Ns/m^2)	S	4 701	1 526	640	324.1	1768.6	120.1	70.1	V	V	V	V	V

Properties of the Saturated Liquid and Vapor

T_{sat}, K	351.45	373	393	413	433	453	473	483	503	513
p_{sat}, kPa	101.3	226	429	753	1 256	1 960	2 940	3 560	5 100	6 020
ρ_ℓ, kg/m^3	757.0	733.7	709.0	680.3	648.5	610.5	564.0	537.6	466.2	420.3
ρ_g, kg/m^3	1.435	3.175	5.841	10.25	17.15	27.65	44.40	56.85	101.1	160.2
h_ℓ, kJ/kg	202.5	271.7	340.0	413.2	491.5	576.5	670.7	722.2	837.4	909.8
h_g, kJ/kg	1 165.5	1 198.7	1 225.5	1 247.2	1 264.4	1 275.3	1 269.0	1 259.0	1 224.6	1 190.3
$\Delta h_{g,\ell}$, kJ/kg	963.0	927.0	885.5	834.0	772.9	698.9	598.3	536.7	387.3	280.5
$c_{p,\ell}$, kJ/(kg K)	3.00	3.30	3.61	3.96	4.65	5.51	6.16	6.61		
$c_{p,g}$, kJ/(kg K)	1.83	1.92	2.02	2.11	2.31	2.80	3.18	3.78	6.55	
η_ℓ, μNs/m^2	428.7	314.3	240.0	185.5	144.6	113.6	89.6	79.7	63.2	56.3
η_g, μNs/m^2	10.4	11.1	11.7	12.3	12.9	13.7	14.5	15.1	16.7	18.5
λ_ℓ, (mW/m^2)/(K/m)	153.6	150.7	146.5	141.9	137.2	134.8	129.1	125.6	108.0	79.11
λ_g, (mW/m^2)/(K/m)	19.9	22.4	24.5	26.8	29.3	32.1	35.3	37.8	43.9	50.7
Pr_ℓ	8.37	6.88	5.91	5.18	4.90	4.64	4.28	4.19		
Pr_g	0.96	0.95	0.96	0.97	1.02	1.20	1.31	1.51	2.49	
σ, mN/m	17.7	15.7	13.6	11.5	9.3	6.9	4.5	3.3	0.9	0.34
$\beta_{e,\ell}$, kK^{-1}	1.41	1.60	1.90	2.41	3.13	4.18	6.06	7.56	16.0	

Transport Properties of the Superheated Vapor

Property (at low pressure)	Temperature, °C												
	−150	−100	−50	0	25	100	200	300	400	600	800	1 000	1 200
	Temperature, K												
	123.15	173.15	223.15	273.15	298.15	373.15	473.15	573.15	673.15	873.15	1 073.15	1 273.15	1 473.15
Specific heat capacity, $c_{p,g}$ (kJ/kg K)	S	L	L	L	L	1.825	2.114	2.370	2.596	2.964	3.245	3.458	3.622
Thermal conductivity, λ_g[(W/m^2)/(K/m)]	S	L	L	L	L	0.023	0.039	0.047	(0.059)	(0.079)	—	—	—
Dynamic viscosity, η_g(10^{-5} Ns/m^2)	S	L	L	L	L	1.09	1.38	1.65	1.88	2.36	2.78	3.17	3.52

1-PROPANOL (PROPYL ALCOHOL)

Chemical formula: $CH_3CH_2CH_2OH$
Molecular weight: 60.1
Melting point: 184.15 K
Normal boiling point: 370.95 K

Critical temperature: 536.85 K
Critical pressure: 5.050 MPa
Critical density: 273 kg/m^3
Normal vapor density: 2.68 kg/m^3
(@ 0°C, 101.3 kPa)

Properties of the Liquid at Temperatures Below the Normal Boiling Point

	Temperature, °C												
	−150	−100	−75	−50	−25	0	20	50	100	150	200	250	300
	Temperature, K												
Property	**123.15**	**173.15**	**198.15**	**223.15**	**248.15**	**273.15**	**293.15**	**323.15**	**373.15**	**423.15**	**473.15**	**523.15**	**573.15**
Density, ρ_l (kg/m^3)	S	(872)	(862)	(849)	(835)	818	804	779	V	V	V	V	V
Specific heat capacity, $c_{p,l}$ (kJ/kg K)	S	1.805	1.855	1.955	2.077	2.219	2.345	2.571	V	V	V	V	V
Thermal conductivity, λ_l [(W/m^2)/(K/m)]	S	0.172	0.170	0.167	0.164	0.160	0.157	0.153	V	V	V	V	V
Dynamic viscosity, η_l (10^{-5} Ns/m^2)	S	(61000)	7 562	2 020	820	413.7	223.1	112.6	V	V	V	V	V

Properties of the Saturated Liquid and Vapor

T_{sat}, K	373.2	393.2	413.2	433.2	453.2	473.2	493.2	513.2	523.2	533.1
p_{sat}, kPa	109.4	218.5	399.2	683.6	1 089	1 662	2 426	3 402	3 998	4 689
ρ_ℓ, kg/m^3	732.5	711	687.5	660	628.5	592.0	548.5	492.0	452.5	390.5
ρ_g, kg/m^3	2.26	4.43	8.05	13.8	22.5	35.3	55.6	90.4	118.0	161.0
h_ℓ, kJ/kg	0.0	65	139	222	315	433	548	691		
h_g, kJ/kg	687	710	733	766	802	860	904	955		
$\Delta h_{g,\ell}$, kJ/kg	687	645	594	544	486	427	356	264	209	138
$c_{p,\ell}$, kJ/(kg K)	3.21	3.47	3.86	4.36	5.02	5.90	6.78	7.79		
$c_{p,g}$, kJ/(kg K)	1.65	1.82	1.93	2.05	2.20	2.36	2.97	3.94		
η_ℓ, μNs/m^2	447	337	250	188	148	119	90.6	70.0	61.4	53.9
η_g, μNs/m^2	9.61	10.3	10.9	11.5	12.2	12.9	14.2	15.7	17.0	19.3
λ_ℓ, (mW/m^2)/(K/m)	142.4	139.2	138.4	133.5	127.9	120.7	111.8	100.6	94.1	89.3
λ_g, (mW/m^2)/(K/m)	20.9	23.0	26.2	28.9	31.4	34.7	38.0	43.9	47.5	53.5
Pr_ℓ	10.1	8.40	6.97	5.14	5.81	5.82	5.50	5.42		
Pr_g	0.76	0.82	0.80	0.82	0.85	0.88	1.11	1.41		
σ, mN/m	17.6	16.15	14.42	12.7	10.77	8.85	6.35	4.04	2.6	0.96
$\beta_{e,\ell}$, kK^{-1}	1.33	1.57	1.91	2.43	3.15	4.14	5.82	9.19	13.6	

Transport Properties of the Superheated Vapor

	Temperature, °C												
	−150	−100	−50	0	25	100	200	300	400	600	800	1 000	1 200
	Temperature, K												
Property (at low pressure)	123.15	173.15	223.15	273.15	298.15	373.15	473.15	573.15	673.15	873.15	1 073.15	1 273.15	1 473.15
Specific heat capacity, $c_{p,g}$ (kJ/kg K)	S	L	L	L	L	1.805	2.173	2.470	2.709	3.086	3.370	3.601	3.760
Thermal conductivity, λ_g[(W/m^2)/(K/m)]	S	L	L	L	L	0.021	0.033	(0.045)	(0.058)	(0.084)	—	—	—
Dynamic viscosity, η_g(10^{-5} Ns/m^2)	S	L	L	L	L	0.931	1.24	1.48	(1.72)	(2.14)	—	—	—

1-BUTANOL (BUTYL ALCOHOL)

Chemical formula: $C_2H_5CH_2CH_2OH$
Molecular weight: 74.12
Melting point: 183.2 K
Normal boiling point: 390.65 K

Critical temperature: 561.15 K
Critical pressure: 4.960 MPa
Critical density: 270.5 kg/m^3
Normal vapor density: 3.31 kg/m^3
(@ 0 °C, 101.3 kPa)

Properties of the Liquid at Temperatures Below the Normal Boiling Point

	Temperature, °C												
	−150	−100	−75	−50	−25	0	20	50	100	150	200	250	300
	Temperature, K												
Property	123.15	173.15	198.15	223.15	248.15	273.15	293.15	323.15	373.15	423.15	473.15	523.15	573.15
Density, ρ_l (kg/m^3)	S	S	(854)	(845)	835	825	810	791	753	V	V	V	V
Specific heat capacity, $c_{p,l}$ (kJ/kg K)	S	S	1.859	1.947	2.056	2.202	2.345	2.621	(2.84)	V	V	V	V
Thermal conductivity, λ_l [(W/m^2)/(K/m)]	S	S	0.170	0.165	0.162	0.158	0.155	0.149	0.138	V	V	V	V
Dynamic viscosity, η_l (10^{-5} Ns/m^2)	S	S	(9400)	3 471	1 250	530	295.1	142.0	53.1	V	V	V	V

Properties of the Saturated Liquid and Vapor

T_{sat}, K	390.65	410.2	429.2	446.5	469.5	485.2	508.3	530.2	545.5	558.9
p_{sat}, kPa	101.3	182	327	482	759	1 190	1 830	2 530	3 210	4 030
ρ_ℓ, kg/m^3	712	688	664	640	606	581	538	487	440	364
ρ_g, kg/m^3	2.30	4.10	7.9	12.5	23.8	27.8	48.2	74.0	102.3	240.2
h_ℓ, kJ/kg	0.0	64.8	135.0	206.8	315.3	399.6	541.9	700.2		
h_g, kJ/kg	591.3	629.8	672.3	716.5	784.1	836.8	924.4	1 015.3		
$\Delta h_{g,\ell}$, kJ/kg	591.3	565.0	537.3	509.7	468.8	437.2	382.5	315.1	248.4	143.0
$c_{p,\ell}$, kJ/(kg K)	3.20	3.54	3.95	4.42	5.15	5.74	6.71	7.76		
$c_{p,g}$, kJ/(kg K)	1.87	1.95	2.03	2.14	2.24	2.37	2.69	3.05	3.97	
η_ℓ, μNs/m^2	403.8	346.1	278.8	230.8	188.5	144.2	130.8	115.4	111.5	105.8
η_g, μNs/m^2	9.29	10.3	10.7	11.4	12.1	12.7	13.9	15.4	17.1	28.3
λ_ℓ, (mW/m^2)/(K/m)	127.1	122.3	117.5	112.6	105.4	101.4	91.7	82.9	74.0	62.8
λ_g, (mW/m^2)/(K/m)	21.7	24.2	26.7	28.2	31.3	33.1	36.9	40.2	43.6	51.5
Pr_ℓ	10.3	9.86	9.17	8.64	10.2	8.10	8.67	9.08		
Pr_g	0.81	0.83	0.81	0.86	0.87	0.91	1.01	1.17	1.56	
σ, mN/m	17.1	15.6	13.9	12.3	10.2	7.50	6.44	4.23	2.11	0.96
$\beta_{e,\ell}$, kK^{-1}	1.69	1.92	2.19	2.46	2.98	3.48	4.90	8.45	14.7	

Transport Properties of the Superheated Vapor

	Temperature, °C												
	−150	−100	−50	0	25	100	200	300	400	600	800	1 000	1 200
	Temperature, K												
Property (at low pressure)	123.15	173.15	223.15	273.15	298.15	373.15	473.15	573.15	673.15	873.15	1 073.15	1 273.15	1 473.15
Specific heat capacity, $c_{p,g}$ (kJ/kg K)	S	S	L	L	L	L	2.103	2.402	2.663	3.088	—	—	—
Thermal conductivity, λ_g[(W/m^2)/(K/m)]	S	S	L	L	L	L	0.030	0.043	(0.058)	(0.089)	—	—	—
Dynamic viscosity, η_g(10^{-5} Ns/m^2)	S	S	L	L	L	L	1.14	1.36	1.61	2.06	—	—	—

1-PENTANOL (AMYL ALCOHOL)

Chemical formula: $C_5H_{12}O$
Molecular weight: 88.15
Melting point: 195.0 K
Boiling point: 411 K

Critical temperature: 586 K
Critical pressure: 3.85 MPa
Critical density: (273) kg/m^3
Normal vapor density: 3.93 kg/m^3
(@ 0°C, 101.3 kPa)

Properties of the Liquid at Temperatures Below the Normal Boiling Point

Property	Temperature, °C: −150	−100	−75	−50	−25	0	20	50	100	150	200	250	300
	Temperature, K: 123.15	173.15	198.15	223.15	248.15	273.15	293.15	323.15	373.15	423.15	473.15	523.15	573.15
Density, ρ_l (kg/m^3)	S	S	(895)	(874)	(853)	829	814	(786)	(734)	V	V	V	V
Specific heat capacity, $c_{p,l}$ (kJ/kg K)	S	S	1.863	1.930	2.026	2.161	2.315	2.625	3.287	V	V	V	V
Thermal conductivity, λ_l [(W/m^2)/(K/m)]	S	S	0.171	0.167	0.163	0.159	0.156	0.151	0.142	V	V	V	V
Dynamic viscosity, η_l (10^{-5} Ns/m^2)	S	S	(20000)	(8000)	(2900)	1 110	510	200.1	60.3	V	V	V	V

Transport Properties of the Superheated Vapor

Property (at low pressure)	Temperature, °C: −150	−100	−50	0	25	100	200	300	400	600	800	1 000	1 200
	Temperature, K: 123.15	173.15	223.15	273.15	298.15	373.15	473.15	573.15	673.15	873.15	1 073.15	1 273.15	1 473.15
Specific heat capacity, $c_{p,g}$ (kJ/kg K)	S	S	L	L	L	L	2.138	2.451	2.722	3.135	3.477	—	—
Thermal conductivity, λ_g[(W/m^2)/(K/m)]	S	S	L	L	L	L	(0.028)	(0.041)	(0.053)	(0.084)	—	—	—
Dynamic viscosity, η_g(10^{-5} Ns/m^2)	S	S	L	L	L	L	(1.09)	(1.32)	(1.57)	(2.03)	—	—	—

1-HEXANOL (HEXYL ALCOHOL)

Chemical formula: $C_6H_{14}O$
Molecular weight: 102.18
Melting point: 229.2 K
Boiling point: 430.2 K

Critical temperature: 610 K
Critical pressure: 3.485 MPa
Critical density: 268 kg/m^3
Normal vapor density: 4.56 kg/m^3
(@ 0 °C, 101.3 kPa)

Properties of the Liquid at Temperatures Below the Normal Boiling Point

	Temperature, °C												
	−150	−100	−75	−50	−25	0	20	50	100	150	200	250	300
	Temperature, K												
Property	123.15	173.15	198.15	223.15	248.15	273.15	293.15	323.15	373.15	423.15	473.15	523.15	573.15
Density, ρ_l (kg/m^3)	S	S	S	S	852	833	820	(797)	(758)	(715)	V	V	V
Specific heat capacity, $c_{p,l}$ (kJ/kg K)	S	S	S	S	1.813	1.976	2.173	(2.57)	(3.10)	(3.46)	V	V	V
Thermal conductivity, λ_l [(W/m^2)/(K/m)]	S	S	S	S	0.150	0.146	0.143	0.140	0.131	0.128	V	V	V
Dynamic viscosity, η_l (10^{-5} Ns/m^2)	S	S	S	S	(3200)	1 079	542.7	216.0	79.4	36.7	V	V	V

Transport Properties of the Superheated Vapor

	Temperature, °C												
	−150	−100	−50	0	25	100	200	300	400	600	800	1 000	1 200
Property (at low pressure)	Temperature, K												
	123.15	173.15	223.15	273.15	298.15	373.15	473.15	573.15	673.15	873.15	1 073.15	1 273.15	1 473.15
Specific heat capacity, $c_{p,g}$ (kJ/kg K)	S	S	S	L	L	L	(2.174)	(2.484)	(2.754)	(3.188)	—	—	—
Thermal conductivity, λ_g[(W/m^2)/(K/m)]	S	S	S	L	L	L	(0.026)	(0.037)	(0.049)	(0.079)	—	—	—
Dynamic viscosity, η_g(10^{-5} Ns/m^2)	S	S	S	L	L	L	(1.04)	(1.29)	(1.53)	(2.01)	—	—	—

1-HEPTANOL (HEPTYL ALCOHOL)

Chemical formula: $C_7H_{16}O$
Molecular weight: 116.20
Melting point: 240.35 K
Boiling point: 449.05 K

Critical temperature: 641.9 K
Critical pressure: 3.04 MPa
Critical density: 267 kg/m^3
Normal vapor density: 5.19 kg/m^3
(@ 0°C, 101.3 kPa)

Properties of the Liquid at Temperatures Below the Normal Boiling Point

	Temperature, °C												
	−150	−100	−75	−50	−25	0	20	50	100	150	200	250	300
	Temperature, K												
Property	123.15	173.15	198.15	223.15	248.15	273.15	293.15	323.15	373.15	423.15	473.15	523.15	573.15
Density, ρ_l (kg/m^3)	S	S	S	S	(863)	839	822	795	756	722	V	V	V
Specific heat capacity, $c_{p,l}$ (kJ/kg K)	S	S	S	S	2.031	2.220	2.358	(2.91)	(3.46)	(3.54)	V	V	V
Thermal conductivity, λ_l [(W/m^2)/(K/m)]	S	S	S	S	0.145	0.140	0.137	0.133	(0.126)	(0.120)	V	V	V
Dynamic viscosity, η_l (10^{-5} Ns/m^2)	S	S	S	S	(4800)	(1490)	701.1	268.1	80.0	41.5	V	V	V

Transport Properties of the Superheated Vapor

	Temperature, °C												
	−150	−100	−50	0	25	100	200	300	400	600	800	1 000	1 200
Property (at low pressure)	Temperature, K												
	123.15	173.15	223.15	273.15	298.15	373.15	473.15	573.15	673.15	873.15	1 073.15	1 273.15	1 473.15
Specific heat capacity, $c_{p,g}$ (kJ/kg K)	S	S	S	L	L	L	2.197	2.510	2.782	3.225	—	—	—
Thermal conductivity, λ_g[(W/m^2)/(K/m)]	S	S	S	L	L	L	(0.027)	(0.038)	(0.051)	(0.079)	—	—	—
Dynamic viscosity, η_g(10^{-5} Ns/m^2)	S	S	S	L	L	L	(1.00)	(1.24)	(1.47)	(1.94)	—	—	—

1-OCTANOL (OCTYL ALCOHOL)

Chemical formula: $C_8H_{18}O$
Molecular weight: 130.23
Melting point: 257.7 K
Boiling point: 452.9 K

Critical temperature: 637 K
Critical pressure: 2.74 MPa
Critical density: 266 kg/m^3
Normal vapor density: 5.81 kg/m^3
(@ 0°C, 101.3 kPa)

Properties of the Liquid at Temperatures Below the Normal Boiling Point

	Temperature, °C												
	−150	−100	−75	−50	−25	0	20	50	100	150	200	250	300
	Temperature, K												
Property	123.15	173.15	198.15	223.15	248.15	273.15	293.15	323.15	373.15	423.15	473.15	523.15	573.15
Density, ρ_l (kg/m^3)	S	S	S	S	S	842	829	809	773	(734)	V	V	V
Specific heat capacity, $c_{p,l}$ (kJ/kg K)	S	S	S	S	S	2.156	2.219	(2.55)	(2.92)	(3.25)	V	V	V
Thermal conductivity, λ_l [(W/m^2)/(K/m)]	S	S	S	S	S	0.140	0.137	0.133	(0.126)	(0.119)	V	V	V
Dynamic viscosity, η_l (10^{-5} Ns/m^2)	S	S	S	S	S	1 980	893.3	321.7	106	46	V	V	V

Transport Properties of the Superheated Vapor

	Temperature, °C												
	− 150	− 100	− 50	0	25	100	200	300	400	600	800	1 000	1 200
Property (at low pressure)	Temperature, K												
	123.15	173.15	223.15	273.15	298.15	373.15	473.15	573.15	673.15	873.15	1 073.15	1 273.15	1 473.15
Specific heat capacity, $c_{p,g}$ (kJ/kg K)	S	S	S	L	L	L	2.217	2.536	2.810	3.242	—	—	—
Thermal conductivity, λ_g[(W/m^2)/(K/m)]	S	S	S	L	L	L	(0.027)	(0.039)	(0.052)	(0.080)	—	—	—
Dynamic viscosity, η_g(10^{-5} Ns/m^2)	S	S	S	L	L	L	(1.01)	(1.24)	(1.48)	(1.95)	—	—	—

ISOPROPANOL (ISOPROPYL ALCOHOL)

Chemical formula: $(CH_3)_2CHOH$
Molecular weight: 60.1
Melting point: 194.15 K
Normal boiling point: 355.65 K

Critical temperature: 508.75 K
Critical pressure: 5.370 MPa
Critical density: 274 kg/m^3
Normal vapor density: 2.68 kg/m^3
(@ 0 °C, 101.3 kPa)

Properties of the Liquid at Temperatures Below the Normal Boiling Point

	Temperature, °C												
	−150	−100	−75	−50	−25	0	20	50	100	150	200	250	300
	Temperature, K												
Property	123.15	173.15	198.15	223.15	248.15	273.15	293.15	323.15	373.15	423.15	473.15	523.15	573.15
Density, ρ_l (kg/m^3)	S	S	(844)	832	818	801	785	764	V	V	V	V	V
Specific heat capacity, $c_{p,l}$ (kJ/kg K)	S	S	1.830	1.930	2.073	2.265	2.496	2.956	V	V	V	V	V
Thermal conductivity, λ_l [(W/m^2)/(K/m)]	S	S	0.158	0.153	0.149	0.144	0.141	0.135	V	V	V	V	V
Dynamic viscosity, η_l (10^{-5} Ns/m^2)	S	S	(11000)	3 761	1 250	460.1	243.2	106.3	V	V	V	V	V

Properties of the Saturated Liquid and Vapor

T_{sat}, K	355.65	373	390	408	425	443	459	478	498	508
p_{sat}, kPa	101.3	200	380	580	925	1 425	2 025	3 039	4 052	5 369
ρ_ℓ, kg/m^3	732.3	712.7	683.0	660.0	630.1	597.4	566.0	514.8	460.5	288.0
ρ_g, kg/m^3	2.06	4.15	7.73	14.3	21.00	32.78	46.40	72.3	108.4	252
h_ℓ, kJ/kg	0.0	60.1	121.8	190.2	257.6	331.8	400.1	484.0		
h_g, kJ/kg	677.8	688.0	736.8	767.9	796.1	822.9	841.7	851.5		
$\Delta h_{g,\ell}$, kJ/kg	677.8	627.9	615.0	577.7	538.5	491.1	441.6	367.5	284.5	82.5
$c_{p,\ell}$, kJ/(kg K)	3.37	3.55	3.71	3.88	4.04	4.20	4.34	4.49		
$c_{p,g}$, kJ/(kg K)	1.63	1.71	1.80	1.94	2.15	2.37	2.83	3.97		
η_ℓ, μNs/m^2	502	376	295	230	184	147	122	93.5	72.5	
η_g, μNs/m^2	9.08	9.80	10.3	10.9	11.4	11.9	12.5	13.7	15.2	28.2
λ_ℓ, (mW/m^2)/(K/m)	131.1	127.5	124.3	122.8	120.1	117.3	115.8	113.2	110.7	107.7
λ_g, (mW/m^2)/(K/m)	19.8	22.2	24.6	27.1	29.3	31.8	34.9	38.9	42.3	47.1
Pr_ℓ	12.9	10.5	8.81	7.27	6.19	5.27	4.57	3.71		
Pr_g	0.75	0.75	0.75	0.78	0.84	0.89	1.01	1.40		
σ, mN/m	18.6	17.2	14.2	11.84	9.4	6.9	4.97	2.6	1.05	0.000
$\beta_{e,\ell}$, kK^{-1}	1.41	1.81	2.27	2.65	3.20	3.95	5.10	7.76	37.3	

Transport Properties of the Superheated Vapor

	Temperature, °C												
	−150	−100	−50	0	25	100	200	300	400	600	800	1 000	1 200
Property (at low pressure)	Temperature, K												
	123.15	173.15	223.15	273.15	298.15	373.15	473.15	573.15	673.15	873.15	1 073.15	1 273.15	1 473.15
Specific heat capacity, $c_{p,g}$ (kJ/kg K)	S	S	L	L	L	1.817	2.165	2.458	2.700	3.086	3.379	3.601	3.772
Thermal conductivity, λ_g[(W/m^2)/(K/m)]	S	S	L	L	L	0.022	0.034	(0.045)	(0.058)	(0.084)	—	—	—
Dynamic viscosity, η_g(10^{-5} Ns/m^2)	S	S	L	L	L	0.951	1.26	1.50	1.76	(2.15)	—	—	—

ISOBUTANOL (ISOBUTYL ALCOHOL)

Chemical formula: $C_4H_{10}O$
Molecular weight: 74.12
Melting point: 165.2 K
Boiling point: 381.0 K

Critical temperature: 547.7 K
Critical pressure: 4.30 MPa
Critical density: 272 kg/m^3
Normal vapor density: 3.31 kg/m^3
(@ 0°C, 101.3 kPa)

Properties of the Liquid at Temperatures Below the Normal Boiling Point

	Temperature, °C												
	–150	–100	–75	–50	–25	0	20	50	100	150	200	250	300
	Temperature, K												
Property	123.15	173.15	198.15	223.15	248.15	273.15	293.15	323.15	373.15	423.15	473.15	523.15	573.15
Density, ρ_l (kg/m^3)	S	(887)	(870)	(853)	(835)	817	806	794	732	V	V	V	V
Specific heat capacity, $c_{p,l}$ (kJ/kg K)	S	1.736	1.786	1.875	2.052	2.156	2.311	2.684	(3.40)	V	V	V	V
Thermal conductivity, λ_l [(W/m^2)/(K/m)]	S	0.157	0.152	0.147	0.142	0.137	0.133	0.127	0.119	V	V	V	V
Dynamic viscosity, η_l (10^{-5} Ns/m^2)	S	—	—	(13000)	2300.6	800	390.1	162.0	52.0	V	V	V	V

Transport Properties of the Superheated Vapor

	Temperature, °C												
	– 150	– 100	– 50	0	25	100	200	300	400	600	800	1 000	1 200
Property (at low pressure)	Temperature, K												
	123.15	173.15	223.15	273.15	298.15	373.15	473.15	573.15	673.15	873.15	1 073.15	1 273.15	1 473.15
Specific heat capacity, $c_{p,g}$ (kJ/kg K)	S	L	L	L	L	L	(2.087)	(2.415)	(2.707)	(3.189)	—	—	—
Thermal conductivity, λ_g[(W/m^2)/(K/m)]	S	L	L	L	L	L	(0.030)	(0.042)	(0.055)	(0.080)	—	—	—
Dynamic viscosity, η_g(10^{-5} Ns/m^2)	S	L	L	L	L	L	(1.15)	(1.39)	(1.61)	(2.02)	—	—	—

tent-BUTANOL (t-BUTYL ALCOHOL)

Chemical formula: $(CH_3)_3COH$
Molecular weight: 74.12
Melting point: 298.8 K
Normal boiling point: 355.6 K

Critical temperature: 506.2 K
Critical pressure: 3.970 MPa
Critical density: 270 kg/m^3
Normal vapor density: 3.31 kg/m^3 (@ 0°C, 101.3 kPa)

Properties of the Liquid at Temperatures Below the Normal Boiling Point

	Temperature, °C												
	−150	−100	−75	−50	−25	0	20	50	100	150	200	250	300
	Temperature, K												
Property	123.15	173.15	198.15	223.15	248.15	273.15	293.15	323.15	373.15	423.15	473.15	523.15	573.15
Density, ρ_l (kg/m^3)	S	S	S	S	S	S	S	750	V	V	V	V	V
Specific heat capacity, $c_{p,l}$ (kJ/kg K)	S	S	S	S	S	S	S	3.18	V	V	V	V	V
Thermal conductivity, λ_l [(W/m^2)/(K/m)]	S	S	S	S	S	S	S	0.106	V	V	V	V	V
Dynamic viscosity, η_l (10^{-5} Ns/m^2)	S	S	S	S	S	S	S	144	V	V	V	V	V

Properties of the Saturated Liquid and Vapor

T_{sat}, K	355.6	375	390	405	420	435	450	465	480	506.2
p_{sat}, kPa	101.3	207	322	483	779	1 010	1 516	1 896	2 619	3 970
ρ_ℓ, kg/m^3	710	688	670	647	621	596	567	533	487	270
ρ_g, kg/m^3	2.64	5.12	8.11	12.4	18.5	27.1	39.1	56.4	82.3	270
h_ℓ, kJ/kg	−182.0	−130.7	−78.4	−35.2	15.0	63.3	117.4	171.2	238.1	351.6
h_g, kJ/kg	324.6	355.0	378.0	400.2	421.1	440.1	456.5	468.5	472.6	351.6
$\Delta h_{g,\ell}$, kJ/kg	506.6	485.7	456.4	435.4	406.1	376.8	339.1	297.3	234.5	
$c_{p,\ell}$, kJ/(kg K)	2.90	3.06	3.19	3.34	3.47	3.62	3.79	4.01	4.38	
$c_{p,g}$, kJ/(kg K)	1.81	1.92	2.02	2.13	2.26	2.42	2.64	3.00	3.73	
η_ℓ, μNs/m^2	531	312	235	180	142	118	98	84	66	65
η_g, μNs/m^2	9.4	10.0	10.4	11.0	11.5	12.1	12.8	13.6	14.8	65
λ_ℓ, (mW/m^2)/(K/m)	109	104	100	96	92	88	83	77	70	53
λ_g, (mW/m^2)/(K/m)	17.9	19.8	21.4	23.1	24.9	26.9	29.1	31.6	34.5	53
Pr_ℓ	14.1	9.18	7.50	6.26	5.36	4.85	4.47	4.37	4.13	
Pr_g	0.95	0.97	0.98	1.01	1.04	1.09	1.16	1.29	1.59	
σ, mN/m	14.5	13.0	11.5	10.0	8.6	7.1	5.5	3.8	2.2	
$\beta_{e,\ell}$, kK^{-1}	1.84	2.09	2.32	3.27	3.47	3.85	4.57	5.98	9.13	

Transport Properties of the Superheated Vapor

	Temperature, °C												
	−150	−100	−50	0	25	100	200	300	400	600	800	1 000	1 200
Property (at low pressure)	Temperature, K												
	123.15	173.15	223.15	273.15	298.15	373.15	473.15	573.15	673.15	873.15	1 073.15	1 273.15	1 473.15
Specific heat capacity, $c_{p,g}$ (kJ/kg K)	S	S	S	S	S	1.832	2.186	2.488	2.743	3.142	—	—	—
Thermal conductivity, λ_g[(W/m^2)/(K/m)]	S	S	S	S	S	0.022	(0.035)	(0.048)	(0.062)	(0.089)	—	—	—
Dynamic viscosity, η_g(10^{-5} Ns/m^2)	S	S	S	S	S	1.02	(1.29)	(1.55)	(1.79)	(2.23)	—	—	—

ISOPENTANOL (3-METHYL-1-BUTANOL)

Chemical formula: $C_5H_{12}O$
Molecular weight: 88.15
Melting point: 155.95 K
Boiling point: 403.75 K

Critical temperature: 579.8 K
Critical pressure: 3.85 MPa
Critical density: (270) kg/m^3
Normal vapor density: 3.93 kg/m^3
(@ 0 °C, 101.3 kPa)

Properties of the Liquid at Temperatures Below the Normal Boiling Point

	Temperature, °C												
	−150	−100	−75	−50	−25	0	20	50	100	150	200	250	300
	Temperature, K												
Property	123.15	173.15	198.15	223.15	248.15	273.15	293.15	323.15	373.15	423.15	473.15	523.15	573.15
Density, ρ_l (kg/m^3)	S	(909)	(891)	(870)	(840)	824	810	(782)	(730)	V	V	V	V
Specific heat capacity, $c_{p,l}$ (kJ/kg K)	S	(1.76)	(1.81)	(1.88)	(2.00)	(2.16)	2.345	(2.66)	(3.17)	V	V	V	V
Thermal conductivity, λ_l [(W/m^2)/(K/m)]	S	0.157	0.152	1.147	0.141	0.137	0.133	0.128	0.120	V	V	V	V
Dynamic viscosity, η_l (10^{-5} Ns/m^2)	S	—	(12000)	(5000)	(2100)	860.2	430.1	185.1	62.0	V	V	V	V

1,3-PROPYLENE GLYCOL

Chemical formula: $C_3H_8O_2$
Molecular weight: 76.10
Melting point: 246.4 K
Boiling point: 487.6 K

Critical temperature: 658 K
Critical pressure: 5.98 MPa
Critical density: 327 kg/m^3
Normal vapor density: 3.40 kg/m^3
(@ 0 °C, 101.3 kPa)

Properties of the Liquid at Temperatures Below the Normal Boiling Point

	Temperature, °C												
	−150	−100	−75	−50	−25	0	20	50	100	150	200	250	300
	Temperature, K												
Property	123.15	173.15	198.15	223.15	248.15	273.15	293.15	323.15	373.15	423.15	473.15	523.15	573.15
Density, ρ_l (kg/m^3)	S	S	S	S	(1083)	(1066)	1 053	1033	1 003	964	(920)	V	V
Specific heat capacity, $c_{p,l}$ (kJ/kg K)	S	S	S	S	2.240	2.361	2.469	2.647	2.950	3.242	(3.51)	V	V
Thermal conductivity, λ_l [(W/m^2)/(K/m)]	S	S	S	S	0.193	0.198	0.201	0.203	0.203	0.195	0.179	V	V
Dynamic viscosity, η_l (10^{-5} Ns/m^2)	S	S	S	S	(97500)	17160	5 260	1 540	360	130	(64)	V	V

ETHYLENE GLYCOL

Chemical formula: C_2H_6O
Molecular weight: 62.07
Melting point: 260.2 K
Boiling point: 470.4 K

Critical temperature: 645 K
Critical pressure: 7.7 MPa
Critical density: (360) kg/m^3
Normal vapor density: 2.77 kg/m^3
(@ 0°C, 101.3 kPa)

Properties of the Liquid at Temperatures Below the Normal Boiling Point

Property	Temperature, °C												
	−150	−100	−75	−50	−25	0	20	50	100	150	200	250	300
	Temperature, K												
	123.15	173.15	198.15	223.15	248.15	273.15	293.15	323.15	373.15	423.15	473.15	523.15	573.15
Density, ρ_l (kg/m^3)	S	S	S	S	S	1 128	1 115	1091	1 055	1 016	V	V	V
Specific heat capacity, $c_{p,l}$ (kJ/kg K)	S	S	S	S	S	2.261	2.357	2.500	2.847	(2.94)	V	V	V
Thermal conductivity, λ_l [(W/m^2)/(K/m)]	S	S	S	S	S	0.254	0.256	0.260	0.265	(0.252)	V	V	V
Dynamic viscosity, η_l (10^{-5} Ns/m^2)	S	S	S	S	S	5 701	2 041	707	202	85.9	V	V	V

Transport Properties of the Superheated Vapor

Property (at low pressure)	Temperature, °C												
	− 150	− 100	− 50	0	25	100	200	300	400	600	800	1 000	1 200
	Temperature, K												
	123.15	173.15	223.15	273.15	298.15	373.15	473.15	573.15	673.15	873.15	1 073.15	1 273.15	1 473.15
Specific heat capacity, $c_{p,g}$ (kJ/kg K)	S	S	S	L	L	L	(1.826)	(2.057)	(2.260)	(2.590)	–	–	–
Thermal conductivity, λ_g[(W/m^2)/(K/m)]	S	S	S	L	L	L	(0.029)	(0.040)	(0.052)	(0.076)	–	–	–
Dynamic viscosity, η_g(10^{-5} Ns/m^2)	S	S	S	L	L	L	(1.31)	(1.59)	(1.86)	(2.35)	–	–	–

GLYCEROL

Chemical formula: $C_3H_8O_3$
Molecular weight: 92.09
Melting point: 291.0 K
Boiling point: 563.2 K

Critical temperature: 726 K
Critical pressure: 6.69 MPa
Critical density: 361 kg/m^3
Normal vapor density: 4.11 kg/m^3
(@ 0 °C, 101.3 kPa)

Properties of the Liquid at Temperatures Below the Normal Boiling Point

	Temperature, °C												
	−150	−100	−75	−50	−25	0	20	50	100	150	200	250	300
	Temperature, K												
Property	123.15	173.15	198.15	223.15	248.15	273.15	293.15	323.15	373.15	423.15	473.15	523.15	573.15
Density, ρ_l (kg/m^3)	S	S	S	S	S	S	1 260	1 242	1 209	154	1 090	(1007)	V
Specific heat capacity, $c_{p,l}$ (kJ/kg K)	S	S	S	S	S	S	2.366	2.512	2.805	3.06	3.34	(3.74)	V
Thermal conductivity, λ_l [(W/m^2)/(K/m)]	S	S	S	S	S	S	0.286	0.290	0.297	0.300	0.295	0.282	V
Dynamic viscosity, η_l (10^{-5} Ns/m^2)	S	S	S	S	S	S	149900	(18000)	1300	170	22.0	(3.0)	V

Transport Properties of the Superheated Vapor

	Temperature, °C												
	− 150	− 100	− 50	0	25	100	200	300	400	600	800	1 000	1 200
Property (at low pressure)	Temperature, K												
	123.15	173.15	223.15	273.15	298.15	373.15	473.15	573.15	673.15	873.15	1 073.15	1 273.15	1 473.15
Specific heat capacity, $c_{p,g}$ (kJ/kg K)	S	S	S	S	L	L	L	(2.15)	(2.29)	(2.53)	—	—	—
Thermal conductivity, λ_g[(W/m^2)/(K/m)]	S	S	S	S	L	L	L	(0.030)	(0.040)	(0.062)	—	—	—
Dynamic viscosity, η_g(10^{-5} Ns/m^2)	S	S	S	S	L	L	L	(1.42)	(1.66)	(2.16)	—	—	—

CYCLOHEXANOL

Chemical formula: $C_6H_{12}O$
Molecular weight: 100.16
Melting point: 298.0 K
Boiling point: 434.3 K

Critical temperature: 624.3 K
Critical pressure: 3.75 MPa
Critical density: (307) kg/m^3
Normal vapor density: 4.47 kg/m^3 (@ 0°C, 101.3 kPa)

Properties of the Liquid at Temperatures Below the Normal Boiling Point

	Temperature, °C												
	−150	−100	−75	−50	−25	0	20	50	100	150	200	250	300
	Temperature, K												
Property	123.15	173.15	198.15	223.15	248.15	273.15	293.15	323.15	373.15	423.15	473.15	523.15	573.15
Density, ρ_l (kg/m^3)	S	S	S	S	S	S	S	925	874	821	V	V	V
Specific heat capacity, $c_{p,l}$ (kJ/kg K)	S	S	S	S	S	S	S	(2.03)	(2.32)	(2.56)	V	V	V
Thermal conductivity, λ_l [(W/m^2)/(K/m)]	S	S	S	S	S	S	S	0.132	0.124	0.117	V	V	V
Dynamic viscosity, η_l (10^{-5} Ns/m^2)	S	S	S	S	S	S	S	1 181	185.1	68.9	V	V	V

Transport Properties of the Superheated Vapor

	Temperature, °C												
	−150	−100	−50	0	25	100	200	300	400	600	800	1 000	1 200
Property (at low pressure)	Temperature, K												
	123.15	173.15	223.15	273.15	298.15	373.15	473.15	573.15	673.15	873.15	1 073.15	1 273.15	1 473.15
Specific heat capacity, $c_{p,g}$ (kJ/kg K)	S	S	S	S	L	L	(2.024)	(2.384)	(2.691)	(3.169)	—	—	—
Thermal conductivity, λ_g[(W/m^2)/(K/m)]	S	S	S	S	L	L	(0.029)	(0.041)	(0.053)	(0.077)	—	—	—
Dynamic viscosity, η_g(10^{-5} Ns/m^2)	S	S	S	S	L	L	(1.11)	(1.35)	(1.57)	(1.99)	—	—	—

BENZYL ALCOHOL

Chemical formula: C_7H_8O
Molecular weight: 108.14
Melting point: 257.8 K
Boiling point: 478.6 K

Critical temperature: 677 K
Critical pressure: 4.66 MPa
Critical density: 324 kg/m^3
Normal vapor density: 4.82 kg/m^3 (@ 0°C, 101.3 kPa)

Properties of the Liquid at Temperatures Below the Normal Boiling Point

	Temperature, °C												
	−150	−100	−75	−50	−25	0	20	50	100	150	200	250	300
	Temperature, K												
Property	123.15	173.15	198.15	223.15	248.15	273.15	293.15	323.15	373.15	423.15	473.15	523.15	573.15
Density, ρ_l (kg/m^3)	S	S	S	S	S	1 061	1 045	1 022	983	940	893	V	V
Specific heat capacity, $c_{p,l}$ (kJ/kg K)	S	S	S	S	S	1.860	1.972	2.135	(2.37)	(2.53)	(2.60)	V	V
Thermal conductivity, λ_l [(W/m^2)/(K/m)]	S	S	S	S	S	(0.161)	0.160	(0.158)	0.156	0.152	(0.151)	V	V
Dynamic viscosity, η_l (10^{-5} Ns/m^2)	S	S	S	S	S	(1200)	558.4	257.4	102	56	35	V	V

PHENOL

Chemical formula: C_6H_5OH
Molecular weight: 94.1
Melting point: 313.90 K
Normal boiling point: 454.95 K

Critical temperature: 693.2 K
Critical pressure: 6.130 MPa
Critical density: 435.7 kg/m^3
Normal vapor density: 4.20 kg/m^3 (@ 0 °C, 101.3 kPa)

Properties of the Liquid at Temperatures Below the Normal Boiling Point

	Temperature, °C												
	−150	−100	−75	−50	−25	0	20	50	100	150	200	250	300
	Temperature, K												
Property	123.15	173.15	198.15	223.15	248.15	273.15	293.15	323.15	373.15	423.15	473.15	523.15	573.15
Density, ρ_l (kg/m^3)	S	S	S	S	S	S	S	1 050	973	931	V	V	V
Specific heat capacity, $c_{p,l}$ (kJ/kg K)	S	S	S	S	S	S	S	2.244	2.382	(2.42)	V	V	V
Thermal conductivity, λ_l [(W/m^2)/(K/m)]	S	S	S	S	S	S	S	0.156	0.135	(0.130)	V	V	V
Dynamic viscosity, η_l (10^{-5} Ns/m^2)	S	S	S	S	S	S	S	342.1	105.0	67.0	V	V	V

Properties of the Saturated Liquid and Vapor

T_{sat}, K	455	480	505	530	555	580	605	635	665	693.15
p_{sat}, kPa	101.3	216	404	693	1 100	1 650	2 360	3 410	4 720	6 129
ρ_ℓ, kg/m^3	955	932	905	877	851	809	772	713	636	436
ρ_g, kg/m^3	2.60	5.36	9.88	16.9	27.2	41.9	62.9	101	170	436
h_ℓ, kJ/kg	−153	−93	−41	20	80	147	206	290	399	514
h_g, kJ/kg	336	374	411	447	482	515	545	575	587	514
$\Delta h_{g,\ell}$, kJ/kg	489	467	452	427	402	368	339	285	188	
$c_{p,\ell}$, kJ/(kg K)	2.55	2.61	2.66	2.76	2.85	3.01	3.10	3.43	3.77	
$c_{p,g}$, kJ/(kg K)	1.63	1.74	1.85	1.95	2.06	2.22	2.44	2.94	5.10	
η_ℓ, μNs/m^2	351	256	219	166	137	113	90.1	75.2	65.3	45.3
η_g, μNs/m^2	12.8	13.5	14.3	15.2	16.2	17.3	18.6	20.2	23.4	45.3
λ_ℓ, (mW/m^2)/(K/m)	175	170	166	162	157	154	149	141	130	92
λ_g, (mW/m^2)/(K/m)	28.9	31.6	34.8	38.4	42.4	46.8	51.4	56.2	64.3	92
Pr_ℓ	5.11	3.93	3.51	2.83	2.59	2.21	1.87	1.57	1.89	
Pr_g	0.72	0.74	0.76	0.77	0.79	0.82	0.88	1.07	1.86	
σ, mN/m	24.5	20.9	18.2	14.6	12.4	10.2	7.6	5.3	2.7	
$\beta_{e,\ell}$, kK^{-1}	0.91	1.05	1.27	1.34	1.83	2.41	3.25	4.72	15.6	

Transport Properties of the Superheated Vapor

	Temperature, °C												
	−150	−100	−50	0	25	100	200	300	400	600	800	1 000	1 200
	Temperature, K												
Property (at low pressure)	123.15	173.15	223.15	273.15	298.15	373.15	473.15	573.15	673.15	873.15	1 073.15	1 273.15	1 473.15
Specific heat capacity, $c_{p,g}$ (kJ/kg K)	S	S	S	S	S	L	1.644	1.886	2.078	2.328	—	—	—
Thermal conductivity, λ_g[(W/m^2)/(K/m)]	S	S	S	S	S	L	(0.025)	(0.035)	(0.045)	(0.065)	—	—	—
Dynamic viscosity, η_g(10^{-5} Ns/m^2)	S	S	S	S	S	L	(1.17)	(1.43)	(1.67)	(2.13)	—	—	—

o-CRESOL

Chemical formula: C_7H_8O
Molecular weight: 108.14
Melting point: 304.1 K
Boiling point: 464.2 K

Critical temperature: 697.6 K
Critical pressure: 5.01 MPa
Critical density: 384 kg/m^3
Normal vapor density: 4.83 kg/m^3
(@ 0°C, 101.3 kPa)

Properties of the Liquid at Temperatures Below the Normal Boiling Point

	Temperature, °C												
	−150	−100	−75	−50	−25	0	20	50	100	150	200	250	300
	Temperature, K												
Property	123.15	173.15	198.15	223.15	248.15	273.15	293.15	323.15	373.15	423.15	473.15	523.15	573.15
Density, ρ_l (kg/m^3)S	S	S	S	S	S	S	S	1 020	979	934	V	V	V
Specific heat capacity, $c_{p,l}$ (kJ/kg K)	S	S	S	S	S	S	S	2.20	2.28	2.35	V	V	V
Thermal conductivity, λ_l [(W/m^2)/(K/m)]	S	S	S	S	S	S	S	0.153	0.143	0.138	V	V	V
Dynamic viscosity, η_l (10^{-5} Ns/m^2)	S	S	S	S	S	S	S	300	100.0	50.4	V	V	V

Transport Properties of the Superheated Vapor

	Temperature, °C												
	−150	−100	−50	0	25	100	200	300	400	600	800	1 000	1 200
Property (at low pressure)	Temperature, K												
	123.15	173.15	223.15	273.15	298.15	373.15	473.15	573.15	673.15	873.15	1 073.15	1 273.15	1 473.15
Specific heat capacity, $c_{p,g}$ (kJ/kg K)	S	S	S	S	S	L	1.73	1.98	2.18	2.48	2.69	2.84	2.96
Thermal conductivity, λ_g[(W/m^2)/(K/m)]	S	S	S	S	S	L	(0.023)	(0.033)	(0.042)	(0.061)	—	—	—
Dynamic viscosity, η_g(10^{-5} Ns/m^2)	S	S	S	S	S	L	(1.05)	(1.27)	(1.49)	(1.93)	—	—	—

m-CRESOL

Chemical formula: C_7H_8O
Molecular weight: 108.14
Melting point: 285.4 K
Boiling point: 475.4 K

Critical temperature: 705.8 K
Critical pressure: 4.56 MPa
Critical density: 346 kg/m^3
Normal vapor density: 4.83 kg/m^3
(@ 0 °C, 101.3 kPa)

Properties of the Liquid at Temperatures Below the Normal Boiling Point

	Temperature, °C												
	−150	−100	−75	−50	−25	0	20	50	100	150	200	250	300
	Temperature, K												
Property	123.15	173.15	198.15	223.15	248.15	273.15	293.15	323.15	373.15	423.15	473.15	523.15	573.15
Density, ρ_l (kg/m^3)	S	S	S	S	S	S	1 034	1 009	973	(930)	(882)	V	V
Specific heat capacity, $c_{p,l}$ (kJ/kg K)	S	S	S	S	S	S	2.010	2.158	2.303	(2.40)	(2.51)	V	V
Thermal conductivity, λ_l [(W/m^2)/(K/m)]	S	S	S	S	S	S	0.150	0.148	0.143	0.135	(0.130)	V	V
Dynamic viscosity, η_l (10^{-5} Ns/m^2)	S	S	S	S	S	S	1 691	420.1	120.0	57.4	34.7	V	V

Transport Properties of the Superheated Vapor

	Temperature, °C												
	−150	−100	−50	0	25	100	200	300	400	600	800	1 000	1 200
Property	Temperature, K												
(at low pressure)	123.15	173.15	223.15	273.15	298.15	373.15	473.15	573.15	673.15	873.15	1 073.15	1 273.15	1 473.15
Specific heat capacity, $c_{p,g}$ (kJ/kg K)	S	S	S	S	L	L	L	1.97	2.18	2.47	2.69	2.84	2.96
Thermal conductivity, λ_g[(W/m^2)/(K/m)]	S	S	S	S	L	L	L	(0.033)	(0.042)	(0.061)	—	—	—
Dynamic viscosity, η_g(10^{-5} Ns/m^2)	S	S	S	S	L	L	L	(1.25)	(1.47)	(1.88)	—	—	—

p-CRESOL

Chemical formula: $C_7H_{11}O$
Molecular weight: 108.14
Melting point: 307.9 K
Boiling point: 475.1 K

Critical temperature: 699.2 K
Critical pressure: 5.15 MPa
Critical density: 391 kg/m^3
Normal vapor density: 4.83 kg/m^3
(@ 0°C, 101.3 kPa)

Properties of the Liquid at Temperatures Below the Normal Boiling Point

Property	Temperature, °C: −150	−100	−75	−50	−25	0	20	50	100	150	200	250	300
	Temperature, K: 123.15	173.15	198.15	223.15	248.15	273.15	293.15	323.15	373.15	423.15	473.15	523.15	573.15
Density, ρ_l (kg/m^3)	S	S	S	S	S	S	S	1 011	973	931	(883)	V	V
Specific heat capacity, $c_{p,l}$ (kJ/kg K)	S	S	S	S	S	S	S	2.162	2.310	(2.43)	(2.50)	V	V
Thermal conductivity, λ_l [(W/m^2)/(K/m)]	S	S	S	S	S	S	S	0.141	0.135	(0.130)	(0.126)	V	V
Dynamic viscosity, η_l (10^{-5} Ns/m^2)	S	S	S	S	S	S	S	465.1	130.0	61.2	37.1	V	V

Transport Properties of the Superheated Vapor

Property (at low pressure)	Temperature, °C: −150	−100	−50	0	25	100	200	300	400	600	800	1 000	1 200
	Temperature, K: 123.15	173.15	223.15	273.15	298.15	373.15	473.15	573.15	673.15	873.15	1 073.15	1 273.15	1 473.15
Specific heat capacity, $c_{p,g}$ (kJ/kg K)	S	S	S	S	S	L	L	1.97	2.18	2.47	2.69	2.84	2.96
Thermal conductivity, λ_g[(W/m^2)/(K/m)]	S	S	S	S	S	L	L	(0.033)	(0.042)	(0.061)	—	—	—
Dynamic viscosity, η_g(10^{-5} Ns/m^2)	S	S	S	S	S	L	L	(1.25)	(1.47)	(1.88)	—	—	—

METHYL FORMATE

Chemical formula: $C_2H_4O_2$
Molecular weight: 60.05
Melting point: 174.2 K
Boiling point: 304.9 K

Critical temperature: 487.2 K
Critical pressure: 6.004 MPa
Critical density: 349 kg/m^3
Normal vapor density: 2.68 kg/m^3
(@ 0 °C, 101.3 kPa)

Properties of the Liquid at Temperatures Below the Normal Boiling Point

	Temperature, °C												
	−150	−100	−75	−50	−25	0	20	50	100	150	200	250	300
	Temperature, K												
Property	123.15	173.15	198.15	223.15	248.15	273.15	293.15	323.15	373.15	423.15	473.15	523.15	573.15
Density, ρ_l (kg/m^3)	S	S	(1.102)	(1.069)	(1.036)	1 003	975	V	V	V	V	V	V
Specific heat capacity, $c_{p,l}$ (kJ/kg K)	S	S	(1.80)	(1.84)	(1.89)	(1.95)	2.026	V	V	V	V	V	V
Thermal conductivity, λ_l [(W/m^2)/(K/m)]	S	S	(0.228)	0.217	0.206	0.195	0.186	V	V	V	V	V	V
Dynamic viscosity, η_l (10^{-5} Ns/m^2)	S	S	(115)	(83)	(59)	43.1	34.5	V	V	V	V	V	V

Transport Properties of the Superheated Vapor

	Temperature, °C												
	−150	−100	−50	0	25	100	200	300	400	600	800	1 000	1 200
	Temperature, K												
Property (at low pressure)	123.15	173.15	223.15	273.15	298.15	373.15	473.15	573.15	673.15	873.15	1 073.15	1 273.15	1 473.15
Specific heat capacity, $c_{p,g}$ (kJ/kg K)	S	S	L	L	L	(1.411)	(1.591)	(1.755)	(1.900)	(2.133)	—	—	—
Thermal conductivity, λ_g[(W/m^2)/(K/m)]	S	S	L	L	L	(0.020)	(0.030)	(0.040)	(0.050)	(0.070)	—	—	—
Dynamic viscosity, η_g(10^{-5} Ns/m^2)	S	S	L	L	L	(1.18)	(1.49)	(1.79)	(2.06)	(2.56)	—	—	—

ETHYL FORMATE

Chemical formula: $C_3H_6O_2$
Molecular weight: 740.08
Melting point: 193.8 K
Boiling point: 327.4 K

Critical temperature: 508.4 K
Critical pressure: 4.74 MPa
Critical density: 323 kg/m^3
Normal vapor density: 3.31 kg/m^3
(@ 0 °C, 101.3 kPa)

Properties of the Liquid at Temperatures Below the Normal Boiling Point

	Temperature, °C												
	−150	−100	−75	−50	−25	0	20	50	100	150	200	250	300
	Temperature, K												
Property	123.15	173.15	198.15	223.15	248.15	273.15	293.15	323.15	373.15	423.15	473.15	523.15	573.15
Density, ρ_l (kg/m^3)	S	S	(1020)	(998)	(976)	948	923	883	V	V	V	V	V
Specific heat capacity, $c_{p,l}$ (kJ/kg K)	S	S	(1.72)	(1.74)	(1.76)	(1.80)	1.989	2.177	V	V	V	V	V
Thermal conductivity, λ_l [(W/m^2)/(K/m)]	S	S	0.195	0.186	0.177	0.168	0.161	0.152	V	V	V	V	V
Dynamic viscosity, η_l (10^{-5} Ns/m^2)	S	S	(115)	(86)	(65)	50.5	40.8	30.6	V	V	V	V	V

Transport Properties of the Superheated Vapor

	Temperature, °C												
	-150	-100	-50	0	25	100	200	300	400	600	800	1 000	1 200
Property (at low pressure)	Temperature, K												
	123.15	173.15	223.15	273.15	298.15	373.15	473.15	573.15	673.15	873.15	1 073.15	1 273.15	1 473.15
Specific heat capacity, $c_{p,g}$ (kJ/kg K)	S	S	L	L	L	(1.474)	(1.703)	(1.908)	(2.087)	(2.373)	—	—	—
Thermal conductivity, λ_g[(W/m^2)/(K/m)]	S	S	L	L	L	(0.019)	(0.028)	(0.039)	(0.049)	(0.070)	—	—	—
Dynamic viscosity, η_g(10^{-5} Ns/m^2)	S	S	L	L	L	(1.06)	(1.34)	(1.61)	(1.86)	(2.32)	—	—	—

PROPYL FORMATE

Chemical formula: $C_4H_8O_2$
Molecular weight: 88.11
Melting point: 180.3 K
Boiling point: 353.7 K

Critical temperature: 538 K
Critical pressure: 4.06 MPa
Critical density: 309 kg/m^3
Normal vapor density: 3.93 kg/m^3
(@ 0 °C, 101.3 kPa)

Properties of the Liquid at Temperatures Below the Normal Boiling Point

	Temperature, °C												
	−150	−100	−75	−50	−25	0	20	50	100	150	200	250	300
	Temperature, K												
Property	123.15	173.15	198.15	223.15	248.15	273.15	293.15	323.15	373.15	423.15	473.15	523.15	573.15
Density, ρ_l (kg/m^3)	S	S	(1013)	(985)	(957)	929	906	871	V	V	V	V	V
Specific heat capacity, $c_{p,l}$ (kJ/kg K)	S	S	(1.91)	(1.92)	(1.93)	(1.96)	(1.98)	(2.05)	V	V	V	V	V
Thermal conductivity, λ_l [(W/m^2)/(K/m)]	S	S	0.175	0.167	0.160	0.153	0.147	0.139	V	V	V	V	V
Dynamic viscosity, η_l (10^{-5} Ns/m^2)	S	S	—	—	(90)	66.7	51.5	37.0	V	V	V	V	V

METHYL ACETATE

Chemical formula: $CH_3CO_2CH_2$
Molecular weight: 74.08
Melting point: 174.45 K
Normal boiling point: 330.3 K

Critical temperature: 506.8 K
Critical pressure: 4.687 MPa
Critical density: 325 kg/m^3
Normal vapor density: 3.30 kg/m^3
(@ 0 °C, 101.3 kPa)

Properties of the Liquid at Temperatures Below the Normal Boiling Point

	Temperature, °C												
	−150	−100	−75	−50	−25	0	20	50	100	150	200	250	300
	Temperature, K												
Property	123.15	173.15	198.15	223.15	248.15	273.15	293.15	323.15	373.15	423.15	473.15	523.15	573.1
Density, ρ_l (kg/m^3)	S	S	(1045)	1 018	989	959	934	894	V	V	V	V	V
Specific heat capacity, $c_{p,l}$ (kJ/kg K)	S	S	(1.757)	(1.822)	(1.887)	1.953	2.119	(2.18)	V	V	V	V	V
Thermal conductivity, λ_l [(W/m^2)/(K/m)]	S	S	0.190	0.182	0.173	0.164	0.157	0.148	V	V	V	V	V
Dynamic viscosity, η_l (10^{-5} Ns/m^2)	S	S	(118)	(87)	(64)	47.7	38.9	28.6	V	V	V	V	V

Properties of the Saturated Liquid and Vapor

T_{sat}, K	331	350	370	390	410	430	450	470	490	506.8
p_{sat}, kPa	101.3	200	359	537	854	1 344	1 930	2 688	3 723	4 687
ρ_ℓ, kg/m^3	875	850	820	780	750	715	680	620	540	325
ρ_g, kg/m^3	2.83	5.16	8.87	14.5	22.7	34.6	52.2	79.3	127.8	325
h_ℓ, kJ/kg	−173.4	−136.1	−96.5	−57.7	−17.8	26.7	76.4	126.7	196.4	254.6
h_g, kJ/kg	228.5	249.1	269.8	289.8	308.8	326.0	340.2	348.6	342.9	254.6
$\Delta h_{g,\ell}$, kJ/kg	401.9	385.2	366.3	347.5	326.6	297.3	263.8	221.9	146.5	
$c_{p,\ell}$, kJ/(kg K)	1.92	1.99	2.08	2.18	2.32	2.46	2.65	2.94	3.58	
$c_{p,g}$, kJ/(kg K)	1.19	1.25	1.35	1.45	1.57	1.72	1.95	2.38	3.86	
η_ℓ, μNs/m^2	260	225	192	168	145	121	99	80	62	56
η_g, μNs/m^2	8.9	9.5	10.1	10.7	11.4	12.1	13.1	14.1	16.1	56
λ_ℓ, (mW/m^2)/(K/m)	157	146	133	122	110	98	86	74	61	50.3
λ_g, (mW/m^2)/(K/m)	14.2	15.7	17.5	19.5	21.7	24.1	27.0	30.4	35.3	50.3
Pr_ℓ	3.18	3.07	3.00	3.00	3.04	3.05	3.06	3.18	3.64	
Pr_g	0.75	0.76	0.78	0.80	0.82	0.86	0.95	1.10	1.76	
σ, mN/m	19.4	17.0	14.7	12.0	9.4	6.9	4.6	2.4	1.1	
$\beta_{e,\ell}$, kK^{-1}	1.64	1.84	2.11	2.46	2.95	3.67	4.86	7.22	14.4	

Transport Properties of the Superheated Vapor

	Temperature, °C												
	−150	−100	−50	0	25	100	200	300	400	600	800	1 000	1 200
	Temperature, K												
Property (at low pressure)	123.15	173.15	223.15	273.15	298.15	373.15	473.15	573.15	673.15	873.15	1 073.15	1 273.15	1 473.15
Specific heat capacity, $c_{p,g}$ (kJ/kg K)	S	S	L	L	L	1.480	1.699	1.897	2.076	2.387	—	—	—
Thermal conductivity, λ_g[(W/m^2)/(K/m)]	S	S	L	L	L	0.019	(0.028)	(0.039)	(0.049)	(0.071)	—	—	—
Dynamic viscosity, η_g(10^{-5} Ns/m^2)	S	S	L	L	L	0.960	1.23	1.56	(1.88)	(2.37)	—	—	—

ETHYL ACETATE

Chemical formula: $CH_3CO_2C_2H_5$
Molecular weight: 88.10
Melting point: 189.55 K
Normal boiling point: 350.25 K

Critical temperature: 523.25 K
Critical pressure: 3.832 MPa
Critical density: 307.7 kg/m^3
Normal vapor density: 3.93 kg/m^3
(@ 0°C, 101.3 kPa)

Properties of the Liquid at Temperatures Below the Normal Boiling Point

	Temperature, °C												
	−150	−100	−75	−50	−25	0	20	50	100	150	200	250	300
	Temperature, K												
Property	123.15	173.15	198.15	223.15	248.15	273.15	293.15	323.15	373.15	423.15	473.15	523.15	573.15
Density, ρ_l (kg/m^3)	S	S	(999)	(976)	(951)	924	901	864	V	V	V	V	V
Specific heat capacity, $c_{p,l}$ (kJ/kg K)	S	S	1.821	1.821	1.838	1.880	1.922	1.968	V	V	V	V	V
Thermal conductivity, λ_l [(W/m^2)/(K/m)]	S	S	0.171	0.164	0.156	0.150	0.144	0.135	V	V	V	V	V
Dynamic viscosity, η_l (10^{-5} Ns/m^2)	S	S	(150)	(109)	(80)	58.1	45.2	34.5	V	V	V	V	V

Properties of the Saturated Liquid and Vapor

T_{sat}, K	350.25	370	390	410	430	450	470	490	510	523.25
p_{sat}, kPa	101.3	193	310	510	792	1 172	1 655	2 275	3 172	3 832
ρ_ℓ, kg/m^3	830	800	770	740	705	670	625	570	475	307.7
ρ_g, kg/m^3	3.20	5.63	9.45	15.1	23.4	35.5	54.2	81.3	134.5	307.7
h_ℓ, kJ/kg	−89.7	−47.3	−6.7	35.7	81.3	131.7	182.0	241.7	315.8	362.8
h_g, kJ/kg	274.6	300.2	326.2	351.8	376.5	399.6	420.6	434.3	433.0	362.8
$\Delta h_{g,\ell}$, kJ/kg	364.3	347.5	332.9	316.1	295.2	267.9	238.6	192.6	117.2	
$c_{p,\ell}$, kJ/(kg K)	2.10	2.17	2.28	2.36	2.50	2.64	2.82	3.09	3.76	
$c_{p,g}$, kJ/(kg K)	1.46	1.54	1.63	1.73	1.85	2.00	2.25	2.68	4.52	
η_ℓ, μNs/m^2	255	221	193	158	134	111	90	72	56	57
η_g, μNs/m^2	8.9	9.5	10.1	10.7	11.4	12.2	13.0	14.3	16.4	57
λ_ℓ, (mW/m^2)/(K/m)	125	118	111	104	98	92	85	77	66	48.2
λ_g, (mW/m^2)/(K/m)	15.8	17.4	19.3	21.5	23.5	25.8	28.5	32.0	36.5	48.2
Pr_ℓ	4.28	4.08	3.96	3.59	3.42	3.19	2.99	2.89	3.19	
Pr_g	0.82	0.84	0.85	0.86	0.90	0.95	1.03	1.20	2.03	
σ, mN/m	17.4	15.0	12.6	10.0	7.8	5.8	3.9	2.3	0.74	
$\beta_{e,\ell}$, kK^{-1}	1.65	1.86	2.14	2.50	3.02	3.79	5.17	7.80	17.4	

Transport Properties of the Superheated Vapor

	Temperature, °C												
	−150	−100	−50	0	25	100	200	300	400	600	800	1 000	1 200
Property	Temperature, K												
(at low pressure)	123.15	173.15	223.15	273.15	298.15	373.15	473.15	573.15	673.15	873.15	1 073.15	1 273.15	1 473.15
Specific heat capacity, $c_{p,g}$ (kJ/kg K)	S	S	L	L	L	1.511	1.729	2.00	2.26	2.73	—	—	—
Thermal conductivity, λ_g[(W/m^2)/(K/m)]	S	S	L	L	L	0.018	0.030	(0.038)	(0.051)	(0.072)	—	—	—
Dynamic viscosity, η_g(10^{-5} Ns/m^2)	S	S	L	L	L	0.943	1.22	1.46	1.70	2.24	—	—	—

PROPYL ACETATE

Chemical formula: $C_5H_{10}O_2$
Molecular weight: 102.13
Melting point: 178.0 K
Boiling point: 374.8 K

Critical temperature: 549.4 K
Critical pressure: 3.33 MPa
Critical density: 296 kg/m^3
Normal vapor density: 4.56 kg/m^3
(@ 0 °C, 101.3 kPa)

Properties of the Liquid at Temperatures Below the Normal Boiling Point

	Temperature, °C												
	−150	−100	−75	−50	−25	0	20	50	100	150	200	250	300
	Temperature, K												
Property	123.15	173.15	198.15	223.15	248.15	273.15	293.15	323.15	373.15	423.15	473.15	523.15	573.15
Density, ρ_l (kg/m^3)	S	S	(979)	(956)	(933)	910	888	855	796	V	V	V	V
Specific heat capacity, $c_{p,l}$ (kJ/kg K)	S	S	1.78	1.79	1.82	1.87	1.91	1.99	2.13	V	V	V	V
Thermal conductivity, λ_l [(W/m^2)/(K/m)]	S	S	(0.162)	0.156	0.150	0.143	0.138	0.132	0.119	V	V	V	V
Dynamic viscosity, η_l (10^{-5} Ns/m^2)	S	S	(285)	(182)	(118)	77.1	53.7	39.0	26.1	V	V	V	V

Transport Properties of the Superheated Vapor

	Temperature, °C												
	−150	−100	−50	0	25	100	200	300	400	600	800	1 000	1 200
Property (at low pressure)	Temperature, K												
	123.15	173.15	223.15	273.15	298.15	373.15	473.15	573.15	673.15	873.15	1 073.15	1 273.15	1 473.15
Specific heat capacity, $c_{p,g}$ (kJ/kg K)	S	S	L	L	L	L	(1.828)	(2.084)	(2.304)	(2.652)	—	—	—
Thermal conductivity, λ_g[(W/m^2)/(K/m)]	S	S	L	L	L	L	(0.026)	(0.037)	(0.049)	(0.078)	—	—	—
Dynamic viscosity, η_g(10^{-5} Ns/m^2)	S	S	L	L	L	L	(1.15)	(1.38)	(1.61)	(2.03)	—	—	—

METHYL PROPIONATE

Chemical formula: $C_4H_8O_2$
Molecular weight: 88.11
Melting point: 185.7 K
Boiling point: 353.0 K

Critical temperature: 530.6 K
Critical pressure: 4.00 MPa
Critical density: 312 kg/m^3
Normal vapor density: 3.93 kg/m^3
(@ 0 °C, 101.3 kPa)

Properties of the Liquid at Temperatures Below the Normal Boiling Point

	Temperature, °C												
	−150	−100	−75	−50	−25	0	20	50	100	150	200	250	300
	Temperature, K												
Property	123.15	173.15	198.15	223.15	248.15	273.15	293.15	323.15	373.15	423.15	473.15	523.15	573.15
Density, ρ_l (kg/m^3)	S	S	(1026)	(997)	(968)	939	915	879	V	V	V	V	V
Specific heat capacity, $c_{p,l}$ (kJ/kg K)	S	S	(1.79)	(1.82)	(1.87)	(1.92)	(1.96)	(2.04)	V	V	V	V	V
Thermal conductivity, λ_l [(W/m^2)/(K/m)]	S	S	0.174	0.167	0.159	0.152	0.146	0.138	V	V	V	V	V
Dynamic viscosity, η_l (10^{-5} Ns/m^2)	S	S	—	—	(110)	77.3	58.6	41.6	V	V	V	V	V

ETHYL PROPIONATE

Chemical formula: $C_5H_{10}O_2$
Molecular weight: 102.13
Melting point: 199.3 K
Boiling point: 372.0 K

Critical temperature: 546 K
Critical pressure: 3.36 MPa
Critical density: 296 kg/m^3
Normal vapor density: 4.56 kg/m^3 (@ 0°C, 101.3 kPa)

Properties of the Liquid at Temperatures Below the Normal Boiling Point

	Temperature, °C												
	−150	−100	−75	−50	−25	0	20	50	100	150	200	250	300
	Temperature, K												
Property	123.15	173.15	198.15	223.15	248.15	273.15	293.15	323.15	373.15	423.15	473.15	523.15	573.15
Density, ρ_l (kg/m^3)	S	S	S	(960)	(935)	912	880	856	V	V	V	V	V
Specific heat capacity, $c_{p,l}$ (kJ/kg K)	S	S	S	(1.78)	(1.83)	(1.88)	1.948	2.034	V	V	V	V	V
Thermal conductivity, λ_l [(W/m^2)/(K/m)]	S	S	S	0.156	0.149	0.143	0.138	0.130	V	V	V	V	V
Dynamic viscosity, η_l (10^{-5} Ns/m^2)	S	S	S	(120)	(81)	58.4	45.6	33.4	V	V	V	V	V

Transport Properties of the Superheated Vapor

	Temperature, °C												
	−150	−100	−50	0	25	100	200	300	400	600	800	1 000	1 200
Property (at low pressure)	Temperature, K												
	123.15	173.15	223.15	273.15	298.15	373.15	473.15	573.15	673.15	873.15	1 073.15	1 273.15	1 473.15
Specific heat capacity, $c_{p,g}$ (kJ/kg K)	S	S	L	L	L	(1.541)	(1.828)	(2.083)	(2.304)	(2.652)	—	—	—
Thermal conductivity, λ_g[(W/m^2)/(K/m)]	S	S	L	L	L	(0.017)	(0.026)	(0.037)	(0.047)	(0.068)	—	—	—
Dynamic viscosity, η_g(10^{-5} Ns/m^2)	S	S	L	L	L	(0.91)	(1.16)	(1.39)	(1.62)	(2.03)	—	—	—

PROPYL PROPIONATE

Chemical formula: $C_6H_{12}O_2$
Molecular weight: 116.16
Melting point: 197.3 K
Boiling point: 395.7 K

Critical temperature: 578 K
Critical pressure: 3.02 MPa
Critical density: (290) kg/m^3
Normal vapor density: 5.18 kg/m^3
(@ 0 °C, 101.3 kPa)

Properties of the Liquid at Temperatures Below the Normal Boiling Point

	Temperature, °C												
	−150	−100	−75	−50	−25	0	20	50	100	150	200	250	300
	Temperature, K												
Property	123.15	173.15	198.15	223.15	248.15	273.15	293.15	323.15	373.15	423.15	473.15	523.15	573.15
Density, ρ_l (kg/m^3)	S	S	(977)	(952)	(927)	902	882	(848)	(787)	V	V	V	V
Specific heat capacity, $c_{p,l}$ (kJ/kg K)	S	S	(1.84)	(1.88)	(1.92)	(1.96)	(2.00)	2.072	(2.40)	V	V	V	V
Thermal conductivity, λ_l [(W/m^2)/(K/m)]	S	S	0.158	0.152	0.147	0.141	0.136	0.132	0.119	V	V	V	V
Dynamic viscosity, η_l (10^{-5} Ns/m^2)	S	S	—	—	(99)	69.0	53.1	37.5	(22)	V	V	V	V

METHYL BUTANOATE

Chemical formula: $C_5H_{10}O_2$
Molecular weight: 102.13
Melting point: 188.4 K
Boiling point: 375.8 K

Critical temperature: 554.4 K
Critical pressure: 3.48 MPa
Critical density: 300 kg/m^3
Normal vapor density: 4.56 kg/m^3
(@ 0 °C, 101.3 kPa)

Properties of the Liquid at Temperatures Below the Normal Boiling Point

	Temperature, °C												
	−150	−100	−75	−50	−25	0	20	50	100	150	200	250	300
	Temperature, K												
Property	123.15	173.15	198.15	223.15	248.15	273.15	293.15	323.15	373.15	423.15	473.15	523.15	573.15
Density, ρ_l (kg/m^3)	S	S	(1004)	(976)	(948)	920	897	865	807	V	V	V	V
Specific heat capacity, $c_{p,l}$ (kJ/kg K)	S	S	(1.92)	(1.95)	(1.98)	(2.00)	(2.02)	2.114	(2.24)	V	V	V	V
Thermal conductivity, λ_l [(W/m^2)/(K/m)]	S	S	(0.164)	0.158	0.152	0.145	0.140	0.133	0.121	V	V	V	V
Dynamic viscosity, η_l (10^{-5} Ns/m^2)	S	S	—	—	(120)	76.3	57.9	41.3	26.5	V	V	V	V

ETHYL BUTANOATE (ETHYL BUTYRATE)

Chemical formula: $C_6H_{12}O_2$
Molecular weight: 116.16
Melting point: 180.0 K
Boiling point: 394.0 K

Critical temperature: 566 K
Critical pressure: 3.14 MPa
Critical density: 294 kg/m^3
Normal vapor density: 5.18 kg/m^3 (@ 0°C, 101.3 kPa)

Properties of the Liquid at Temperatures Below the Normal Boiling Point

	Temperature, °C												
	−150	−100	−75	−50	−25	0	20	50	100	150	200	250	300
	Temperature, K												
Property	123.15	173.15	198.15	223.15	248.15	273.15	293.15	323.15	373.15	423.15	473.15	523.15	573.15
Density, ρ_l (kg/m^3)	S	S	(981)	(954)	(927)	900	879	847	(794)	V	V	V	V
Specific heat capacity, $c_{p,l}$ (kJ/kg K)	S	S	(1.76)	(1.80)	(1.84)	(1.86)	1.888	1.96	2.11	V	V	V	V
Thermal conductivity, λ_l [(W/m^2)/(K/m)]	S	S	0.159	0.153	0.147	0.141	0.137	0.130	0.119	V	V	V	V
Dynamic viscosity, η_l (10^{-5} Ns/m^2)	S	S	—	—	(120)	76.0	57.7	40.6	25.5	V	V	V	V

METHYL BENZOATE

Chemical formula: $C_8H_8O_2$
Molecular weight: 136.15
Melting point: 260.8 K
Boiling point: 472.2 K

Critical temperature: 692 K
Critical pressure: 3.65 MPa
Critical density: 344 kg/m^3
Normal vapor density: 6.07 kg/m^3 (@ 0°C, 101.3 kPa)

Properties of the Liquid at Temperatures Below the Normal Boiling Point

	Temperature, °C												
	−150	−100	−75	−50	−25	0	20	50	100	150	200	250	300
	Temperature, K												
Property	123.15	173.15	198.15	223.15	248.15	273.15	293.15	323.15	373.15	423.15	473.15	523.15	573.15
Density, ρ_l (kg/m^3)	S	S	S	S	S	1108	1088	(1059)	(1011)	(963)	V	V	V
Specific heat capacity, $c_{p,l}$ (kJ/kg K)	S	S	S	S	S	1.520	(1.58)	(1.66)	(1.80)	(1.95)	V	V	V
Thermal conductivity, λ_l [(W/m^2)/(K/m)]	S	S	S	S	S	0.151	0.147	0.141	0.131	0.121	V	V	V
Dynamic viscosity, η_l (10^{-5} Ns/m^2)	S	S	S	S	S	(86)	66.7	46.4	(31)	—	V	V	V

ETHYL BENZOATE

Chemical formula: $C_9H_{10}O_2$
Molecular weight: 150.18
Melting point: 238.3 K
Boiling point: 485.9 K

Critical temperature: 697 K
Critical pressure: 3.24 MPa
Critical density: 333 kg/m^3
Normal vapor density: 6.70 kg/m^3
(@ 0 °C, 101.3 kPa)

Properties of the Liquid at Temperatures Below the Normal Boiling Point

	Temperature, °C												
	−150	−100	−75	−50	−25	0	20	50	100	150	200	250	300
	Temperature, K												
Property	123.15	173.15	198.15	223.15	248.15	273.15	293.15	323.15	373.15	423.15	473.15	523.15	573.15
Density, ρ_l (kg/m^3)	S	S	S	S	(1083)	1 064	1 046	1 020	(974)	(930)	(884)	V	V
Specific heat capacity, $c_{p,l}$ (kJ/kg K)	S	S	S	S	(1.46)	(1.54)	(1.60)	(1.68)	(1.84)	(1.99)	(2.14)	V	V
Thermal conductivity, λ_l [(W/m^2)/(K/m)]	S	S	S	S	0.151	0.146	0.142	0.136	0.127	0.118	0.109	V	V
Dynamic viscosity, η_l (10^{-5} Ns/m^2)	S	S	S	S	—	—	206.7	110	67	49.7	(42)	V	V

METHYL SALICYLATE

Chemical formula: $C_8H_8O_3$
Molecular weight: 152.14
Melting point: 264.55 K
Boiling point: 496.45 K

Critical temperature: 709.2 K
Critical pressure: 4.09 MPa
Critical density: (416) kg/m^3
Normal vapor density: 6.79 kg/m^3
(@ 0 °C, 101.3 kPa)

Properties of the Liquid at Temperatures Below the Normal Boiling Point

	Temperature, °C												
	−150	−100	−75	−50	−25	0	20	50	100	150	200	250	300
	Temperature, K												
Property	123.15	173.15	198.15	223.15	248.15	273.15	293.15	323.15	373.15	423.15	473.15	523.15	573.15
Density, ρ_l (kg/m^3)	S	S	S	S	S	(1162)	1 180	1 152	1 105	1 055	1 004	V	V
Specific heat capacity, $c_{p,l}$ (kJ/kg K)	S	S	S	S	S	(1.54)	(1.58)	(1.63)	(1.76)	(1.83)	(1.92)	V	V
Thermal conductivity, λ_l [(W/m^2)/(K/m)]	S	S	S	S	S	(0.152)	(0.149)	(0.143)	(0.133)	(0.124)	(0.115)	V	V
Dynamic viscosity, η_l (10^{-5} Ns/m^2)	S	S	S	S	S	348.0	222.0	110.0	82.0	50.2	34.7	V	V

FORMALDEHYDE

Chemical formula: CH_2O
Molecular weight: 30.03
Melting point: 156.0 K
Boiling point: 254.0 K

Critical temperature: 408 K
Critical pressure: 6.59 MPa
Critical density: (300) kg/m^3
Normal vapor density: 1.34 kg/m^3 (@ 0 °C, 101.3 kPa)

Properties of the Liquid at Temperatures Below the Normal Boiling Point

Property	Temperature, °C												
	−150	−100	−75	−50	−25	0	20	50	100	150	200	250	300
	Temperature, K												
	123.15	173.15	198.15	223.15	248.15	273.15	293.15	323.15	373.15	423.15	473.15	523.15	573.15
Density, ρ_l (kg/m^3)	S	(932)	(895)	(858)	821	V	V	V	V	V	V	V	V
Specific heat capacity, $c_{p,l}$ (kJ/kg K)	S	(2.30)	(2.39)	(2.47)	(2.62)	V	V	V	V	V	V	V	V
Thermal conductivity, λ_l [(W/m^2)/(K/m)]	S	0.315	0.293	0.271	0.250	V	V	V	V	V	V	V	V
Dynamic viscosity, η_l (10^{-5} Ns/m^2)	S	—	—	(24)	(19)	V	V	V	V	V	V	V	V

Transport Properties of the Superheated Vapor

Property (at low pressure)	Temperature, °C												
	−150	−100	−50	0	25	100	200	300	400	600	800	1 000	1 200
	Temperature, K												
	123.15	173.15	223.15	273.15	298.15	373.15	473.15	573.15	673.15	873.15	1 073.15	1 273.15	1 473.15
Specific heat capacity, $c_{p,g}$ (kJ/kg K)	S	L	L	1.151	1.172	1.264	1.411	1.562	1.700	1.943	2.119	2.257	2.357
Thermal conductivity, λ_g[(W/m^2)/(K/m)]	S	L	L	(0.012)	(0.014)	(0.019)	(0.027)	(0.036)	(0.045)	(0.064)	—	—	—
Dynamic viscosity, η_g(10^{-5} Ns/m^2)	S	L	L	(0.88)	(0.96)	(1.20)	(1.50)	(1.78)	(2.04)	(2.51)	—	—	—

ACETALDEHYDE

Chemical formula: C_2H_4O
Molecular weight: 44.05
Melting point: 149.15 K
Boiling point: 293.6 K

Critical temperature: 461 K
Critical pressure: 5.57 MPa
Critical density: (262) kg/m^3
Normal vapor density: 1.97 kg/m^3
(@ 0 °C, 101.3 kPa)

Properties of the Liquid at Temperatures Below the Normal Boiling Point

	Temperature, °C												
	−150	−100	−75	−50	−25	0	20	50	100	150	200	250	300
	Temperature, K												
Property	123.15	173.15	198.15	223.15	248.15	273.15	293.15	323.15	373.15	423.15	473.15	523.15	573.15
Density, ρ_l (kg/m^3)	S	(920)	(892)	(863)	(834)	804	783	V	V	V	V	V	V
Specific heat capacity, $c_{p,l}$ (kJ/kg K)	S	(1.97)	(2.00)	(2.05)	(2.13)	2.20	2.28	V	V	V	V	V	V
Thermal conductivity, λ_l [(W/m^2)/(K/m)]	S	0.238	0.224	0.211	0.197	0.184	0.180	V	V	V	V	V	V
Dynamic viscosity, η_l (10^{-5} Ns/m^2)	S	—	—	(46)	(35)	26.8	22.2	V	V	V	V	V	V

Transport Properties of the Superheated Vapor

	Temperature, °C												
	−150	−100	−50	0	25	100	200	300	400	600	800	1 000	1 200
Property (at low pressure)	Temperature, K												
	123.15	173.15	223.15	273.15	298.15	373.15	473.15	573.15	673.15	873.15	1 073.15	1 273.15	1 473.15
Specific heat capacity, $c_{p,g}$ (kJ/kg K)	L	L	L	L	1.239	1.423	1.675	1.892	2.089	2.403	—	—	—
Thermal conductivity, λ_g[(W/m^2)/(K/m)]	L	L	L	L	0.012	0.018	(0.026)	(0.037)	(0.049)	(0.079)	—	—	—
Dynamic viscosity, η_g(10^{-5} Ns/m^2)	L	L	L	L	0.899	1.028	(1.36)	(1.62)	(1.86)	(2.30)	—	—	—

PARALDEHYDE

Chemical formula: $C_6H_{12}O_3$
Molecular weight: 132.15
Melting point: 285.75 K
Boiling point: 397.15 K

Critical temperature: 563.2 K
Critical pressure: 5.51 MPa
Critical density: 359 kg/m^3
Normal vapor density: 5.90 kg/m^3 (@ 0°C, 101.3 kPa)

Properties of the Liquid at Temperatures Below the Normal Boiling Point

	Temperature, °C												
	−150	−100	−75	−50	−25	0	20	50	100	150	200	250	300
	Temperature, K												
Property	123.15	173.15	198.15	223.15	248.15	273.15	293.15	323.15	373.15	423.15	473.15	523.15	573.15
Density, ρ_l (kg/m^3)	S	S	S	S	S	S	994	956	899	V	V	V	V
Specific heat capacity, $c_{p,l}$ (kJ/kg K)	S	S	S	S	S	S	1.933	(2.03)	(2.22)	V	V	V	V
Thermal conductivity, λ_l [(W/m^2)/(K/m)]	S	S	S	S	S	S	0.145	0.142	0.135	V	V	V	V
Dynamic viscosity, η_l (10^{-5} Ns/m^2)	S	S	S	S	S	S	117.9	69.0	37.0	V	V	V	V

Transport Properties of the Superheated Vapor

	Temperature, °C												
	−150	−100	−50	0	25	100	200	300	400	600	800	1 000	1 200
	Temperature, K												
Property (at low pressure)	123.15	173.15	223.15	273.15	298.15	373.15	473.15	573.15	673.15	873.15	1 073.15	1 273.15	1 473.15
Specific heat capacity, $c_{p,g}$ (kJ/kg K)	S	S	S	S	L	L	(1.859)	(2.116)	(2.335)	(2.676)	—	—	—
Thermal conductivity, λ_g[(W/m^2)/(K/m)]	S	S	S	S	L	L	(0.028)	(0.038)	(0.049)	(0.071)	—	—	—
Dynamic viscosity, η_g(10^{-5} Ns/m^2)	S	S	S	S	L	L	(1.17)	(1.41)	(1.64)	(2.06)	—	—	—

FURFURAL

Chemical formula: $C_5H_4O_2$
Molecular weight: 96.08
Melting point: 236.65 K
Boiling point: 433.7 K

Critical temperature: 670.15 K
Critical pressure: 5.89 MPa
Critical density: 346 kg/m^3
Normal vapor density: 4.29 kg/m^3 (@ 0 °C, 101.3 kPa)

Properties of the Liquid at Temperatures Below the Normal Boiling Point

	Temperature, °C												
	−150	−100	−75	−50	−25	0	20	50	100	150	200	250	300
	Temperature, K												
Property	123.15	173.15	198.15	223.15	248.15	273.15	293.15	323.15	373.15	423.15	473.15	523.15	573.15
Density, ρ_l (kg/m^3)	S	S	S	S	(1201)	1 181	1 160	1 128	1 077	(1020)	V	V	V
Specific heat capacity, $c_{p,l}$ (kJ/kg K)	S	S	S	S	(1.47)	1.541	1.636	1.750	(1.87)	(2.00)	V	V	V
Thermal conductivity, λ_l [(W/m^2)/(K/m)]	S	S	S	S	(0.184)	(0.178)	0.172	0.167	(0.155)	(0.142)	V	V	V
Dynamic viscosity, η_l (10^{-5} Ns/m^2)	S	S	S	S	(410)	247.6	163.1	(120)	(64)	(38)	V	V	V

Transport Properties of the Superheated Vapor

	Temperature, °C												
	− 150	− 100	− 50	0	25	100	200	300	400	600	800	1 000	1 200
Property (at low pressure)	Temperature, K												
	123.15	173.15	223.15	273.15	298.15	373.15	473.15	573.15	673.15	873.15	1 073.15	1 273.15	1 473.15
Specific heat capacity, $c_{p,g}$ (kJ/kg K)	S	S	S	L	L	L	(1.512)	(1.673)	(1.808)	(2.019)	—	—	—
Thermal conductivity, λ_g[(W/m^2)/(K/m)]	S	S	S	L	L	L	(0.023)	(0.032)	(0.040)	(0.057)	—	—	—
Dynamic viscosity, η_g(10^{-5} Ns/m^2)	S	S	S	L	L	L	(1.20)	(1.46)	(1.70)	(2.16)	—	—	—

BENZALDEHYDE

Chemical formula: C_7H_6O
Molecular weight: 106.12
Melting point: 216.0 K
Boiling point: 452.0 K

Critical temperature: 695 K
Critical pressure: 4.66 MPa
Critical density: 327 kg/m^3
Normal vapor density: 4.73 kg/m^3
(@ 0 °C, 101.3 kPa)

Properties of the Liquid at Temperatures Below the Normal Boiling Point

	Temperature, °C												
	−150	−100	−75	−50	−25	0	20	50	100	150	200	250	300
	Temperature, K												
Property	123.15	173.15	198.15	223.15	248.15	273.15	293.15	323.15	373.15	423.15	473.15	523.15	573.15
Density, ρ_l (kg/m^3)	S	S	S	(1106)	(1084)	1 062	1 046	1 017	971	924	V	V	V
Specific heat capacity, $c_{p,l}$ (kJ/kg K)	S	S	S	1.474	1.515	1.565	1.610	1.632	1.804	1.926	V	V	V
Thermal conductivity, λ_l [(W/m^2)/(K/m)]	S	S	S	0.167	0.162	0.157	0.154	0.148	0.139	0.129	V	V	V
Dynamic viscosity, η_l (10^{-5} Ns/m^2)	S	S	S	—	(330)	230.1	160.1	(95)	—	—	V	V	V

SALICYALDEHYDE

Chemical formula: $C_7H_6O_7$
Molecular weight: 122.12
Melting point: 274.75 K
Boiling point: 469.65 K

Critical temperature: (696) K
Critical pressure: (5.4) MPa
Critical density: —
Normal vapor density: 5.45 kg/m^3
(@ 0 °C, 101.3 kPa)

Properties of the Liquid at Temperatures Below the Normal Boiling Point

	Temperature, °C												
	−150	−100	−75	−50	−25	0	20	50	100	150	200	250	300
	Temperature, K												
Property	123.15	173.15	198.15	223.15	248.15	273.15	293.15	323.15	373.15	423.15	473.15	523.15	573.15
Density, ρ_l (kg/m^3)	S	S	S	S	S	S	1 150	1 132	1 084	1 034	V	V	V
Specific heat capacity, $c_{p,l}$ (kJ/kg K)	S	S	S	S	S	S	1.608	(1.71)	(1.89)	(2.07)	V	V	V
Thermal conductivity, λ_l [(W/m^2)/(K/m)]	S	S	S	S	S	S	(0.154)	(0.148)	(0.139)	(0.129)	V	V	V
Dynamic viscosity, η_l (10^{-5} Ns/m^2)	S	S	S	S	S	S	320.1	150.0	80.0	49.1	V	V	V

KETENE

Chemical formula: C_2H_2O
Molecular weight: 42.04
Melting point: 138.0 K
Boiling point: 232.0 K

Critical temperature: 380 K
Critical pressure: 6.48 MPa
Critical density: (290) kg/m^3
Normal vapor density: 1.88 kg/m^3
(@ 0 °C, 101.3 kPa)

Properties of the Liquid at Temperatures Below the Normal Boiling Point

	Temperature, °C												
	−150	−100	−75	−50	−25	0	20	50	100	150	200	250	300
	Temperature, K												
Property	123.15	173.15	198.15	223.15	248.15	273.15	293.15	323.15	373.15	423.15	473.15	523.15	573.15
Density, ρ_l (kg/m^3)	S	(1080)	(1030)	(979)	V	V	V	V	V	V	V	V	V
Specific heat capacity, $c_{p,l}$ (kJ/kg K)	S	(1.79)	(1.92)	(2.02)	V	V	V	V	V	V	V	V	V
Thermal conductivity, λ_l [(W/m^2)/(K/m)]	S	(0.267)	(0.250)	(0.233)	V	V	V	V	V	V	V	V	V
Dynamic viscosity, η_l (10^{-5} Ns/m^2)	S	—	—	(110)	V	V	V	V	V	V	V	V	V

Transport Properties of the Superheated Vapor

	Temperature, °C												
	−150	−100	−50	0	25	100	200	300	400	600	800	1 000	1 200
Property (at low pressure)	Temperature, K												
	123.15	173.15	223.15	273.15	298.15	373.15	473.15	573.15	673.15	873.15	1 073.15	1 273.15	1 473.15
Specific heat capacity, $c_{p,g}$ (kJ/kg K)	S	L	L	1.093	1.143	1.290	1.461	1.599	1.717	1.905	2.043	2.148	2.227
Thermal conductivity, λ_g[(W/m^2)/(K/m)]	S	L	L	(0.015)	(0.017)	(0.024)	(0.034)	(0.045)	(0.055)	(0.070)	—	—	—
Dynamic viscosity, η_g(10^{-5} Ns/m^2)	S	L	L	(1.05)	(1.15)	(1.43)	(1.78)	(2.10)	(2.40)	(2.94)	—	—	—

ACETONE

Chemical formula: CH_3COCH_3
Molecular weight: 58.1
Melting point: 179.95 K
Normal boiling point: 329.25 K

Critical temperature: 508.15 K
Critical pressure: 4.761 MPa
Critical density: 273 kg/m^3
Normal vapor density: 2.59 kg/m^3
(@ 0°C, 101.3 kPa)

Properties of the Liquid at Temperatures Below the Normal Boiling Point

	Temperature, °C												
	−150	−100	−75	−50	−25	0	20	50	100	150	200	250	300
	Temperature, K												
Property	123.15	173.15	198.15	223.15	248.15	273.15	293.15	323.15	373.15	423.15	473.15	523.15	573.15
Density, ρ_l (kg/m^3)	S	S	893	868	840	812	791	756	V	V	V	V	V
Specific heat capacity, $c_{p,l}$ (kJ/kg K)	S	S	2.010	2.039	2.072	2.102	2.156	2.252	V	V	V	V	V
Thermal conductivity, λ_l [(W/m^2)/(K/m)]	S	S	0.179	0.175	0.170	0.165	0.160	0.154	V	V	V	V	V
Dynamic viscosity, η_l (10^{-5} Ns/m^2)	S	S	134.1	82.0	56.0	39.8	32.5	24.9	V	V	V	V	V

Properties of the Saturated Liquid and Vapor

T_{sat}, K	329.25	340	360	380	400	420	440	460	480	508.15
p_{sat}, kPa	101.3	152	274	452	731	1 082	1 637	2 279	3 252	4 761
ρ_ℓ, kg/m^3	750	736	710	683	655	625	590	553	504	273
ρ_g, kg/m^3	2.23	3.11	5.49	9.13	14.5	22.3	33.6	50.3	77.2	273
h_ℓ, kJ/kg	−258	−233	−180	−131	−74	−32	23	79	138	257
h_g, kJ/kg	248	261	285	308	330	350	367	379	380	257
$\Delta h_{g,\ell}$, kJ/kg	506	494	465	439	414	382	344	300	242	
$c_{p,\ell}$, kJ/(kg K)	2.28	2.32	2.42	2.53	2.65	2.83	3.03	3.29	3.76	
$c_{p,g}$, kJ/(kg K)	1.41	1.46	1.55	1.66	1.79	1.95	2.18	2.54	3.38	
η_ℓ, μNs/m^2	235	213	188	165	141	119	99	80	64	49
η_g, μNs/m^2	9.4	9.8	10.4	11.1	11.8	12.6	13.5	14.4	15.8	49
λ_ℓ, (mW/m^2)/(K/m)	142	137	129	121	112	104	96	87	77	58
λ_g, (mW/m^2)/(K/m)	12.7	14.1	16.1	18.5	21.2	24.2	27.2	31.0	36.0	58
Pr_ℓ	3.77	3.61	3.53	3.49	3.34	3.24	3.12	3.03	3.13	
Pr_g	1.04	1.01	1.00	1.00	1.00	1.02	1.08	1.18	1.48	
σ, mN/m	18.4	17.0	14.5	12.1	9.6	7.1	4.6	3.1	1.6	
$\beta_{e,\ell}$, kK^{-1}	1.65	1.81	1.92	2.18	2.50	3.04	3.82	5.13	9.22	

Transport Properties of the Superheated Vapor

	Temperature, °C												
	−150	−100	−50	0	25	100	200	300	400	600	800	1 000	1 200
Property (at low pressure)	Temperature, K												
	123.15	173.15	223.15	273.15	298.15	373.15	473.15	573.15	673.15	873.15	1 073.15	1 273.15	1 473.15
Specific heat capacity, $c_{p,g}$ (kJ/kg K)	S	S	L	L	L	1.557	1.838	2.093	2.311	2.659	2.906	3.098	3.006
Thermal conductivity, λ_g[(W/m^2)/(K/m)]	S	S	L	L	L	0.018	(0.027)	(0.038)	(0.051)	(0.076)	—	—	—
Dynamic viscosity, η_g(10^{-5} Ns/m^2)	S	S	L	L	L	0.931	1.21	1.46	1.72	2.20	2.64	3.05	3.42

METHYL ETHYL KETONE (2-BUTANONE)

Chemical formula: C_4H_8O
Molecular weight: 72.11
Melting point: 186.5 K
Boiling point: 352.8 K

Critical temperature: 535.6 K
Critical pressure: 4.15 MPa
Critical density: 270 kg/m^3
Normal vapor density: 3.22 kg/m^3 (@ 0 °C, 101.3 kPa)

Properties of the Liquid at Temperatures Below the Normal Boiling Point

	Temperature, °C												
	−150	−100	−75	−50	−25	0	20	50	100	150	200	250	300
	Temperature, K												
Property	123.15	173.15	198.15	223.15	248.15	273.15	293.15	323.15	373.15	423.15	473.15	523.15	573.15
Density, ρ_l (kg/m^3)	S	S	(893)	(870)	(847)	826	803	(773)	V	V	V	V	V
Specific heat capacity, $c_{p,l}$ (kJ/kg K)	S	S	2.071	2.091	2.120	2.155	2.219	2.311	V	V	V	V	V
Thermal conductivity, λ_l [(W/m^2)/(K/m)]	S	S	0.169	0.163	0.156	0.150	0.145	0.137	V	V	V	V	V
Dynamic viscosity, η_l (10^{-5} Ns/m^2)	S	S	(125)	(94)	(70)	54.0	42.3	31.0	V	V	V	V	V

Transport Properties of the Superheated Vapor

	Temperature, °C												
	−150	−100	−50	0	25	100	200	300	400	600	800	1 000	1 200
	Temperature, K												
Property (at low pressure)	123.15	173.15	223.15	273.15	298.15	373.15	473.15	573.15	673.15	873.15	1 073.15	1 273.15	1 473.15
Specific heat capacity, $c_{p,g}$ (kJ/kg K)	S	S	L	L	L	1.659	1.942	2.196	2.423	2.802	—	—	—
Thermal conductivity, λ_g[(W/m^2)/(K/m)]	S	S	L	L	L	(0.017)	(0.027)	(0.038)	(0.050)	(0.072)	—	—	—
Dynamic viscosity, η_g(10^{-5} Ns/m^2)	S	S	L	L	L	(0.92)	(1.17)	(1.40)	(1.63)	(2.04)	—	—	—

DIETHYL KETONE (3-PENTANONE)

Chemical formula: $C_5H_{10}O$
Molecular weight: 86.13
Melting point: 234.2 K
Boiling point: 375.1 K

Critical temperature: 561 K
Critical pressure: 3.74 MPa
Critical density: 256 kg/m^3
Normal vapor density: 3.84 kg/m^3
(@ 0°C, 101.3 kPa)

Properties of the Liquid at Temperatures Below the Normal Boiling Point

Property	Temperature, °C / Temperature, K												
	−150	−100	−75	−50	−25	0	20	50	100	150	200	250	300
	123.15	173.15	198.15	223.15	248.15	273.15	293.15	323.15	373.15	423.15	473.15	523.15	573.15
Density, ρ_l (kg/m^3)	S	S	S	S	(858)	834	914	787	(740)	V	V	V	V
Specific heat capacity, $c_{p,l}$ (kJ/kg K)	S	S	S	S	2.139	2.171	2.214	2.268	2.418	V	V	V	V
Thermal conductivity, λ_l [(W/m^2)/(K/m)]	S	S	S	S	0.153	0.147	0.142	0.135	(0.123)	V	V	V	V
Dynamic viscosity, η_l (10^{-5} Ns/m^2)	S	S	S	S	(77)	59.4	47.0	34.4	22.5	V	V	V	V

Transport Properties of the Superheated Vapor

Property (at low pressure)	Temperature, °C / Temperature, K												
	−150	−100	−50	0	25	100	200	300	400	600	800	1 000	1 200
	123.15	173.15	223.15	273.15	298.15	373.15	473.15	573.15	673.15	873.15	1 073.15	1 273.15	1 473.15
Specific heat capacity, $c_{p,g}$ (kJ/kg K)	S	S	S	L	L	L	2.032	2.305	2.528	2.912	3.122	—	—
Thermal conductivity, λ_g[(W/m^2)/(K/m)]	S	S	S	L	L	L	(0.027)	(0.039)	(0.055)	—	—	—	—
Dynamic viscosity, η_g(10^{-5} Ns/m^2)	S	S	S	L	L	L	(1.04)	(1.25)	(1.43)	—	—	—	—

DIPROPYL KETONE

Chemical formula: $C_7H_{14}O$
Molecular weight: 114.18
Melting point: 239.15 K
Boiling point: 416.65 K

Critical temperature: 604.2 K
Critical pressure: 3.0 MPa
Critical density: 256 kg/m^3
Normal vapor density: 5.09 kg/m^3
(@ 0 °C, 101.3 kPa)

Properties of the Liquid at Temperatures Below the Normal Boiling Point

	Temperature, °C												
	−150	−100	−75	−50	−25	0	20	50	100	150	200	250	300
	Temperature, K												
Property	123.15	173.15	198.15	223.15	248.15	273.15	293.15	323.15	373.15	423.15	473.15	523.15	573.15
Density, ρ_l (kg/m^3)	S	S	S	S	(853)	832	817	790	748	V	V	V	V
Specific heat capacity, $c_{p,l}$ (kJ/kg K)	S	S	S	S	(2.10)	(2.14)	(2.18)	2.198	2.407	V	V	V	V
Thermal conductivity, λ_l [(W/m^2)/(K/m)]	S	S	S	S	0.148	0.144	0.140	0.133	(0.122)	V	V	V	V
Dynamic viscosity, η_l (10^{-5} Ns/m^2)	S	S	S	S	—	(100)	75.1	(52)	(33)	V	V	V	V

BENZOPHENONE

Chemical formula: $C_{13}H_{10}O$
Molecular weight: 128.21
Melting point: 321.25 K
Boiling point: 579.15 K

Critical temperature: (827) K
Critical pressure: (2.07) MPa
Critical density: —
Normal vapor density: 5.72 kg/m^3
(@ 0 °C, 101.3 kPa)

Properties of the Liquid at Temperatures Below the Normal Boiling Point

	Temperature, °C												
	−150	−100	−75	−50	−25	0	20	50	100	150	200	250	300
	Temperature, K												
Property	123.15	173.15	198.15	223.15	248.15	273.15	293.15	323.15	373.15	423.15	473.15	523.15	573.15
Density, ρ_l (kg/m^3)	S	S	S	S	S	S	S	1 085	1 042	(998)	(955)	(911)	(866)
Specific heat capacity, $c_{p,l}$ (kJ/kg K)	S	S	S	S	S	S	S	(1.55)	(1.68)	(1.81)	(1.96)	(2.10)	(2.24)
Thermal conductivity, λ_l [(W/m^2)/(K/m)]	S	S	S	S	S	S	S	(0.136)	(0.130)	(0.124)	(0.119)	(0.113)	(0.107)
Dynamic viscosity, η_l (10^{-5} Ns/m^2)	S	S	S	S	S	S	S	560.2	180.1	96.0	62.0	42.0	(28)

ACETOPHENONE

Chemical formula: C_8H_8O
Molecular weight: 120.14
Melting point: 253.55 K
Boiling point: 476.85 K

Critical temperature: 704.2 K
Critical pressure: 3.8 MPa
Critical density: (324) kg/m^3
Normal vapor density: 5.36 kg/m^3
(@ 0°C, 101.3 kPa)

Properties of the Liquid at Temperatures Below the Normal Boiling Point

	Temperature, °C												
	−150	−100	−75	−50	−25	0	20	50	100	150	200	250	300
	Temperature, K												
Property	123.15	173.15	198.15	223.15	248.15	273.15	293.15	323.15	373.15	423.15	473.15	523.15	573.15
Density, ρ_l (kg/m^3)	S	S	S	S	S	S	1 028	1 002	958	914	(865)	V	V
Specific heat capacity, $c_{p,l}$ (kJ/kg K)	S	S	S	S	S	S	1.88	1.93	1.968	(2.03)	(2.09)	V	V
Thermal conductivity, λ_l [(W/m^2)/(K/m)]	S	S	S	S	S	S	0.142	0.140	0.132	(0.123)	(0.115)	V	V
Dynamic viscosity, η_l (10^{-5} Ns/m^2)	S	S	S	S	S	S	185.1	124.1	62.0	(41)	(30)	V	V

Transport Properties of the Superheated Vapor

	Temperature, °C												
	− 150	− 100	− 50	0	25	100	200	300	400	600	800	1 000	1 200
Property (at low pressure)	Temperature, K												
	123.15	173.15	223.15	273.15	298.15	373.15	473.15	573.15	673.15	873.15	1 073.15	1 273.15	1 473.15
Specific heat capacity, $c_{p,g}$ (kJ/kg K)	S	S	S	S	L	L	L	(1.835)	(2.043)	(2.366)	—	—	—
Thermal conductivity, λ_g[(W/m^2)/(K/m)]	S	S	S	S	L	L	L	(0.030)	(0.040)	(0.059)	—	—	—
Dynamic viscosity, η_g(10^{-5} Ns/m^2)	S	S	S	S	L	L	L	(1.28)	(1.50)	(1.91)	—	—	—

FORMIC ACID

Chemical formula: CH_2O_2
Molecular weight: 46.02
Melting point: 281.5 K
Boiling point: 373.8 K

Critical temperature: 579.2 K
Critical pressure: 5.5 MPa
Critical density: 400 kg/m^3
Normal vapor density: 2.05 kg/m^3 (@ 0 °C, 101.3 kPa)

Properties of the Liquid at Temperatures Below the Normal Boiling Point

	Temperature, °C												
	−150	−100	−75	−50	−25	0	20	50	100	150	200	250	300
	Temperature, K												
Property	123.15	173.15	198.15	223.15	248.15	273.15	293.15	323.15	373.15	423.15	473.15	523.15	573.15
Density, ρ_l (kg/m^3)	S	S	S	S	S	S	1 220	1 184	1 108	V	V	V	V
Specific heat capacity, $c_{p,l}$ (kJ/kg K)	S	S	S	S	S	S	2.169	2.202	2.282	V	V	V	V
Thermal conductivity, λ_l [(W/m^2)/(K/m)]	S	S	S	S	S	S	0.261	0.250	0.232	V	V	V	V
Dynamic viscosity, η_l (10^{-5} Ns/m^2)	S	S	S	S	S	S	179.1	103.1	54.2	V	V	V	V

Transport Properties of the Superheated Vapor

	Temperature, °C												
	−150	−100	−50	0	25	100	200	300	400	600	800	1 000	1 200
Property (at low pressure)	Temperature, K												
	123.15	173.15	223.15	273.15	298.15	373.15	473.15	573.15	673.15	873.15	1 073.15	1 273.15	1 473.15
Specific heat capacity, $c_{p,g}$ (kJ/kg K)	S	S	S	S	L	L	1.348	1.480	1.589	1.757	1.870	1.953	2.037
Thermal conductivity, λ_g[(W/m^2)/(K/m)]	S	S	S	S	L	L	(0.027)	(0.036)	(0.045)	(0.062)	—	—	—
Dynamic viscosity, η_g(10^{-5} Ns/m^2)	S	S	S	S	L	L	(1.53)	(1.84)	(2.15)	(2.69)	—	—	—

ACETIC ACID

Chemical formula: CH_3CO_2H
Molecular weight: 60.05
Melting point: 289.85 K
Normal boiling point: 391.15 K

Critical temperature: 594.75 K
Critical pressure: 5.790 MPa
Critical density: 350.6 kg/m^3
Normal vapor density: 2.68 kg/m^3
(@ 0°C, 101.3 kPa)

Properties of the Liquid at Temperatures Below the Normal Boiling Point

	Temperature, °C												
	−150	−100	−75	−50	−25	0	20	50	100	150	200	250	300
	Temperature, K												
Property	123.15	173.15	198.15	223.15	248.15	273.15	293.15	323.15	373.15	423.15	473.15	523.15	573.15
Density, ρ_l (kg/m^3)	S	S	S	S	S	S	1 049	1 018	960	V	V	V	V
Specific heat capacity, $c_{p,l}$ (kJ/kg K)	S	S	S	S	S	S	1.997	2.156	2.349	V	V	V	V
Thermal conductivity, λ_l [(W/m^2)/(K/m)]	S	S	S	S	S	S	0.161	0.155	0.142	V	V	V	V
Dynamic viscosity, η_l (10^{-5} Ns/m^2)	S	S	S	S	S	S	121.0	79.2	45.8	V	V	V	V

Properties of the Saturated Liquid and Vapor

T_{sat}, K	391.15	420	440	460	480	500	520	540	560	594.75
p_{sat}, kPa	101.3	230	382	427	898	1 320	1 890	2 630	3 590	5 790
ρ_ℓ, kg/m^3	939	900	874	846	815	782	743	697	642	350.6
ρ_g, kg/m^3	1.93	4.53	7.56	12.0	18.4	27.3	39.9	57.8	85.0	350.6
h_ℓ, kJ/kg	260	326	372	420	473	524	480	643	710	854
h_g, kJ/kg	642	703	740	775	807	834	850	874	882	854
$\Delta h_{g,\ell}$, kJ/kg	382	377	368	355	334	310	270	231	172	
$c_{p,\ell}$, kJ/(kg K)	2.42	2.55	2.66	2.76	2.91	3.04	3.21	3.43	3.82	
$c_{p,g}$, kJ/(kg K)	1.39	1.49	1.58	1.69	1.82	1.99	2.24	2.66	3.59	
η_ℓ, μNs/m^2	372	276	232	194	166	138	115	95	76	
η_g, μNs/m^2	10.4	11.4	11.9	12.4	13.0	13.7	14.4	15.3	16.5	
λ_ℓ, (mW/m^2)/(K/m)	158	150	143	137	131	125	118	112	105	93
λ_g, (mW/m^2)/(K/m)	20.7	23.8	26.3	29.0	32.2	35.9	40.3	45.9	53.6	93
Pr_ℓ	5.70	4.69	4.32	3.91	3.69	3.36	3.13	2.91	2.76	
Pr_g	0.70	0.71	0.71	0.72	0.73	0.76	0.80	0.89	1.11	
σ, mN/m	18.1	15.3	13.5	11.6	9.7	7.9	6.0	4.28	2.47	
$\beta_{e,\ell}$, kK^{-1}	1.40	1.52	1.63	1.87	2.24	2.75	3.62	5.15	8.0	

Transport Properties of the Superheated Vapor

	Temperature, °C												
	− 150	− 100	− 50	0	25	100	200	300	400	600	800	1 000	1 200
Property (at low pressure)	Temperature, K												
	123.15	173.15	223.15	273.15	298.15	373.15	473.15	573.15	673.15	873.15	1 073.15	1 273.15	1 473.15
Specific heat capacity, $c_{p,g}$ (kJ/kg K)	S	S	S	S	L	L	1.503	1.699	1.852	2.100	2.274	2.398	2.523
Thermal conductivity, λ_g[(W/m^2)/(K/m)]	S	S	S	S	L	L	(0.026)	(0.037)	(0.049)	(0.074)	—	—	—
Dynamic viscosity, η_g(10^{-5} Ns/m^2)	S	S	S	S	L	L	1.35	1.66	1.98	(2.63)	—	—	—

PROPIONIC ACID

Chemical formula: $C_3H_6O_2$
Molecular weight: 74.08
Melting point: 252.5 K
Boiling point: 414.0 K

Critical temperature: 612 K
Critical pressure: 5.37 MPa
Critical density: 322 kg/m^3
Normal vapor density: 3.31 kg/m^3
(@ 0 °C, 101.3 kPa)

Properties of the Liquid at Temperatures Below the Normal Boiling Point

	Temperature, °C												
	−150	−100	−75	−50	−25	0	20	50	100	150	200	250	300
	Temperature, K												
Property	123.15	173.15	198.15	223.15	248.15	273.15	293.15	323.15	373.15	423.15	473.15	523.15	573.15
Density, ρ_l (kg/m^3)	S	S	S	S	S	1 015	993	963	911	V	V	V	V
Specific heat capacity, $c_{p,l}$ (kJ/kg K)	S	S	S	S	S	2.077	2.165	2.299	2.516	V	V	V	V
Thermal conductivity, λ_l [(W/m^2)/(K/m)]	S	S	S	S	S	0.153	0.150	0.144	0.136	V	V	V	V
Dynamic viscosity, η_l (10^{-5} Ns/m^2)	S	S	S	S	S	154.1	169.4	73.8	45.0	V	V	V	V

Transport Properties of the Superheated Vapor

	Temperature, °C												
	−150	−100	−50	0	25	100	200	300	400	600	800	1 000	1 200
Property (at low pressure)	Temperature, K												
	123.15	173.15	223.15	273.15	298.15	373.15	473.15	573.15	673.15	873.15	1 073.15	1 273.15	1 473.15
Specific heat capacity, $c_{p,g}$ (kJ/kg K)	S	S	S	L	L	L	3.245	3.647	3.973	4.494	4.905	5.121	5.337
Thermal conductivity, λ_g[(W/m^2)/(K/m)]	S	S	S	L	L	L	(0.026)	(0.036)	(0.046)	(0.067)	—	—	—
Dynamic viscosity, η_g(10^{-5} Ns/m^2)	S	S	S	L	L	L	(1.22)	(1.48)	(1.72)	(2.17)	—	—	—

BUTYRIC ACID

Chemical formula: $C_4H_8O_2$
Molecular weight: 88.15
Melting point: 267.9 K
Boiling point: 436.4 K

Critical temperature: 628 K
Critical pressure: 5.27 MPa
Critical density: 304 kg/m^3
Normal vapor density: 3.93 kg/m^3
(@ 0°C, 101.3 kPa)

Properties of the Liquid at Temperatures Below the Normal Boiling Point

	Temperature, °C												
	−150	−100	−75	−50	−25	0	20	50	100	150	200	250	300
	Temperature, K												
Property	123.15	173.15	198.15	223.15	248.15	273.15	293.15	323.15	373.15	423.15	473.15	523.15	573.15
Density, ρ_l (kg/m^3)	S	S	S	S	S	977	958	927	882	829	V	V	V
Specific heat capacity, $c_{p,l}$ (kJ/kg K)	S	S	S	S	S	1.947	2.010	2.156	2.36	(2.64)	V	V	V
Thermal conductivity, λ_l [(W/m^2)/(K/m)]	S	S	S	S	S	0.153	0.150	0.144	0.136	0.127	V	V	V
Dynamic viscosity, η_l (10^{-5} Ns/m^2)	S	S	S	S	S	228.1	154.1	98.0	55.0	34.0	V	V	V

Transport Properties of the Superheated Vapor

	Temperature, °C												
	−150	−100	−50	0	25	100	200	300	400	600	800	1 000	1 200
Property (at low pressure)	Temperature, K												
	123.15	173.15	223.15	273.15	298.15	373.15	473.15	573.15	673.15	873.15	1 073.15	1 273.15	1 473.15
Specific heat capacity, $c_{p,g}$ (kJ/kg K)	S	S	S	L	L	L	(1.779)	(2.020)	(2.229)	(2.565)	—	—	—
Thermal conductivity, λ_g[(W/m^2)/(K/m)]	S	S	S	L	L	L	(0.025)	(0.035)	(0.045)	(0.065)	—	—	—
Dynamic viscosity, η_g(10^{-5} Ns/m^2)	S	S	S	L	L	L	(1.13)	(1.36)	(1.59)	(2.01)	—	—	—

PENTANOIC ACID (VALERIC ACID)

Chemical formula: $C_5H_{10}O_2$
Molecular weight: 102.13
Melting point: 239.0 K
Boiling point: 458.7 K

Critical temperature: 651 K
Critical pressure: 3.85 MPa
Critical density: 300 kg/m^3
Normal vapor density: 4.56 kg/m^3 (@ 0 °C, 101.3 kPa)

Properties of the Liquid at Temperatures Below the Normal Boiling Point

	Temperature, °C												
	−150	−100	−75	−50	−25	0	20	50	100	150	200	250	300
	Temperature, K												
Property	123.15	173.15	198.15	223.15	248.15	273.15	293.15	323.15	373.15	423.15	473.15	523.15	573.15
Density, ρ_l (kg/m^3)	S	S	S	S	(980)	957	942	(915)	(868)	(823)	V	V	V
Specific heat capacity, $c_{p,l}$ (kJ/kg K)	S	S	S	S	1.87	1.95	2.04	2.17	(2.38)	(2.60)	V	V	V
Thermal conductivity, λ_l [(W/m^2)/(K/m)]	S	S	S	S	(0.153)	0.149	0.146	0.140	0.132	0.123	V	V	V
Dynamic viscosity, η_l (10^{-5} Ns/m^2)	S	S	S	S	(600)	355.1	223.1	131.0	66.0	(40)	V	V	V

HEXANOIC ACID (CAPROIC ACID)

Chemical formula: $C_6H_{12}O_2$
Molecular weight: 116.15
Melting point: 269.25 K
Boiling point: 478.35 K

Critical temperature: 665.2 K
Critical pressure: 4.0 MPa
Critical density: 294 kg/m^3
Normal vapor density: 5.18 kg/m^3 (@ 0 °C, 101.3 kPa)

Thermodynamic Properties of the Superheated Vapor

	Temperature, °C												
	−150	−100	−75	−50	−25	0	20	50	100	150	200	250	300
	Temperature, K												
Property	123.15	173.15	198.15	223.15	248.15	273.15	293.15	323.15	373.15	423.15	473.15	523.15	573.15
Density, ρ_l (kg/m^3)	S	S	S	S	S	945	929	900	(861)	(816)	(771)	V	V
Specific heat capacity, $c_{p,l}$ (kJ/kg K)	S	S	S	S	S	(2.09)	2.15	2.156	(2.43)	(2.60)	(2.76)	V	V
Thermal conductivity, λ_l [(W/m^2)/(K/m)]	S	S	S	S	S	0.149	0.146	0.140	0.132	0.123	0.117	V	V
Dynamic viscosity, η_l (10^{-5} Ns/m^2)	S	S	S	S	S	540.2	319.9	175.0	86.0	(50)	(34)	V	V

Transport Properties of the Superheated Vapor

	Temperature, °C												
	−150	−100	−50	0	25	100	200	300	400	600	800	1 000	1 200
Property (at low pressure)	Temperature, K												
	123.15	173.15	223.15	273.15	298.15	373.15	473.15	573.15	673.15	873.15	1 073.15	1 273.15	1 473.15
Specific heat capacity, $c_{p,g}$ (kJ/kg K)	S	S	S	L	L	L	L	(2.139)	(2.372)	(2.737)	—	—	—
Thermal conductivity, λ_g[(W/m^2)/(K/m)]	S	S	S	L	L	L	L	(0.033)	(0.043)	(0.063)	—	—	—
Dynamic viscosity, η_g(10^{-5} Ns/m^2)	S	S	S	L	L	L	L	(1.24)	(1.45)	(1.84)	—	—	—

ACETIC ANHYDRIDE

Chemical formula: $C_4H_6O_3$
Molecular weight: 102.09
Melting point: 199.0 K
Boiling point: 412.0 K

Critical temperature: 569 K
Critical pressure: 4.67 MPa
Critical density: (337) kg/m^3
Normal vapor density: 4.56 kg/m^3
(@ 0 °C, 101.3 kPa)

Properties of the Liquid at Temperatures Below the Normal Boiling Point

	Temperature, °C												
	−150	−100	−75	−50	−25	0	20	50	100	150	200	250	300
Property	Temperature, K												
	123.15	173.15	198.15	223.15	248.15	273.15	293.15	323.15	373.15	423.15	473.15	523.15	573.15
Density, ρ_l (kg/m^3)	S	S	S	(1161)	(1133)	1 105	1 082	1 044	(979)	V	V	V	V
Specific heat capacity, $c_{p,l}$ (kJ/kg K)	S	S	S	(1.61)	(1.64)	(1.68)	(1.71)	1.750	1.905	V	V	V	V
Thermal conductivity, λ_l [(W/m^2)/(K/m)]	S	S	S	0.186	0.178	0.171	0.165	0.158	0.143	V	V	V	V
Dynamic viscosity, η_l (10^{-5} Ns/m^2)	S	S	S	(270)	(180)	123.5	90.5	62.0	38.6	V	V	V	V

Transport Properties of the Superheated Vapor

	Temperature, °C												
	−150	−100	−50	0	25	100	200	300	400	600	800	1 000	1 200
Property (at low pressure)	Temperature, K												
	123.15	173.15	223.15	273.15	298.15	373.15	473.15	573.15	673.15	873.15	1 073.15	1 273.15	1 473.15
Specific heat capacity, $c_{p,g}$ (kJ/kg K)	S	S	L	L	L	L	1.442	1.655	1.829	2.086	—	—	—
Thermal conductivity, λ_g[(W/m^2)/(K/m)]	S	S	L	L	L	L	(0.024)	(0.033)	(0.042)	(0.060)	—	—	—
Dynamic viscosity, η_g(10^{-5} Ns/m^2)	S	S	L	L	L	L	(1.24)	(1.49)	(1.74)	(2.18)	—	—	—

PROPIONIC ANHYDRIDE

Chemical formula: $C_6H_{10}O_3$
Molecular weight: 130.14
Melting point: 228.15 K
Boiling point: 440.15 K

Critical temperature: 630.2 K
Critical pressure: 3.40 MPa
Critical density: 310 kg/m^3
Normal vapor density: 5.81 kg/m^3
(@ 0 °C, 101.3 kPa)

Properties of the Liquid at Temperatures Below the Normal Boiling Point

	Temperature, °C												
	−150	−100	−75	−50	−25	0	20	50	100	150	200	250	300
	Temperature, K												
Property	123.15	173.15	198.15	223.15	248.15	273.15	293.15	323.15	373.15	423.15	473.15	523.15	573.15
Density, ρ_l (kg/m^3)	S	S	S	S	(1062)	(1034)	1 012	978	921	864	V	V	V
Specific heat capacity, $c_{p,l}$ (kJ/kg K)	S	S	S	S	(1.71)	(1.74)	(1.77)	(1.83)	(1.95)	(2.10)	V	V	V
Thermal conductivity, λ_l [(W/m^2)/(K/m)]	S	S	S	S	(0.135)	(0.133)	(0.131)	(0.129)	(0.127)	(0.125)	V	V	V
Dynamic viscosity, η_l (10^{-5} Ns/m^2)	S	S	S	S	(220)	160.0	112.0	73.1	43.0	28.0	V	V	V

Transport Properties of the Superheated Vapor

	Temperature, °C												
	− 150	− 100	− 50	0	25	100	200	300	400	600	800	1 000	1 200
Property (at low pressure)	Temperature, K												
	123.15	173.15	223.15	273.15	298.15	373.15	473.15	573.15	673.15	873.15	1 073.15	1 273.15	1 473.15
Specific heat capacity, $c_{p,g}$ (kJ/kg K)	S	S	S	L	L	L	(1.67)	(1.90)	(2.41)	(2.26)	—	—	—
Thermal conductivity, λ_g[(W/m^2)/(K/m)]	S	S	S	L	L	L	(0.025)	(0.036)	(0.047)	(0.059)	—	—	—
Dynamic viscosity, η_g(10^{-5} Ns/m^2)	S	S	S	L	L	L	(1.13)	(1.37)	(1.61)	(1.83)	—	—	—

DIMETHYL ETHER

Chemical formula: C_2H_6O
Molecular weight: 46.07
Melting point: 131.65 K
Boiling point: 248.3 K

Critical temperature: 400 K
Critical pressure: 5.27 MPa
Critical density: 259 kg/m^3
Normal vapor density: 2.06 kg/m^3 (@ 0°C, 101.3 kPa)

Properties of the Liquid at Temperatures Below the Normal Boiling Point

	Temperature, °C												
	−150	−100	−75	−50	−25	0	20	50	100	150	200	250	300
	Temperature, K												
Property	123.15	173.15	198.15	223.15	248.15	273.15	293.15	323.15	373.15	423.15	473.15	523.15	573.15
Density, ρ_l (kg/m^3)	S	(846)	(812)	(773)	735	V	V	V	V	V	V	V	V
Specific heat capacity, $c_{p,l}$ (kJ/kg K)	S	2.139	2.148	2.181	2.219	V	V	V	V	V	V	V	V
Thermal conductivity, λ_l [(W/m^2)/(K/m)]	S	0.228	0.211	0.194	0.178	V	V	V	V	V	V	V	V
Dynamic viscosity, η_l (10^{-5} Ns/m^2)	S	—	—	(30)	(23)	V	V	V	V	V	V	V	V

Transport Properties of the Superheated Vapor

	Temperature, °C												
	− 150	− 100	− 50	0	25	100	200	300	400	600	800	1 000	1 200
Property (at low pressure)	Temperature, K												
	123.15	173.15	223.15	273.15	298.15	373.15	473.15	573.15	673.15	873.15	1 073.15	1 273.15	1 473.15
Specific heat capacity, $c_{p,g}$ (kJ/kg K)	S	L	L	1.352	1.424	1.642	1.933	2.210	2.463	2.952	3.287	3.538	3.720
Thermal conductivity, λ_g[(W/m^2)/(K/m)]	S	L	L	0.015	0.017	0.025	0.038	0.054	0.068	0.100	0.133	0.163	0.190
Dynamic viscosity, η_g(10^{-5} Ns/m^2)	S	L	L	0.850	0.931	1.17	1.48	1.74	2.02	2.52	2.98	3.40	3.80

DIETHYL ETHER

Chemical formula: $(CH_3CH_2)_2O$
Molecular weight: 74.10
Melting point: 158.9 K
Normal boiling point: 307.8 K

Critical temperature: 467 K
Critical pressure: 3.610 MPa
Critical density: 265 kg/m^3
Normal vapor density: 3.31 kg/m^3
(@ 0 °C, 101.3 kPa)

Properties of the Liquid at Temperatures Below the Normal Boiling Point

	Temperature, °C												
	−150	−100	−75	−50	−25	0	20	50	100	150	200	250	300
	Temperature, K												
Property	123.15	173.15	198.15	223.15	248.15	273.15	293.15	323.15	373.15	423.15	473.15	523.15	573.15
Density, ρ_l (kg/m^3)	S	842	816	790	764	736	714	V	V	V	V	V	V
Specific heat capacity, $c_{p,l}$ (kJ/kg K)	S	2.014	2.093	2.135	2.190	2.261	2.336	V	V	V	V	V	V
Thermal conductivity, λ_l [(W/m^2)/(K/m)]	S	0.180	0.170	0.159	0.150	0.140	0.138	V	V	V	V	V	V
Dynamic viscosity, η_l (10^{-5} Ns/m^2)	S	171.0	88.1	55.0	38.7	29.6	24.3	V	V	V	V	V	V

Properties of the Saturated Liquid and Vapor

T_{sat}, K	307.75	323	343	363	383	403	423	443	458	463
p_{sat}, kPa	101.3	170	307	511	811	1 220	1 770	2 490	3 150	3 490
ρ_ℓ, kg/m^3	696.2	676.4	653.2	625.0	594.2	558.0	517.9	465.8	401.8	366.3
ρ_g, kg/m^3	3.16	5.08	8.92	14.77	23.49	36.38	55.51	87.31	132.0	162.0
h_ℓ, kJ/kg	0.0	36.6	85.9	137.0	190.3	248.6	305.0	367.1	416.1	433.0
h_g, kJ/kg	349.9	373.6	404.1	434.3	464.5	497.7	520.6	532.5	523.9	524.6
$\Delta h_{g,\ell}$, kJ/kg	349.9	337.0	318.2	297.3	274.2	249.1	215.6	165.4	107.8	81.6
$c_{p,\ell}$, kJ/(kg K)	2.37	2.43	2.51	2.61	2.72	2.86	3.01	3.20	3.75	4.07
$c_{p,g}$, kJ/(kg K)	1.40	1.96	2.05	2.14	2.26	2.43	2.75	3.44	4.15	4.50
η_ℓ, μNs/m^2	210	177	148	127	109	95	84	77	70	67
η_g, μNs/m^2	7.86	8.28	8.86	9.45	10.1	10.9	11.7	13.1	15.0	16.4
λ_ℓ, (mW/m^2)/(K/m)	126	120	112	104	95.6	87.5	79.9	71.4	65.3	63.3
λ_g, (mW/m^2)/(K/m)	15.8	17.4	19.5	22.4	24.2	27.0	30.2	34.2	38.6	41.5
Pr_ℓ	3.95	3.58	3.32	3.19	3.10	3.11	3.16	3.45	4.02	4.31
Pr_g	0.70	0.93	0.93	0.90	0.94	0.98	1.07	1.32	1.61	1.78
σ, mN/m	15.25	13.5	11.3	9.1	7.0	4.9	3.1	1.5	0.5	0.2
$\beta_{e,\ell}$, kK^{-1}	1.93	1.80	2.00	2.48	3.27	4.38	6.10	8.82	25.8	45.7

Transport Properties of the Superheated Vapor

	Temperature, °C												
	−150	−100	−50	0	25	100	200	300	400	600	800	1 000	1 200
	Temperature, K												
Property (at low pressure)	123.15	173.15	223.15	273.15	298.15	373.15	473.15	573.15	673.15	873.15	1 073.15	1 273.15	1 473.15
Specific heat capacity, $c_{p,g}$ (kJ/kg K)	S	L	L	L	L	1.842	2.173	2.424	2.583	(3.09)	—	—	—
Thermal conductivity, λ_g[(W/m^2)/(K/m)]	S	L	L	L	L	0.025	0.033	0.050	0.067	0.108	—	—	—
Dynamic viscosity, η_g(10^{-5} Ns/m^2)	S	L	L	L	L	0.939	1.18	1.41	1.65	2.12	—	—	—

METHYLPROPYL ETHER

Chemical formula: $C_4H_{10}O$
Molecular weight: 74.12
Melting point: 123.15 K
Boiling point: 312.25 K

Critical temperature: 476.2 K
Critical pressure: 3.80 MPa
Critical density: 269 kg/m^3
Normal vapor density: 3.31 kg/m^3
(@ 0 °C, 101.3 kPa)

Properties of the Liquid at Temperatures Below the Normal Boiling Point

	Temperature, °C												
	−150	−100	−75	−50	−25	0	20	50	100	150	200	250	300
	Temperature, K												
Property	123.15	173.15	198.15	223.15	248.15	273.15	293.15	323.15	373.15	423.15	473.15	523.15	573.15
Density, ρ_l (kg/m^3)	(875)	(850)	(825)	(800)	(775)	747	(725)	V	V	V	V	V	V
Specific heat capacity, $c_{p,l}$ (kJ/kg K)	(1.90)	1.942	1.980	2.028	2.085	2.154	2.216	V	V	V	V	V	V
Thermal conductivity, λ_l [(W/m^2)/(K/m)]	0.202	0.181	0.171	0.161	0.151	0.141	0.134	V	V	V	V	V	V
Dynamic viscosity, η_l (10^{-5} Ns/m^2)	S	—	—	—	(40)	31.4	26.0	V	V	V	V	V	V

ETHYLPROPYL ETHER

Chemical formula: $C_5H_{12}O$
Molecular weight: 88.15
Melting point: 146.4 K
Boiling point: 336.8 K

Critical temperature: 500.6 K
Critical pressure: 3.25 MPa
Critical density: (263) kg/m^3
Normal vapor density: 3.93 kg/m^3
(@ 0 °C, 101.3 kPa)

Properties of the Liquid at Temperatures Below the Normal Boiling Point

	Temperature, °C												
	−150	−100	−75	−50	−25	0	20	50	100	150	200	250	300
	Temperature, K												
Property	123.15	173.15	198.15	223.15	248.15	273.15	293.15	323.15	373.15	423.15	473.15	523.15	573.15
Density, ρ_l (kg/m^3)	S	(844)	(821)	(798)	(775)	(752)	733	(704)	V	V	V	V	V
Specific heat capacity, $c_{p,l}$ (kJ/kg K)	S	1.958	2.000	2.060	2.112	2.169	2.223	(2.30)	V	V	V	V	V
Thermal conductivity, λ_l [(W/m^2)/(K/m)]	S	0.170	0.161	0.152	0.144	0.136	0.129	0.120	V	V	V	V	V
Dynamic viscosity, η_l (10^{-5} Ns/m^2)	S	—	—	—	(52)	40.2	32.4	24.5	V	V	V	V	V

METHYL-t-BUTYL ETHER

Chemical formula: $CH_3OC_4H_9$
Molecular weight: 88.1
Melting point: 162.4 K
Normal boiling point: 331.2 K

Critical temperature: 503.4 K
Critical pressure: 3.411 MPa
Critical density: 275 kg/m^3
Normal vapor density: 3.93 kg/m^3 (@ 0 °C, 101.3 kPa)

Properties of the Liquid at Temperatures Below the Normal Boiling Point

	Temperature, °C												
	−150	−100	−75	−50	−25	0	20	50	100	150	200	250	300
	Temperature, K												
Property	123.15	173.15	198.15	223.15	248.15	273.15	293.15	323.15	373.15	423.15	473.15	523.15	573.15
Density, ρ_l (kg/m^3)	S	(860)	(835)	(810)	(785)	(760)	741	(710)	V	V	V	V	V
Specific heat capacity, $c_{p,l}$ (kJ/kg K)	S	1.766	1.821	1.886	1.949	2.039	2.105	(2.22)	V	V	V	V	V
Thermal conductivity, λ_l [(W/m^2)/(K/m)]	S	(0.171)	(0.163)	(0.154)	(0.146)	(0.138)	(0.132)	(0.122)	V	V	V	V	V
Dynamic viscosity, η_l (10^{-5} Ns/m^2)	S	—	—	(70)	(53)	(38.7)	(31.3)	(23.9)	V	V	V	V	V

Properties of the Saturated Liquid and Vapor

T_{sat}, K	331.2	340	360	380	400	420	440	460	480	503.4
p_{sat}, kPa	101.3	133	234	386	602	897	1 290	1 800	2 440	3 411
ρ_ℓ, kg/m^3	706	697	673	649	623	591	556	517	467	275
ρ_g, kg/m^3	3.40	4.39	7.51	12.2	18.9	28.5	42.3	62.7	96.2	275
h_ℓ, kJ/kg	−22.2	−1.7	46.5	96.6	148.7	203.1	259.8	319.2	382	488.1
h_g, kJ/kg	292.2	306.6	339.8	373.7	407.8	441.8	474.9	505.4	529.3	488.1
$\Delta h_{g,\ell}$, kJ/kg	314.4	308.3	293.3	277.1	259.1	238.7	215.1	186.3	147.3	
$c_{p,\ell}$, kJ/(kg K)	2.30	2.35	2.46	2.57	2.69	2.83	3.01	3.27	3.83	
$c_{p,g}$, kJ/(kg K)	1.78	1.83	1.95	2.08	2.23	2.39	2.61	2.95	3.76	
η_ℓ, μNs/m^2	231	214	184	151	129	109	91	74	59	44
η_g, μNs/m^2	8.50	8.76	9.38	10.0	10.7	11.5	12.3	13.3	14.9	44
λ_ℓ, (mW/m^2)/(K/m)	108	106	100	94	89	83	78	71	62	47
λ_g, (mW/m^2)/(K/m)	15.4	16.2	18.2	20.3	22.5	24.9	27.6	30.7	34.4	47
Pr_ℓ	4.90	4.76	4.53	4.12	3.92	3.73	3.52	3.36	3.53	
Pr_g	0.98	0.99	1.01	1.02	1.06	1.10	1.16	1.28	1.63	
σ, mN/m	14.7	13.8	11.7	9.8	7.9	6.1	4.3	2.7	1.3	
$\beta_{e,\ell}$, kK^{-1}	1.56	1.64	1.85	2.13	2.46	3.05	4.09	5.62	10.1	

DIPROPYL ETHER

Chemical formula: $C_6H_{14}O$
Molecular weight: 102.18
Melting point: 187.7 K
Boiling point: 341.45 K

Critical temperature: 500 K
Critical pressure: 2.88 MPa
Critical density: 259 kg/m^3
Normal vapor density: 4.56 kg/m^3 (@ 0°C, 101.3 kPa)

Properties of the Liquid at Temperatures Below the Normal Boiling Point

	Temperature, °C												
	−150	−100	−75	−50	−25	0	20	50	100	150	200	250	300
	Temperature, K												
Property	123.15	173.15	198.15	223.15	248.15	273.15	293.15	323.15	373.15	423.15	473.15	523.15	573.15
Density, ρ_l (kg/m^3)	S	S	(822)	(800)	(777)	756	736	(707)	V	V	V	V	V
Specific heat capacity, $c_{p,l}$ (kJ/kg K)	S	S	1.825	1.892	1.968	2.052	2.119	2.244	V	V	V	V	V
Thermal conductivity, λ_l [(W/m^2)/(K/m)]	S	S	0.144	0.139	0.134	0.129	0.128	0.120	V	V	V	V	V
Dynamic viscosity, η_l (10^{-5} Ns/m^2)	S	S	—	—	(72)	54.2	42.0	30.3	V	V	V	V	V

Transport Properties of the Superheated Vapor

	Temperature, °C												
	−150	−100	−50	0	25	100	200	300	400	600	800	1 000	1 200
Property (at low pressure)	Temperature, K												
	123.15	173.15	223.15	273.15	298.15	373.15	473.15	573.15	673.15	873.15	1 073.15	1 273.15	1 473.15
Specific heat capacity, $c_{p,g}$ (kJ/kg K)	S	S	L	L	L	1.820	2.172	2.484	2.754	3.184	—	—	—
Thermal conductivity, λ_g[(W/m^2)/(K/m)]	S	S	L	L	L	0.019	(0.031)	(0.045)	(0.063)	—	—	—	—
Dynamic viscosity, η_g(10^{-5} Ns/m^2)	S	S	L	L	L	0.789	(1.02)	(1.21)	(1.40)	—	—	—	—

ETHYLENE OXIDE

Chemical formula: $\langle (CH_2)_2 \rangle O$
Molecular weight: 44.054
Melting point: 161 K
Normal boiling point: 283.5 K

Critical temperature: 469 K
Critical pressure: 7.194 MPa
Critical density: 315 kg/m^3
Normal vapor density: 1.97 kg/m^3
(@ 0 °C, 101.3 kPa)

Properties of the Liquid at Temperatures Below the Normal Boiling Point

	Temperature, °C												
	−150	−100	−75	−50	−25	0	20	50	100	150	200	250	300
	Temperature, K												
Property	123.15	173.15	198.15	223.15	248.15	273.15	293.15	323.15	373.15	423.15	473.15	523.15	573.15
Density, ρ_l (kg/m^3)	S	(1026)	(994)	962	929	897	V	V	V	V	V	V	
Specific heat capacity, $c_{p,l}$ (kJ/kg K)	S	1.859	1.859	1.859	1.892	1.951	V	V	V	V	V	V	V
Thermal conductivity, λ_l [(W/m^2)/(K/m)]	S	(0.261)	(0.248)	0.235	0.222	0.208	V	V	V	V	V	V	V
Dynamic viscosity, η_l (10^{-5} Ns/m^2)	S	—	(82)	57.0	40.0	32.0	V	V	V	V	V	V	V

Properties of the Saturated Liquid and Vapor

T_{sat}, K	283.5	300	320	340	360	380	400	420	440	469
p_{sat}, kPa	101.3	186	359	621	1 030	1 660	2 480	3 450	4 830	7 194
ρ_ℓ, kg/m^3	889	866	835	804	760	721	691	682	584	315
ρ_g, kg/m^3	1.94	3.44	6.33	10.8	17.5	27.2	41.1	61.6	93.9	315
h_ℓ, kJ/kg	−440	−409	−367	−333	−289	−238	−183	−129	−65	66
h_g, kJ/kg	129	144	161	178	193	206	215	219	211	66
$\Delta h_{g,\ell}$, kJ/kg	569	553	528	511	482	444	398	348	276	
$c_{p,\ell}$, kJ/(kg K)	1.96	2.01	2.09	2.19	2.32	2.48	2.68	2.92	3.21	
$c_{p,g}$, kJ/(kg K)	1.09	1.17	1.28	1.40	1.55	1.73	1.99	2.47	3.40	
η_ℓ, μNs/m^2	284	245	210	182	160	142	128	116	105	47
η_g, μNs/m^2	9.0	9.6	10.3	11.0	11.7	12.5	13.4	14.4	15.9	47
λ_ℓ, (mW/m^2)/(K/m)	158	152	144	135	127	119	110	102	93	69
λ_g, (mW/m^2)/(K/m)	11.5	13.6	16.3	19.3	22.5	25.1	29.9	34.5	40.4	69
Pr_ℓ	3.52	3.24	3.04	2.93	2.92	2.97	3.12	3.34	3.63	
Pr_g	0.85	0.83	0.81	0.80	0.81	0.86	0.89	1.01	1.34	
σ, mN/m	25.87	23.24	20.10	17.03	14.03	11.11	8.29	5.59	3.05	
$\beta_{e,\ell}$, kK^{-1}	1.60	1.76	1.98	2.30	2.71	3.30	4.20	5.76	9.26	

Transport Properties of the Superheated Vapor

	Temperature, °C												
	−150	−100	−50	0	25	100	200	300	400	600	800	1 000	1 200
	Temperature, K												
Property (at low pressure)	123.15	173.15	223.15	273.15	298.15	373.15	473.15	573.15	673.15	873.15	1 073.15	1 273.15	1 473.15
Specific heat capacity, $c_{p,g}$ (kJ/kg K)	S	L	L	L	1.097	1.315	1.637	1.901	2.114	2.445	—	—	—
Thermal conductivity, λ_g[(W/m^2)/(K/m)]	S	L	L	L	0.012	0.019	(0.032)	(0.044)	(0.057)	(0.082)	—	—	—
Dynamic viscosity, η_g(10^{-5} Ns/m^2)	S	L	L	L	(0.94)	(1.18)	(1.49)	(1.78)	(2.06)	(2.55)	—	—	—

PROPYLENE OXIDE

Chemical formula: $CH_3(CHCH_2)O$
Molecular weight: 58.08
Melting point: 161 K
Normal boiling point: 307.5 K

Critical temperature: 482.2 K
Critical pressure: 4.920 MPa
Critical density: 312 kg/m^3
Normal vapor density: 2.59 kg/m^3
(@ 0°C, 101.3 kPa)

Properties of the Liquid at Temperatures Below the Normal Boiling Point

	Temperature, °C												
	−150	−100	−75	−50	−25	0	20	50	100	150	200	250	300
	Temperature, K												
Property	123.15	173.15	198.15	223.15	248.15	273.15	293.15	323.15	373.15	423.15	473.15	523.15	573.15
Density, ρ_l (kg/m^3)	S	(970)	(941)	912	883	854	829	V	V	V	V	V	V
Specific heat capacity, $c_{p,l}$ (kJ/kg K)	S	1.890	1.880	1.906	1.947	1.997	2.051	V	V	V	V	V	V
Thermal conductivity, λ_l [(W/m^2)/(K/m)]	S	(0.205)	(0.194)	0.183	0.172	0.162	(0.153)	V	V	V	V	V	V
Dynamic viscosity, η_l (10^{-5} Ns/m^2)	S	—	—	(140)	(77)	45.5	31.5	V	V	V	V	V	V

Properties of the Saturated Liquid and Vapor

T_{sat}, K	307.5	320	340	360	380	400	420	440	460	482.2
p_{sat}, kPa	101.3	159	297	496	814	1 230	1 790	2 480	3 450	4 920
ρ_ℓ, kg/m^3	812	796	769	740	709	675	636	491	531	312
ρ_g, kg/m^3	2.38	3.57	6.39	10.7	17.2	26.6	40.5	61.1	95.9	312
h_ℓ, kJ/kg	−293	−271	−220	−177	−135	−86	−37	33	106	200
h_g, kJ/kg	184	198	220	242	263	282	298	309	307	200
$\Delta h_{g,\ell}$, kJ/kg	477	469	440	419	398	368	335	276	201	
$c_{p,\ell}$, kJ/(kg K)	2.06	2.10	2.18	2.29	2.42	2.60	2.81	3.07	3.38	
$c_{p,g}$, kJ/(kg K)	1.32	1.39	1.49	1.62	1.76	1.94	2.19	2.63	3.83	
η_ℓ, μNs/m^2	278	251	217	191	171	155	143	133	125	48
η_g, μNs/m^2	9.1	9.5	10.2	10.8	11.5	12.3	13.1	14.1	15.5	48
λ_ℓ, (mW/m^2)/(K/m)	147	143	135	128	120	113	105	96	88	71
λ_g, (mW/m^2)/(K/m)	12.1	13.3	15.5	17.8	20.5	23.5	26.9	31.3	47.4	71
Pr_ℓ	3.90	3.69	3.49	3.41	3.44	3.57	3.84	4.23	4.79	
Pr_g	0.99	0.99	0.98	0.98	0.99	1.02	1.07	1.18	1.25	
σ, mN/m	19.9	18.2	15.5	12.9	10.4	8.0	5.7	3.5	1.6	
$\beta_{e,\ell}$, kK^{-1}	1.59	1.71	1.9	2.24	2.64	3.27	4.26	6.22	19.4	

Transport Properties of the Superheated Vapor

	Temperature, °C												
	−150	−100	−50	0	25	100	200	300	400	600	800	1 000	1 200
Property (at low pressure)	Temperature, K												
	123.15	173.15	223.15	273.15	298.15	373.15	473.15	573.15	673.15	873.15	1 073.15	1 273.15	1 473.15
Specific heat capacity, $c_{p,g}$ (kJ/kg K)	S	L	L	L	L	1.513	1.828	2.109	2.348	2.689			
Thermal conductivity, λ_g[(W/m^2)/(K/m)]	S	L	L	L	L	(0.021)	(0.033)	(0.045)	(0.058)	(0.084)			
Dynamic viscosity, η_g(10^{-5} Ns/m^2)	S	L	L	L	L	(1.11)	(1.40)	(1.67)	(1.93)	(2.40)			

FURAN

Chemical formula: C_4H_4O
Molecular weight: 68.07
Melting point: 187.5 K
Boiling point: 304.5 K

Critical temperature: 490.2 K
Critical pressure: 5.50 MPa
Critical density: 312 kg/m^3
Normal vapor density: 3.04 kg/m^3
(@ 0 °C, 101.3 kPa)

Properties of the Liquid at Temperatures Below the Normal Boiling Point

	Temperature, °C												
	−150	−100	−75	−50	−25	0	20	50	100	150	200	250	300
	Temperature, K												
Property	123.15	173.15	198.15	223.15	248.15	273.15	293.15	323.15	373.15	423.15	473.15	523.15	573.15
Density, ρ_l (kg/m^3)	S	S	(1040)	(1018)	(996)	965	937	V	V	V	V	V	V
Specific heat capacity, $c_{p,l}$ (kJ/kg K)	S	S	1.463	1.512	1.550	1.616	1.670	V	V	V	V	V	V
Thermal conductivity, λ_l [(W/m^2)/(K/m)]	S	S	(0.186)	(0.178)	(0.169)	(0.161)	(0.154)	V	V	V	V	V	V
Dynamic viscosity, η_l (10^{-5} Ns/m^2)	S	S	—	(65)	(53)	(44.7)	38.2	V	V	V	V	V	V

Transport Properties of the Superheated Vapor

	Temperature, °C												
	−150	−100	−50	0	25	100	200	300	400	600	800	1 000	1 200
Property (at low pressure)	Temperature, K												
	123.15	173.15	223.15	273.15	298.15	373.15	473.15	573.15	673.15	873.15	1 073.15	1 273.15	1 473.15
Specific heat capacity, $c_{p,g}$ (kJ/kg K)	S	S	L	L	L	1.218	1.507	1.742	1.928	2.200	2.374	2.499	2.623
Thermal conductivity, λ_g[(W/m^2)/(K/m)]	S	S	L	L	L	(0.017)	(0.027)	(0.037)	(0.048)	(0.067)	—	—	—
Dynamic viscosity,	S	S	L	L	L	(1.07)	(1.35)	(1.62)	(1.87)	(2.33)	—	—	—

1,4-DIOXAN

Chemical formula: $C_4H_8O_2$
Molecular weight: 88.11
Melting point: 284.25 K
Boiling point: 374.5 K

Critical temperature: 587 K
Critical pressure: 5.21 MPa
Critical density: 370 kg/m^3
Normal vapor density: 3.93 kg/m^3
(@ 0°C, 101.3 kPa)

Properties of the Liquid at Temperatures Below the Normal Boiling Point

	Temperature, °C												
	−150	−100	−75	−50	−25	0	20	50	100	150	200	250	300
	Temperature, K												
Property	123.15	173.15	198.15	223.15	248.15	273.15	293.15	323.15	373.15	423.15	473.15	523.15	573.15
Density, ρ_l (kg/m^3)	S	S	S	S	S	S	1 034	1 008	(965)	V	V	V	V
Specific heat capacity, $c_{p,l}$ (kJ/kg K)	S	S	S	S	S	S	1.721	1.800	(1.96)	V	V	V	V
Thermal conductivity, λ_l [(W/m^2)/(K/m)]	S	S	S	S	S	S	0.163	(0.161)	(0.159)	V	V	V	V
Dynamic viscosity, η_l (10^{-5} Ns/m^2)	S	S	S	S	S	S	126.0	77.9	(42)	V	V	V	V

Transport Properties of the Superheated Vapor

	Temperature, °C												
	−150	−100	−50	0	25	100	200	300	400	600	800	1 000	1 200
Property	Temperature, K												
(at low pressure)	123.15	173.15	223.15	273.15	298.15	373.15	473.15	573.15	673.15	873.15	1 073.15	1 273.15	1 473.15
Specific heat capacity, $c_{p,g}$ (kJ/kg K)	S	S	S	S	L	L	1.692	1.996	2.243	2.590	—	—	—
Thermal conductivity, λ_g[(W/m^2)/(K/m)]	S	S	S	S	L	L	(0.030)	(0.043)	(0.059)	—	—	—	—
Dynamic viscosity, η_g(10^{-5} Ns/m^2)	S	S	S	S	L	L	(1.24)	(1.51)	(1.77)	—	—	—	—

FLUOROMETHANE

Chemical formula: CH_3F
Molecular weight: 34.03
Melting point: 131.4 K
Boiling point: 194.8 K

Critical temperature: 317.8 K
Critical pressure: 5.88 MPa
Critical density: 274 kg/m^3
Normal vapor density: 1.52 kg/m^3 (@ 0 °C, 101.3 kPa)

Transport Properties of the Superheated Vapor

Property (at low pressure)	Temperature, °C: −150	−100	−50	0	25	100	200	300	400	500	1 000	1 500	2 000
	Temperature, K: 123.15	173.15	223.15	273.15	298.15	373.15	473.15	573.15	673.15	773.15	1 273.15	1 773.15	2 273.15
Specific heat capacity, $c_{p,g}$ (kJ/kg K)	S	L	1.019	1.070	1.099	1.245	1.450	1.645	1.823	1.980	2.509	2.772	2.912
Thermal conductivity, λ_g[(W/m^2)/(K/m)]	S	L	0.012	0.016	0.018	0.025	0.035	0.048	0.062	0.076	0.139	0.194	0.242
Dynamic viscosity, η_g(10^{-5} Ns/m^2)	S	L	0.86	1.06	1.16	1.44	1.80	2.16	2.47	2.80	4.08	5.17	6.14

DIFLUOROMETHANE

Chemical formula: CH_2F
Molecular weight: 52.03
Melting point: —
Boiling point: 221.15 K

Critical temperature: 351.6 K
Critical pressure: 5.82 MPa
Critical density: (423) kg/m^3
Normal vapor density: 2.32 kg/m^3 (@ 0 °C, 101.3 kPa)

Transport Properties of the Superheated Vapor

Property (at low pressure)	Temperature, °C: −150	−100	−50	0	25	100	200	300	400	500	1 000	1 500	2 000
	Temperature, K: 123.15	173.15	223.15	273.15	298.15	373.15	473.15	573.15	673.15	773.15	1 273.15	1 773.15	2 273.15
Specific heat capacity, $c_{p,g}$ (kJ/kg K)	S	L	0.733	0.794	0.824	0.941	1.093	1.229	1.344	1.440	1.737	1.876	1.948
Thermal conductivity, λ_g[(W/m^2)/(K/m)]	S	L	0.010	0.013	0.014	0.020	0.028	0.038	0.048	0.057	0.101	0.137	0.168
Dynamic viscosity, η_g(10^{-5} Ns/m^2)	S	L	0.92	1.13	1.23	1.51	1.90	2.26	2.60	2.93	4.24	5.38	6.36

TRIFLUOROMETHANE (FLUOROFORM) (REFRIGERANT 14)

Chemical formula: CHF_3
Molecular weight: 70.02
Melting point: 110.15 K
Boiling point: 191.15 K

Critical temperature: 299.5 K
Critical pressure: 4.87 MPa
Critical density: (511) kg/m^3
Normal vapor density: 3.13 kg/m^3
(@ 0 °C, 101.3 kPa)

Transport Properties of the Superheated Vapor

Property (at low pressure)	Temperature, °C: −150	−100	−50	0	25	100	200	300	400	500	1 000	1 500	2 000
	Temperature, K: 123.15	173.15	223.15	273.15	298.15	373.15	473.15	573.15	673.15	773.15	1 273.15	1 773.15	2 273.15
Specific heat capacity, $c_{p,g}$ (kJ/kg K)	L	L	0.621	0.708	0.730	0.843	0.956	1.057	1.138	1.240	1.373	1.445	1.481
Thermal conductivity, λ_g[(W/m^2)/(K/m)]	L	L	0.010	0.013	0.015	0.020	0.028	0.037	0.045	0.053	0.087	0.114	0.137
Dynamic viscosity, η_g(10^{-5} Ns/m^2)	L	L	1.06	1.35	1.48	1.83	2.22	2.60	2.94	3.24	4.64	5.84	6.89

TETRAFLUOROMETHANE (CARBON TETRAFLUORIDE)

Chemical formula: CF_4
Molecular weight: 88.00
Melting point: 86.4 K
Boiling point: 145.2 K

Critical temperature: 227.8 K
Critical pressure: 3.79 MPa
Critical density: 628 kg/m^3
Normal vapor density: 3.93 kg/m^3
(@ 0 °C, 101.3 kPa)

Transport Properties of the Superheated Vapor

Property (at low pressure)	Temperature, °C: −150	−100	−50	0	25	100	200	300	400	500	1 000	1 500	2 000
	Temperature, K: 123.15	173.15	223.15	273.15	298.15	373.15	473.15	573.15	673.15	773.15	1 273.15	1 773.15	2 273.15
Specific heat capacity, $c_{p,g}$ (kJ/kg K)	L	0.501	0.577	0.664	0.694	0.789	0.895	0.986	1.045	1.090	1.192	1.258	1.291
Thermal conductivity, λ_g[(W/m^2)/(K/m)]	L	0.006	0.010	0.015	0.017	0.023	0.031	0.039	0.046	0.050	0.082	0.104	0.126
Dynamic viscosity, η_g(10^{-5} Ns/m^2)	L	1.07	1.38	1.58	1.70	2.04	2.52	2.94	3.24	3.57	4.98	6.17	7.25

CHLOROMETHANE

Chemical formula: CH_3Cl
Molecular weight: 50.49
Melting point: 175.45 K
Boiling point: 248.9 K

Critical temperature: 416.3 K
Critical pressure: 6.68 MPa
Critical density: 363 kg/m^3
Normal vapor density: 2.25 kg/m^3 (@ 0 °C, 101.3 kPa)

Properties of the Liquid at Temperatures Below the Normal Boiling Point

	Temperature, °C												
	−200	−180	−160	−140	−120	−100	−50	0	20	50	100	150	200
	Temperature, K												
Property	73.15	93.15	113.15	133.15	153.15	173.15	223.15	273.15	293.15	323.15	373.15	423.15	473.15
Density, ρ_l (kg/m^3)	S	S	S	S	S	S	1 050	V	V	V	V	V	V
Specific heat capacity, $c_{p,l}$ (kJ/kg K)	S	S	S	S	S	S	1.499	V	V	V	V	V	V
Thermal conductivity, λ_l [(W/m^2)/(K/m)]	S	S	S	S	S	S	0.199	V	V	V	V	V	V
Dynamic viscosity, η_l (10^{-5} Ns/m^2)	S	S	S	S	S	S	39.7	V	V	V	V	V	V

Transport Properties of the Superheated Vapor

	Temperature, °C												
	−150	−100	−50	0	25	100	200	300	400	500	1 000	1 500	2 000
Property	Temperature, K												
(at low pressure)	123.15	173.15	223.15	273.15	298.15	373.15	473.15	573.15	673.15	773.15	1 273.15	1 773.15	2 273.15
Specific heat capacity, $c_{p,g}$ (kJ/kg K)	S	S	L	0.775	0.808	0.917	1.055	1.181	1.290	1.382	1.717	1.881	1.970
Thermal conductivity, λ_g[(W/m^2)/(K/m)]	S	S	L	0.009	0.011	0.016	0.023	0.030	0.042	0.057	0.092	0.127	0.157
Dynamic viscosity, η_g(10^{-5} Ns/m^2)	S	S	L	0.970	1.08	1.36	1.76	2.11	2.40	2.66	3.92	4.97	5.89

DICHLOROMETHANE

Chemical formula: CH_2Cl_2
Molecular weight: 84.93
Melting point: 177.2 K
Boiling point: 313.0 K

Critical temperature: 510 K
Critical pressure: 6.08 MPa
Critical density: 440 kg/m^3
Normal vapor density: 3.79 kg/m^3
(@ 0 °C, 101.3 kPa)

Properties of the Liquid at Temperatures Below the Normal Boiling Point

	Temperature, °C												
	−200	−180	−160	−140	−120	−100	−50	0	20	50	100	150	200
	Temperature, K												
Property	73.15	93.15	113.15	133.15	153.15	173.15	223.15	273.15	293.15	323.15	373.15	423.15	473.15
Density, ρ_l (kg/m^3)	S	S	S	S	S	S	1 450	1 362	1 336	V	V	V	V
Specific heat capacity, $c_{p,l}$ (kJ/kg K)	S	S	S	S	S	S	1.126	1.143	1.156	V	V	V	V
Thermal conductivity, λ_l [(W/m^2)/(K/m)]	S	S	S	S	S	S	0.163	0.148	0.141	V	V	V	V
Dynamic viscosity, η_l (10^{-5} Ns/m^2)	S	S	S	S	S	S	118	53.9	44.2	V	V	V	V

Transport Properties of the Superheated Vapor

	Temperature, °C												
	−150	−100	−50	0	25	100	200	300	400	500	1 000	1 500	2 000
Property (at low pressure)	Temperature, K												
	123.15	173.15	223.15	273.15	298.15	373.15	473.15	573.15	673.15	773.15	1 273.15	1 773.15	2 273.15
Specific heat capacity, $c_{p,g}$ (kJ/kg K)	S	S	L	L	L	0.691	0.783	0.854	0.909	0.950	1.097	1.324	1.367
Thermal conductivity, λ_g[(W/m^2)/(K/m)]	S	S	L	L	L	0.012	0.017	0.022	0.027	0.032	0.055	0.074	0.090
Dynamic viscosity, η_g(10^{-5} Ns/m^2)	S	S	L	L	L	1.27	1.60	1.94	2.20	2.49	3.69	4.68	5.55

TRICHLOROMETHANE (CHLOROFORM)

Chemical formula: $CHCl_3$
Molecular weight: 119.4
Melting point: 209.9 K
Normal boiling point: 334.5 K

Critical temperature: 536.4 K
Critical pressure: 5.470 MPa
Critical density: 498 kg/m^3
Normal vapor density: 5.33 kg/m^3
(@ 0°C, 101.3 kPa)

Properties of the Liquid at Temperatures Below the Normal Boiling Point

	Temperature, °C												
	−200	−180	−160	−140	−120	−100	−50	0	20	50	100	150	200
	Temperature, K												
Property	73.15	93.15	113.15	133.15	153.15	173.15	223.15	273.15	293.15	323.15	373.15	423.15	473.15
Density, ρ_l (kg/m^3)	S	S	S	S	S	S	1 618	1 526	1 490	1 433	V	V	V
Specific heat capacity, $c_{p,l}$ (kJ/kg K)	S	S	S	S	S	S	0.959	0.971	0.992	1.034	V	V	V
Thermal conductivity, λ_l [(W/m^2)/(K/m)]	S	S	S	S	S	S	0.156	0.145	0.128	0.115	V	V	V
Dynamic viscosity, η_l (10^{-5} Ns/m^2)	S	S	S	S	S	S	(120)	70.6	56.3	42.4	V	V	V

Properties of the Saturated Liquid and Vapor

T_{sat}, K	334.5	360	380	400	420	440	460	480	505	536.4
p_{sat}, kPa	101.3	221	374	599	914	1 321	1 893	2 598	3 725	5 470
ρ_ℓ, kg/m^3	1 415	1 361	1 333	1 282	1 248	1 184	1 114	1 050	969	498
ρ_g, kg/m^3	4.50	9.33	15.3	23.4	36.1	53.0	77.3	109	158	498
h_ℓ, kJ/kg	−108	−79	−61	−41	−17	4	27	45	67	136
h_g, kJ/kg	141	154	164	173	181	188	194	197	197	136
$\Delta h_{g,\ell}$, kJ/kg	249	233	225	214	198	184	167	152	130	
$c_{p,\ell}$, kJ/(kg K)	1.00	1.03	1.07	1.11	1.15	1.21	1.32	1.43	1.59	
$c_{p,g}$, kJ/(kg K)	0.60	0.63	0.66	0.69	0.73	0.79	0.87	1.00	1.28	
η_ℓ, μNs/m^2	400	342	299	267	220	197	181	165	150	65
η_g, μNs/m^2	11.2	12.2	13.0	13.7	14.6	15.5	16.5	17.9	19.8	65
λ_ℓ, (mW/m^2)/(K/m)	111	107	103	98.4	94.2	91.3	81.6	73.3	69.1	26.8
λ_g, (mW/m^2)/(K/m)	8.71	9.74	10.6	11.5	12.5	13.5	13.6	15.9	17.5	26.8
Pr_ℓ	3.60	3.29	3.11	3.01	2.69	2.61	2.93	3.22	3.45	
Pr_g	0.77	0.79	0.81	0.81	0.85	0.91	1.06	1.13	1.45	
σ, mN/m	22.5	18.6	16.0	13.6	11.2	9.5	6.8	5.9	3.1	
$\beta_{e,\ell}$, kK^{-1}	1.38	1.58	1.78	2.03	2.36	2.82	3.54	4.63	6.86	

Transport Properties of the Superheated Vapor

	Temperature, °C												
	−150	−100	−50	0	25	100	200	300	400	500	1 000	1 500	2 000
Property (at low pressure)	Temperature, K												
	123.15	173.15	223.15	273.15	298.15	373.15	473.15	573.15	673.15	773.15	1 273.15	1 773.15	2 273.15
Specific heat capacity, $c_{p,g}$ (kJ/kg K)	S	S	L	L	L	0.607	0.666	0.708	0.737	0.762	0.829	0.861	0.877
Thermal conductivity, λ_g[(W/m^2)/(K/m)]	S	S	L	L	L	0.010	0.014	0.018	0.021	0.025	0.041	0.054	0.065
Dynamic viscosity, η_g(10^{-5} Ns/m^2)	S	S	L	L	L	1.29	1.60	1.92	2.25	2.54	3.66	4.63	5.48

TETRACHLOROMETHANE (CARBON TETRACHLORIDE)

Chemical formula: CCl_4
Molecular weight: 153.8
Melting point: 250.25 K
Normal boiling point: 349.85 K

Critical temperature: 556.35 K
Critical pressure: 4.560 MPa
Critical density: 588 kg/m^3
Normal vapor density: 6.87 kg/m^3
(@ 0 °C, 101.3 kPa)

Properties of the Liquid at Temperatures Below the Normal Boiling Point

	Temperature, °C												
	−200	−180	−160	−140	−120	−100	−50	0	20	50	100	150	200
	Temperature, K												
Property	73.15	93.15	113.15	133.15	153.15	173.15	223.15	273.15	293.15	323.15	373.15	423.15	473.15
Density, ρ_l (kg/m^3)	S	S	S	S	S	S	S	1 633	1 594	1 534	V	V	V
Specific heat capacity, $c_{p,l}$ (kJ/kg K)	S	S	S	S	S	S	S	0.842	0.850	0.862	V	V	V
Thermal conductivity, λ_l [(W/m^2)/(K/m)]	S	S	S	S	S	S	S	0.107	0.106	0.105	V	V	V
Dynamic viscosity, η_l (10^{-5} Ns/m^2)	S	S	S	S	S	S	S	134.9	96.1	65.4	V	V	V

Properties of the Saturated Liquid and Vapor

T_{sat}, K	349.9	370	390	410	430	450	470	495	525	556.35
p_{sat}, kPa	101.3	184	307	473	701	1 020	1 390	2 020	3 160	4 560
ρ_ℓ, kg/m^3	1 484	1 442	1 397	1 351	1 303	1 250	1 199	1 107	989	588
ρ_g, kg/m^3	5.44	9.40	15.2	23.4	34.8	50.3	71.2	108.5	184.5	588
h_ℓ, kJ/kg	−36	−17	−1	16	38	53	72	92	123	177
h_g, kJ/kg	159	169	179	188	197	205	212	218	221	177
$\Delta h_{g,\ell}$, kJ/kg	195	188	180	172	159	152	140	126	98	
$c_{p,\ell}$, kJ/(kg K)	0.92	0.94	0.97	1.01	1.06	1.14	1.24	1.36	1.57	
$c_{p,g}$, kJ/(kg K)	0.58	0.60	0.62	0.65	0.68	0.73	0.80	0.91	1.30	
η_ℓ, μNs/m^2	494	407	352	309	274	241	205	154	98	63
η_g, μNs/m^2	11.9	12.5	13.3	14.1	14.9	15.7	16.7	18.9	21.0	63
λ_ℓ, (mW/m^2)/(K/m)	92	87	83	78	74	70	65	57	45	25
λ_g, (mW/m^2)/(K/m)	8.6	9.3	10.0	10.7	11.5	12.3	13.2	14.3	16.3	25
Pr_ℓ	4.94	4.40	4.16	4.08	3.93	3.92	3.91	3.67	3.42	
Pr_g	0.80	0.81	0.82	0.85	0.88	0.93	1.01	1.20	1.67	
σ, mN/m	20.2	17.6	15.4	13.1	10.9	8.8	6.9	4.4	2.0	
$\beta_{e,\ell}$, kK^{-1}	1.36	1.50	1.68	1.90	2.19	2.58	3.14	4.30	7.83	

Transport Properties of the Superheated Vapor

	Temperature, °C												
	− 150	− 100	− 50	0	25	100	200	300	400	500	1 000	1 500	2 000
Property	Temperature, K												
(at low pressure)	123.15	173.15	223.15	273.15	298.15	373.15	473.15	573.15	673.15	773.15	1 273.15	1 773.15	2 273.15
Specific heat capacity, $c_{p,g}$ (kJ/kg K)	S	S	S	L	L	0.586	0.624	0.645	0.657	0.670	0.691	0.696	0.699
Thermal conductivity, λ_g[(W/m^2)/(K/m)]	S	S	S	L	L	0.009	0.012	0.015	0.019	0.021	0.032	0.041	0.049
Dynamic viscosity, η_g(10^{-5} Ns/m^2)	S	S	S	L	L	1.23	1.53	1.83	2.12	2.38	3.45	4.35	5.15

BROMOMETHANE

Chemical formula: CH_3Br
Molecular weight: 94.94
Melting point: 179.5 K
Boiling point: 276.7 K

Critical temperature: 464 K
Critical pressure: 6.61 MPa
Critical density: (586) kg/m^3
Normal vapor density: 4.24 kg/m^3
(@ 0°C, 101.3 kPa)

Properties of the Liquid at Temperatures Below the Normal Boiling Point

	Temperature, °C												
	−200	−180	−160	−140	−120	−100	−50	0	20	50	100	150	200
	Temperature, K												
Property	73.15	93.15	113.15	133.15	153.15	173.15	223.15	273.15	293.15	323.15	373.15	423.15	473.15
Density, ρ_l (kg/m^3)	S	S	S	S	S	S	1 857	1 729	V	V	V	V	V
Specific heat capacity, $c_{p,l}$ (kJ/kg K)	S	S	S	S	S	S	0.816	0.829	V	V	V	V	V
Thermal conductivity, λ_l [(W/m^2)/(K/m)]	S	S	S	S	S	S	0.114	0.103	V	V	V	V	V
Dynamic viscosity, η_l (10^{-5} Ns/m^2)	S	S	S	S	S	S	(61)	37.7	V	V	V	V	V

Transport Properties of the Superheated Vapor

	Temperature, °C												
	−150	−100	−50	0	25	100	200	300	400	500	1 000	1 500	2 000
Property (at low pressure)	Temperature, K												
	123.15	173.15	223.15	273.15	298.15	373.15	473.15	573.15	673.15	773.15	1 273.15	1 773.15	2 273.15
Specific heat capacity, $c_{p,g}$ (kJ/kg K)	S	S	L	L	0.452	0.507	0.579	0.643	0.697	0.748	0.914	1.087	1.105
Thermal conductivity, λ_g[(W/m^2)/(K/m)]	S	S	L	L	0.080	0.011	0.016	0.023	0.029	0.035	0.065	0.091	0.113
Dynamic viscosity, η_g(10^{-5} Ns/m^2)	S	S	L	L	1.34	1.70	2.16	2.60	3.03	3.43	5.20	6.67	7.95

DIBROMOMETHANE

Chemical formula: CH_2Br_2
Molecular weight: 173.84
Melting point: 220.6 K
Boiling point: 371.8 K

Critical temperature: 583 K
Critical pressure: 7.17 MPa
Critical density: (759) kg/m^3
Normal vapor density: 7.76 kg/m^3
(@ 0°C, 101.3 kPa)

Properties of the Liquid at Temperatures Below the Normal Boiling Point

	Temperature, °C												
	−200	−180	−160	−140	−120	−100	−50	0	20	50	100	150	200
	Temperature, K												
Property	73.15	93.15	113.15	133.15	153.15	173.15	223.15	273.15	293.15	323.15	373.15	423.15	473.15
Density, ρ_l (kg/m^3)	S	S	S	S	S	S	(2680)	2 549	2 497	2 413	V	V	V
Specific heat capacity, $c_{p,l}$ (kJ/kg K)	S	S	S	S	S	S	0.600	0.601	0.605	0.610	V	V	V
Thermal conductivity, λ_l [(W/m^2)/(K/m)]	S	S	S	S	S	S	0.120	0.111	0.108	0.102	V	V	V
Dynamic viscosity, η_l (10^{-5} Ns/m^2)	S	S	S	S	S	S	(260)	134.0	102.5	78.7	V	V	V

Transport Properties of the Superheated Vapor

	Temperature, °C												
	−150	−100	−50	0	25	100	200	300	400	500	1 000	1 500	2 000
Property (at low pressure)	Temperature, K												
	123.15	173.15	223.15	273.15	298.15	373.15	473.15	573.15	673.15	773.15	1 273.15	1 773.15	2 273.15
Specific heat capacity, $c_{p,g}$ (kJ/kg K)	S	S	L	L	L	0.356	0.394	0.420	0.453	0.471	0.533	(0.571)	(0.588)
Thermal conductivity, λ_g[(W/m^2)/(K/m)]	S	S	L	L	L	0.008	(0.013)	(0.016)	(0.019)	(0.023)	(0.037)	(0.051)	(0.066)
Dynamic viscosity, η_g(10^{-5} Ns/m^2)	S	S	L	L	L	(1.60)	(2.02)	(2.44)	(2.84)	(3.28)	(5.00)	(6.42)	(7.70)

TRIBROMOMETHANE (BROMOFORM)

Chemical formula: $CHBr_3$
Molecular weight: 252.77
Melting point: 281.05 K
Boiling point: 423.65 K

Critical temperature: 685.1 K
Critical pressure: 5.65 MPa
Critical density: (854) kg/m^3
Normal vapor density: 11.3 kg/m^3 (@ 0 °C, 101.3 kPa)

Properties of the Liquid at Temperatures Below the Normal Boiling Point

	Temperature, °C												
	−200	−180	−160	−140	−120	−100	−50	0	20	50	100	150	200
	Temperature, K												
Property	73.15	93.15	113.15	133.15	153.15	173.15	223.15	273.15	293.15	323.15	373.15	423.15	473.15
Density, ρ_l (kg/m^3)	S	S	S	S	S	S	S	S	2 890	2 793	2 676	(2540)	V
Specific heat capacity, $c_{p,l}$ (kJ/kg K)	S	S	S	S	S	S	S	S	(0.48)	(0.50)	(0.52)	(0.56)	V
Thermal conductivity, λ_l [(W/m^2)/(K/m)]	S	S	S	S	S	S	S	S	0.099	0.095	0.088	0.081	V
Dynamic viscosity, η_l (10^{-5} Ns/m^2)	S	S	S	S	S	S	S	S	210.9	150.0	89.5	(62)	V

Transport Properties of the Superheated Vapor

	Temperature, °C												
	− 150	− 100	− 50	0	25	100	200	300	400	500	1 000	1 500	2 000
Property (at low pressure)	Temperature, K												
	123.15	173.15	223.15	273.15	298.15	373.15	473.15	573.15	673.15	773.15	1 273.15	1 773.15	2 273.15
Specific heat capacity, $c_{p,g}$ (kJ/kg K)	S	S	S	S	L	L	0.325	0.343	0.356	0.367	0.396	0.408	0.415
Thermal conductivity, λ_g[(W/m^2)/(K/m)]	S	S	S	S	L	L	0.009	0.011	0.013	0.015	0.025	0.034	0.041
Dynamic viscosity, η_g(10^{-5} Ns/m^2)	S	S	S	S	L	L	1.88	2.28	2.66	3.04	4.73	6.16	7.39

TETRABROMOMETHANE

Chemical formula: CBr_4
Molecular weight: 331.67
Melting point: 363.25 K
Boiling point: 460.15 K

Critical temperature: 712.15 K
Critical pressure: 4.25 MPa
Critical density: —
Normal vapor density: 14.8 kg/m^3
(@ 0°C, 101.3 kPa)

Properties of the Liquid at Temperatures Below the Normal Boiling Point

	Temperature, °C												
	−200	−180	−160	−140	−120	−100	−50	0	20	50	100	150	200
	Temperature, K												
Property	73.15	93.15	113.15	133.15	153.15	173.15	223.15	273.15	293.15	323.15	373.15	423.15	473.15
Density, ρ_l (kg/m^3)	S	S	S	S	S	S	S	S	S	S	2 955	2 828	V
Specific heat capacity, $c_{p,l}$ (kJ/kg K)	S	S	S	S	S	S	S	S	S	S	0.452	(0.48)	V
Thermal conductivity, λ_l [$(W/m^2)/(K/m)$]	S	S	S	S	S	S	S	S	S	S	0.083	0.077	V
Dynamic viscosity, η_l (10^{-5} Ns/m^2)	S	S	S	S	S	S	S	S	S	S	24.3	13.0	V

Transport Properties of the Superheated Vapor

	Temperature, °C												
	−150	−100	−50	0	25	100	200	300	400	500	1 000	1 500	2 000
Property	Temperature, K												
(at low pressure)	123.15	173.15	223.15	273.15	298.15	373.15	473.15	573.15	673.15	773.15	1 273.15	1 773.15	2 273.15
Specific heat capacity, $c_{p,g}$ (kJ/kg K)	S	S	S	S	S	L	0.300	0.308	0.312	0.315	0.321	0.324	0.325
Thermal conductivity, λ_g[$(W/m^2)/(K/m)$]	S	S	S	S	S	L	0.008	0.009	0.011	0.015	0.019	0.025	0.029
Dynamic viscosity, η_g(10^{-5} Ns/m^2)	S	S	S	S	S	L	1.84	2.21	2.59	2.93	4.41	5.67	6.76

TRICHLOROFLUOROMETHANE (REFRIGERANT 11)

Chemical formula: CCl_3F
Molecular weight: 137.37
Melting point: 162.0 K
Boiling point: 296.95 K

Critical temperature: 471.2 K
Critical pressure: 4.40 MPa
Critical density: 554 kg/m^3
Normal vapor density: 6.13 kg/m^3
(@ 0°C, 101.3 kPa)

Properties of the Liquid at Temperatures Below the Normal Boiling Point

	Temperature, °C												
	−200	−180	−160	−140	−120	−100	−50	0	20	50	100	150	200
	Temperature, K												
Property	73.15	93.15	113.15	133.15	153.15	173.15	223.15	273.15	293.15	323.15	373.15	423.15	473.15
Density, ρ_l (kg/m^3)	S	S	S	S	S	1 760	1 642	1 534	1 488	V	V	V	V
Specific heat capacity, $c_{p,l}$ (kJ/kg K)	S	S	S	S	S	0.800	0.829	0.862	0.883	V	V	V	V
Thermal conductivity, λ_l [(W/m^2)/(K/m)]	S	S	S	S	S	0.123	0.109	0.0945	0.0889	V	V	V	V
Dynamic viscosity, η_l (10^{-5} Ns/m^2)	S	S	S	S	S	(210)	108.2	54.3	44.1	V	V	V	V

Transport Properties of the Superheated Vapor

	Temperature, °C												
	− 150	− 100	− 50	0	25	100	200	300	400	500	1 000	1 500	2 000
Property	Temperature, K												
(at low pressure)	123.15	173.15	223.15	273.15	298.15	373.15	473.15	573.15	673.15	773.15	1 273.15	1 773.15	2 273.15
Specific heat capacity, $c_{p,g}$ (kJ/kg K)	S	L	L	L	0.569	0.619	0.666	0.697	0.719	0.733	0.765	0.776	0.781
Thermal conductivity, λ_g[(W/m^2)/(K/m)]	S	L	L	L	0.008	0.011	0.016	0.019	0.023	0.026	0.040	0.051	0.061
Dynamic viscosity, η_g(10^{-5} Ns/m^2)	S	L	L	L	1.13	1.35	1.71	2.03	2.34	2.63	3.86	4.89	5.79

REFRIGERANT 12

Chemical formula: CCl_2F_2
Molecular weight: 120.92
Melting point: 118 K
Normal boiling point: 243.2 K

Critical temperature: 384.8 K
Critical pressure: 4.132 MPa
Critical density: 561.8 kg/m^3
Normal vapor density: 5.40 kg/m^3
(@ 0°C, 101.3 kPa)

Properties of the Liquid at Temperatures Below the Normal Boiling Point

	Temperature, °C												
	−200	−180	−160	−140	−120	−100	−50	0	20	50	100	150	200
	Temperature, K												
Property	73.15	93.15	113.15	133.15	153.15	173.15	223.15	273.15	293.15	323.15	373.15	423.15	473.15
Density, ρ_l (kg/m^3)	S	S	S	1 798	1 743	1 688	1 546	V	V	V	V	V	V
Specific heat capacity, $c_{p,l}$ (kJ/kg K)	S	S	S	0.842	0.842	0.846	0.871	V	V	V	V	V	V
Thermal conductivity, λ_l [(W/m^2)/(K/m)]	S	S	S	0.129	0.122	0.115	0.0966	V	V	V	V	V	V
Dynamic viscosity, η_l (10^{-5} Ns/m^2)	S	S	S	(130)	(100)	(80)	46.7	V	V	V	V	V	V

Properties of the Saturated Liquid and Vapor

T_{sat}, K	243.2	260	275	290	305	320	335	350	365	384.8
p_{sat}, kPa	101.3	200	333	528	793	1 145	1 602	2 183	2 907	4 132
ρ_ℓ, kg/m^3	1 486	1 436	1 388	1 338	1 284	1 225	1 157	1 075	969.7	561.8
ρ_g, kg/m^3	6.33	11.8	19.2	29.9	44.8	65.4	94.6	136.4	203.2	561.8
h_ℓ, kJ/kg	473.6	488.3	501.9	516.3	531.1	546.8	563.8	582.3	603.2	649.8
h_g, kJ/kg	641.9	649.8	656.6	662.9	668.8	674.0	677.8	679.9	679.0	649.8
$\Delta h_{g,\ell}$, kJ/kg	168.3	161.5	154.7	146.6	137.7	127.2	114.0	97.6	75.8	
$c_{p,\ell}$, kJ/(kg K)	0.896	0.911	0.932	0.957	0.990	1.03	1.08	1.13	1.22	
$c_{p,g}$, kJ/(kg K)	0.569	0.614	0.646	0.689	0.746	0.825	0.920	1.22	1.68	
η_ℓ, μNs/m^2	373	303	262	231	208	187	167	144	119	
η_g, μNs/m^2	10.3	11.0	11.7	12.5	13.3	14.2	15.2	16.5	18.1	
λ_ℓ, (mW/m^2)/(K/m)	95.1	87.4	80.5	73.3	66.8	59.8	53.0	46.2	39.2	15.4
λ_g, (mW/m^2)/(K/m)	6.9	7.7	8.4	9.2	10.0	10.8	11.6	12.3	13.4	15.4
Pr_ℓ	3.51	3.16	3.03	3.02	3.14	3.22	3.40	3.52	3.70	
Pr_g	0.85	0.88	0.90	0.94	0.99	1.08	1.21	1.64	2.27	
σ, mN/m	15.5	13.5	11.4	9.4	7.7	5.9	4.2	2.8	1.3	
$\beta_{e,\ell}$, kK^{-1}	1.96	2.22	1.50	2.87	3.36	4.17	5.48	8.50	14.5	

Transport Properties of the Superheated Vapor

	Temperature, °C												
	−150	−100	−50	0	25	100	200	300	400	500	1 000	1 500	2 000
Property	Temperature, K												
(at low pressure)	123.15	173.15	223.15	273.15	298.15	373.15	473.15	573.15	673.15	773.15	1 273.15	1 773.15	2 273.15
Specific heat capacity, $c_{p,g}$ (kJ/kg K)	L	L	L	0.571	0.600	0.658	0.724	0.762	0.793	0.814	0.861	0.878	0.884
Thermal conductivity, λ_g[(W/m^2)/(K/m)]	L	L	L	0.009	0.010	0.014	0.019	0.023	0.027	0.031	0.049	0.061	0.073
Dynamic viscosity, η_g(10^{-5} Ns/m^2)	L	L	L	1.15	1.25	1.54	1.89	2.26	2.59	2.87	4.11	5.19	6.09

REFRIGERANT 13

Chemical formula: $CClF_3$
Molecular weight: 104.47
Melting point: 93.2 K
Normal boiling point: 191.7 K

Critical temperature: 302.28 K
Critical pressure: 3.900 MPa
Critical density: 571 kg/m^3
Normal vapor density: 4.66 kg/m^3
(@ 0 °C, 101.3 kPa)

Properties of the Liquid at Temperatures Below the Normal Boiling Point

	Temperature, °C												
	−200	−180	−160	−140	−120	−100	−50	0	20	50	100	150	200
	Temperature, K												
Property	73.15	93.15	113.15	133.15	153.15	173.15	223.15	273.15	293.15	323.15	373.15	423.15	473.15
Density, ρ_l (kg/m^3)	S	(1870)	(1860)	(1730)	1 664	1 593	V	V	V	V	V	V	V
Specific heat capacity, $c_{p,l}$ (kJ/kg K)	S	(0.81)	(0.82)	(0.84)	(0.86)	(0.88)	V	V	V	V	V	V	V
Thermal conductivity, λ_l [(W/m^2)/(K/m)]	S	0.141	0.130	0.118	0.107	0.097	V	V	V	V	V	V	V
Dynamic viscosity, η_l (10^{-5} Ns/m^2)	S	—	—	—	S	(40)	V	V	V	V	V	V	V

Properties of the Saturated Liquid and Vapor

T_{sat}, K	191.7	200	210	220	235	250	265	280	295	302.28
p_{sat}, kPa	101.3	155	245	371	641	1 060	1 590	2 340	3 320	3 900
ρ_ℓ, kg/m^3	1 521	1 489	1 450	1 408	1 339	1 257	1 175	1 066	893	571
ρ_g, kg/m^3	6.89	10.3	15.8	23.4	39.9	67.2	103	166	290	571
h_ℓ, kJ/kg	417.1	424.6	433.9	443.3	458.1	474.5	489.8	509.1	532.8	562.0
h_g, kJ/kg	566.4	569.8	573.7	577.5	582.6	587.2	589.8	590.3	583.9	562.0
$\Delta h_{g,\ell}$, kJ/kg	149.3	145.2	139.8	134.2	124.5	112.7	100.0	81.2	51.1	
$c_{p,\ell}$, kJ/(kg K)	1.11	1.13	1.16	1.21	1.27	1.36	1.56	1.96	3.92	
$c_{p,g}$, kJ/(kg K)	.529	.552	.588	.633	.696	.854	.962	1.30	2.52	
η_ℓ, μNs/m^2	318	282	248	220	188	163	142	114	69.5	
η_g, μNs/m^2	9.83	10.3	10.8	11.3	12.2	13.2	14.3	15.8	19.9	
λ_ℓ, (mW/m^2)/(K/m)	99.3	95.0	89.8	84.6	77.0	69.2	60.7	48.6	34.0	23.6
λ_g, (mW/m^2)/(K/m)	6.4	7.0	7.6	8.3	9.0	10.0	11.5	13.2	16.2	23.6
Pr_ℓ	3.55	3.35	3.20	3.15	3.10	3.20	3.65	4.56	8.01	
Pr_g	0.81	0.82	0.84	0.86	0.94	1.13	1.20	1.56	3.10	
σ, mN/m	13.5	12.3	10.8	9.41	7.36	5.41	3.58	1.91	0.49	
$\beta_{e,\ell}$, kK^{-1}	2.51	2.66	2.92	3.30	4.20	5.55	7.50	12.3	76.0	

Transport Properties of the Superheated Vapor

	Temperature, °C												
	−150	−100	−50	0	25	100	200	300	400	500	1 000	1 500	2 000
Property (at low pressure)	Temperature, K												
	123.15	173.15	223.15	273.15	298.15	373.15	473.15	573.15	673.15	773.15	1 273.15	1 773.15	2 273.15
Specific heat capacity, $c_{p,g}$ (kJ/kg K)	L	L	L	L	L	0.717	0.795	0.851	0.891	0.921	0.983	1.011	1.018
Thermal conductivity, λ_g[(W/m^2)/(K/m)]	L	L	L	L	L	0.017	0.023	0.029	0.035	0.041	0.061	0.077	0.092
Dynamic viscosity, η_g(10^{-5} Ns/m^2)	L	L	L	L	L	1.79	2.21	2.60	2.93	3.30	4.55	5.65	6.63

REFRIGERANT 21

Chemical formula: $CHCl_2F$
Molecular weight: 102.92
Normal boiling point: 281.9 K
Melting point: 138 K

Critical temperature: 451.25 K
Critical pressure: 5.181 MPa
Critical density: 525.0 kg/m^3
Normal vapor density: 4.59 kg/m^3
(@ 0°C, 101.3 kPa)

Properties of the Liquid at Temperatures Below the Normal Boiling Point

Property	Temperature, °C												
	−200	−180	−160	−140	−120	−100	−50	0	20	50	100	150	200
	Temperature, K												
	73.15	93.15	113.15	133.15	153.15	173.15	223.15	273.15	293.15	323.15	373.15	423.15	473.15
Density, ρ_l (kg/m^3)	S	S	S	S	(1700)	(1624)	1 534	1 426	V	V	V	V	V
Specific heat capacity, $c_{p,l}$ (kJ/kg K)	S	S	S	S	0.842	0.846	0.892	1.017	V	V	V	V	V
Thermal conductivity, λ_l [(W/m^2)/(K/m)]	S	S	S	S	0.155	0.150	0.131	0.112	V	V	V	V	V
Dynamic viscosity, η_l (10^{-5} Ns/m^2)	S	S	S	S	(270)	186.7	72.4	39.7	V	V	V	V	V

Properties of the Saturated Liquid and Vapor

T_{sat}, K	281.9	300	320	340	360	380	400	420	440	451.25
p_{sat}, kPa	101.3	196	364	626	1 010	1 540	2 240	3 160	4 350	5 181
ρ_ℓ, kg/m^3	1 406	1 360	1 311	1 258	1 199	1 134	1 057	962.8	823.0	525
ρ_g, kg/m^3	4.62	8.55	15.48	26.14	41.93	64.94	98.91	151.1	124.1	525
h_ℓ, kJ/kg	509.1	528.1	549.9	572.3	595.2	618.4	642.4	668.6	701.4	748.7
h_g, kJ/kg	748.1	756.5	765.5	773.6	780.5	786.6	790.7	791.5	785.3	748.7
$\Delta h_{g,\ell}$, kJ/kg	239.0	228.4	215.6	201.3	185.3	168.2	148.3	122.9	83.9	
$c_{p,\ell}$, kJ/(kg K)	1.04	1.05	1.07	1.10	1.17	1.26	1.37	1.52	1.89	
$c_{p,g}$, kJ/(kg K)	0.721	.733	.780	.832	.902	.982	1.05	1.43	2.50	
η_ℓ, μNs/m^2	366	311	269	235	209	187	165	135	99	
η_g, μNs/m^2	11.0	11.7	12.5	13.3	14.3	15.4	16.5	18.3	25.0	
λ_ℓ, (mW/m^2)/(K/m)	114	108	101	94.2	87.5	80.7	73.8	66.7	59.9	
λ_g, (mW/m^2)/(K/m)	7.9	8.6	9.5	10.4	11.4	12.5	13.7	15.2	16.7	
Pr_ℓ	3.34	3.02	2.85	2.74	2.79	2.89	3.06	3.08	3.12	
Pr_g	1.00	1.01	1.03	1.06	1.13	1.21	1.26	1.72	3.74	
σ, mN/m	20.1	17.6	14.7	12.1	9.4	7.0	4.7	2.6	0.73	
$\beta_{e,\ell}$, kK^{-1}	1.64	1.85	2.11	2.47	2.95	3.78	5.32	8.50	54.0	

Transport Properties of the Superheated Vapor

Property (at low pressure)	Temperature, °C												
	−150	−100	−50	0	25	100	200	300	400	500	1 000	1 500	2 000
	Temperature, K												
	123.15	173.15	223.15	273.15	298.15	373.15	473.15	573.15	673.15	773.15	1 273.15	1 773.15	2 273.15
Specific heat capacity, $c_{p,g}$ (kJ/kg K)	S	L	L	L	0.593	0.671	0.732	0.787	0.830	0.867	0.952	—	—
Thermal conductivity, λ_g[(W/m^2)/(K/m)]	S	L	L	L	0.009	0.012	(0.021)	(0.029)	(0.039)	—	—	—	—
Dynamic viscosity, η_g(10^{-5} Ns/m^2)	S	L	L	L	1.13	1.36	(1.52)	(1.80)	(2.00)	—	—	—	—

REFRIGERANT 22

Chemical formula: $CHClF_2$
Molecular weight: 86.48
Melting point: 113.2 K
Normal boiling point: 242.4 K

Critical temperature: 369.3 K
Critical pressure: 4.986 MPa
Critical density: 513 kg/m^3
Normal vapor density: 3.86 kg/m^3
(@ 0°C, 101.3 kPa)

Properties of the Liquid at Temperatures Below the Normal Boiling Point

	Temperature, °C												
	−200	−180	−160	−140	−120	−100	−50	0	20	50	100	150	200
	Temperature, K												
Property	73.15	93.15	113.15	133.15	153.15	173.15	223.15	273.15	293.15	323.15	373.15	423.15	473.15
Density, ρ_l (kg/m^3)	S	S	S	S	S	1 565	1 438	V	V	V	V	V	V
Specific heat capacity, $c_{p,l}$ (kJ/kg K)	S	S	S	S	S	1.061	1.093	V	V	V	V	V	V
Thermal conductivity, λ_l [(W/m^2)/(K/m)]	S	S	S	S	S	0.150	0.125	V	V	V	V	V	V
Dynamic viscosity, η_l (10^{-5} Ns/m^2)	S	S	S	S	S	72.8	36.5	V	V	V	V	V	V

Properties of the Saturated Liquid and Vapor

T_{sat}, K	242.4	250	265	280	295	310	325	340	355	369.3
p_{sat}, kPa	101.3	218	376	619	958	1 420	2 020	2 800	3 800	4 986
ρ_ℓ, kg/m^3	1 413	1 360	1 313	1 260	1 206	1 146	1 076	991	877	513
ρ_g, kg/m^3	4.70	9.59	16.1	26.3	40.6	60.9	90.2	134	208	513
h_ℓ, kJ/kg	453.6	469.3	490.2	508.1	526.3	545.2	565.3	587.3	613.2	667.3
h_g, kJ/kg	687.0	694.9	701.0	706.7	711.5	715.0	716.9	716.0	708.9	667.3
$\Delta h_{g,\ell}$, kJ/kg	233.4	225.6	210.8	198.6	185.2	169.8	151.6	128.7	95.7	
$c_{p,\ell}$, kJ/(kg K)	1.10	1.13	1.16	1.19	1.24	1.30	1.41	1.65	2.43	
$c_{p,g}$, kJ/(kg K)	.599	.646	.691	.747	.820	.930	1.09	1.40	2.31	
η_ℓ, μNs/m^2	332	282	251	225	204	187	172	150	119	
η_g, μNs/m^2	10.1	10.9	11.7	12.3	13.2	14.2	15.7	16.4	18.8	
λ_ℓ, (mW/m^2)/(K/m)	119	109	101	94.2	86.6	78.8	70.2	59.2	44.0	31.9
λ_g, (mW/m^2)/(K/m)	7.15	8.22	9.10	10.1	11.2	12.4	14.0	16.0	18.8	31.9
Pr_ℓ	3.07	2.92	2.88	2.84	2.92	3.09	3.45	4.18	6.89	
Pr_g	.85	.86	.89	.91	.97	1.07	1.22	1.69	2.31	
σ, mN/m	18.3	15.5	13.0	10.6	8.4	6.2	4.3	2.5	1.0	
$\beta_{e,\ell}$, kK^{-1}	2.08	2.37	2.68	3.10	3.67	4.65	6.34	9.56	25.6	

Transport Properties of the Superheated Vapor

	Temperature, °C												
	−150	−100	−50	0	25	100	200	300	400	500	1 000	1 500	2 000
Property	Temperature, K												
(at low pressure)	123.15	173.15	223.15	273.15	298.15	373.15	473.15	573.15	673.15	773.15	1 273.15	1 773.15	2 273.15
Specific heat capacity, $c_{p,g}$ (kJ/kg K)	L	L	L	0.616	0.647	0.728	0.821	0.894	0.952	0.998	1.123	1.176	1.203
Thermal conductivity, λ_g[(W/m^2)/(K/m)]	L	L	L	0.010	0.011	0.015	0.023	0.028	0.034	0.040	0.067	0.087	1.051
Dynamic viscosity, η_g(10^{-5} Ns/m^2)	L	L	L	1.18	1.28	1.59	2.00	2.35	2.70	3.00	4.35	5.45	6.46

1,1,1-TRIFLUOROETHANE

Chemical formula: $C_2H_3F_3$
Molecular weight: 84.04
Melting point: 161.9 K
Boiling point: 225.5 K

Critical temperature: 346.2 K
Critical pressure: 3.76 MPa
Critical density: 380 kg/m^3
Normal vapor density: 3.75 kg/m^3
(@ 0°C, 101.3 kPa)

Properties of the Liquid at Temperatures Below the Normal Boiling Point

	Temperature, °C												
	−200	−180	−160	−140	−120	−100	−50	0	20	50	100	150	200
	Temperature, K												
Property	73.15	93.15	113.15	133.15	153.15	173.15	223.15	273.15	293.15	323.15	373.15	423.15	473.15
Density, ρ_l (kg/m^3)	S	S	S	S	S	(1300)	1 176	V	V	V	V	V	V
Specific heat capacity, $c_{p,l}$ (kJ/kg K)	S	S	S	S	S	1.227	1.315	V	V	V	V	V	V
Thermal conductivity, λ_l [(W/m^2)/(K/m)]	S	S	S	S	S	0.145	0.118	V	V	V	V	V	V
Dynamic viscosity, η_l (10^{-5} Ns/m^2)	S	S	S	S	S	—	(43)	V	V	V	V	V	V

CHLOROETHANE

Chemical formula: C_2H_5Cl
Molecular weight: 64.51
Melting point: 136.8 K
Boiling point: 285.4 K

Critical temperature: 460.4 K
Critical pressure: 5.27 MPa
Critical density: 324 kg/m^3
Normal vapor density: 2.88 kg/m^3
(@ 0°C, 101.3 kPa)

Properties of the Liquid at Temperatures Below the Normal Boiling Point

	Temperature, °C												
	−200	−180	−160	−140	−120	−100	−50	0	20	50	100	150	200
	Temperature, K												
Property	73.15	93.15	113.15	133.15	153.15	173.15	223.15	273.15	293.15	323.15	373.15	423.15	473.15
Density, ρ_l (kg/m^3)	S	S	S	S	1 082	1 056	991	924	V	V	V	V	V
Specific heat capacity, $c_{p,l}$ (kJ/kg K)	S	S	S	S	1.344	1.486	1.495	1.566	V	V	V	V	V
Thermal conductivity, λ_l [(W/m^2)/(K/m)]	S	S	S	S	0.164	0.158	0.143	0.128	V	V	V	V	V
Dynamic viscosity, η_l (10^{-5} Ns/m^2)	S	S	S	S	—	—	(50)	32.3	V	V	V	V	V

Transport Properties of the Superheated Vapor

	Temperature, °C												
	− 150	− 100	− 50	0	25	100	200	300	400	500	1 000	1 500	2 000
Property (at low pressure)	Temperature, K												
	123.15	173.15	223.15	273.15	298.15	373.15	473.15	573.15	673.15	773.15	1 273.15	1 773.15	2 273.15
Specific heat capacity, $c_{p,g}$ (kJ/kg K)	S	L	L	L	0.971	1.132	1.349	1.527	1.674	1.798	2.229	—	—
Thermal conductivity, λ_g[(W/m^2)/(K/m)]	S	L	L	L	0.011	0.018	0.025	(0.034)	(0.043)	(0.058)	—	—	—
Dynamic viscosity, η_g(10^{-5} Ns/m^2)	S	L	L	L	0.975	1.21	1.51	1.79	2.05	2.36	3.35	4.21	4.98

1,1-DICHLOROETHANE

Chemical formula: $C_2H_4Cl_2$
Molecular weight: 98.96
Melting point: 176.2 K
Boiling point: 330.4 K

Critical temperature: 523 K
Critical pressure: 5.07 MPa
Critical density: 412 kg/m^3
Normal vapor density: 4.42 kg/m^3
(@ 0°C, 101.3 kPa)

Properties of the Liquid at Temperatures Below the Normal Boiling Point

	Temperature, °C												
	−200	−180	−160	−140	−120	−100	−50	0	20	50	100	150	200
	Temperature, K												
Property	73.15	93.15	113.15	133.15	153.15	173.15	223.15	273.15	293.15	323.15	373.15	423.15	473.15
Density, ρ_l (kg/m^3)	S	S	S	S	S	S	(1276)	1 209	1 174	(1128)	V	V	V
Specific heat capacity, $c_{p,l}$ (kJ/kg K)	S	S	S	S	S	S	1.226	1.257	1.276	(1.32)	V	V	V
Thermal conductivity, λ_l [(W/m^2)/(K/m)]	S	S	S	S	S	S	0.131	0.118	0.114	0.107	V	V	V
Dynamic viscosity, η_l (10^{-5} Ns/m^2)	S	S	S	S	S	S	(120)	(61.7)	48.0	(35)	V	V	V

Transport Properties of the Superheated Vapor

	Temperature, °C												
	− 150	− 100	− 50	0	25	100	200	300	400	500	1 000	1 500	2 000
Property (at low pressure)	Temperature, K												
	123.15	173.15	223.15	273.15	298.15	373.15	473.15	573.15	673.15	773.15	1 273.15	1 773.15	2 273.15
Specific heat capacity, $c_{p,g}$ (kJ/kg K)	S	S	L	L	L	0.885	1.015	1.124	1.212	1.285	1.505	—	—
Thermal conductivity, λ_g[(W/m^2)/(K/m)]	S	S	L	L	L	0.014	0.020	(0.026)	(0.032)	(0.039)	—	—	—
Dynamic viscosity, η_g(10^{-5} Ns/m^2)	S	S	L	L	L	(1.17)	(1.48)	(1.78)	(2.07)	(2.33)	—	—	—

1,2-DICHLOROETHANE

Chemical formula: $C_2H_4Cl_2$
Molecular weight: 98.96
Melting point: 237.5 K
Boiling point: 356.6 K

Critical temperature: 561 K
Critical pressure: 5.37 MPa
Critical density: 449 kg/m^3
Normal vapor density: 4.42 kg/m^3 (@ 0°C, 101.3 kPa)

Properties of the Liquid at Temperatures Below the Normal Boiling Point

	Temperature, °C												
	−200	−180	−160	−140	−120	−100	−50	0	20	50	100	150	200
	Temperature, K												
Property	73.15	93.15	113.15	133.15	153.15	173.15	223.15	273.15	293.15	323.15	373.15	423.15	473.15
Density, ρ_l (kg/m^3)	S	S	S	S	S	S	S	1 287	1 258	1 209	V	V	V
Specific heat capacity, $c_{p,l}$ (kJ/kg K)	S	S	S	S	S	S	S	1.223	1.260	1.319	V	V	V
Thermal conductivity, λ_l [(W/m^2)/(K/m)]	S	S	S	S	S	S	S	0.140	0.135	0.128	V	V	V
Dynamic viscosity, η_l (10^{-5} Ns/m^2)	S	S	S	S	S	S	S	113.7	84.0	58.4	V	V	V

Transport Properties of the Superheated Vapor

	Temperature, °C												
	−150	−100	−50	0	25	100	200	300	400	500	1 000	1 500	2 000
Property (at low pressure)	Temperature, K												
	123.15	173.15	223.15	273.15	298.15	373.15	473.15	573.15	673.15	773.15	1 273.15	1 773.15	2 273.15
Specific heat capacity, $c_{p,g}$ (kJ/kg K)	S	S	S	K	K	0.900	1.012	1.111	1.195	1.268	1.492	—	—
Thermal conductivity, λ_g[(W/m^2)/(K/m)]	S	S	S	L	L	0.013	(0.020)	(0.028)	(0.037)	(0.047)	—	—	—
Dynamic viscosity, η_g(10^{-5} Ns/m^2)	S	S	S	L	L	(1.20)	(1.52)	(1.83)	(2.12)	(2.41)	—	—	—

1,1,1-TRICHLOROETHANE

Chemical formula: $C_2H_3Cl_3$
Molecular weight: 133.40
Melting point: 240.15 K
Boiling point: 347.15 K

Critical temperature: 528.2 K
Critical pressure: 4.15 MPa
Critical density: (455) kg/m^3
Normal vapor density: 5.95 kg/m^3
(@ 0°C, 101.3 kPa)

Properties of the Liquid at Temperatures Below the Normal Boiling Point

	Temperature, °C												
	−200	−180	−160	−140	−120	−100	−50	0	20	50	100	150	200
	Temperature, K												
Property	73.15	93.15	113.15	133.15	153.15	173.15	223.15	273.15	293.15	323.15	373.15	423.15	473.15
Density, ρ_l (kg/m^3)	S	S	S	S	S	S	S	1 371	1 338	(1290)	V	V	V
Specific heat capacity, $c_{p,l}$ (kJ/kg K)	S	S	S	S	S	S	S	1.063	1.076	(1.11)	V	V	V
Thermal conductivity, λ_l [(W/m^2)/(K/m)]	S	S	S	S	S	S	S	0.106	0.102	0.096	V	V	V
Dynamic viscosity, η_l (10^{-5} Ns/m^2)	S	S	S	S	S	S	S	(130)	84.0	(48)	V	V	V

Transport Properties of the Superheated Vapor

	Temperature, °C												
	−150	−100	−50	0	25	100	200	300	400	500	1 000	1 500	2 000
Property (at low pressure)	Temperature, K												
	123.15	173.15	223.15	273.15	298.15	373.15	473.15	573.15	673.15	773.15	1 273.15	1 773.15	2 273.15
Specific heat capacity, $c_{p,g}$ (kJ/kg K)	S	S	S	L	L	0.787	0.875	0.942	(1.001)	(1.047)	—	—	—
Thermal conductivity, λ_g[(W/m^2)/(K/m)]	S	S	S	L	L	(0.012)	(0.017)	(0.022)	(0.027)	(0.032)	—	—	—
Dynamic viscosity, η_g(10^{-5} Ns/m^2)	S	S	S	L	L	(1.18)	(1.50)	(1.81)	(2.09)	(2.37)	—	—	—

1,1,2,2-TETRACHLOROETHANE

Chemical formula: $C_2H_2Cl_4$
Molecular weight: 167.88
Melting point: 230.65 K
Boiling point: 419.35 K

Critical temperature: 650.2 K
Critical pressure: 5.13 MPa
Critical density: (491) kg/m^3
Normal vapor density: 7.49 kg/m^3
(@ 0 °C, 101.3 kPa)

Properties of the Liquid at Temperatures Below the Normal Boiling Point

	Temperature, °C												
	−200	−180	−160	−140	−120	−100	−50	0	20	50	100	150	200
	Temperature, K												
Property	73.15	93.15	113.15	133.15	153.15	173.15	223.15	273.15	293.15	323.15	373.15	423.15	473.15
Density, ρ_l (kg/m^3)	S	S	S	S	S	S	S	1 633	1 600	(1550)	(1470)	V	V
Specific heat capacity, $c_{p,l}$ (kJ/kg K)	S	S	S	S	S	S	S	0.944	0.952	0.973	0.985	V	V
Thermal conductivity, λ_l [(W/m^2)/(K/m)]	S	S	S	S	S	S	S	0.118	0.114	0.108	0.099	V	V
Dynamic viscosity, η_l (10^{-5} Ns/m^2)	S	S	S	S	S	S	S	265.7	177.0	113.2	62.0	V	V

PENTACHLOROETHANE

Chemical formula: C_2HCl_5
Molecular weight: 202.33
Melting point: 244.15 K
Boiling point: 433.65 K

Critical temperature: 669.2 K
Critical pressure: 4.58 MPa
Critical density: (516) kg/m^3
Normal vapor density: 9.03 kg/m^3
(@ 0 °C, 101.3 kPa)

Properties of the Liquid at Temperatures Below the Normal Boiling Point

	Temperature, °C												
	−200	−180	−160	−140	−120	−100	−50	0	20	50	100	150	200
	Temperature, K												
Property	73.15	93.15	113.15	133.15	153.15	173.15	223.15	273.15	293.15	323.15	373.15	423.15	473.15
Density, ρ_l (kg/m^3)	S	S	S	S	S	S	S	1 709	1 678	(1630)	(1555)	(1480)	V
Specific heat capacity, $c_{p,l}$ (kJ/kg K)	S	S	S	S	S	S	S	(0.85)	(0.86)	(0.87)	(0.89)	(0.92)	V
Thermal conductivity, λ_l [(W/m^2)/(K/m)]	S	S	S	S	S	S	S	0.104	0.100	0.095	0.086	0.077	V
Dynamic viscosity, η_l (10^{-5} Ns/m^2)	S	S	S	S	S	S	S	(390)	258	(180)	(110)	(80)	V

BROMOETHANE

Chemical formula: C_2H_5Br
Molecular weight: 108.97
Melting point: 154.6 K
Boiling point: 311.5 K

Critical temperature: 503.8 K
Critical pressure: 6.23 MPa
Critical density: 507 kg/m^3
Normal vapor density: 4.86 kg/m^3 (@ 0 °C, 101.3 kPa)

Properties of the Liquid at Temperatures Below the Normal Boiling Point

	Temperature, °C												
	−200	−180	−160	−140	−120	−100	−50	0	20	50	100	150	200
	Temperature, K												
Property	73.15	93.15	113.15	133.15	153.15	173.15	223.15	273.15	293.15	323.15	373.15	423.15	473.15
Density, ρ_l (kg/m^3)	S	S	S	S	S	1 697	1 600	1 501	1 462	V	V	V	V
Specific heat capacity, $c_{p,l}$ (kJ/kg K)	S	S	S	S	S	0.816	0.846	0.888	0.909	V	V	V	V
Thermal conductivity, λ_l [(W/m^2)/(K/m)]	S	S	S	S	S	0.126	0.115	0.105	0.100	V	V	V	V
Dynamic viscosity, η_l (10^{-5} Ns/m^2)	S	S	S	S	S	289	185	81	39.6	V	V	V	V

Transport Properties of the Superheated Vapor

	Temperature, °C												
	−150	−100	−50	0	25	100	200	300	400	500	1 000	1 500	2 000
	Temperature, K												
Property (at low pressure)	123.15	173.15	223.15	273.15	298.15	373.15	473.15	573.15	673.15	773.15	1 273.15	1 773.15	2 273.15
Specific heat capacity, $c_{p,g}$ (kJ/kg K)	S	L	L	L	L	0.693	0.817	0.918	1.009	1.083	1.321	—	—
Thermal conductivity, λ_g[(W/m^2)/(K/m)]	S	L	L	L	L	(0.013)	(0.019)	(0.027)	(0.035)	(0.043)	—	—	—
Dynamic viscosity, η_g(10^{-5} Ns/m^2)	S	L	L	L	L	(1.27)	(1.58)	(1.90)	(2.23)	(2.55)	—	—	—

1,2-DIBROMOETHANE

Chemical formula: $C_2H_4Br_2$
Molecular weight: 187.87
Melting point: 283.15 K
Boiling point: 404.85 K

Critical temperature: 650.2 K
Critical pressure: 5.69 MPa
Critical density: (662) kg/m^3
Normal vapor density: 8.39 kg/m^3 (@ 0°C, 101.3 kPa)

Properties of the Liquid at Temperatures Below the Normal Boiling Point

	Temperature, °C												
	−200	−180	−160	−140	−120	−100	−50	0	20	50	100	150	200
	Temperature, K												
Property	73.15	93.15	113.15	133.15	153.15	173.15	223.15	273.15	293.15	323.15	373.15	423.15	473.15
Density, ρ_l (kg/m^3)	S	S	S	S	S	S	S	S	2 180	2 118	(2040)	V	V
Specific heat capacity, $c_{p,l}$ (kJ/kg K)	S	S	S	S	S	S	S	S	0.729	0.733	(0.77)	V	V
Thermal conductivity, λ_l [(W/m^2)/(K/m)]	S	S	S	S	S	S	S	S	0.102	0.098	0.089	V	V
Dynamic viscosity, η_l (10^{-5} Ns/m^2)	S	S	S	S	S	S	S	S	173	115.2	64.4	V	V

Transport Properties of the Superheated Vapor

	Temperature, °C												
	− 150	− 100	− 50	0	25	100	200	300	400	500	1 000	1 500	2 000
Property (at low pressure)	Temperature, K												
	123.15	173.15	223.15	273.15	298.15	373.15	473.15	573.15	673.15	773.15	1 273.15	1 773.15	2 273.15
Specific heat capacity, $c_{p,g}$ (kJ/kg K)	S	S	S	S	L	L	(0.548)	(0.604)	(0.649)	(0.686)	(0.803)	—	—
Thermal conductivity, λ_g[(W/m^2)/(K/m)]	S	S	S	S	L	L	(0.020)	(0.028)	(0.036)	(0.046)	—	—	—
Dynamic viscosity, η_g(10^{-5} Ns/m^2)	S	S	S	S	L	L	(1.83)	(2.15)	(2.55)	(2.93)	—	—	—

TETRACHLORODIFLUOROETHANE (REFRIGERANT 112)

Chemical formula: $C_2Cl_4F_2$
Molecular weight: 203.83
Melting point: 298.0 K
Boiling point: 366.15 K

Critical temperature: 551.2 K
Critical pressure: 3.86 MPa
Critical density: (552) kg/m^3
Normal vapor density: 9.09 kg/m^3 (@ 0 °C, 101.3 kPa)

Properties of the Liquid at Temperatures Below the Normal Boiling Point

	Temperature, °C												
	−200	−180	−160	−140	−120	−100	−50	0	20	50	100	150	200
	Temperature, K												
Property	73.15	93.15	113.15	133.15	153.15	173.15	223.15	273.15	293.15	323.15	373.15	423.15′	473.15
Density, ρ_l (kg/m^3)	S	S	S	S	S	S	S	S	S	(1596)	V	V	V
Specific heat capacity, $c_{p,l}$ (kJ/kg K)	S	S	S	S	S	S	S	S	S	(0.91)	V	V	V
Thermal conductivity, λ_l [(W/m^2)/(K/m)]	S	S	S	S	S	S	S	S	S	0.078	V	V	V
Dynamic viscosity, η_l (10^{-5} Ns/m^2)	S	S	S	S	S	S	S	S	S	90.8	V	V	V

TRICHLOROTRIFLUOROETHANE (REFRIGERANT 113)

Chemical formula: $C_2Cl_3F_3$
Molecular weight: 187.38
Melting point: 238.19 K
Boiling point: 320.75 K

Critical temperature: 487.3 K
Critical pressure: 3.412 MPa
Critical density: 616 kg/m^3
Normal vapor density: 8.36 kg/m^3 (@ 0 °C, 101.3 kPa)

Properties of the Liquid at Temperatures Below the Normal Boiling Point

	Temperature, °C												
	−200	−180	−160	−140	−120	−100	−50	0	20	50	100	150	200
	Temperature, K												
Property	73.15	93.15	113.15	133.15	153.15	173.15	223.15	273.15	293.15	323.15	373.15	423.15	473.15
Density, ρ_l (kg/m^3)	S	S	S	S	S	S	S	1 621	1 575	V	V	V	V
Specific heat capacity, $c_{p,l}$ (kJ/kg K)	S	S	S	S	S	S	S	0.858	0.900	V	V	V	V
Thermal conductivity, λ_l [(W/m^2)/(K/m)]	S	S	S	S	S	S	S	0.0802	0.0761	V	V	V	V
Dynamic viscosity, η_l (10^{-5} Ns/m^2)	S	S	S	S	S	S	S	97.8	72.7	V	V	V	V

Transport Properties of the Superheated Vapor

	Temperature, °C												
	− 150	− 100	− 50	0	25	100	200	300	400	500	1 000	1 500	2 000
Property	Temperature, K												
(at low pressure)	123.15	173.15	223.15	273.15	298.15	373.15	473.15	573.15	673.15	773.15	1 273.15	1 773.15	2 273.15
Specific heat capacity, $c_{p,g}$ (kJ/kg K)	S	S	S	L	L	0.740	0.805	0.851	0.883	0.904	0.952	—	—
Thermal conductivity, λ_g[(W/m^2)/(K/m)]	S	S	S	L	L	0.011	(0.017)	(0.023)	(0.031)	—	—	—	—
Dynamic viscosity, η_g(10^{-5} Ns/m^2)	S	S	S	L	L	1.09	1.40	1.60	1.77	—	—	—	—

DICHLOROTETRAFLUOROETHANE (REFRIGERANT 114)

Chemical formula: $C_2Cl_2F_4$
Molecular weight: 170.92
Melting point: 179.3 K
Boiling point: 276.9 K

Critical temperature: 418.9 K
Critical pressure: 3.36 MPa
Critical density: 581 kg/m^3
Normal vapor density: 7.63 kg/m^3
(@ 0°C, 101.3 kPa)

Properties of the Liquid at Temperatures Below the Normal Boiling Point

	Temperature, °C												
	−200	−180	−160	−140	−120	−100	−50	0	20	50	100	150	200
	Temperature, K												
Property	73.15	93.15	113.15	133.15	153.15	173.15	223.15	273.15	293.15	323.15	373.15	423.15	473.15
Density, ρ_l (kg/m^3)	S	S	S	S	S	S	1 678	1 538	V	V	V	V	V
Specific heat capacity, $c_{p,l}$ (kJ/kg K)	S	S	S	S	S	S	0.929	0.963	V	V	V	V	V
Thermal conductivity, λ_l [(W/m^2)/(K/m)]	S	S	S	S	S	S	0.0841	0.0710	V	V	V	V	V
Dynamic viscosity, η_l (10^{-5} Ns/m^2)	S	S	S	S	S	S	107.2	46.7	V	V	V	V	V

Transport Properties of the Superheated Vapor

	Temperature, °C												
	−150	−100	−50	0	25	100	200	300	400	500	1 000	1 500	2 000
Property (at low pressure)	Temperature, K												
	123.15	173.15	223.15	273.15	298.15	373.15	473.15	573.15	673.15	773.15	1 273.15	1 773.15	2 273.15
Specific heat capacity, $c_{p,g}$ (kJ/kg K)	S	S	L	L	0.677	0.757	0.835	0.894	0.933	0.960	1.019	—	—
Thermal conductivity, λ_g[(W/m^2)/(K/m)]	S	S	L	L	0.010	0.015	(0.024)	(0.033)	(0.044)	—	—	—	—
Dynamic viscosity, η_g(10^{-5} Ns/m^2)	S	S	L	L	1.13	1.23	1.72	2.03	2.32	—	—	—	—

CHLOROPROPANE

Chemical formula: C_3H_7Cl
Molecular weight: 78.54
Melting point: 150.4 K
Boiling point: 319.6 K

Critical temperature: 503 K
Critical pressure: 4.58 MPa
Critical density: 309 kg/m^3
Normal vapor density: 3.50 kg/m^3
(@ 0 °C, 101.3 kPa)

Properties of the Liquid at Temperatures Below the Normal Boiling Point

	Temperature, °C												
	−200	−180	−160	−140	−120	−100	−50	0	20	50	100	150	200
	Temperature, K												
Property	73.15	93.15	113.15	133.15	153.15	173.15	223.15	273.15	293.15	323.15	373.15	423.15	473.15
Density, ρ_l (kg/m^3)	S	S	S	S	S	(1036)	(975)	916	890	V	V	V	V
Specific heat capacity, $c_{p,l}$ (kJ/kg K)	S	S	S	S	S	(1.50)	(1.55)	1.574	1.650	V	V	V	V
Thermal conductivity, λ_l [(W/m^2)/(K/m)]	S	S	S	S	S	0.147	0.135	0.122	0.117	V	V	V	V
Dynamic viscosity, η_l (10^{-5} Ns/m^2)	S	S	S	S	S	—	(74)	40.5	32.9	V	V	V	V

Transport Properties of the Superheated Vapor

	Temperature, °C												
	−150	−100	−50	0	25	100	200	300	400	500	1 000	1 500	2 000
Property (at low pressure)	Temperature, K												
	123.15	173.15	223.15	273.15	298.15	373.15	473.15	573.15	673.15	773.15	1 273.15	1 773.15	2 273.15
Specific heat capacity, $c_{p,g}$ (kJ/kg K)	S	L	L	L	L	1.297	1.534	1.744	1.942	2.075	2.559	—	—
Thermal conductivity, λ_g[(W/m^2)/(K/m)]	S	L	L	L	L	0.018	(0.023)	(0.030)	(0.037)	(0.045)	—	—	—
Dynamic viscosity, η_g(10^{-5} Ns/m^2)	S	L	L	L	L	(0.98)	(1.26)	(1.50)	(1.74)	(2.05)	—	—	—

CHLOROBUTANE

Chemical formula: C_4H_9Cl
Molecular weight: 92.57
Melting point: 247.8 K
Boiling point: 323.95 K

Critical temperature: 507 K
Critical pressure: 3.95 MPa
Critical density: 297 kg/m^3
Normal vapor density: 4.13 kg/m^3
(@ 0 °C, 101.3 kPa)

Properties of the Liquid at Temperatures Below the Normal Boiling Point

	Temperature, °C												
	−200	−180	−160	−140	−120	−100	−50	0	20	50	100	150	200
	Temperature, K												
Property	73.15	93.15	113.15	133.15	153.15	173.15	223.15	273.15	293.15	323.15	373.15	423.15	473.15
Density, ρ_l (kg/m^3)	S	S	S	S	S	S	S	907	897	852	V	V	V
Specific heat capacity, $c_{p,l}$ (kJ/kg K)	S	S	S	S	S	S	S	1.69	1.74	1.80	V	V	V
Thermal conductivity, λ_l [(W/m^2)/(K/m)]	S	S	S	S	S	S	S	0.122	0.118	0.111	V	V	V
Dynamic viscosity, η_l (10^{-5} Ns/m^2)	S	S	S	S	S	S	S	(54)	44.0	32.8	V	V	V

CHLOROPENTANE

Chemical formula: $C_5H_{11}Cl$
Molecular weight: 106.59
Melting point: 174.15 K
Boiling point: 381.45 K

Critical temperature: 562.15 K
Critical pressure: 3.42 MPa
Critical density: (291) kg/m^3
Normal vapor density: 4.76 kg/m^3
(@ 0 °C, 101.3 kPa)

Properties of the Liquid at Temperatures Below the Normal Boiling Point

	Temperature, °C												
	−200	−180	−160	−140	−120	−100	−50	0	20	50	100	150	200
	Temperature, K												
Property	73.15	93.15	113.15	133.15	153.15	173.15	223.15	273.15	293.15	323.15	373.15	423.15	473.15
Density, ρ_l (kg/m^3)	S	S	S	S	S	S	(951)	902	882	853	(803)	V	V
Specific heat capacity, $c_{p,l}$ (kJ/kg K)	S	S	S	S	S	S	(1.60)	(1.71)	(1.77)	(1.86)	(2.00)	V	V
Thermal conductivity, λ_l [(W/m^2)/(K/m)]	S	S	S	S	S	S	0.135	0.124	0.120	0.113	0.103	V	V
Dynamic viscosity, η_l (10^{-5} Ns/m^2)	S	S	S	S	S	S	(122)	(72)	58.0	(43)	(29)	V	V

VINYL CHLORIDE

Chemical formula: C_2H_3Cl
Molecular weight: 62.60
Melting point: 119.4 K
Boiling point: 259.8 K

Critical temperature: 429.7 K
Critical pressure: 5.60 MPa
Critical density: 370 kg/m^3
Normal vapor density: 2.79 kg/m^3
(@ 0 °C, 101.3 kPa)

Properties of the Liquid at Temperatures Below the Normal Boiling Point

	Temperature, °C												
	−200	−180	−160	−140	−120	−100	−50	0	20	50	100	150	200
	Temperature, K												
Property	73.15	93.15	113.15	133.15	153.15	173.15	223.15	273.15	293.15	323.15	373.15	423.15	473.15
Density, ρ_l (kg/m^3)	S	S	S	(1192)	(1156)	(1120)	(1031)	V	V	V	V	V	V
Specific heat capacity, $c_{p,l}$ (kJ/kg K)	S	S	S	1.363	1.359	1.356	1.358	V	V	V	V	V	V
Thermal conductivity, λ_l [(W/m^2)/(K/m)]	S	S	S	0.168	0.161	0.154	0.138	V	V	V	V	V	V
Dynamic viscosity, η_l (10^{-5} Ns/m^2)	S	S	S	—	—	—	(43)	V	V	V	V	V	V

Transport Properties of the Superheated Vapor

	Temperature, °C												
	− 150	− 100	− 50	0	25	100	200	300	400	500	1 000	1 500	2 000
Property (at low pressure)	Temperature, K												
	123.15	173.15	223.15	273.15	298.15	373.15	473.15	573.15	673.15	773.15	1 273.15	1 773.15	2 273.15
Specific heat capacity, $c_{p,g}$ (kJ/kg K)	L	L	L	0.812	0.858	0.996	1.156	1.285	1.386	1.470	1.540	—	—
Thermal conductivity, λ_g[(W/m^2)/(K/m)]	L	L	L	0.009	0.011	0.016	0.025	0.035	0.046	0.060	0.074	—	—
Dynamic viscosity, η_g(10^{-5} Ns/m^2)	L	L	L	0.92	1.10	1.25	1.57	1.87	2.15	2.41	2.65	—	—

1,1-DICHLOROETHYLENE

Chemical formula: $C_2H_2Cl_2$
Molecular weight: 96.95
Melting point: 150.65 K
Boiling point: 304.85 K

Critical temperature: 489.1 K
Critical pressure: 4.39 MPa
Critical density: 443 kg/m^3
Normal vapor density: 4.33 kg/m^3
(@ 0°C, 101.3 kPa)

Properties of the Liquid at Temperatures Below the Normal Boiling Point

	Temperature, °C												
	−200	−180	−160	−140	−120	−100	−50	0	20	50	100	150	200
	Temperature, K												
Property	73.15	93.15	113.15	133.15	153.15	173.15	223.15	273.15	293.15	323.15	373.15	423.15	473.15
Density, ρ_l (kg/m^3)	S	S	S	S	(1494)	(1412)	(1330)	(1248)	(1213)	V	V	V	V
Specific heat capacity, $c_{p,l}$ (kJ/kg K)	S	S	S	S	1.027	1.034	1.065	1.117	1.172	V	V	V	V
Thermal conductivity, λ_l [(W/m^2)/(K/m)]	S	S	S	S	0.146	0.140	0.127	0.113	0.108	V	V	V	V
Dynamic viscosity, η_l (10^{-5} Ns/m^2)	S	S	S	S	—	(220)	116	57.8	46.7	V	V	V	V

Transport Properties of the Superheated Vapor

	Temperature, °C												
	−150	−100	−50	0	25	100	200	300	400	500	1 000	1 500	2 000
Property (at low pressure)	Temperature, K												
	123.15	173.15	223.15	273.15	298.15	373.15	473.15	573.15	673.15	773.15	1 273.15	1 773.15	2 273.15
Specific heat capacity, $c_{p,g}$ (kJ/kg K)	S	L	L	L	L	0.796	0.888	0.959	1.013	1.059	1.202	—	—
Thermal conductivity, λ_g[(W/m^2)/(K/m)]	S	L	L	L	L	(0.013)	(0.019)	(0.024)	(0.030)	(0.035)	—	—	—
Dynamic viscosity, η_g(10^{-5} Ns/m^2)	S	L	L	L	L	(1.28)	(1.61)	(1.93)	(2.23)	(2.51)	—	—	—

TRICHLOROETHYLENE

Chemical formula: C_2HCl_3
Molecular weight: 131.39
Melting point: 186.15 K
Boiling point: 360.4 K

Critical temperature: 571 K
Critical pressure: 4.91 MPa
Critical density: 513 kg/m^3
Normal vapor density: 5.86 kg/m^3 (@ 0 °C, 101.3 kPa)

Properties of the Liquid at Temperatures Below the Normal Boiling Point

	Temperature, °C												
	−200	−180	−160	−140	−120	−100	−50	0	20	50	100	150	200
	Temperature, K												
Property	73.15	93.15	113.15	133.15	153.15	173.15	223.15	273.15	293.15	323.15	373.15	423.15	473.15
Density, ρ_l (kg/m^3)	S	S	S	S	S	S	1 660	1 550	1 463	1 440	V	V	V
Specific heat capacity, $c_{p,l}$ (kJ/kg K)	S	S	S	S	S	S	0.900	0.938	0.950	(0.98)	V	V	V
Thermal conductivity, λ_l [(W/m^2)/(K/m)]	S	S	S	S	S	S	0.133	0.121	0.116	0.110	V	V	V
Dynamic viscosity, η_l (10^{-5} Ns/m^2)	S	S	S	S	S	S	140	71.0	60.0	44.7	V	V	V

Transport Properties of the Superheated Vapor

	Temperature, °C												
	−150	−100	−50	0	25	100	200	300	400	500	1 000	1 500	2 000
Property	Temperature, K												
(at low pressure)	123.15	173.15	223.15	273.15	298.15	373.15	473.15	573.15	673.15	773.15	1 273.15	1 773.15	2 273.15
Specific heat capacity, $c_{p,g}$ (kJ/kg K)	S	S	L	L	L	0.695	0.758	0.804	0.837	0.862	0.938	—	—
Thermal conductivity, λ_g[(W/m^2)/(K/m)]	S	S	L	L	L	(0.012)	(0.018)	(0.025)	(0.033)	(0.043)	—	—	—
Dynamic viscosity, η_g(10^{-5} Ns/m^2)	S	S	L	L	L	(1.37)	(1.74)	(2.08)	(2.42)	(2.73)	—	—	—

TETRACHLOROETHYLENE

Chemical formula: C_2Cl_4
Molecular weight: 165.83
Melting point: 251.0 K
Boiling point: 394.3 K

Critical temperature: 620 K
Critical pressure: 4.46 MPa
Critical density: 572 kg/m^3
Normal vapor density: 7.40 kg/m^3 (@ 0°C, 101.3 kPa)

Properties of the Liquid at Temperatures Below the Normal Boiling Point

	Temperature, °C												
	−200	−180	−160	−140	−120	−100	−50	0	20	50	100	150	200
	Temperature, K												
Property	73.15	93.15	113.15	133.15	153.15	173.15	223.15	273.15	293.15	323.15	373.15	423.15	473.15
Density, ρ_l (kg/m^3)	S	S	S	S	S	S	S	1 656	1 621	1 574	1 490	V	V
Specific heat capacity, $c_{p,l}$ (kJ/kg K)	S	S	S	S	S	S	S	0.829	0.879	0.946	1.043	V	V
Thermal conductivity, λ_l [(W/m^2)/(K/m)]	S	S	S	S	S	S	S	0.114	0.110	0.104	0.094	V	V
Dynamic viscosity, η_l (10^{-5} Ns/m^2)	S	S	S	S	S	S	S	114.4	90.4	65.7	45.4	V	V

Transport Properties of the Superheated Vapor

	Temperature, °C												
	−150	−100	−50	0	25	100	200	300	400	500	1 000	1 500	2 000
Property (at low pressure)	Temperature, K												
	123.15	173.15	223.15	273.15	298.15	373.15	473.15	573.15	673.15	773.15	1 273.15	1 773.15	2 273.15
Specific heat capacity, $c_{p,g}$ (kJ/kg K)	S	S	S	L	L	L	0.687	0.712	0.733	0.745	0.779	—	—
Thermal conductivity, λ_g[(W/m^2)/(K/m)]	S	S	S	L	L	L	(0.013)	(0.017)	(0.021)	(0.024)	—	—	—
Dynamic viscosity, η_g(10^{-5} Ns/m^2)	S	S	S	L	L	L	(1.84)	(2.16)	(2.46)	(2.75)	—	—	—

FLUOROBENZENE

Chemical formula: C_6H_5F
Molecular weight: 96.10
Melting point: 234.0 K
Boiling point: 358.5 K

Critical temperature: 560.1 K
Critical pressure: 4.55 MPa
Critical density: 355 kg/m^3
Normal vapor density: 4.29 kg/m^3
(@ 0 °C, 101.3 kPa)

Properties of the Liquid at Temperatures Below the Normal Boiling Point

	Temperature, °C												
	−200	−180	−160	−140	−120	−100	−50	0	20	50	100	150	200
	Temperature, K												
Property	73.15	93.15	113.15	133.15	153.15	173.15	223.15	273.15	293.15	323.15	373.15	423.15	473.15
Density, ρ_l (kg/m^3)	S	S	S	S	S	S	S	1 047	1 024	985	V	V	V
Specific heat capacity, $c_{p,l}$ (kJ/kg K)	S	S	S	S	S	S	S	1.503	1.524	1.570	V	V	V
Thermal conductivity, λ_l [(W/m^2)/(K/m)]	S	S	S	S	S	S	S	0.133	0.127	0.119	V	V	V
Dynamic viscosity, η_l (10^{-5} Ns/m^2)	S	S	S	S	S	S	S	75.4	59.8	42.8	V	V	V

Transport Properties of the Superheated Vapor

	Temperature, °C												
	−150	−100	−50	0	25	100	200	300	400	500	1 000	1 500	2 000
Property (at low pressure)	Temperature, K												
	123.15	173.15	223.15	273.15	298.15	373.15	473.15	573.15	673.15	773.15	1 273.15	1 773.15	2 273.15
Specific heat capacity, $c_{p,g}$ (kJ/kg K)	S	S	S	L	L	1.225	1.499	1.723	1.901	2.046	2.454	—	—
Thermal conductivity, λ_g[(W/m^2)/(K/m)]	S	S	S	L	L	(0.020)	(0.030)	(0.046)	(0.061)	(0.079)	—	—	—
Dynamic viscosity, η_g(10^{-5} Ns/m^2)	S	S	S	L	L	(0.88)	(1.24)	(1.50)	(1.77)	(2.00)	—	—	—

CHLOROBENZENE

Chemical formula: C_6H_5Cl
Molecular weight: 112.56
Melting point: 227.6 K
Boiling point: 404.9 K

Critical temperature: 632.4 K
Critical pressure: 4.52 MPa
Critical density: 365 kg/m^3
Normal vapor density: 5.02 kg/m^3
(@ 0°C, 101.3 kPa)

Properties of the Liquid at Temperatures Below the Normal Boiling Point

	Temperature, °C												
	−200	−180	−160	−140	−120	−100	−50	0	20	50	100	150	200
	Temperature, K												
Property	73.15	93.15	113.15	133.15	153.15	173.15	223.15	273.15	293.15	323.15	373.15	423.15	473.15
Density, ρ_l (kg/m^3)	S	S	S	S	S	S	S	1 128	1 106	1 074	1 019	V	V
Specific heat capacity, $c_{p,l}$ (kJ/kg K)	S	S	S	S	S	S	S	1.264	1.298	1.348	1.507	V	V
Thermal conductivity, λ_l [(W/m^2)/(K/m)]	S	S	S	S	S	S	S	0.133	0.129	0.122	0.112	V	V
Dynamic viscosity, η_l (10^{-5} Ns/m^2)	S	S	S	S	S	S	S	105	80	58	37	V	V

Transport Properties of the Superheated Vapor

	Temperature, °C												
	−150	−100	−50	0	25	100	200	300	400	500	1 000	1 500	2 000
Property (at low pressure)	Temperature, K												
	123.15	173.15	223.15	273.15	298.15	373.15	473.15	573.15	673.15	773.15	1 273.15	1 773.15	2 273.15
Specific heat capacity, $c_{p,g}$ (kJ/kg K)	S	S	S	L	L	L	1.285	1.470	1.619	1.738	—	—	—
Thermal conductivity, λ_g[(W/m^2)/(K/m)]	S	S	S	L	L	L	(0.026)	(0.038)	(0.053)	(0.068)	—	—	—
Dynamic viscosity, η_g(10^{-5} Ns/m^2)	S	S	S	L	L	L	(1.21)	(1.46)	(1.72)	(1.94)	—	—	—

BROMOBENZENE

Chemical formula: C_6H_5Br
Molecular weight: 157.01
Melting point: 242.45 K
Boiling point: 429.35 K

Critical temperature: 670.2 K
Critical pressure: 4.56 MPa
Critical density: 485 kg/m^3
Normal vapor density: 7.01 kg/m^3 (@ 0 °C, 101.3 kPa)

Properties of the Liquid at Temperatures Below the Normal Boiling Point

	Temperature, °C												
	−200	−180	−160	−140	−120	−100	−50	0	20	50	100	150	200
	Temperature, K												
Property	73.15	93.15	113.15	133.15	153.15	173.15	223.15	273.15	293.15	323.15	373.15	423.15	473.15
Density, ρ_l (kg/m^3)	S	S	S	S	S	S	S	1 522	1 495	1 455	1 386	1 315	V
Specific heat capacity, $c_{p,l}$ (kJ/kg K)	S	S	S	S	S	S	S	0.950	0.963	0.988	1.068	(1.12)	V
Thermal conductivity, λ_l [(W/m^2)/(K/m)]	S	S	S	S	S	S	S	0.116	0.112	0.106	0.098	0.088	V
Dynamic viscosity, η_l (10^{-5} Ns/m^2)	S	S	S	S	S	S	S	156.4	112.4	80.0	51.8	(34)	V

Transport Properties of the Superheated Vapor

	Temperature, °C												
	−150	−100	−50	0	25	100	200	300	400	500	1 000	1 500	2 000
Property (at low pressure)	Temperature, K												
	123.15	173.15	223.15	273.15	298.15	373.15	473.15	573.15	673.15	773.15	1 273.15	1 773.15	2 273.15
Specific heat capacity, $c_{p,g}$ (kJ/kg K)	S	S	S	L	L	L	0.933	1.064	1.168	1.251	—	—	—
Thermal conductivity, λ_g[(W/m^2)/(K/m)]	S	S	S	L	L	L	0.025	0.039	(0.055)	(0.069)	—	—	—
Dynamic viscosity, η_g(10^{-5} Ns/m^2)	S	S	S	L	L	L	(1.32)	(1.61)	(1.90)	(2.15)	—	—	—

IODOBENZENE

Chemical formula: C_6H_5J
Molecular weight: 204.01
Melting point: 241.8 K
Boiling point: 461.4 K

Critical temperature: 721 K
Critical pressure: 4.52 MPa
Critical density: 581 kg/m^3
Normal vapor density: 9.10 kg/m^3
(@ 0 °C, 101.3 kPa)

Properties of the Liquid at Temperatures Below the Normal Boiling Point

	Temperature, °C												
	−200	−180	−160	−140	−120	−100	−50	0	20	50	100	150	200
	Temperature, K												
Property	73.15	93.15	113.15	133.15	153.15	173.15	223.15	273.15	293.15	323.15	373.15	423.15	473.15
Density, ρ_l (kg/m^3)	S	S	S	S	S	S	S	1 861	1 832	1 785	1 708	1 627	V
Specific heat capacity, $c_{p,l}$ (kJ/kg K)	S	S	S	S	S	S	S	0.770	0.775	0.791	(0.84)	(0.87)	V
Thermal conductivity, λ_l [(W/m^2)/(K/m)]	S	S	S	S	S	S	S	0.103	0.101	0.098	0.092	0.081	V
Dynamic viscosity, η_l (10^{-5} Ns/m^2)	S	S	S	S	S	S	S	242	170.5	112.1	69.0	(41)	V

m-CHLOROTOLUENE

Chemical formula: C_7H_7Cl
Molecular weight: 126.58
Melting point: 225.35 K
Boiling point: 434.75 K

Critical temperature: 662.2 K
Critical pressure: 3.85 MPa
Critical density: (348) kg/m^3
Normal vapor density: 5.65 kg/m^3
(@ 0 °C, 101.3 kPa)

Properties of the Liquid at Temperatures Below the Normal Boiling Point

	Temperature, °C												
	−200	−180	−160	−140	−120	−100	−50	0	20	50	100	150	200
	Temperature, K												
Property	73.15	93.15	113.15	133.15	153.15	173.15	223.15	273.15	293.15	323.15	373.15	423.15	473.15
Density, ρ_l (kg/m^3)	S	S	S	S	S	S	S	1 088	1 072	1 038	987	937	V
Specific heat capacity, $c_{p,l}$ (kJ/kg K)	S	S	S	S	S	S	S	(1.35)	(1.39)	(1.46)	(1.58)	(1.70)	V
Thermal conductivity, λ_l [(W/m^2)/(K/m)]	S	S	S	S	S	S	S	0.130	0.126	0.120	0.110	0.100	V
Dynamic viscosity, η_l (10^{-5} Ns/m^2)	S	S	S	S	S	S	S	117.7	87.7	61.6	39.2	27.8	V

BENZYL CHLORIDE

Chemical formula: C_7H_7Cl
Molecular weight: 126.58
Melting point: 233.95 K
Boiling point: 452.55 K

Critical temperature: 685.2 K
Critical pressure: 3.85 MPa
Critical density: —
Normal vapor density: 5.65 kg/m^3
(@ 0°C, 101.3 kPa)

Properties of the Liquid at Temperatures Below the Normal Boiling Point

	Temperature, °C												
	−200	−180	−160	−140	−120	−100	−50	0	20	50	100	150	200
	Temperature, K												
Property	73.15	93.15	113.15	133.15	153.15	173.15	223.15	273.15	293.15	323.15	373.15	423.15	473.15
Density, ρ_l (kg/m^3)	S	S	S	S	S	S	S	1 103	1 082	(1053)	(1003)	(954)	V
Specific heat capacity, $c_{p,l}$ (kJ/kg K)	S	S	S	S	S	S	S	1.403	1.432	1.495	1.675	(1.88)	V
Thermal conductivity, λ_l [(W/m^2)/(K/m)]	S	S	S	S	S	S	S	0.141	0.137	0.131	0.120	0.110	V
Dynamic viscosity, η_l (10^{-5} Ns/m^2)	S	S	S	S	S	S	S	(180)	138.0	(96)	(56)	(40)	V

Transport Properties of the Superheated Vapor

	Temperature, °C												
	− 150	− 100	− 50	0	25	100	200	300	400	500	1 000	1 500	2 000
Property (at low pressure)	Temperature, K												
	123.15	173.15	223.15	273.15	298.15	373.15	473.15	573.15	673.15	773.15	1 273.15	1 773.15	2 273.15
Specific heat capacity, $c_{p,g}$ (kJ/kg K)	S	S	S	L	L	L	1.360	1.569	1.745	1.996	—	—	—
Thermal conductivity, λ_g[(W/m^2)/(K/m)]	S	S	S	L	L	L	(0.022)	(0.031)	(0.042)	(0.056)	—	—	—
Dynamic viscosity, η_g(10^{-5} Ns/m^2)	S	S	S	L	L	L	(1.05)	(1.28)	(1.52)	(1.72)	—	—	—

METHYLAMINE

Chemical formula: CH_5N
Molecular weight: 31.06
Melting point: 179.7 K
Boiling point: 266.8 K

Critical temperature: 430 K
Critical pressure: 7.46 MPa
Critical density: 222 kg/m^3
Normal vapor density: 1.39 kg/m^3
(@ 0°C, 101.3 kPa)

Properties of the Liquid at Temperatures Below the Normal Boiling Point

	Temperature, °C												
	−150	−100	−75	−50	−25	0	20	50	100	150	200	250	300
	Temperature, K												
Property	123.15	173.15	198.15	223.15	248.15	273.15	293.15	323.15	373.15	423.15	473.15	523.15	573.15
Density, ρ_l (kg/m^3)	S	S	769	740	712	V	V	V	V	V	V	V	V
Specific heat capacity, $c_{p,l}$ (kJ/kg K)	S	S	3.207	3.236	3.282	V	V	V	V	V	V	V	V
Thermal conductivity, λ_l [(W/m^2)/(K/m)]	S	S	0.233	0.224	0.216	V	V	V	V	V	V	V	V
Dynamic viscosity, η_l (10^{-5} Ns/m^2)	S	S	—	(38)	(30)	V	V	V	V	V	V	V	V

Transport Properties of the Superheated Vapor

	Temperature, °C												
	− 150	− 100	− 50	0	25	100	200	300	400	600	800	1 000	1 200
Property (at low pressure)	Temperature, K												
	123.15	173.15	223.15	273.15	298.15	373.15	473.15	573.15	673.15	873.15	1 073.15	1 273.15	1 473.15
Specific heat capacity, $c_{p,g}$ (kJ/kg K)	S	S	L	(1.51)	1.665	1.923	2.243	2.534	2.792	3.222	3.537	3.779	4.021
Thermal conductivity, λ_g[(W/m^2)/(K/m)]	S	S	L	0.015	(0.018)	(0.027)	(0.041)	(0.059)	(0.084)	—	—	—	—
Dynamic viscosity, η_g(10^{-5} Ns/m^2)	S	S	L	(0.88)	(0.96)	(1.19)	(1.47)	(1.76)	(2.02)	—	—	—	—

DIMETHYLAMINE

Chemical formula: C_2H_7N
Molecular weight: 45.08
Melting point: 181.0 K
Boiling point: 280.0 K

Critical temperature: 437.6 K
Critical pressure: 5.31 MPa
Critical density: 241 kg/m^3
Normal vapor density: 2.01 kg/m^3
(@ 0 °C, 101.3 kPa)

Properties of the Liquid at Temperatures Below the Normal Boiling Point

	Temperature, °C												
	−150	−100	−75	−50	−25	0	20	50	100	150	200	250	300
	Temperature, K												
Property	123.15	173.15	198.15	223.15	248.15	273.15	293.15	323.15	373.15	423.15	473.15	523.15	573.15
Density, ρ_l (kg/m^3)	S	S	(763)	(735)	(707)	679	V	V	V	V	V	V	V
Specific heat capacity, $c_{p,l}$ (kJ/kg K)	S	S	2.797	2.935	2.989	3.023	V	V	V	V	V	V	V
Thermal conductivity, λ_l [(W/m^2)/(K/m)]	S	S	0.181	0.171	0.160	0.151	V	V	V	V	V	V	V
Dynamic viscosity, η_l (10^{-5} Ns/m^2)	S	S	—	(48)	30.0	24.8	V	V	V	V	V	V	V

Transport Properties of the Superheated Vapor

	Temperature, °C												
	−150	−100	−50	0	25	100	200	300	400	600	800	1 000	1 200
Property (at low pressure)	Temperature, K												
	123.15	173.15	223.15	273.15	298.15	373.15	473.15	573.15	673.15	873.15	1 073.15	1 273.15	1 473.15
Specific heat capacity, $c_{p,g}$ (kJ/kg K)	S	S	L	L	1.531	1.837	2.213	2.551	2.842	3.306	3.633	3.876	4.118
Thermal conductivity, λ_g[(W/m^2)/(K/m)]	S	S	L	L	0.016	0.023	(0.037)	(0.055)	(0.080)	—	—	—	—
Dynamic viscosity, η_g(10^{-5} Ns/m^2)	S	S	L	L	(0.91)	(1.12)	(1.39)	(1.64)	(1.88)	—	—	—	—

TRIMETHYLAMINE

Chemical formula: C_3H_9N
Molecular weight: 59.11
Melting point: 156.0 K
Boiling point: 276.1 K

Critical temperature: 433.2 K
Critical pressure: 4.07 MPa
Critical density: 233 kg/m^3
Normal vapor density: 2.64 kg/m^3
(@ 0°C, 101.3 kPa)

Properties of the Liquid at Temperatures Below the Normal Boiling Point

	Temperature, °C												
	−150	−100	−75	−50	−25	0	20	50	100	150	200	250	300
	Temperature, K												
Property	123.15	173.15	198.15	223.15	248.15	273.15	293.15	323.15	373.15	423.15	473.15	523.15	573.15
Density, ρ_l (kg/m^3)	S	(767)	(740)	(713)	685	658	V	V	V	V	V	V	V
Specific heat capacity, $c_{p,l}$ (kJ/kg K)	S	1.964	1.989	2.052	2.131	2.223	V	V	V	V	V	V	V
Thermal conductivity, λ_l [(W/m^2)/(K/m)]	S	0.168	0.157	0.146	0.135	0.126	V	V	V	V	V	V	V
Dynamic viscosity, η_l (10^{-5} Ns/m^2)	S	—	—	(38)	28.9	19.5	V	V	V	V	V	V	V

Transport Properties of the Superheated Vapor

	Temperature, °C												
	−150	−100	−50	0	25	100	200	300	400	600	800	1 000	1 200
Property (at low pressure)	Temperature, K												
	123.15	173.15	223.15	273.15	298.15	373.15	473.15	573.15	673.15	873.15	1 073.15	1 273.15	1 473.15
Specific heat capacity, $c_{p,g}$ (kJ/kg K)	S	L	L	L	1.552	1.879	2.275	2.624	2.918	3.378	3.694	3.928	4.163
Thermal conductivity, λ_g[(W/m^2)/(K/m)]	S	L	L	L	(0.016)	(0.024)	(0.036)	(0.049)	(0.063)	(0.090)	—	—	—
Dynamic viscosity, η_g(10^{-5} Ns/m^2)	S	L	L	L	(0.76)	(0.96)	(1.20)	(1.43)	(1.65)	(2.03)	—	—	—

ETHYLAMINE

Chemical formula: C_2H_7N
Molecular weight: 45.08
Melting point: 192.0 K
Boiling point: 289.7 K

Critical temperature: 456 K
Critical pressure: 5.62 MPa
Critical density: 253 kg/m^3
Normal vapor density: 2.01 kg/m^3
(@ 0 °C, 101.3 kPa)

Properties of the Liquid at Temperatures Below the Normal Boiling Point

	Temperature, °C												
	−150	−100	−75	−50	−25	0	20	50	100	150	200	250	300
	Temperature, K												
Property	123.15	173.15	198.15	223.15	248.15	273.15	293.15	323.15	373.15	423.15	473.15	523.15	573.15
Density, ρ_l (kg/m^3)	S	S	(788)	761	734	707	V	V	V	V	V	V	V
Specific heat capacity, $c_{p,l}$ (kJ/kg K)	S	S	(2.91)	(2.95)	(2.99)	(3.03)	V	V	V	V	V	V	V
Thermal conductivity, λ_l [(W/m^2)/(K/m)]	S	S	0.212	0.204	0.197	0.191	V	V	V	V	V	V	V
Dynamic viscosity, η_l (10^{-5} Ns/m^2)	S	S	—	(58)	40.9	(32)	V	V	V	V	V	V	V

Transport Properties of the Superheated Vapor

	Temperature, °C												
	−150	−100	−50	0	25	100	200	300	400	600	800	1 000	1 200
Property (at low pressure)	Temperature, K												
	123.15	173.15	223.15	273.15	298.15	373.15	473.15	573.15	673.15	873.15	1 073.15	1 273.15	1 473.15
Specific heat capacity, $c_{p,g}$ (kJ/kg K)	S	S	L	L	1.613	1.914	2.266	2.592	2.861	3.288	—	—	—
Thermal conductivity, λ_g[(W/m^2)/(K/m)]	S	S	L	L	(0.015)	(0.023)	(0.036)	(0.053)	(0.075)	—	—	—	—
Dynamic viscosity, η_g(10^{-5} Ns/m^2)	S	S	L	L	(0.80)	(1.01)	(1.26)	(1.50)	(1.72)	—	—	—	—

DIETHYLAMINE

Chemical formula: $C_4H_{11}N$
Molecular weight: 73.14
Melting point: 225.15 K
Boiling point: 328.6 K

Critical temperature: 496.6 K
Critical pressure: 3.71 MPa
Critical density: 243 kg/m^3
Normal vapor density: 3.26 kg/m^3
(@ 0°C, 101.3 kPa)

Properties of the Liquid at Temperatures Below the Normal Boiling Point

	Temperature, °C												
	−150	−100	−75	−50	−25	0	20	50	100	150	200	250	300
	Temperature, K												
Property	123.15	173.15	198.15	223.15	248.15	273.15	293.15	323.15	373.15	423.15	473.15	523.15	573.15
Density, ρ_l (kg/m^3)	S	S	S	S	(756)	730	709	677	V	V	V	V	V
Specific heat capacity, $c_{p,l}$ (kJ/kg K)	S	S	S	S	(2.27)	(2.35)	2.417	2.525	V	V	V	V	V
Thermal conductivity, λ_l [(W/m^2)/(K/m)]	S	S	S	S	0.149	0.141	0.135	0.127	V	V	V	V	V
Dynamic viscosity, η_l (10^{-5} Ns/m^2)	S	S	S	S	75.4	44.0	33.0	24.0	V	V	V	V	V

Transport Properties of the Superheated Vapor

	Temperature, °C												
	−150	−100	−50	0	25	100	200	300	400	600	800	1 000	1 200
Property (at low pressure)	Temperature, K												
	123.15	173.15	223.15	273.15	298.15	373.15	473.15	573.15	673.15	873.15	1 073.15	1 273.15	1 473.15
Specific heat capacity, $c_{p,g}$ (kJ/kg K)	S	S	S	L	L	1.901	2.279	2.622	2.909	3.361	—	—	—
Thermal conductivity, λ_g[(W/m^2)/(K/m)]	S	S	S	L	L	(0.022)	(0.034)	(0.054)	(0.078)	—	—	—	—
Dynamic viscosity, η_g(10^{-5} Ns/m^2)	S	S	S	L	L	0.920	(1.17)	(1.39)	(1.61)	—	—	—	—

TRIETHYLAMINE

Chemical formula: $C_6H_{15}N$
Molecular weight: 101.19
Melting point: 158.4 K
Boiling point: 362.7 K

Critical temperature: 535 K
Critical pressure: 3.03 MPa
Critical density: 259 kg/m^3
Normal vapor density: 4.52 kg/m^3
(@ 0 °C, 101.3 kPa)

Properties of the Liquid at Temperatures Below the Normal Boiling Point

Property	123.15 (−150 °C)	173.15 (−100 °C)	198.15 (−75 °C)	223.15 (−50 °C)	248.15 (−25 °C)	273.15 (0 °C)	293.15 (20 °C)	323.15 (50 °C)	373.15 (100 °C)	423.15 (150 °C)	473.15 (200 °C)	523.15 (250 °C)	573.15 (300 °C)
Density, ρ_l (kg/m^3)	S	(830)	(810)	(789)	768	746	726	699	V	V	V	V	V
Specific heat capacity, $c_{p,l}$ (kJ/kg K)	S	(1.56)	(1.70)	(1.79)	(1.88)	2.092	2.208	2.367	V	V	V	V	V
Thermal conductivity, λ_l [(W/m^2)/(K/m)]	S	0.151	0.145	0.140	0.133	0.126	0.121	0.111	V	V	V	V	V
Dynamic viscosity, η_l (10^{-5} Ns/m^2)	S	—	(150)	(102)	70.5	51.0	36.2	27.0	V	V	V	V	V

Temperature row headings: Temperature, °C / Temperature, K

Transport Properties of the Superheated Vapor

Property (at low pressure)	123.15 (−150 °C)	173.15 (−100 °C)	223.15 (−50 °C)	273.15 (0 °C)	298.15 (25 °C)	373.15 (100 °C)	473.15 (200 °C)	573.15 (300 °C)	673.15 (400 °C)	873.15 (600 °C)	1 073.15 (800 °C)	1 273.15 (1 000 °C)	1 473.15 (1 200 °C)
Specific heat capacity, $c_{p,g}$ (kJ/kg K)	S	L	L	L	L	(1.863)	(2.258)	(2.601)	(2.896)	(3.369)	—	—	—
Thermal conductivity, λ_g[(W/m^2)/(K/m)]	S	L	L	L	L	(0.020)	(0.031)	(0.043)	(0.056)	(0.080)	—	—	—
Dynamic viscosity, η_g(10^{-5} Ns/m^2)	S	L	L	L	L	(0.85)	(1.07)	(1.29)	(1.50)	(1.87)	—	—	—

Temperature row headings: Temperature, °C / Temperature, K

ISOPROPYLAMINE

Chemical formula: C_3H_9N
Molecular weight: 59.11
Melting point: 177.9 K
Boiling point: 305.6 K

Critical temperature: 476 K
Critical pressure: 5.07 MPa
Critical density: 258 kg/m^3
Normal vapor density: 2.64 kg/m^3 (@ 0 °C, 101.3 kPa)

Properties of the Liquid at Temperatures Below the Normal Boiling Point

	Temperature, °C												
	−150	−100	−75	−50	−25	0	20	50	100	150	200	250	300
	Temperature, K												
Property	123.15	173.15	198.15	223.15	248.15	273.15	293.15	323.15	373.15	423.15	473.15	523.15	573.15
Density, ρ_l (kg/m^3)	S	S	(822)	(796)	(768)	(740)	719	V	V	V	V	V	V
Specific heat capacity, $c_{p,l}$ (kJ/kg K)	S	S	2.507	2.642	2.690	2.737	2.770	V	V	V	V	V	V
Thermal conductivity, λ_l [(W/m^2)/(K/m)]	S	S	0.161	0.155	0.150	0.146	0.143	V	V	V	V	V	V
Dynamic viscosity, η_l (10^{-5} Ns/m^2)	S	S	—	—	(67)	46.05	35.0	V	V	V	V	V	V

Transport Properties of the Superheated Vapor

	Temperature, °C												
	− 150	− 100	− 50	0	25	100	200	300	400	600	800	1 000	1 200
Property (at low pressure)	Temperature, K												
	123.15	173.15	223.15	273.15	298.15	373.15	473.15	573.15	673.15	873.15	1 073.15	1 273.15	1 473.15
Specific heat capacity, $c_{p,g}$ (kJ/kg K)	S	S	L	L	L	1.989	2.386	2.742	3.001	3.391	3.684	3.810	—
Thermal conductivity, λ_g[(W/m^2)/(K/m)]	S	S	L	L	L	(0.021)	(0.035)	(0.054)	(0.078)	—	—	—	—
Dynamic viscosity, η_g(10^{-5} Ns/m^2)	S	S	L	L	L	(0.95)	(1.19)	(1.42)	(1.65)	—	—	—	—

BUTYLAMINE

Chemical formula: $C_4H_{11}N$
Molecular weight: 73.14
Melting point: 224.15 K
Boiling point: 350.6 K

Critical temperature: 524 K
Critical pressure: 4.15 MPa
Critical density: 254 kg/m^3
Normal vapor density: 3.26 kg/m^3
(@ 0°C, 101.3 kPa)

Transport Properties of the Superheated Vapor

	Temperature, °C												
	−150	−100	−50	0	25	100	200	300	400	500	1 000	1 500	2 000
Property (at low pressure)	Temperature, K												
	123.15	173.15	223.15	273.15	298.15	373.15	473.15	573.15	673.15	773.15	1 273.15	1 773.15	2 273.15
Specific heat capacity, $c_{p,g}$ (kJ/kg K)	S	S	S	L	L	(1.99)	(2.39)	(2.74)	(3.00)	(3.39)	(3.68)	(3.81)	—
Thermal conductivity, λ_g[(W/m^2)/(K/m)]	S	S	S	L	L	(0.023)	(0.036)	(0.053)	(0.072)	—	—	—	—
Dynamic viscosity, η_g(10^{-5} Ns/m^2)	S	S	S	L	L	0.820	(1.06)	(1.28)	(1.48)	—	—	—	—

PIPERIDINE

Chemical formula: $C_5H_{11}N$
Molecular weight: 85.15
Melting point: 262.7 K
Boiling point: 379.7 K

Critical temperature: 594 K
Critical pressure: 4.76 MPa
Critical density: 295 kg/m^3
Normal vapor density: 3.80 kg/m^3
(@ 0°C, 101.3 kPa)

Properties of the Liquid at Temperatures Below the Normal Boiling Point

	Temperature, °C												
	−150	−100	−75	−50	−25	0	20	50	100	150	200	250	300
Property	Temperature, K												
	123.15	173.15	198.15	223.15	248.15	273.15	293.15	323.15	373.15	423.15	473.15	523.15	573.15
Density, ρ_l (kg/m^3)	S	S	S	S	S	880	861	834	787	V	V	V	V
Specific heat capacity, $c_{p,l}$ (kJ/kg K)	S	S	S	S	S	(2.09)	2.136	2.202	(2.33)	V	V	V	V
Thermal conductivity, λ_l [(W/m^2)/(K/m)]	S	S	S	S	S	0.131	0.125	0.118	0.105	V	V	V	V
Dynamic viscosity, η_l (10^{-5} Ns/m^2)	S	S	S	S	S	(400)	148.6	84.5	47.5	V	V	V	V

PYRIDINE

Chemical formula: C_5H_5N
Molecular weight: 79.10
Melting point: 231.5 K
Boiling point: 388.5 K

Critical temperature: 630 K
Critical pressure: 5.63 MPa
Critical density: 311 kg/m^3
Normal vapor density: 3.53 kg/m^3
(@ 0°C, 101.3 kPa)

Properties of the Liquid at Temperatures Below the Normal Boiling Point

	Temperature, °C												
	−150	−100	−75	−50	−25	0	20	50	100	150	200	250	300
	Temperature, K												
Property	123.15	173.15	198.15	223.15	248.15	273.15	293.15	323.15	373.15	423.15	473.15	523.15	573.15
Density, ρ_l (kg/m^3)	S	S	S	S	1 028	1 003	983	953	901	V	V	V	V
Specific heat capacity, $c_{p,l}$ (kJ/kg K)	S	S	S	S	(1.60)	1.645	1.696	1.767	(1.90)	V	V	V	V
Thermal conductivity, λ_l [(W/m^2)/(K/m)]	S	S	S	S	(0.176)	0.168	0.162	0.152	0.138	V	V	V	V
Dynamic viscosity, η_l (10^{-5} Ns/m^2)	S	S	S	S	(223)	136	95.8	63.9	40.0	V	V	V	V

Transport Properties of the Superheated Vapor

	Temperature, °C												
	− 150	− 100	− 50	0	25	100	200	300	400	600	800	1 000	1 200
Property (at low pressure)	Temperature, K												
	123.15	173.15	223.15	273.15	298.15	373.15	473.15	573.15	673.15	873.15	1 073.15	1 273.15	1 473.15
Specific heat capacity, $c_{p,g}$ (kJ/kg K)	S	S	S	L	L	L	1.565	1.825	2.033	2.346	2.549	2.694	2.839
Thermal conductivity, λ_g[(W/m^2)/(K/m)]	S	S	S	L	L	L	(0.024)	(0.034)	(0.044)	(0.065)	—	—	—
Dynamic viscosity, η_g(10^{-5} Ns/m^2)	S	S	S	L	L	L	1.22	1.42	(1.66)	(2.09)	—	—	—

ANILINE

Chemical formula: $C_6H_5NH_2$
Molecular weight: 93.06
Melting point: 267.05 K
Normal boiling point: 457.55 K

Critical temperature: 699.0 K
Critical pressure: 5.301 MPa
Critical density: 340 kg/m^3
Normal vapor density: 4.16 kg/m^3
(@ 0°C, 101.3 kPa)

Properties of the Liquid at Temperatures Below the Normal Boiling Point

	Temperature, °C												
	−150	−100	−75	−50	−25	0	20	50	100	150	200	250	300
	Temperature, K												
Property	123.15	173.15	198.15	223.15	248.15	273.15	293.15	323.15	373.15	423.15	473.15	523.15	573.15
Density, ρ_l (kg/m^3)	S	S	S	S	S	1 039	1 022	996	951	905	V	V	V
Specific heat capacity, $c_{p,l}$ (kJ/kg K)	S	S	S	S	S	2.030	2.068	2.121	2.209	2.293	V	V	V
Thermal conductivity, λ_l [(W/m^2)/(K/m)]	S	S	S	S	S	0.200	0.195	0.174	0.167	0.159	V	V	V
Dynamic viscosity, η_l (10^{-5} Ns/m^2)	S	S	S	S	S	1 020	440	185	82	49	V	V	V

Properties of the Saturated Liquid and Vapor

T_{sat}, K	457.5	500	525	550	575	600	625	650	675	699.0
p_{sat}, kPa	101.3	276	456	716	1 080	1 560	2 200	3 010	4 050	5 300
ρ_ℓ, kg/m^3	875	828	800	769	736	699	658	608	541	340
ρ_g, kg/m^3	2.56	6.62	10.7	16.7	25.3	37.5	55.1	81.7	128	340
h_ℓ, kJ/kg	−114	−10	53	137	184	254	323	396	473	584
h_g, kJ/kg	357	427	468	509	549	589	623	653	669	584
$\Delta h_{g,\ell}$, kJ/kg	471	437	415	372	365	335	300	257	196	
$c_{p,\ell}$, kJ/(kg K)	2.37	2.52	2.61	2.71	2.84	2.97	3.13	3.36	3.84	
$c_{p,g}$, kJ/(kg K)	1.74	1.89	1.99	2.09	2.20	2.35	2.55	2.90	3.95	
η_ℓ, μNs/m^2	303	213	178	152	132	117	104	94	86	71
η_g, μNs/m^2	11.8	13.1	13.9	14.7	15.6	16.5	17.6	19.0	21.1	71
λ_ℓ, (mW/m^2)/(K/m)	154	146	141	136	131	126	121	116	111	70
λ_g, (mW/m^2)/(K/m)	23.5	28.1	30.9	33.8	36.8	40.1	43.6	47.8	53.1	70
Pr_ℓ	4.66	3.68	3.29	3.03	2.86	2.76	2.69	2.72	2.98	
Pr_g	0.87	0.88	0.89	0.91	0.93	0.97	1.03	1.15	1.57	
σ, mN/m	25.2	19.9	16.9	14.0	11.2	8.5	5.9	3.6	1.5	
$\beta_{e,\ell}$, kK^{-1}	1.17	1.42	1.62	1.88	2.24	2.77	3.62	5.26	9.82	

Transport Properties of the Superheated Vapor

	Temperature, °C												
	− 150	− 100	− 50	0	25	100	200	300	400	600	800	1 000	1 200
Property (at low pressure)	Temperature, K												
	123.15	173.15	223.15	273.15	298.15	373.15	473.15	573.15	673.15	873.15	1 073.15	1 273.15	1 473.15
Specific heat capacity, $c_{p,g}$ (kJ/kg K)	S	S	S	L	L	L	1.763	2.019	2.217	2.514	2.707	2.782	—
Thermal conductivity, λ_g[(W/m^2)/(K/m)]	S	S	S	L	L	L	(0.013)	(0.020)	(0.027)	—	—	—	—
Dynamic viscosity, η_g(10^{-5} Ns/m^2)	S	S	S	L	L	L	(1.03)	(1.23)	(1.44)	—	—	—	—

METHYLANILINE

Chemical formula: C_7H_9N
Molecular weight: 107.15
Melting point: 216.15 K
Boiling point: 469.45 K

Critical temperature: 701.8 K
Critical pressure: 5.20 MPa
Critical density: (312) kg/m^3
Normal vapor density: 4.78 kg/m^3 (@ 0°C, 101.3 kPa)

Properties of the Liquid at Temperatures Below the Normal Boiling Point

	Temperature, °C												
	−150	−100	−75	−50	−25	0	20	50	100	150	200	250	300
	Temperature, K												
Property	123.15	173.15	198.15	223.15	248.15	273.15	293.15	323.15	373.15	423.15	473.15	523.15	573.15
Density, ρ_l (kg/m^3)	S	S	S	(1042)	(1022)	1 002	986	963	922	880	V	V	V
Specific heat capacity, $c_{p,l}$ (kJ/kg K)	S	S	S	(1.98)	(2.00)	(2.02)	(2.04)	(2.07)	2.127	(2.19)	V	V	V
Thermal conductivity, λ_l [(W/m^2)/(K/m)]	S	S	S	0.173	0.167	0.163	0.159	0.153	0.143	0.134	V	V	V
Dynamic viscosity, η_l (10^{-5} Ns/m^2)	S	S	S	—	(920)	435.1	230	124	63.1	39.1	V	V	V

DIMETHYLANILINE

Chemical formula: $C_8H_{11}N$
Molecular weight: 121.18
Melting point: 270.7 K
Boiling point: 466.7 K

Critical temperature: 687 K
Critical pressure: 3.63 MPa
Critical density: (304) kg/m^3
Normal vapor density: 5.41 kg/m^3 (@ 0°C, 101.3 kPa)

Properties of the Liquid at Temperatures Below the Normal Boiling Point

	Temperature, °C												
	−150	−100	−75	−50	−25	0	20	50	100	150	200	250	300
	Temperature, K												
Property	123.15	173.15	198.15	223.15	248.15	273.15	293.15	323.15	373.15	423.15	473.15	523.15	573.15
Density, ρ_l (kg/m^3)	S	S	S	S	S	S	956	930	891	850	V	V	V
Specific heat capacity, $c_{p,l}$ (kJ/kg K)	S	S	S	S	S	S	1.792	1.876	2.022	(2.14)	V	V	V
Thermal conductivity, λ_l [(W/m^2)/(K/m)]	S	S	S	S	S	S	0.143	0.138	0.129	0.120	V	V	V
Dynamic viscosity, η_l (10^{-5} Ns/m^2)	S	S	S	S	S	S	138.7	91.0	57.2	35.0	V	V	V

DIETHYLANILINE

Chemical formula: $C_{10}H_{15}N$
Molecular weight: 149.24
Melting point: 251.85 K
Boiling point: 490.25 K

Critical temperature: 695.2 K
Critical pressure: 3.2 MPa
Critical density: (294) kg/m^3
Normal vapor density: 6.66 kg/m^3
(@ 0°C, 101.3 kPa)

Properties of the Liquid at Temperatures Below the Normal Boiling Point

Property	Temperature, °C												
	−150	−100	−75	−50	−25	0	20	50	100	150	200	250	300
	Temperature, K												
	123.15	173.15	198.15	223.15	248.15	273.15	293.15	323.15	373.15	423.15	473.15	523.15	573.15
Density, ρ_l (kg/m^3)	S	S	S	S	S	951	935	910	871	832	(798)	V	V
Specific heat capacity, $c_{p,l}$ (kJ/kg K)	S	S	S	S	S	(1.77)	(1.82)	2.010	2.024	(2.16)	(2.33)	V	V
Thermal conductivity, λ_l [(W/m^2)/(K/m)]	S	S	S	S	S	0.140	0.137	0.132	0.129	0.120	0.112	V	V
Dynamic viscosity, η_l (10^{-5} Ns/m^2)	S	S	S	S	S	390.1	218.6	119.1	62.6	40.7	29.0	V	V

ACETONITRILE

Chemical formula: C_2H_3N
Molecular weight: 41.05
Melting point: 229.3 K
Boiling point: 354.8 K

Critical temperature: 548 K
Critical pressure: 4.83 MPa
Critical density: 237 kg/m^3
Normal vapor density: 1.83 kg/m^3
(@ 0°C, 101.3 kPa)

Properties of the Liquid at Temperatures Below the Normal Boiling Point

Property	Temperature, °C												
	−150	−100	−75	−50	−25	0	20	50	100	150	200	250	300
	Temperature, K												
	123.15	173.15	198.15	223.15	248.15	273.15	293.15	323.15	373.15	423.15	473.15	523.15	573.15
Density, ρ_l (kg/m^3)	S	S	S	S	829	803	783	750	V	V	V	V	V
Specific heat capacity, $c_{p,l}$ (kJ/kg K)	S	S	S	S	2.17	2.20	2.256	2.361	V	V	V	V	V
Thermal conductivity, λ_l [(W/m^2)/(K/m)]	S	S	S	S	0.227	0.218	0.211	0.201	V	V	V	V	V
Dynamic viscosity, η_l (10^{-5} Ns/m^2)	S	S	S	S	(63)	44.5	26.0	27.2	V	V	V	V	V

Transport Properties of the Superheated Vapor

Property (at low pressure)	Temperature, °C												
	−150	−100	−50	0	25	100	200	300	400	600	800	1 000	1 200
	Temperature, K												
	123.15	173.15	223.15	273.15	298.15	373.15	473.15	573.15	673.15	873.15	1 073.15	1 273.15	1 473.15
Specific heat capacity, $c_{p,g}$ (kJ/kg K)	S	S	S	L	L	1.444	1.654	1.842	2.001	2.261	2.474	—	—
Thermal conductivity, λ_g[(W/m^2)/(K/m)]	S	S	S	L	L	0.015	0.021	0.030	0.040	0.063	—	—	—
Dynamic viscosity, η_g(10^{-5} Ns/m^2)	S	S	S	L	L	0.84	1.16	1.40	1.63	2.06	—	—	—

PROPIONITRILE

Chemical formula: C_3H_5N
Molecular weight: 55.08
Melting point: 180.3 K
Boiling point: 370.5 K

Critical temperature: 564.4 K
Critical pressure: 4.18 MPa
Critical density: 239 kg/m^3
Normal vapor density: 2.46 kg/m^3
(@ 0°C, 101.3 kPa)

Properties of the Liquid at Temperatures Below the Normal Boiling Point

	Temperature, °C												
	−150	−100	−75	−50	−25	0	20	50	100	150	200	250	300
	Temperature, K												
Property	123.15	173.15	198.15	223.15	248.15	273.15	293.15	323.15	373.15	423.15	473.15	523.15	573.15
Density, ρ_l (kg/m^3)	S	S	(874)	850	826	802	782	750	V	V	V	V	V
Specific heat capacity, $c_{p,l}$ (kJ/kg K)	S	S	2.012	2.036	2.069	2.117	2.163	2.251	V	V	V	V	V
Thermal conductivity, λ_l [(W/m^2)/(K/m)]	S	S	0.200	0.192	0.185	0.177	0.172	0.164	V	V	V	V	V
Dynamic viscosity, η_l (10^{-5} Ns/m^2)	S	S	—	(113)	(79)	54.1	43.2	32.7	V	V	V	V	V

Transport Properties of the Superheated Vapor

	Temperature, °C												
	−150	−100	−50	0	25	100	200	300	400	600	800	1 000	1 200
Property	Temperature, K												
(at low pressure)	123.15	173.15	223.15	273.15	298.15	373.15	473.15	573.15	673.15	873.15	1 073.15	1 273.15	1 473.15
Specific heat capacity, $c_{p,g}$ (kJ/kg K)	S	S	L	L	L	1.531	1.799	2.035	2.240	2.567	—	—	—
Thermal conductivity, λ_g[(W/m^2)/(K/m)]	S	S	L	L	L	(0.017)	(0.025)	(0.035)	(0.045)	(0.064)	—	—	—
Dynamic viscosity, η_g(10^{-5} Ns/m^2)	S	S	L	L	L	(0.87)	(1.10)	(1.33)	(1.54)	(1.93)	—	—	—

BUTYRONITRILE

Chemical formula: C_4H_7N
Molecular weight: 69.11
Melting point: 161.0 K
Boiling point: 391.0 K

Critical temperature: 582.2 K
Critical pressure: 3.79 MPa
Critical density: 242 kg/m^3
Normal vapor density: 3.08 kg/m^3
(@ 0 °C, 101.3 kPa)

Properties of the Liquid at Temperatures Below the Normal Boiling Point

	Temperature, °C												
	−150	−100	−75	−50	−25	0	20	50	100	150	200	250	300
	Temperature, K												
Property	123.15	173.15	198.15	223.15	248.15	273.15	293.15	323.15	373.15	423.15	473.15	523.15	573.15
Density, ρ_l (kg/m^3)	S	(905)	(881)	(857)	(833)	809	790	761	712	V	V	V	V
Specific heat capacity, $c_{p,l}$ (kJ/kg K)	S	(2.05)	(2.07)	(2.08)	(2.10)	(2.12)	(2.14)	2.219	(2.33)	V	V	V	V
Thermal conductivity, λ_l [(W/m^2)/(K/m)]	S	0.186	0.178	0.172	0.166	0.159	0.154	0.148	0.137	V	V	V	V
Dynamic viscosity, η_l (10^{-5} Ns/m^2)	S	—	—	(170)	(115)	79.0	58.1	42.0	26.7	V	V	V	V

Transport Properties of the Superheated Vapor

	Temperature, °C												
	− 150	− 100	− 50	0	25	100	200	300	400	600	800	1 000	1 200
Property (at low pressure)	Temperature, K												
	123.15	173.15	223.15	273.15	298.15	373.15	473.15	573.15	673.15	873.15	1 073.15	1 273.15	1 473.15
Specific heat capacity, $c_{p,g}$ (kJ/kg K)	S	L	L	L	L	L	1.927	2.187	2.406	2.745	—	—	—
Thermal conductivity, λ_g[(W/m^2)/(K/m)]	S	L	L	L	L	L	(0.023)	(0.033)	(0.043)	(0.062)	—	—	—
Dynamic viscosity, η_g(10^{-5} Ns/m^2)	S	L	L	L	L	L	(1.10)	(1.33)	(1.54)	(1.93)	—	—	—

FORMAMIDE

Chemical formula: CH_3NO
Molecular weight: 45.04
Melting point: 275.35 K
Boiling point: 483.65 K

Critical temperature: 745.2 K
Critical pressure: 9.5 MPa
Critical density: (319) kg/m^3
Normal vapor density: 2.01 kg/m^3
(@ 0°C, 101.3 kPa)

Properties of the Liquid at Temperatures Below the Normal Boiling Point

	Temperature, °C												
	−150	−100	−75	−50	−25	0	20	50	100	150	200	250	300
	Temperature, K												
Property	123.15	173.15	198.15	223.15	248.15	273.15	293.15	323.15	373.15	423.15	473.15	523.15	573.15
Density, ρ_l (kg/m^3)	S	S	S	S	S	S	1 112	(1100)	(1070)	(1020)	(980)	V	V
Specific heat capacity, $c_{p,l}$ (kJ/kg K)	S	S	S	S	S	S	2.382	2.412	(2.56)	(2.67)	(2.74)	V	V
Thermal conductivity, λ_l [(W/m^2)/(K/m)]	S	S	S	S	S	S	0.362	0.352	(0.335)	(0.318)	(0.301)	V	V
Dynamic viscosity, η_l (10^{-5} Ns/m^2)	S	S	S	S	S	S	376.4	184.0	85.1	(44)	(23)	V	V

Transport Properties of the Superheated Vapor

	Temperature, °C												
	−150	−100	−50	0	25	100	200	300	400	600	800	1 000	1 200
Property (at low pressure)	Temperature, K												
	123.15	173.15	223.15	273.15	298.15	373.15	473.15	573.15	673.15	873.15	1 073.15	1 273.15	1 473.15
Specific heat capacity, $c_{p,g}$ (kJ/kg K)	S	S	S	S	L	L	L	1.601	(1.79)	(2.05)	—	—	—
Thermal conductivity, λ_g[(W/m^2)/(K/m)]	S	S	S	S	L	L	L	(0.030)	(0.039)	(0.058)	—	—	—
Dynamic viscosity, η_g(10^{-5} Ns/m^2)	S	S	S	S	L	L	L	(1.44)	(1.70)	(2.17)	—	—	—

NITROMETHANE

Chemical formula: CH_3NO_2
Molecular weight: 61.04
Melting point: 244.6 K
Boiling point: 374.4 K

Critical temperature: 588 K
Critical pressure: 6.31 MPa
Critical density: 353 kg/m^3
Normal vapor density: 2.72 kg/m^3
(@ 0 °C, 101.3 kPa)

Properties of the Liquid at Temperatures Below the Normal Boiling Point

	Temperature, °C												
	−150	−100	−75	−50	−25	0	20	50	100	150	200	250	300
	Temperature, K												
Property	123.15	173.15	198.15	223.15	248.15	273.15	293.15	323.15	373.15	423.15	473.15	523.15	573.15
Density, ρ_l (kg/m^3)	S	S	S	S	(1198)	(1164)	1136	1095	1026	V	V	V	V
Specific heat capacity, $c_{p,l}$ (kJ/kg K)	S	S	S	S	(1.66)	(1.70)	1.729	1.775	1.861	V	V	V	V
Thermal conductivity, λ_l [(W/m^2)/(K/m)]	S	S	S	S	0.217	0.210	0.204	0.194	0.179	V	V	V	V
Dynamic viscosity, η_l (10^{-5} Ns/m^2)	S	S	S	S	(115)	84.4	65.7	47.9	29.8	V	V	V	V

Transport Properties of the Superheated Vapor

	Temperature, °C												
	−150	−100	−50	0	25	100	200	300	400	600	800	1 000	1 200
Property (at low pressure)	Temperature, K												
	123.15	173.15	223.15	273.15	298.15	373.15	473.15	573.15	673.15	873.15	1 073.15	1 273.15	1 473.15
Specific heat capacity, $c_{p,g}$ (kJ/kg K)	S	S	S	L	L	L	1.290	1.453	1.600	1.822	1.970	2.077	2.183
Thermal conductivity, λ_g[(W/m^2)/(K/m)]	S	S	S	L	L	L	(0.022)	(0.033)	(0.045)	—	—	—	—
Dynamic viscosity, η_g(10^{-5} Ns/m^2)	S	S	S	L	L	L	(1.20)	(1.44)	(1.69)	—	—	—	—

NITROBENZENE

Chemical formula: $C_6H_5NO_2$
Molecular weight: 123.11
Melting point: 278.85 K
Boiling point: 484.05 K

Critical temperature: 712.2 K
Critical pressure: 3.5 MPa
Critical density: (365) kg/m^3
Normal vapor density: 5.49 kg/m^3
(@ 0°C, 101.3 kPa)

Properties of the Liquid at Temperatures Below the Normal Boiling Point

	Temperature, °C												
	−150	−100	−75	−50	−25	0	20	50	100	150	200	250	300
	Temperature, K												
Property	123.15	173.15	198.15	223.15	248.15	273.15	293.15	323.15	373.15	423.15	473.15	523.15	573.15
Density, ρ_l (kg/m^3)	S	S	S	S	S	S	1 204	1 174	1 124	1 073	1 020	V	V
Specific heat capacity, $c_{p,l}$ (kJ/kg K)	S	S	S	S	S	S	1.465	1.507	1.617	(1.77)	(1.85)	V	V
Thermal conductivity, λ_l [(W/m^2)/(K/m)]	S	S	S	S	S	S	0.149	0.146	0.138	0.130	0.117	V	V
Dynamic viscosity, η_l (10^{-5} Ns/m^2)	S	S	S	S	S	S	201.0	124.1	70.0	50.5	39.0	V	V

o-NITROTOLUENE

Chemical formula: $C_7H_7NO_2$
Molecular weight: 137.13
Melting point: 269.95 K
Boiling point: 494.85 K

Critical temperature: 720.2 K
Critical pressure: 3.4 MPa
Critical density: (370) kg/m^3
Normal vapor density: 6.12 kg/m^3
(@ 0°C, 101.3 kPa)

Properties of the Liquid at Temperatures Below the Normal Boiling Point

	Temperature, °C												
	−150	−100	−75	−50	−25	0	20	50	100	150	200	250	300
	Temperature, K												
Property	123.15	173.15	198.15	223.15	248.15	273.15	293.15	323.15	373.15	423.15	473.15	523.15	573.15
Density, ρ_l (kg/m^3)	S	S	S	S	S	(1180)	1 163	1 132	1 082	1 034	985	V	V
Specific heat capacity, $c_{p,l}$ (kJ/kg K)	S	S	S	S	S	1.072	1.139	1.252	1.432	1.641	(1.86)	V	V
Thermal conductivity, λ_l [(W/m^2)/(K/m)]	S	S	S	S	S	0.144	0.141	0.136	0.128	0.120	0.112	V	V
Dynamic viscosity, η_l (10^{-5} Ns/m^2)	S	S	S	S	S	383.1	237.0	142.0	74.2	47.8	33.0	V	V

m-NITROTOLUENE

Chemical formula: $C_7H_7NO_2$
Molecular weight: 137.13
Melting point: 289.25 K
Boiling point: 505.75 K

Critical temperature: 725.2 K
Critical pressure: 3.1 MPa
Critical density: (370) kg/m^3
Normal vapor density: 6.12 kg/m^3
(@ 0 °C, 101.3 kPa)

Properties of the Liquid at Temperatures Below the Normal Boiling Point

Temperature, °C	−150	−100	−75	−50	−25	0	20	50	100	150	200	250	300
Property / Temperature, K	123.15	173.15	198.15	223.15	248.15	273.15	293.15	323.15	373.15	423.15	473.15	523.15	573.15
Density, ρ_l (kg/m^3)	S	S	S	S	S	S	1 157	(1130)	(1082)	(1036)	(988)	V	V
Specific heat capacity, $c_{p,l}$ (kJ/kg K)	S	S	S	S	S	S	1.097	1.227	(1.40)	1.60)	(1.83)	V	V
Thermal conductivity, λ_l [(W/m^2)/(K/m)]	S	S	S	S	S	S	0.141	0.136	0.128	0.120	0.112	V	V
Dynamic viscosity, η_l (10^{-5} Ns/m^2)	S	S	S	S	S	S	233.1	139.8	75.0	(46)	(27)	V	V

p-NITROTOLUENE

Chemical formula: $C_7H_7NO_2$
Molecular weight: 137.13
Melting point: 324.95 K
Boiling point: 511.75 K

Critical temperature: 735.2 K
Critical pressure: 3.0 MPa
Critical density: (370) kg/m^3
Normal vapor density: 6.12 kg/m^3
(@ 0 °C, 101.3 kPa)

Properties of the Liquid at Temperatures Below the Normal Boiling Point

Temperature, °C	−150	−100	−75	−50	−25	0	20	50	100	150	200	250	300
Property / Temperature, K	123.15	173.15	198.15	223.15	248.15	273.15	293.15	323.15	373.15	423.15	473.15	523.15	573.15
Density, ρ_l (kg/m^3)	S	S	S	S	S	S	S	S	1 082	1 036	988	V	V
Specific heat capacity, $c_{p,l}$ (kJ/kg K)	S	S	S	S	S	S	S	S	1.357	1.453	1.687	V	V
Thermal conductivity, λ_l [(W/m^2)/(K/m)]	S	S	S	S	S	S	S	S	0.137	0.128	0.120	V	V
Dynamic viscosity, η_l (10^{-5} Ns/m^2)	S	S	S	S	S	S	S	S	76.0	48.2	34.7	V	V

METHYL MERCAPTAN (METHANETHIOL)

Chemical formula: CH_4S
Molecular weight: 48.11
Melting point: 151.2 K
Boiling point: 279.95 K

Critical temperature: 470 K
Critical pressure: 7.23 MPa
Critical density: 332 kg/m^3
Normal vapor density: 2.15 kg/m^3 (@ 0 °C, 101.3 kPa)

Properties of the Liquid at Temperatures Below the Normal Boiling Point

	Temperature, °C												
	−200	−180	−160	−140	−120	−100	−50	0	20	50	100	150	200
	Temperature, K												
Property	73.15	93.15	113.15	133.15	153.15	173.15	223.15	273.15	293.15	323.15	373.15	423.15	473.15
Density, ρ_l (kg/m^3)	S	S	S	S	(1046)	(1019)	(957)	(895)	V	V	V	V	V
Specific heat capacity, $c_{p,l}$ (kJ/kg K)	S	S	S	S	(1.84)	1.825	1.800	1.842	V	V	V	V	V
Thermal conductivity, λ_l [(W/m^2)/(K/m)]	S	S	S	S	—	—	(0.247)	(0.207)	V	V	V	V	V
Dynamic viscosity, η_l (10^{-5} Ns/m^2)	S	S	S	S	—	—	(49)	(29)	V	V	V	V	V

Transport Properties of the Superheated Vapor

	Temperature, °C												
	−150	−100	−50	0	25	100	200	300	400	500	1 000	1 500	2 000
Property (at low pressure)	Temperature, K												
	123.15	173.15	223.15	273.15	298.15	373.15	473.15	573.15	673.15	773.15	1 273.15	1 773.15	2 273.15
Specific heat capacity, $c_{p,g}$ (kJ/kg K)	S	L	L	L	1.054	1.184	1.345	1.494	1.625	1.742	2.124	—	—
Thermal conductivity, λ_g[(W/m^2)/(K/m)]	S	L	L	L	(0.012)	(0.018)	(0.025)	(0.034)	(0.042)	(0.051)	—	—	—
Dynamic viscosity, η_g(10^{-5} Ns/m^2)	S	L	L	L	(0.96)	(1.20)	(1.52)	(1.81)	(2.09)	(2.37)	—	—	—

ETHYL MERCAPTAN (ETHANETHIOL)

Chemical formula: C_2H_6S
Molecular weight: 62.13
Melting point: 126.2 K
Boiling point: 307.55 K

Critical temperature: 498.7 K
Critical pressure: 5.49 MPa
Critical density: 300 kg/m^3
Normal vapor density: 2.77 kg/m^3
(@ 0 °C, 101.3 kPa)

Properties of the Liquid at Temperatures Below the Normal Boiling Point

	Temperature, °C												
	−200	−180	−160	−140	−120	−100	−50	0	20	50	100	150	200
	Temperature, K												
Property	73.15	93.15	113.15	133.15	153.15	173.15	223.15	273.15	293.15	323.15	373.15	423.15	473.15
Density, ρ_l (kg/m^3)	S	S	S	(994)	(972)	(950)	(896)	842	820	V	V	V	V
Specific heat capacity, $c_{p,l}$ (kJ/kg K)	S	S	S	1.840	1.815	1.805	1.803	1.851	1.885	V	V	V	V
Thermal conductivity, λ_l [(W/m^2)/(K/m)]	S	S	S	—	—	—	(0.172)	(0.143)	0.133	V	V	V	V
Dynamic viscosity, η_l (10^{-5} Ns/m^2)	S	S	S	—	—	—	(62)	36.3	30.0	V	V	V	V

Transport Properties of the Superheated Vapor

	Temperature, °C												
	−150	−100	−50	0	25	100	200	300	400	500	1 000	1 500	2 000
Property (at low pressure)	Temperature, K												
	123.15	173.15	223.15	273.15	298.15	373.15	473.15	573.15	673.15	773.15	1 273.15	1 773.15	2 273.15
Specific heat capacity, $c_{p,g}$ (kJ/kg K)	S	L	L	L	L	1.357	1.581	1.780	1.954	2.102	—	—	—
Thermal conductivity, λ_g[(W/m^2)/(K/m)]	S	L	L	L	L	(0.017)	(0.025)	(0.033)	(0.042)	(0.051)	—	—	—
Dynamic viscosity, η_g(10^{-5} Ns/m^2)	S	L	L	L	L	(1.05)	(1.32)	(1.58)	(1.83)	(2.07)	—	—	—

DIMETHYL SULFIDE

Chemical formula: C_2H_6S
Molecular weight: 62.13
Melting point: 174.85 K
Boiling point: 310.5 K

Critical temperature: 503 K
Critical pressure: 5.53 MPa
Critical density: 309 kg/m^3
Normal vapor density: 2.77 kg/m^3
(@ 0°C, 101.3 kPa)

Properties of the Liquid at Temperatures Below the Normal Boiling Point

	Temperature, °C												
	−200	−180	−160	−140	−120	−100	−50	0	20	50	100	150	200
	Temperature, K												
Property	73.15	93.15	113.15	133.15	153.15	173.15	223.15	273.15	293.15	323.15	373.15	423.15	473.15
Density, ρ_l (kg/m^3)	S	S	S	S	S	S	(931)	872	850	V	V	V	V
Specific heat capacity, $c_{p,l}$ (kJ/kg K)	S	S	S	S	S	S	1.813	1.863	1.897	V	V	V	V
Thermal conductivity, λ_l [(W/m^2)/(K/m)]	S	S	S	S	S	S	(0.151)	(0.133)	(0.126)	V	V	V	V
Dynamic viscosity, η_l (10^{-5} Ns/m^2)	S	S	S	S	S	S	(59)	35.5	29.4	V	V	V	V

Transport Properties of the Superheated Vapor

	Temperature, °C												
	−150	−100	−50	0	25	100	200	300	400	500	1 000	1 500	2 000
Property (at low pressure)	Temperature, K												
	123.15	173.15	223.15	273.15	298.15	373.15	473.15	573.15	673.15	773.15	1 273.15	1 773.15	2 273.15
Specific heat capacity, $c_{p,g}$ (kJ/kg K)	S	S	L	L	L	1.363	1.576	1.768	1.937	2.086	—	—	—
Thermal conductivity, λ_g[(W/m^2)/(K/m)]	S	S	L	L	L	(0.022)	(0.033)	(0.045)	(0.060)	(0.077)	—	—	—
Dynamic viscosity, η_g(10^{-5} Ns/m^2)	S	S	L	L	L	(1.05)	(1.30)	(1.59)	(1.86)	(2.10)	—	—	—

DIETHYL SULFIDE

Chemical formula: $C_4H_{10}S$
Molecular weight: 90.18
Melting point: 169.85 K
Boiling point: 363.15 K

Critical temperature: 557 K
Critical pressure: 3.96 MPa
Critical density: 284 kg/m^3
Normal vapor density: 4.02 kg/m^3
(@ 0 °C, 101.3 kPa)

Properties of the Liquid at Temperatures Below the Normal Boiling Point

	Temperature, °C												
	−200	−180	−160	−140	−120	−100	−50	0	20	50	100	150	200
	Temperature, K												
Property	73.15	93.15	113.15	133.15	153.15	173.15	223.15	273.15	293.15	323.15	373.15	423.15	473.15
Density, ρ_l (kg/m^3)	S	S	S	S	S	(954)	(905)	856	837	807	V	V	V
Specific heat capacity, $c_{p,l}$ (kJ/kg K)	S	S	S	S	S	(1.74)	1.767	1.847	1.889	1.899	V	V	V
Thermal conductivity, λ_l [(W/m^2)/(K/m)]	S	S	S	S	S	(0.176)	(0.160)	(0.142)	0.137	(0.126)	V	V	V
Dynamic viscosity, η_l (10^{-5} Ns/m^2)	S	S	S	S	S	—	(104)	57.4	44.6	33.4	V	V	V

THIOPHENE

Chemical formula: C_4H_4S
Molecular weight: 84.14
Melting point: 234.85 K
Boiling point: 357.25 K

Critical temperature: 570.2 K
Critical pressure: 4.95 MPa
Critical density: 384 kg/m^3
Normal vapor density: 3.75 kg/m^3
(@ 0 °C, 101.3 kPa)

Properties of the Liquid at Temperatures Below the Normal Boiling Point

	Temperature, °C												
	−200	−180	−160	−140	−120	−100	−50	0	20	50	100	150	200
	Temperature, K												
Property	73.15	93.15	113.15	133.15	153.15	173.15	223.15	273.15	293.15	323.15	373.15	423.15	473.15
Density, ρ_l (kg/m^3)	S	S	S	S	S	S	S	1 086	1 065	(1013)	V	V	V
Specific heat capacity, $c_{p,l}$ (kJ/kg K)	S	S	S	S	S	S	S	1.43	1.46	(1.49)	V	V	V
Thermal conductivity, λ_l [(W/m^2)/(K/m)]	S	S	S	S	S	S	S	(0.142)	0.137	(0.128)	V	V	V
Dynamic viscosity, η_l (10^{-5} Ns/m^2)	S	S	S	S	S	S	S	87.5	66.4	47.4	V	V	V

HYDROGEN FLUORIDE

Chemical formula: HF
Molecular weight: 20.063
Melting point: 190 K
Normal boiling point: 292.69 K

Critical temperature: 461.15 K
Critical pressure: 6.485 MPa
Critical density: 290.0 kg/m^3
Normal vapor density: 0.89 kg/m^3
(@ 0 °C, 101.3 kPa)

Properties of the Liquid at Temperatures Below the Normal Boiling Point

	Temperature, °C												
	−200	−180	−160	−140	−120	−100	−50	0	20	50	100	150	200
	Temperature, K												
Property	73.15	93.15	113.15	133.15	153.15	173.15	223.15	273.15	293.15	323.15	373.15	423.15	473.15
Density, ρ_l (kg/m^3)	S	S	S	S	S	S	1 123	1 002	V	V	V	V	V
Specific heat capacity, $c_{p,l}$ (kJ/kg K)	S	S	S	S	S	S	2.97	3.60	V	V	V	V	V
Thermal conductivity, λ_l [(W/m^2)/(K/m)]	S	S	S	S	S	S	0.51	0.455	V	V	V	V	V
Dynamic viscosity, η_l (10^{-5} Ns/m^2)	S	S	S	S	S	S	57	25.6	V	V	V	V	V

Properties of the Saturated Liquid and Vapor

T_{sat}, K	292.69	305	325	345	365	385	405	425	445	461.15
p_{sat}, kPa	101.3	152	285	500	820	1 320	2 100	3 150	4 800	6 490
ρ_ℓ, kg/m^3	968	945	905	862	816	765	710	640	545	290
ρ_g, kg/m^3	2.0	3.5	5.0	10.0	14.0	20	28	45	88	290
h_ℓ, kJ/kg	0.0	37.9	101.6	168	239	316	400	493	598	
h_g, kJ/kg	330	407.9	536.6	653	769	896	1 010	1 068	993	
$\Delta h_{g,\ell}$, kJ/kg	330	370	435	485	530	580	610	575	395	
$c_{p,\ell}$, kJ/(kg K)	3.04	3.12	3.26	3.44	3.68	4.00	4.41	4.92	5.56	
$c_{p,g}$, kJ/(kg K)	1.46	1.46	1.46	1.46	1.46	1.46	1.46	1.46	1.46	
η_ℓ, μNs/m^2	215	191	161	139	121	106	93	81.6	71.2	39.8
η_g, μNs/m^2	10.9	12.2	13.5	14.5	15.4	16.3	17.2	18.1	18.9	39.8
λ_ℓ, (mW/m^2)/(K/m)	402	387	362	335	310	283	255	227	199	
λ_g, (mW/m^2)/(K/m)	21.0	21.8	23.0	24.3	25.6	26.9	28.3	29.6	30.9	32.0
Pr_ℓ	1.63	1.54	1.45	1.43	1.44	1.50	1.61	1.77	1.99	2.22
Pr_g	0.76	0.82	0.85	0.87	0.88	0.88	0.88	0.89	0.89	
σ, mN/m	8.65	7.85	6.75	5.6	4.6	3.5	2.5	1.6	0.7	
$\beta_{e,\ell}$, kK^{-1}	1.93	2.13	2.45	2.91	3.54	4.40	6.08	9.93	27.6	

Transport Properties of the Superheated Vapor

	Temperature, °C												
	−150	−100	−50	0	25	100	200	300	400	500	1 000	1 500	2 000
Property	Temperature, K												
(at low pressure)	123.15	173.15	223.15	273.15	298.15	373.15	473.15	573.15	673.15	773.15	1 273.15	1 773.15	2 273.15
Specific heat capacity, $c_{p,g}$ (kJ/kg K)	S	S	L	L	1.457	1.457	1.457	1.457	1.457	1.470	1.566	1.675	1.750
Thermal conductivity, λ_g[(W/m^2)/(K/m)]	S	S	L	L	0.026	0.033	0.041	0.049	0.057	0.064	0.099	0.132	0.164
Dynamic viscosity, η_g(10^{-5} Ns/m^2)	S	S	L	L	1.25	1.56	1.96	2.34	2.70	3.02	4.40	5.57	6.61

HYDROGEN CHLORIDE

Chemical formula: HCl
Molecular weight: 36.461
Melting point: 158.93 K
Normal boiling point: 188.05 K

Critical temperature: 324.6 K
Critical pressure: 8.309 MPa
Critical density: 450 kg/m^3
Normal vapor density: 1.63 kg/m^3 (@ 0 °C, 101.3 kPa)

Properties of the Liquid at Temperatures Below the Normal Boiling Point

	Temperature, °C												
	−200	−180	−160	−140	−120	−100	−50	0	20	50	100	150	200
Property	Temperature, K												
	73.15	93.15	113.15	133.15	153.15	173.15	223.15	273.15	293.15	323.15	373.15	423.15	473.15
Density, ρ_l (kg/m^3)	S	S	S	S	S	1 235	V	V	V	V	V	V	V
Specific heat capacity, $c_{p,l}$ (kJ/kg K)	S	S	S	S	S	1.60	V	V	V	V	V	V	V
Thermal conductivity, λ_l [(W/m^2)/(K/m)]	S	S	S	S	S	(0.421)	V	V	V	V	V	V	V
Dynamic viscosity, η_l (10^{-5} Ns/m^2)	S	S	S	S	S	(58)	V	V	V	V	V	V	V

Properties of the Saturated Liquid and Vapor

T_{sat}, K	188.05	200	215	230	245	260	275	290	305	324.65
p_{sat}, kPa	101.3	180	370	670	1 100	1 800	2 700	3 800	5 500	8 309
ρ_ℓ, kg/m^3	1 190	1 155	1 115	1 070	1 020	970	925	845	755	450
ρ_g, kg/m^3	2.5	5	10	15	25	40	55	90	140	450
h_ℓ, kJ/kg	0.0	20	45	71	99	130	164	203	247	
h_g, kJ/kg	442	452	461	467	473	478	480	478	465	
$\Delta h_{g,\ell}$, kJ/kg	442	432	416	396	374	348	316	275	218	
$c_{p,\ell}$, kJ/(kg K)	1.61	1.66	1.74	1.84	1.95	2.15	2.34	2.67	3.28	
$c_{p,g}$, kJ/(kg K)	0.85	0.87	0.91	0.96	1.04	1.16	1.36	1.74	2.74	
η_ℓ, μNs/m^2	407	332	259	204	160	126	101	77	60	34
η_g, μNs/m^2	9.0	9.6	10.4	11.2	12.1	13.0	14.1	15.1	16.8	34
λ_ℓ, (mW/m^2)/(K/m)	337	323	305	285	264	242	219	195	169	61
λ_g, (mW/m^2)/(K/m)	8.6	9.3	10.5	12.0	13.5	15.6	17.8	21.6	26.9	61
Pr_ℓ	1.94	1.71	1.48	1.32	1.18	1.12	1.08	1.05	1.16	
Pr_g	0.89	0.90	0.90	0.90	0.93	0.97	1.06	1.22	1.71	
σ, mN/m	23.2	21.0	18.3	15.5	12.9	10.2	7.7	5.2	2.8	
$\beta_{e,\ell}$, kK^{-1}	2.31	2.49	2.72	3.30	3.81	4.00	5.91	9.52	30.55	

Transport Properties of the Superheated Vapor

	Temperature, °C												
	−150	−100	−50	0	25	100	200	300	400	500	1 000	1 500	2 000
Property (at low pressure)	Temperature, K												
	123.15	173.15	223.15	273.15	298.15	373.15	473.15	573.15	673.15	773.15	1 273.15	1 773.15	2 273.15
Specific heat capacity, $c_{p,g}$ (kJ/kg K)	S	L	0.795	0.795	0.795	0.795	0.795	0.808	0.821	0.837	0.913	0.959	1.407
Thermal conductivity, λ_g[(W/m^2)/(K/m)]	S	L	0.010	0.013	0.014	0.018	0.024	0.030	0.036	0.043	0.068	0.091	0.111
Dynamic viscosity, η_g(10^{-5} Ns/m^2)	S	L	1.08	1.31	1.45	1.83	2.30	2.75	3.17	3.56	5.25	6.64	7.87

HYDROGEN BROMIDE

Chemical formula: HBr
Molecular weight: 80.91
Melting point: 186.25 K
Boiling point: 206.1 K

Critical temperature: 362.6 K
Critical pressure: 8.4 MPa
Critical density: 809 kg/m^3
Normal vapor density: 3.64 kg/m^3
(@ 0 °C, 101.3 kPa)

Properties of the Liquid at Temperatures Below the Normal Boiling Point

	Temperature, °C												
	−200	−180	−160	−140	−120	−100	−50	0	20	50	100	150	200
	Temperature, K												
Property	73.15	93.15	113.15	133.15	153.15	173.15	223.15	273.15	293.15	323.15	373.15	423.15	473.15
Density, ρ_l (kg/m^3)	S	S	S	S	S	S	2 132	V	V	V	V	V	V
Specific heat capacity, $c_{p,l}$ (kJ/kg K)	S	S	S	S	S	S	0.749	V	V	V	V	V	V
Thermal conductivity, λ_l [(W/m^2)/(K/m)]	S	S	S	S	S	S	0.131	V	V	V	V	V	V
Dynamic viscosity, η_l (10^{-5} Ns/m^2)	S	S	S	S	S	S	(50)	V	V	V	V	V	V

Transport Properties of the Superheated Vapor

	Temperature, °C												
	− 150	− 100	− 50	0	25	100	200	300	400	500	1 000	1 500	2 000
Property (at low pressure)	Temperature, K												
	123.15	173.15	223.15	273.15	298.15	373.15	473.15	573.15	673.15	773.15	1 273.15	1 773.15	2 273.15
Specific heat capacity, $c_{p,g}$ (kJ/kg K)	S	S	0.360	0.360	0.360	0.360	0.360	0.368	0.377	0.381	0.419	0.444	0.448
Thermal conductivity, λ_g[(W/m^2)/(K/m)]	S	S	0.006	0.008	0.009	0.011	0.014	0.018	0.022	0.026	0.043	0.058	0.071
Dynamic viscosity, η_g(10^{-5} Ns/m^2)	S	S	1.41	1.70	1.85	2.34	3.00	3.62	4.22	4.78	7.24	9.29	11.02

HYDROGEN IODIDE

Chemical formula: HI
Molecular weight: 127.91
Melting point: 222.4 K
Boiling point: 237.6 K

Critical temperature: 424 K
Critical pressure: 8.3 MPa
Critical density: 976 kg/m^3
Normal vapor density: 5.71 kg/m^3
(@ 0 °C, 101.3 kPa)

Properties of the Liquid at Temperatures Below the Normal Boiling Point

	Temperature, °C												
	−200	−180	−160	−140	−120	−100	−50	0	20	50	100	150	200
	Temperature, K												
Property	73.15	93.15	113.15	133.15	153.15	173.15	223.15	273.15	293.15	323.15	373.15	423.15	473.15
Density, ρ_l (kg/m^3)	S	S	S	S	S	S	2 863	V	V	V	V	V	V
Specific heat capacity, $c_{p,l}$ (kJ/kg K)	S	S	S	S	S	S	0.460	V	V	V	V	V	V
Thermal conductivity, λ_l [(W/m^2)/(K/m)]	S	S	S	S	S	S	(0.173)	V	V	V	V	V	V
Dynamic viscosity, η_l (10^{-5} Ns/m^2)	S	S	S	S	S	S	141.8	V	V	V	V	V	V

Transport Properties of the Superheated Vapor

	Temperature, °C												
	−150	−100	−50	0	25	100	200	300	400	500	1 000	1 500	2 000
Property	Temperature, K												
(at low pressure)	123.15	173.15	223.15	273.15	298.15	373.15	473.15	573.15	673.15	773.15	1 273.15	1 773.15	2 273.15
Specific heat capacity, $c_{p,g}$ (kJ/kg K)	S	S	L	0.226	0.226	0.230	0.234	0.239	0.243	0.247	0.272	0.281	0.285
Thermal conductivity, λ_g[(W/m^2)/(K/m)]	S	S	L	0.006	0.006	0.008	0.010	0.012	0.014	0.016	0.025	0.033	0.039
Dynamic viscosity, η_g(10^{-5} Ns/m^2)	S	S	L	1.73	1.89	2.38	2.93	3.46	3.96	4.41	6.42	8.10	9.48

HYDROGEN CYANIDE

Chemical formula: HCN
Molecular weight: 27.03
Melting point: 258.95 K
Boiling point: 298.85 K

Critical temperature: 456.7 K
Critical pressure: 5.39 MPa
Critical density: 194 kg/m^3
Normal vapor density: 1.21 kg/m^3 (@ 0°C, 101.3 kPa)

Properties of the Liquid at Temperatures Below the Normal Boiling Point

	Temperature, °C												
	−200	−180	−160	−140	−120	−100	−50	0	20	50	100	150	200
	Temperature, K												
Property	73.15	93.15	113.15	133.15	153.15	173.15	223.15	273.15	293.15	323.15	373.15	423.15	473.15
Density, ρ_l (kg/m^3)	S	S	S	S	S	S	S	715	688	V	V	V	V
Specific heat capacity, $c_{p,l}$ (kJ/kg K)	S	S	S	S	S	S	S	2.61	2.63	V	V	V	V
Thermal conductivity, λ_l [(W/m^2)/(K/m)]	S	S	S	S	S	S	S	(0.312)	(0.294)	V	V	V	V
Dynamic viscosity, η_l (10^{-5} Ns/m^2)	S	S	S	S	S	S	S	23.55	19.2	V	V	V	V

Transport Properties of the Superheated Vapor

	Temperature, °C												
	−150	−100	−50	0	25	100	200	300	400	500	1 000	1 500	2 000
Property (at low pressure)	Temperature, K												
	123.15	173.15	223.15	273.15	298.15	373.15	473.15	573.15	673.15	773.15	1 273.15	1 773.15	2 273.15
Specific heat capacity, $c_{p,g}$ (kJ/kg K)	S	S	S	L	L	1.411	1.516	1.599	1.687	1.750	1.985	2.093	2.156
Thermal conductivity, λ_g[(W/m^2)/(K/m)]	S	S	S	L	L	0.017	0.023	0.032	0.040	0.048	0.092	0.127	0.159
Dynamic viscosity, η_g(10^{-5} Ns/m^2)	S	S	S	L	L	0.981	1.30	1.62	1.94	2.15	3.33	4.32	5.20

HYDROGEN SULFIDE

Chemical formula: H_2S
Molecular weight: 34.08
Melting point: 187.6 K
Normal boiling point: 212.8 K

Critical temperature: 373.15 K
Critical pressure: 8.937 MPa
Critical density: 346 kg/m^3
Normal vapor density: 1.52 kg/m^3
(@ 0 °C, 101.3 kPa)

Properties of the Liquid at Temperatures Below the Normal Boiling Point

	Temperature, °C												
	−200	−180	−160	−140	−120	−100	−50	0	20	50	100	150	200
	Temperature, K												
Property	73.15	93.15	113.15	133.15	153.15	173.15	223.15	273.15	293.15	323.15	373.15	423.15	473.15
Density, ρ_l (kg/m^3)	S	S	S	S	S	S	980	V	V	V	V	V	V
Specific heat capacity, $c_{p,l}$ (kJ/kg K)	S	S	S	S	S	S	(1.875)	V	V	V	V	V	V
Thermal conductivity, λ_l [(W/m^2)/(K/m)]	S	S	S	S	S	S	(0.211)	V	V	V	V	V	V
Dynamic viscosity, η_l (10^{-5} Ns/m^2)	S	S	S	S	S	S	(24)	V	V	V	V	V	V

Properties of the Saturated Liquid and Vapor

T_{sat}, K	212.8	220	240	260	280	300	320	340	360	373.15
p_{sat}, kPa	101.3	140	325	680	1 020	2 000	3 250	4 890	7 050	8 937
ρ_ℓ, kg/m^3	965	955	915	875	830	780	720	650	565	346
ρ_g, kg/m^3	2.0	2.6	5.5	11.0	21.0	35.0	55.0	95.0	160.0	346
h_ℓ, kJ/kg	−356	−341	−301	−256	−207	−161	−104	−42	45	68
h_g, kJ/kg	199	204	219	230	239	244	241	228	190	68
$\Delta h_{g,\ell}$, kJ/kg	555	545	520	485	445	405	345	270	145	
$c_{p,\ell}$, kJ/(kg K)	1.83	1.85	1.91	2.00	2.13	2.35	2.64	3.10	4.38	
$c_{p,g}$, kJ/(kg K)	1.02	1.03	1.08	1.16	1.28	1.45	1.77	2.48	6.45	
η_ℓ, μNs/m^2	423	378	272	205	162	130	110	87	66	40.5
η_g, μNs/m^2	9.2	9.6	10.5	11.4	12.4	13.5	14.8	16.5	19.2	40.5
λ_ℓ, (mW/m^2)/(K/m)	233	224	199	175	153	131	107	85	62	49.5
λ_g, (mW/m^2)/(K/m)	9.1	9.6	11.1	12.9	14.7	17.0	19.8	24.1	30.5	49.5
Pr_ℓ	3.32	3.12	2.61	2.34	2.26	2.33	2.71	3.17	4.66	
Pr_g	1.03	1.03	1.02	1.03	1.08	1.15	1.32	1.70	4.06	
σ, mN/m	29.0	27.5	23.5	19.6	16.0	12.5	9.2	5.5	2.2	
$\beta_{e,\ell}$, kK^{-1}	1.86	1.98	2.31	2.69	3.34	4.40	6.10	8.99	31.5	

Transport Properties of the Superheated Vapor

	Temperature, °C												
	−150	−100	−50	0	25	100	200	300	400	500	1 000	1 500	2 000
Property (at low pressure)	Temperature, K												
	123.15	173.15	223.15	273.15	298.15	373.15	473.15	573.15	673.15	773.15	1 273.15	1 773.15	2 273.15
Specific heat capacity, $c_{p,g}$ (kJ/kg K)	S	S	0.976	0.992	1.005	1.030	1.076	1.122	1.172	1.223	1.499	1.541	1.608
Thermal conductivity,	S	S	0.014	0.017	0.018	0.023	0.031	0.038	0.045	0.053	0.089	0.122	0.150
Dynamic viscosity, η_g(10^{-5} Ns/m^2)	S	S	0.97	1.17	1.27	1.59	1.99	2.38	2.72	3.05	4.44	5.59	6.61

AIR

Chemical formula: N_2 (78.1%); O_2 (20.9%); Ar (0.9%)
Molecular weight: 28.96
Melting point: 60.2 K
Boiling point: 78.9 K

Critical temperature: 132.6 K
Critical pressure: 3.769 MPa
Critical density: 313 kg/m^3
Normal vapor density: 1.29 kg/m^3 (@ 0°C, 101.3 kPa)

Properties of the Liquid at Temperatures Below the Normal Boiling Point

Property	Temperature, °C: −200	−180	−160	−140	−120	−100	−50	0	20	50	100	150	200
	Temperature, K: 73.15	93.15	113.15	133.15	153.15	173.15	223.15	273.15	293.15	323.15	373.15	423.15	473.15
Density, ρ_l (kg/m^3)	897	V	V	V	V	V	V	V	V	V	V	V	V
Specific heat capacity, $c_{p,l}$ (kJ/kg K)	1.957	V	V	V	V	V	V	V	V	V	V	V	V
Thermal conductivity, λ_l [(W/m^2)/(K/m)]	0.152	V	V	V	V	V	V	V	V	V	V	V	V
Dynamic viscosity, η_l (10^{-5} Ns/m^2)	20.3	V	V	V	V	V	V	V	V	V	V	V	V

Properties of the Saturated Liquid and Vapor

T_{sat}, K	78.9	85	90	95	100	110	115	120	125	132.6
p_{sat}, kPa	101.3	192	304	457	662	1 260	1 670	2 160	2 740	
p_{con}, kPa	0.721	1.45	2.40	3.75	6.60	11.2	15.2	20.1	26.1	37.69
ρ_ℓ, kg/m^3	876	847	822	796	768	705	669	627	569	
ρ_g, kg/m^3	3.27	6.26	9.98	15.2	22.4	45.1	62.8	87.3	123	313
h_ℓ, kJ/kg	−124.6	−113.1	−103.5	−93.5	−83.3	−61.9	−50.3	−37.5	−22.0	
h_g, kJ/kg	76.9	81.6	84.8	87.4	89.3	90.1	88.4	84.8	78.2	37.4
$\Delta h_{g,\ell}$, kJ/kg	201.5	194.7	188.3	180.9	172.6	152.0	138.7	122.3	100.2	
$c_{p,\ell}$, kJ/(kg K)	1.87	1.91	1.94	1.99	2.05	2.14	2.48	2.92	4.59	
$c_{p,g}$, kJ/(kg K)	1.05	1.07	1.09	1.13	1.26	1.56	1.92	2.46	3.38	
η_ℓ, μNs/m^2	183	142	116	97	82	63	55	48	41	
η_g, μNs/m^2	5.6	6.0	6.4	6.8	7.3	8.5	9.0	9.8	10.9	
λ_ℓ, (mW/m^2)/(K/m)	148	137	128	120	111	94	85	77	66	
λ_g, (mW/m^2)/(K/m)	7.4	7.8	8.4	9.2	10.1	12.5	13.9	15.2	17.4	
Pr_ℓ	2.31	1.98	1.76	1.61	1.51	1.43	1.60	1.82	2.85	
Pr_g	0.79	0.82	0.83	0.84	0.91	1.06	1.24	1.59	2.12	
σ, mN/m	9.64	8.29	7.26	6.22	5.22	3.34	2.45	1.62	0.88	
$\beta_{e,\ell}$, kK^{-1}	5.5	6.0	6.6	7.3	8.4	12.5	16.0	21.5	33.0	

Transport Properties of the Superheated Vapor

Property (at low pressure)	Temperature, °C: −150	−100	−50	0	25	100	200	300	400	500	1 000	1 500	2 000
	Temperature, K: 123.15	173.15	223.15	273.15	298.15	373.15	473.15	573.15	673.15	773.15	1 273.15	1 773.15	2 273.15
Specific heat capacity, $c_{p,g}$ (kJ/kg K)	1.019	1.008	1.005	1.005	1.005	1.005	1.022	1.043	1.051	1.089	1.189	1.239	1.269
Thermal conductivity, λ_g[(W/m^2)/(K/m)]	0.011	0.016	0.020	0.024	0.026	0.031	0.039	0.044	0.050	0.056	0.079	0.106	0.128
Dynamic viscosity, η_g(10^{-5} Ns/m^2)	0.853	1.17	1.46	1.72	1.82	2.17	2.57	2.93	3.25	3.55	4.79	5.78	6.81

AIR (*Continued*)

Pressure, MPa	Property	Value at different temperatures, K								
		200	300	400	500	600	700	800	900	1 000
0.050	v, m³/kg	1.147	1.722	2.297	2.871	3.445	4.020	4.594	5.168	5.742
(76.24)	h, kJ/kg	359.56	459.96	560.79	662.81	766.71	872.87	981.41	1 092.23	1 205.16
	s, kJ/(kg K)	4.329 8	4.736 8	5.026 9	5.254 4	5.443 8	5.607 4	5.752 3	5.882 8	6.001 8
	u, kJ/kg	302.22	373.86	445.95	519.25	594.44	671.89	751.72	833.83	918.04
0.101 325	v, m³/kg	0.565 3	0.849 7	1.133	1.417	1.701	1.984	2.267	2.551	2.834
(81.82)	h, kJ/kg	359.31	459.85	560.73	662.79	766.70	872.88	981.43	1 092.26	1 205.19
	s, kJ/(kg K)	4.126 1	4.533 7	4.823 9	5.051 6	5.241 0	5.404 6	5.549 5	5.680 0	5.799 0
	u, kJ/kg	302.03	373.75	445.88	519.20	594.40	671.86	751.69	833.81	918.03
0.20	v, m³/kg	0.285 7	0.430 4	0.574 4	0.718 2	0.861 9	1.006	1.149	1.293	1.436
(88.14)	h, kJ/kg	358.83	459.63	560.62	662.74	766.70	872.90	981.47	1 092.31	1 205.25
	s, kJ/(kg K)	3.929 1	4.337 9	4.628 4	4.856 2	5.045 7	5.209 3	5.354 3	5.484 8	5.603 8
	u, kJ/kg	301.68	373.55	445.75	519.11	594.32	671.80	751.65	833.77	918.00
0.40	v, m³/kg	0.142 2	0.215 1	0.287 3	0.359 3	0.431 3	0.503 1	0.575 0	0.646 8	0.718 6
(95.78)	h, kJ/kg	357.85	459.18	560.40	662.64	766.68	872.94	981.54	1 092.41	1 205.37
	s, kJ/(kg K)	3.726 6	4.137 5	4.428 7	4.656 8	4.846 4	5.010 2	5.155 2	5.285 7	5.404 7
	u, kJ/kg	300.97	373.15	445.48	518.91	594.17	671.68	751.55	833.70	917.94
0.70	v, m³/kg	0.080 69	0.122 8	0.164 3	0.205 6	0.246 7	0.287 8	0.328 9	0.370 0	0.411 0
(103.0)	h, kJ/kg	356.37	458.50	560.07	662.50	766.65	872.99	981.65	1 092.56	1 205.55
	s, kJ/(kg K)	3.560 6	3.974 9	4.267 1	4.495 6	4.685 4	4.849 3	4.994 3	5.125 0	5.244 0
	u, kJ/kg	299.89	372.54	445.07	518.61	593.95	671.51	751.41	833.58	917.84
1.0	v, m³/kg	0.056 09	0.085 90	0.115 1	0.144 0	0.172 9	0.201 7	0.230 5	0.259 2	0.288 0
(108.2)	h, kJ/kg	354.89	457.83	559.74	662.36	766.63	873.04	981.76	1 092.72	1 205.74
	s, kJ/(kg K)	3.452 7	3.870 5	4.163 7	4.392 6	4.582 6	4.746 6	4.891 8	5.022 4	5.141 5
	u, kJ/kg	298.80	371.94	444.66	518.32	593.72	671.33	751.27	833.47	917.75
2.0	v, m³/kg	0.027 39	0.042 85	0.057 67	0.072 28	0.086 79	0.101 2	0.115 7	0.130 1	0.144 5
(119.8)	h, kJ/kg	349.87	455.63	558.67	661.90	766.56	873.23	982.14	1 093.23	1 206.36
	s, kJ/(kg K)	3.235 3	3.664 9	3.961 4	4.191 7	4.382 4	4.546 8	4.692 2	4.823 0	4.942 2
	u, kJ/kg	295.09	369.92	443.32	517.34	592.98	670.75	750.81	833.10	917.45
5.0	v, m³/kg	0.010 22	0.017 07	0.023 26	0.029 24	0.035 13	0.040 97	0.046 77	0.052 56	0.058 34
	h, kJ/kg	334.39	449.26	555.63	660.66	766.43	873.86	983.31	1 094.80	1 208.24
	s, kJ/(kg K)	2.914 2	3.382 4	3.688 7	3.923 0	4.115 8	4.281 4	4.427 5	4.558 8	4.678 3
	u, kJ/kg	283.27	363.91	439.34	514.46	590.79	669.04	749.44	832.00	916.55
7.0	v, m³/kg	0.007 01	0.012 19	0.016 72	0.021 05	0.025 29	0.029 49	0.033 65	0.037 80	0.041 94
	h, kJ/kg	323.93	445.27	553.77	659.93	766.41	874.33	984.12	1 095.88	1 209.51
	s, kJ/(kg K)	2.777 6	3.273 4	3.585 9	3.822 8	4.016 9	4.183 2	4.329 8	4.461 4	4.581 1
	u, kJ/kg	274.89	359.96	436.75	512.58	589.36	667.91	748.55	831.27	915.96
10	v, m³/kg	0.004 68	0.008 55	0.011 83	0.014 92	0.017 92	0.020 88	0.023 82	0.026 73	0.029 64
	h, kJ/kg	308.87	439.71	551.23	659.00	766.50	875.11	985.40	1 097.53	1 211.45
	s, kJ/(kg K)	2.616 9	3.153 4	3.474 7	3.715 2	3.911 2	4.078 6	4.225 8	4.357 9	4.477 9
	u, kJ/kg	262.09	354.16	432.96	509.82	587.26	666.26	747.22	830.20	915.09
20	v, m³/kg	0.002 44	0.004 44	0.006 18	0.007 80	0.009 35	0.010 86	0.012 35	0.013 83	0.015 29
	h, kJ/kg	277.13	425.39	544.95	657.23	767.64	878.29	990.07	1 103.34	1 218.14
	s, kJ/(kg K)	2.296 7	2.905 3	3.250 1	3.500 8	3.702 1	3.872 7	4.021 9	4.155 3	4.276 3
	u, kJ/kg	228.33	336.54	421.27	501.25	580.66	661.04	743.02	826.78	912.29
50	v, m³/kg	0.001 59	0.002 26	0.002 95	0.003 62	0.004 26	0.004 89	0.005 50	0.006 10	0.006 70
	h, kJ/kg	269.73	414.72	542.54	661.80	777.54	892.28	1 007.28	1 123.10	1 239.97
	s, kJ/(kg K)	1.979 7	2.570 6	2.939 2	3.205 5	3.416 6	3.593 5	3.747 0	3.883 4	4.006 5
	u, kJ/kg	190.47	301.91	395.05	480.91	564.45	647.89	732.23	817.86	904.89
70	v, m³/kg	0.001 43	0.001 89	0.002 37	0.002 85	0.003 31	0.003 76	0.004 20	0.004 64	0.005 07
	h, kJ/kg	279.48	421.23	549.50	670.41	787.92	904.27	1 020.66	1 137.61	1 255.56
	s, kJ/(kg K)	1.878 4	2.455 4	2.825 2	3.095 2	3.309 5	3.488 8	3.644 2	3.782 1	3.906 2
	u, kJ/kg	179.21	288.78	383.32	471.01	556.16	640.94	726.38	812.93	900.73
100	v, m³/kg	0.001 31	0.001 62	0.001 95	0.002 28	0.002 60	0.002 92	0.003 23	0.003 54	0.003 84
	h, kJ/kg	298.89	438.15	566.11	687.97	806.89	924.73	1 042.54	1 160.85	1 279.90
	s, kJ/(kg K)	1.771 0	2.337 7	2.706 4	2.978 5	3.195 3	3.377 0	3.534 3	3.673 6	3.799 1
	u, kJ/kg	168.38	275.91	370.94	459.94	546.51	632.61	719.21	806.78	895.47

AMMONIA

Chemical formula: NH_3
Molecular weight: 17.032
Melting point: 195.45 K
Normal boiling point: 239.75 K

Critical temperature: 405.55 K
Critical pressure: 11.290 MPa
Critical density: 235 kg/m³
Normal vapor density: 0.76 kg/m³
(@ 0°C, 101.3 kPa)

Properties of the Liquid at Temperatures Below the Normal Boiling Point

	Temperature, °C												
	−200	−180	−160	−140	−120	−100	−50	0	20	50	100	150	200
	Temperature, K												
Property	73.15	93.15	113.15	133.15	153.15	173.15	223.15	273.15	293.15	323.15	373.15	423.15	473.15
Density, ρ_l (kg/m³)	S	S	S	S	S	S	695	V	V	V	V	V	V
Specific heat capacity, $c_{p,l}$ (kJ/kg K)	S	S	S	S	S	S	4.45	V	V	V	V	V	V
Thermal conductivity, λ_l [(W/m²)/(K/m)]	S	S	S	S	S	S	0.547	V	V	V	V	V	V
Dynamic viscosity, η_l (10^{-5} Ns/m²)	S	S	S	S	S	S	31.7	V	V	V	V	V	V

Properties of the Saturated Liquid and Vapor

T_{sat}, K	239.75	250	270	290	310	330	350	370	390	400
p_{sat}, kPa	101.3	165.4	381.9	775.3	1 424.9	2 422	3 870	5 891	8 606	10 280
ρ_ℓ, kg/m³	682	669	643	615	584	551	512	466	400	344
ρ_g, kg/m³	0.86	1.41	3.09	6.08	11.0	18.9	31.5	52.6	93.3	137
h_ℓ, kJ/kg	808.0	854.0	945.7	1 039.6	1 135.7	1 235.7	1 341.9	1 457.5	1 591.4	1 675.3
h_g, kJ/kg	2 176	2 192	2 219	2 240	2 251	2 255	2 251	2 202	2 099	1 982
$\Delta h_{g,\ell}$, kJ/kg	1 368	1 338	1 273	1 200	1 115	1 019	899	744	508	307
$c_{p,\ell}$, kJ/(kg K)	4.472	4.513	4.585	4.649	4.857	5.066	5.401	5.861	7.74	
$c_{p,g}$, kJ/(kg K)	2.12	2.32	2.69	3.04	3.44	3.90	4.62	6.21	8.07	
η_ℓ, μNs/m²	285	246	190	152	125	105	88.5	70.2	50.7	39.5
η_g, μNs/m²	9.25	9.59	10.30	11.05	11.86	12.74	13.75	15.06	17.15	19.5
λ_ℓ, (mW/m²)/(K/m)	614	592	569	501	456	411	365	320	275	252
λ_g, (mW/m²)/(K/m)	18.8	19.8	22.7	25.2	28.9	34.3	39.5	50.4	69.2	79.4
Pr_ℓ	2.06	1.88	1.58	1.39	1.36	1.32	1.34	1.41	1.43	
Pr_g	1.04	1.11	1.17	1.25	1.31	1.34	1.49	1.70	1.86	
σ, mN/m	33.9	31.5	26.9	22.4	18.0	13.7	9.60	5.74	2.21	0.68
$\beta_{e,\ell}$, kK⁻¹	1.90	1.98	2.22	2.63	3.18	4.01	5.50	8.75	19.7	29.2

Transport Properties of the Superheated Vapor

	Temperature, °C												
	−150	−100	−50	0	25	100	200	300	400	500	1 000	1 500	2 000
Property (at low pressure)	Temperature, K												
	123.15	173.15	223.15	273.15	298.15	373.15	473.15	573.15	673.15	773.15	1 273.15	1 773.15	2 273.15
Specific heat capacity, $c_{p,g}$ (kJ/kg K)	S	S	L	2.056	2.093	2.219	2.366	2.516	2.663	2.805	3.538	4.099	4.509
Thermal conductivity, λ_g[(W/m²)/(K/m)]	S	S	L	0.022	0.024	0.033	0.047	0.067	0.088	0.109	0.209	0.304	0.388
Dynamic viscosity, η_g(10^{-5} Ns/m²)	S	S	L	0.930	1.00	1.28	1.65	1.99	2.34	2.67	4.16	5.40	6.49

NITRIC OXIDE

Chemical formula: NO
Molecular weight: 30.01
Melting point: 109.6 K
Boiling point: 121.4 K

Critical temperature: 180 K
Critical pressure: 6.48 MPa
Critical density: 517 kg/m^3
Normal vapor density: 1.34 kg/m^3
(@ 0 °C, 101.3 kPa)

Transport Properties of the Superheated Vapor

	Temperature, °C												
	−150	−100	−50	0	25	100	200	300	400	500	1 000	1 500	2 000
Property (at low pressure)	Temperature, K												
	123.15	173.15	223.15	273.15	298.15	373.15	473.15	573.15	673.15	773.15	1 273.15	1 773.15	2 273.15
Specific heat capacity, $c_{p,g}$ (kJ/kg K)	0.971	0.971	0.971	0.971	0.971	0.980	1.005	1.030	1.059	1.089	1.176	1.218	1.239
Thermal conductivity, λ_g[(W/m^2)/(K/m)]	0.013	0.018	0.021	0.024	0.026	0.031	0.038	0.046	0.053	0.059	0.088	0.113	0.135
Dynamic viscosity, η_g(10^{-5} Ns/m^2)	0.85	1.21	1.49	1.79	1.92	2.27	2.68	3.12	3.47	3.85	5.29	6.55	7.72

NITROUS OXIDE

Chemical formula: N_2O
Molecular weight: 44.01
Melting point: 182.3 K
Boiling point: 184.7 K

Critical temperature: 309.58 K
Critical pressure: 7.159 MPa
Critical density: 452 kg/m^3
Normal vapor density: 1.96 kg/m^3
(@ 0 °C, 101.3 kPa)

Transport Properties of the Superheated Vapor

	Temperature, °C												
	−150	−100	−50	0	25	100	200	300	400	500	1 000	1 500	2 000
Property (at low pressure)	Temperature, K												
	123.15	173.15	223.15	273.15	298.15	373.15	473.15	573.15	673.15	773.15	1 273.15	1 773.15	2 273.15
Specific heat capacity, $c_{p,g}$ (kJ/kg K)	S	S	0.816	0.858	0.879	0.950	1.017	1.076	1.130	1.176	1.285	1.331	1.361
Thermal conductivity, λ_g[(W/m^2)/(K/m)]	S	S	0.011	0.015	0.017	0.024	0.032	0.041	0.050	0.057	0.086	0.113	0.135
Dynamic viscosity, η_g(10^{-5} Ns/m^2)	S	S	1.11	1.36	1.49	1.83	2.25	2.65	3.02	3.33	4.76	5.96	7.01

NITROGEN DIOXIDE

Chemical formula: NO_2
Molecular weight: 46.01
Melting point: 261.9 K
Boiling point: 294.3 K

Critical temperature: 431.4 K
Critical pressure: 1.013 MPa
Critical density: 271 kg/m^3
Normal vapor density: 2.05 kg/m^3 (@ 0°C, 101.3 kPa)

Properties of the Liquid at Temperatures Below the Normal Boiling Point

	Temperature, °C												
	−200	−180	−160	−140	−120	−100	−50	0	20	50	100	150	200
	Temperature, K												
Property	73.15	93.15	113.15	133.15	153.15	173.15	223.15	273.15	293.15	323.15	373.15	423.15	473.15
Density, ρ_l (kg/m^3)	S	S	S	S	S	S	S	1 494	1 446	V	V	V	V
Specific heat capacity, $c_{p,l}$ (kJ/kg K)	S	S	S	S	S	S	S	1.505	1.535	V	V	V	V
Thermal conductivity, λ_l [(W/m^2)/(K/m)]	S	S	S	S	S	S	S	0.140	0.130	V	V	V	V
Dynamic viscosity, η_l (10^{-5} Ns/m^2)	S	S	S	S	S	S	S	49.4	4.21	V	V	V	V

Transport Properties of the Superheated Vapor

	Temperature, °C												
	− 150	− 100	− 50	0	25	100	200	300	400	500	1 000	1 500	2 000
Property (at low pressure)	Temperature, K												
	123.15	173.15	223.15	273.15	298.15	373.15	473.15	573.15	673.15	773.15	1 273.15	1 773.15	2 273.15
Specific heat capacity, $c_{p,g}$ (kJ/kg K)	S	S	S	L	0.808	0.858	0.929	0.984	1.034	1.080	1.193	1.256	1.281
Thermal conductivity, λ_g[(W/m^2)/(K/m)]	S	S	S	L	1.18	0.065	0.033	0.040	0.047	0.055	0.085	—	—
Dynamic viscosity, η_g(10^{-5} Ns/m^2)	S	S	S	L	(1.49)	1.84	2.26	2.65	2.99	3.32	4.55	—	—

NITROGEN PEROXIDE

Chemical formula: N_2O_4
Molecular weight: 92.02
Melting point: 261.9 K
Boiling point: 294.25 K

Critical temperature: 431.4 K
Critical pressure: 10.13 MPa
Critical density: 271 kg/m^3
Normal vapor density: 4.11 kg/m^3 (@ 0°C, 101.3 kPa)

Transport Properties of the Superheated Vapor

	Temperature, °C												
	− 150	− 100	− 50	0	25	100	200	300	400	500	1 000	1 500	2 000
Property (at low pressure)	Temperature, K												
	123.15	173.15	223.15	273.15	298.15	373.15	473.15	573.15	673.15	773.15	1 273.15	1 773.15	2 273.15
Specific heat capacity, $c_{p,g}$ (kJ/kg K)	S	S	S	L	0.858	0.946	1.047	1.118	1.175	1.221	1.350	1.388	1.411
Thermal conductivity, λ_g[(W/m^2)/(K/m)]	S	S	S	L	1.18	0.065	0.033	0.040	0.047	0.055	0.085	—	—
Dynamic viscosity, η_g(10^{-5} Ns/m^2)	S	S	S	L	1.70	2.05	2.47	2.88	3.26	3.61	5.14	—	—

CARBON MONOXIDE

Chemical formula: CO
Molecular weight: 28.011
Melting point: 68.16 K
Normal boiling point: 81.66 K

Critical temperature: 133.16 K
Critical pressure: 3.498 MPa
Critical density: 301 kg/m^3
Normal vapor density: 1.25 kg/m^3
(@ 0°C, 101.3 kPa)

Properties of the Liquid at Temperatures Below the Normal Boiling Point

	Temperature, °C												
	−200	−180	−160	−140	−120	−100	−50	0	20	50	100	150	200
	Temperature, K												
Property	73.15	93.15	113.15	133.15	153.15	173.15	223.15	273.15	293.15	323.15	373.15	423.15	473.15
Density, ρ_l (kg/m^3)	837	V	V	V	V	V	V	V	V	V	V	V	V
Specific heat capacity, $c_{p,l}$ (kJ/kg K)	2.16	V	V	V	V	V	V	V	V	V	V	V	V
Thermal conductivity, λ_l [(W/m^2)/(K/m)]	0.160	V	V	V	V	V	V	V	V	V	V	V	V
Dynamic viscosity, η_l (10^{-5} Ns/m^2)	23.0	V	V	V	V	V	V	V	V	V	V	V	V

Properties of the Saturated Liquid and Vapor

T_{sat}, K	81.66	90	95	100	105	110	115	120	125	133.16
p_{sat}, kPa	101.3	245	437	548	776	1 070	1 433	1 875	2 423	3 498
ρ_ℓ, kg/m^3	789	751	728	702	675	646	613	574	526	301
ρ_g, kg/m^3	4.40	9.99	15.4	22.4	31.0	42.6	57.2	77.5	153	301
h_ℓ, kJ/kg	150.4	168.6	179.8	192.3	204.7	216.1	227.1	239.3	258.1	314.33
h_g, kJ/kg	366.0	369.7	371.2	372.9	374.3	374.6	374.0	370.9	363.7	314.33
$\Delta h_{g,\ell}$, kJ/kg	215.6	201.1	191.4	180.6	169.6	158.5	146.9	131.6	105.6	
$c_{p,\ell}$, kJ/(kg K)	2.15	2.17	2.20	2.26	2.34	2.46	2.61	2.79	3.02	
$c_{p,g}$, kJ/(kg K)	1.22	1.35	1.52	1.60	1.79	2.03	2.42	3.18	5.25	
η_ℓ, μNs/m^2	154	120	105	93.5	83.9	75.9	69.1	63.3	58.3	
η_g, μNs/m^2	7.08	7.51	7.80	8.11	8.44	8.80	9.23	9.78	11.7	
λ_ℓ, (mW/m^2)/(K/m)	141	125	116	107	98.2	89.3	80.3	71.4	62.5	
λ_g, (mW/m^2)/(K/m)	6.89	8.22	8.93	9.71	10.5	11.5	12.6	14.0	18.4	
Pr_ℓ	2.35	2.08	1.99	1.97	2.00	2.09	2.25	2.47	2.82	
Pr_g	1.25	1.23	1.33	1.34	1.44	1.55	1.77	2.22	3.34	
σ, mN/m	9.47	7.73	6.71	5.71	4.73	3.78	2.86	1.97	1.13	
$\beta_{e,\ell}$, kK^{-1}	5.46	6.65	7.52	8.44	9.63	11.6	14.9	20.7	31	

Transport Properties of the Superheated Vapor

	Temperature, °C												
	−150	−100	−50	0	25	100	200	300	400	500	1 000	1 500	2 000
Property (at low pressure)	Temperature, K												
	123.15	173.15	223.15	273.15	298.15	373.15	473.15	573.15	673.15	773.15	1 273.15	1 773.15	2 273.15
Specific heat capacity, $c_{p,g}$ (kJ/kg K)	1.038	1.038	1.038	1.038	1.038	1.038	1.055	1.080	1.105	1.130	1.235	1.285	1.315
Thermal conductivity, λ_g[(W/m^2)/(K/m)]	0.011	0.015	0.019	0.023	0.025	0.030	0.037	0.043	0.049	0.055	0.083	0.107	0.128
Dynamic viscosity, η_g(10^{-5} Ns/m^2)	0.81	1.12	1.40	1.66	1.78	2.11	2.51	2.87	3.20	3.52	4.90	5.89	6.92

CARBON DIOXIDE

Chemical formula: CO_2
Molecular weight: 44.011
Melting point: 216.55 K
Normal boiling point: 194.65 K

Critical temperature: 304.19 K
Critical pressure: 7.382 MPa
Critical density: 468 kg/m^3
Normal vapor density: 1.96 kg/m^3
(@ 0 °C, 101.3 kPa)

Properties of the Liquid at Temperatures Below the Normal Boiling Point

	Temperature, °C												
	−200	−180	−160	−140	−120	−100	−78.5	0	20	50	100	150	200
	Temperature, K												
Property	73.15	93.15	113.15	133.15	153.15	173.15	194.65	273.15	293.15	323.15	373.15	423.15	473.15
Density, ρ_l (kg/m^3)	S	S	S	S	S	S	1 179	V	V	V	V	V	V
Specific heat capacity, $c_{p,l}$ (kJ/kg K)	S	S	S	S	S	S	2.15	V	V	V	V	V	V
Thermal conductivity, λ_l [(W/m^2)/(K/m)]	S	S	S	S	S	S	177	V	V	V	V	V	V
Dynamic viscosity, η_l (10^{-5} Ns/m^2)	S	S	S	S	S	S	25.0	V	V	V	V	V	V

Properties of the Saturated Liquid and Vapor

T_{sat}, K	216.55	230	240	250	260	270	280	290	300	304.19
p_{sat}, kPa	518	891	1 282	1 787	2 421	3 203	4 159	5 315	6 712	7 382
ρ_ℓ, kg/m^3	1 179	1 130	1 089	1 046	998	944	883	805	676	468
ρ_g, kg/m^3	15.8	20.8	32.7	45.9	63.6	88.6	121	172	268	468
h_ℓ, kJ/kg	−206.2	−181.5	−162.5	−142.6	−121.9	−99.6	−75.7	−47.6	−10.8	42.8
h_g, kJ/kg	141.1	148.5	151.7	151.1	148.6	142.9	134.9	122.8	96.3	42.8
$\Delta h_{g,\ell}$, kJ/kg	347.3	330.0	314.2	293.7	270.5	242.5	210.6	170.4	107.1	
$c_{p,\ell}$, kJ/(kg K)	2.15	2.08	2.09	2.13	2.24	2.42	2.76	3.63	7.69	
$c_{p,g}$, kJ/(kg K)	0.89	0.98	1.10	1.20	1.42	1.64	1.94	3.03	9.25	
η_ℓ, μNs/m^2	250	200	166	138	117	102	90.8	79.0	59.6	31.6
η_g, μNs/m^2	11.0	12.0	12.7	13.5	14.3	15.2	16.5	18.7	22.8	31.6
λ_ℓ, (mW/m^2)/(K/m)	177	160	146	134	122	110	98	86.1	74.1	47.5
λ_g, (mW/m^2)/(K/m)	11.5	12.9	14.2	15.7	17.4	19.7	22.9	28.0	39.2	47.5
Pr_ℓ	3.04	2.60	2.38	2.19	2.15	2.24	2.56	3.33	6.19	
Pr_g	0.85	0.91	0.98	1.03	1.17	1.27	1.40	2.02	5.38	
σ, mN/m	17.1	13.8	11.4	9.16	7.02	5.01	3.19	1.61	0.33	
$\beta_{e,\ell}$, kK^{-1}	2.86	3.60	4.18	4.91	6.00	7.63	10.2	18.2	57	

Transport Properties of the Superheated Vapor

	Temperature, °C												
	−150	−100	−50	0	25	100	200	300	400	500	1 000	1 500	2 000
Property	Temperature, K												
(at low pressure)	123.15	173.15	223.15	273.15	298.15	373.15	473.15	573.15	673.15	773.15	1 273.15	1 773.15	2 273.15
Specific heat capacity, $c_{p,g}$ (kJ/kg K)	S	S	0.775	0.816	0.846	0.934	1.001	1.063	1.114	1.156	1.269	1.319	1.352
Thermal conductivity, λ_g[(W/m^2)/(K/m)]	S	S	0.011	0.015	0.016	0.022	0.030	0.038	0.045	0.052	0.083	0.108	0.132
Dynamic viscosity, η_g(10^{-5} Ns/m^2)	S	S	1.13	1.37	1.49	1.82	2.22	2.59	2.93	3.24	4.59	5.70	6.57

Thermodynamic Properties of the Superheated Vapor

Pressure, MPa	Property	Value at different temperatures, K							
		300	400	500	600	700	800	900	1 000
1.0	v, m³/kg	0.053 79	0.074 18	0.093 76	0.113 0	0.132 2	0.151 3	0.170 3	0.189 3
(233.0)	h, kJ/kg	419.95	513.65	613.22	718.90	829.90	945.34	1 064.46	1 186.61
	s, kJ/(kg K)	1.773 7	2.043 0	2.264 9	2.457 4	2.628 4	2.782 5	2.922 8	3.051 5
2.0	v, m³/kg	0.025 35	0.036 40	0.046 54	0.056 38	0.066 08	0.075 71	0.085 29	0.094 84
(253.6)	h, kJ/kg	409.41	508.45	610.01	716.73	828.37	944.25	1 063.68	1 186.07
	s, kJ/(kg K)	1.617 4	1.902 5	2.128 9	2.323 3	2.495 4	2.650 0	2.790 7	2.919 6
5.0	v, m³/kg	0.007 79	0.013 74	0.018 24	0.022 41	0.026 43	0.030 39	0.034 29	0.038 17
(287.5)	h, kJ/kg	366.98	492.26	600.43	710.37	823.93	941.10	1 061.45	1 184.53
	s, kJ/(kg K)	1.335 1	1.699 3	1.940 7	2.141 0	2.316 0	2.472 4	2.614 1	2.743 8
10	v, m³/kg		0.006 20	0.008 85	0.011 12	0.013 25	0.015 30	0.017 31	0.019 29
	h, kJ/kg		463.19	584.73	700.24	816.97	936.20	1 058.02	1 182.21
	s, kJ/(kg K)		1.512 7	1.784 6	1.995 2	2.175 1	2.334 3	2.477 7	2.608 6
20	v, m³/kg		0.002 62	0.004 26	0.005 54	0.006 70	0.007 79	0.008 85	0.009 88
	h, kJ/kg		403.03	555.30	681.94	804.67	927.73	1 052.23	1 178.42
	s, kJ/(kg K)		1.263 4	1.605 4	1.836 6	2.025 8	2.190 1	2.336 7	2.469 7

CYANOGEN

Chemical formula: C_2N_2
Molecular weight: 52.04
Melting point: 245.3 K
Boiling point: 251.95 K

Critical temperature: 399.7 K
Critical pressure: 5.89 MPa
Critical density: (260) kg/m³
Normal vapor density: 2.32 kg/m³
(@ 0 °C, 101.3 kPa)

Transport Properties of the Superheated Vapor

Property (at low pressure)	Temperature, °C												
	−150	−100	−50	0	25	100	200	300	400	500	1 000	1 500	2 000
	Temperature, K												
	123.15	173.15	223.15	273.15	298.15	373.15	473.15	573.15	673.15	773.15	1 273.15	1 773.15	2 273.15
Specific heat capacity, $c_{p,g}$ (kJ/kg K)	S	S	S	1.068	1.097	1.168	1.168	1.302	1.348	1.394	1.529	1.592	1.623
Thermal conductivity, λ_g[(W/m²)/(K/m)]	S	S	S	0.013	0.015	0.020	0.027	0.033	0.040	0.047	0.076	0.100	0.121
Dynamic viscosity, η_g(10⁻⁵ Ns/m²)	S	S	S	0.928	1.10	1.27	1.60	1.91	2.20	2.48	3.66	4.64	5.50

CYANOGEN FLUORIDE

Chemical formula: CNF
Molecular weight: 45.02
Melting point: —
Boiling point: 200.55 K

Critical temperature: —
Critical pressure: —
Critical density: —
Normal vapor density: 2.01 kg/m^3 (@ 0°C, 101.3 kPa)

Transport Properties of the Superheated Vapor

	Temperature, °C												
	−150	−100	−50	0	25	100	200	300	400	500	1 000	1 500	2 000
Property (at low pressure)	Temperature, K												
	123.15	173.15	223.15	273.15	298.15	373.15	473.15	573.15	673.15	773.15	1 273.15	1 773.15	2 273.15
Specific heat capacity, $c_{p,g}$ (kJ/kg K)	S	S	0.887	0.946	0.965	1.004	1.063	1.110	1.149	1.180	1.272	1.316	1.340
Thermal conductivity, λ_g[(W/m^2)/(K/m)]	S	S	0.019	0.027	0.028	0.035	0.044	0.053	0.062	0.070	0.103	0.132	0.158
Dynamic viscosity, η_g(10^{-5} Ns/m^2)	S	S	1.50	1.80	1.95	2.35	2.81	3.32	3.74	4.13	5.85	7.20	8.50

CYANOGEN CHLORIDE

Chemical formula: CNCl
Molecular weight: 61.48
Melting point: 266.25 K
Boiling point: 286.15 K

Critical temperature: —
Critical pressure: —
Critical density: —
Normal vapor density: 2.74 kg/m^3 (@ 0°C, 101.3 kPa)

Transport Properties of the Superheated Vapor

	Temperature, °C												
	−150	−100	−50	0	25	100	200	300	400	500	1 000	1 500	2 000
Property (at low pressure)	Temperature, K												
	123.15	173.15	223.15	273.15	298.15	373.15	473.15	573.15	673.15	773.15	1 273.15	1 773.15	2 273.15
Specific heat capacity, $c_{p,g}$ (kJ/kg K)	S	S	S	L	0.729	0.775	0.816	0.846	0.871	0.888	0.950	0.979	0.992
Thermal conductivity, λ_g[(W/m^2)/(K/m)]	S	S	S	L	0.013	0.018	0.023	0.028	0.034	0.039	0.062	0.080	0.096
Dynamic viscosity, η_g(10^{-5} Ns/m^2)	S	S	S	L	1.31	1.65	2.08	2.45	2.83	3.20	4.68	5.87	6.98

CARBONYL SULFIDE (CARBON OXYSULFIDE)

Chemical formula: COS
Molecular weight: 60.07
Melting point: 134.3 K
Boiling point: 222.9 K

Critical temperature: 375.4 K
Critical pressure: 6.5 MPa
Critical density: 429 kg/m^3
Normal vapor density: 2.68 kg/m^3
(@ 0°C, 101.3 kPa)

Properties of the Liquid at Temperatures Below the Normal Boiling Point

	Temperature, °C												
	−200	−180	−160	−140	−120	−100	−50	0	20	50	100	150	200
	Temperature, K												
Property	73.15	93.15	113.15	133.15	153.15	173.15	223.15	273.15	293.15	323.15	373.15	423.15	473.15
Density, ρ_l (kg/m^3)	S	S	S	S	(1316)	1 276	1 180	V	V	V	V	V	V
Specific heat capacity, $c_{p,l}$ (kJ/kg K)	S	S	S	S	1.20	1.19	1.22	V	V	V	V	V	V
Thermal conductivity, λ_l [(W/m^2)/(K/m)]	S	S	S	S	(0.274)	(0.251)	(0.199)	V	V	V	V	V	V
Dynamic viscosity, η_l (10^{-5} Ns/m^2)	S	S	S	S	54.0	43.7	25.8	V	V	V	V	V	V

Transport Properties of the Superheated Vapor

	Temperature, °C												
	− 150	− 100	− 50	0	25	100	200	300	400	500	1 000	1 500	2 000
Property (at low pressure)	Temperature, K												
	123.15	173.15	223.15	273.15	298.15	373.15	473.15	573.15	673.15	773.15	1 273.15	1 773.15	2 273.15
Specific heat capacity, $c_{p,g}$ (kJ/kg K)	S	L	0.636	0.674	0.691	0.745	0.800	0.842	0.875	0.900	0.967	0.992	1.013
Thermal conductivity, λ_g[(W/m^2)/(K/m)]	S	L	0.008	0.011	0.012	0.017	0.022	0.027	0.033	0.038	0.050	0.079	0.095
Dynamic viscosity, η_g(10^{-5} Ns/m^2)	S	L	0.93	1.13	1.23	1.35	1.96	2.32	2.67	3.01	4.42	5.59	6.63

CARBON DISULFIDE

Chemical formula: CS
Molecular weight: 76.13
Melting point: 161.3 K
Boiling point: 319.4 K

Critical temperature: 552.0 K
Critical pressure: 7.90 MPa
Critical density: 475 kg/m^3
Normal vapor density: 3.40 kg/m^3
(@ 0 °C, 101.3 kPa)

Properties of the Liquid at Temperatures Below the Normal Boiling Point

	Temperature, °C												
	−200	−180	−160	−140	−120	−100	−50	0	20	50	100	150	200
	Temperature, K												
Property	73.15	93.15	113.15	133.15	153.15	173.15	223.15	273.15	293.15	323.15	373.15	423.15	473.15
Density, ρ_l (kg/m^3)	S	S	S	S	S	1 432	1 362	1 292	1 262	V	V	V	V
Specific heat capacity, $c_{p,l}$ (kJ/kg K)	S	S	S	S	S	0.988	0.988	0.996	1.00	V	V	V	V
Thermal conductivity, λ_l [(W/m^2)/(K/m)]	S	S	S	S	S	(0.212)	(0.194)	0.177	0.170	V	V	V	V
Dynamic viscosity, η_l (10^{-5} Ns/m^2)	S	S	S	S	S	(138)	(68)	44.0	36.7	V	V	V	V

Transport Properties of the Superheated Vapor

	Temperature, °C												
	− 150	− 100	− 50	0	25	100	200	300	400	500	1 000	1 500	2 000
Property (at low pressure)	Temperature, K												
	123.15	173.15	223.15	273.15	298.15	373.15	473.15	573.15	673.15	773.15	1 273.15	1 773.15	2 273.15
Specific heat capacity, $c_{p,g}$ (kJ/kg K)	S	L	L	L	L	0.641	0.678	0.708	0.729	0.745	0.779	0.804	0.816
Thermal conductivity, λ_g[(W/m^2)/(K/m)]	S	L	L	L	L	0.011	0.015	0.019	0.023	0.026	0.043	0.055	0.068
Dynamic viscosity, η_g(10^{-5} Ns/m^2)	S	L	L	L	L	1.26	1.60	1.94	2.27	2.54	3.88	5.00	5.96

SULFUR DIOXIDE

Chemical formula: SO
Molecular weight: 64.06
Melting point: 198.7 K
Boiling point: 263.0 K

Critical temperature: 430.8 K
Critical pressure: 7.884 MPa
Critical density: 525 kg/m^3
Normal vapor density: 2.86 kg/m^3
(@ 0 °C, 101.3 kPa)

Properties of the Liquid at Temperatures Below the Normal Boiling Point

	Temperature, °C												
	−200	−180	−160	−140	−120	−100	−50	0	20	50	100	150	200
	Temperature, K												
Property	73.15	93.15	113.15	133.15	153.15	173.15	223.15	273.15	293.15	323.15	373.15	423.15	473.15
Density, ρ_l (kg/m^3)	S	S	S	S	S	S	1 557	V	V	V	V	V	V
Specific heat capacity, $c_{p,l}$ (kJ/kg K)	S	S	S	S	S	S	(1.381)	V	V	V	V	V	V
Thermal conductivity, λ_l [(W/m^2)/(K/m)]	S	S	S	S	S	S	0.242	V	V	V	V	V	V
Dynamic viscosity, η_l (10^{-5} Ns/m^2)	S	S	S	S	S	S	68.1	V	V	V	V	V	V

Transport Properties of the Superheated Vapor

	Temperature, °C												
	−150	−100	−50	0	25	100	200	300	400	500	1 000	1 500	2 000
Property (at low pressure)	Temperature, K												
	123.15	173.15	223.15	273.15	298.15	373.15	473.15	573.15	673.15	773.15	1 273.15	1 773.15	2 273.15
Specific heat capacity, $c_{p,g}$ (kJ/kg K)	S	S	L	0.586	0.607	0.662	0.712	0.754	0.787	0.816	0.883	0.921	0.946
Thermal conductivity, λ_g[(W/m^2)/(K/m)]	S	S	L	0.009	0.010	0.014	0.019	0.024	0.029	0.034	0.057	0.073	0.087
Dynamic viscosity, η_g(10^{-5} Ns/m^2)	S	S	L	1.17	1.28	1.63	2.07	2.46	2.80	3.15	4.57	5.84	6.91

SULFUR HEXAFLUORIDE

Chemical formula: SF_6
Molecular weight: 146.05
Melting point: 222.5 K
Boiling point: 209.3 K

Critical temperature: 318.7 K
Critical pressure: 3.76 MPa
Critical density: 738 kg/m^3
Normal vapor density: 6.52 kg/m^3 (@ 0°C, 101.3 kPa)

Transport Properties of the Superheated Vapor

Property (at low pressure)	Temperature, °C: −150	−100	−50	0	25	100	200	300	400	500	1 000	1 500	2 000
	Temperature, K: 123.15	173.15	223.15	273.15	298.15	373.15	473.15	573.15	673.15	773.15	1 273.15	1 773.15	2 273.15
Specific heat capacity, $c_{p,g}$ (kJ/kg K)	S	S	0.502	0.599	0.641	0.755	0.858	0.918	0.968	0.986	1.045	1.063	1.070
Thermal conductivity, λ_g[(W/m^2)/(K/m)]	S	S	0.008	0.012	0.014	0.020	0.027	0.034	0.040	0.045	0.068	0.087	0.103
Dynamic viscosity, η_g(10^{-5} Ns/m^2)	S	S	1.15	1.43	1.56	1.89	2.30	2.71	3.07	3.42	4.87	6.09	7.17

PHOSGENE

Chemical formula: $COCl_2$
Molecular weight: 98.92
Melting point: 145.0 K
Boiling point: 280.8 K

Critical temperature: 455 K
Critical pressure: 5.67 MPa
Critical density: 521 kg/m^3
Normal vapor density: 4.41 kg/m^3 (@ 0°C, 101.3 kPa)

Properties of the Liquid at Temperatures Below the Normal Boiling Point

Property	Temperature, °C: −200	−180	−160	−140	−120	−100	−50	0	20	50	100	150	200
	Temperature, K: 73.15	93.15	113.15	133.15	153.15	173.15	223.15	273.15	293.15	323.15	373.15	423.15	473.15
Density, ρ_l (kg/m^3)	S	S	S	S	(1715)	1 667	1 549	1 436	V	V	V	V	V
Specific heat capacity, $c_{p,l}$ (kJ/kg K)	S	S	S	S	1.05	1.03	1.00	1.02	V	V	V	V	V
Thermal conductivity, λ_l [(W/m^2)/(K/m)]	S	S	S	S	S	S	(0.150)	(0.138)	V	V	V	V	V
Dynamic viscosity, η_l (10^{-5} Ns/m^2)	S	S	S	S	S	S	(220)	(114)	V	V	V	V	V

Transport Properties of the Superheated Vapor

Property (at low pressure)	Temperature, °C: −150	−100	−50	0	25	100	200	300	400	500	1 000	1 500	2 000
	Temperature, K: 123.15	173.15	223.15	273.15	298.15	373.15	473.15	573.15	673.15	773.15	1 273.15	1 773.15	2 273.15
Specific heat capacity, $c_{p,g}$ (kJ/kg K)	S	L	L	L	0.615	0.657	0.695	0.724	0.745	0.779	—	—	—
Thermal conductivity, λ_g[(W/m^2)/(K/m)]	S	L	L	L	(0.010)	(0.015)	(0.023)	(0.032)	(0.044)	(0.056)	—	—	—
Dynamic viscosity, η_g(10^{-5} Ns/m^2)	S	L	L	L	(0.92)	(1.17)	(1.48)	(1.74)	(2.00)	(2.25)	—	—	—

HELIUM

Chemical formula: He
Molecular weight: 4.002 6
Melting point: 0.95 K
Normal boiling point: 4.21 K

Critical temperature: 5.19 K
Critical pressure: 0.229 MPa
Critical density: 69.3 kg/m^3
Normal vapor density: 0.18 kg/m^3
(@ 0°C, 101.3 kPa)

Properties of the Saturated Liquid and Vapor

T_{sat}, K	4.21	4.3	4.4	4.5	4.6	4.7	4.8	4.9	5.0	5.19
p_{sat}, kPa	101.3	111	120	132	144	157	169	184	199	229
ρ_ℓ, kg/m^3	125.0	123.6	122.0	119.5	117.0	114.4	111.0	106.5	101.0	69.3
ρ_g, kg/m^3	11.58	12.43	13.13	14.12	15.07	16.08	16.95	18.08	19.16	
h_ℓ, kJ/kg	0.00	0.42	0.50	0.53	0.57	0.62	0.69	0.82	1.02	
h_g, kJ/kg	20.9	20.72	20.20	19.33	18.57	17.42	16.29	14.62	13.62	
$\Delta h_{g,\ell}$, kJ/kg	20.9	20.3	19.7	18.8	18.0	16.8	15.6	13.8	12.0	
$c_{p,\ell}$, kJ/(kg K)	4.48	4.77	5.11	5.53	5.94	6.57	7.53	9.08	11.5	
$c_{p,g}$, kJ/(kg K)	5.19	5.19	5.19	5.19	5.19	5.19	5.19	5.19	5.19	
η_ℓ, μNs/m^2	36.5	36.3	36.1	36.0	35.8	35.5	35.4	35.1	34.9	
η_g, μNs/m^2	1.09	1.11	1.13	1.16	1.18	1.20	1.22	1.25	1.27	1.31
λ_ℓ, (mW/m^2)/(K/m)	31.2	32.4	33.9	35.4	37.0	38.7	40.5	42.3	44.3	
λ_g, (mW/m^2)/(K/m)	8.3	8.45	8.60	8.79	8.95	9.1	9.28	9.43	9.62	
Pr_ℓ	5.23	5.37	5.44	5.62	5.75	6.19	6.58	7.95	9.06	
Pr_g	0.68	0.68	0.68	0.68	0.68	0.68	0.68	0.69	0.69	
σ, mN/m	0.111	0.099	0.087	0.075	0.063	0.061	0.040	0.028	0.019	
$\beta_{e,\ell}$, kK^{-1}	117	144	177	214	256	306	384	528	806	

Transport Properties of the Superheated Vapor

	Temperature, °C												
	−150	−100	−50	0	25	100	200	300	400	500	1 000	1 500	2 000
Property (at low pressure)	Temperature, K												
	123.15	173.15	223.15	273.15	298.15	373.15	473.15	573.15	673.15	773.15	1 273.15	1 773.15	2 273.15
Specific heat capacity, $c_{p,g}$ (kJ/kg K)	5.200	5.200	5.200	5.200	5.200	5.200	5.200	5.200	5.200	5.200	5.200	5.200	5.200
Thermal conductivity, λ_g[(W/m^2)/(K/m)]	0.083	0.104	0.124	0.143	0.150	0.174	0.205	0.237	0.270	0.302	0.423	0.538	0.587
Dynamic viscosity, η_g(10^{-5} Ns/m^2)	1.09	1.35	1.63	1.89	1.96	2.28	2.67	3.06	3.41	3.75	5.19	6.61	7.84

Thermodynamic Properties of the Superheated Vapor

Pressure, MPa	Property	Value at different temperatures, K							
		40	70	100	300	500	700	1 000	1 500
0.101 325	v, m^3/kg	0.821 9	1.438	2.053	6.153	10.25	14.35	20.50	30.75
	h, kJ/kg	219.62	375.63	531.50	1 570.19	2 608.80	3 647.41	5 205.32	7 801.84
	s, kJ/(kg K)	19.465 0	22.375 5	24.228 6	29.934 3	32.587 1	34.334 4	36.186 6	38.292 2
0.20	v, m^3/kg	0.417 3	0.729 8	1.042	3.119	5.196	7.273	10.39	15.58
	h, kJ/kg	219.59	375.81	531.75	1 570.51	2 609.12	3 647.72	5 205.62	7 802.12
	s, kJ/(kg K)	18.047 0	20.961 6	22.815 7	28.522 0	31.174 7	32.922 0	34.774 2	36.879 8
0.50	v, m^3/kg	0.168 1	0.293 6	0.418 4	1.249	2.080	2.911	4.157	6.234
	h, kJ/kg	219.49	376.35	532.51	1 571.51	2 610.10	3 648.68	5 206.54	7 802.99
	s, kJ/(kg K)	16.127 1	19.054 2	20.911 0	26.619 0	29.271 7	31.018 9	32.871 1	34.976 6
1.0	v, m^3/kg	0.085 00	0.148 1	0.210 7	0.626 1	1.041	1.457	2.080	3.118
	h, kJ/kg	219.35	377.25	533.79	1 573.16	2 611.74	3 650.27	5 208.07	7 804.45
	s, kJ/(kg K)	14.660 0	17.607 6	19.469 0	25.179 7	27.832 3	29.579 5	31.431 6	33.537 1
2.0	v, m^3/kg	0.043 50	0.075 44	0.106 8	0.314 6	0.522 1	0.729 6	1.041	1.560
	h, kJ/kg	219.24	379.06	536.33	1 576.46	2 615.00	3 653.45	5 211.14	7 807.34
	s, kJ/(kg K)	13.169 0	16.154 6	18.024 7	23.740 9	26.393 5	28.140 6	29.992 5	32.097 9
4.0	v, m^3/kg	0.022 88	0.039 10	0.054 90	0.158 8	0.262 4	0.366 1	0.521 6	0.781 0
	h, kJ/kg	219.93	382.80	541.40	1 583.04	2 621.51	3 659.80	5 217.25	7 813.13
	s, kJ/(kg K)	11.645 0	14.689 7	16.576 2	22.303 2	24.955 7	26.702 5	28.554 1	30.659 2
7.0	v, m^3/kg	0.014 21	0.023 55	0.032 64	0.091 97	0.151 1	0.210 3	0.299 1	0.447 1
	h, kJ/kg	223.35	388.84	549.09	1 592.86	2 631.24	3 669.28	5 226.37	7 821.77
	s, kJ/(kg K)	10.401 4	13.495 3	15.401 6	21.144 0	23.796 2	25.542 5	27.393 8	29.498 5
10	v, m^3/kg	0.010 83	0.017 35	0.023 74	0.065 25	0.106 6	0.148 0	0.210 0	0.313 6
	h, kJ/kg	228.94	395.47	556.93	1 602.63	2 640.91	3 678.71	5 235.45	7 830.37
	s, kJ/(kg K)	9.617 3	12.729 3	14.650 2	20.406 2	23.058 3	24.804 2	26.655 0	28.759 3
20	v, m^3/kg	0.007 00	0.010 17	0.013 36	0.034 05	0.054 62	0.075 21	0.106 1	0.157 8
	h, kJ/kg	255.08	420.70	584.25	1 634.95	2 672.86	3 709.83	5 265.41	7 858.75
	s, kJ/(kg K)	8.152 8	11.240 9	13.186 6	18.977 0	21.628 2	23.372 8	25.222 2	27.325 2
50	v, m^3/kg	0.004 65	0.005 83	0.007 08	0.015 27	0.023 36	0.031 49	0.043 74	0.064 21
	h, kJ/kg	350.04	506.54	670.41	1 730.11	2 766.57	3 800.87	5 352.83	7 941.55
	s, kJ/(kg K)	6.404 2	9.309 5	11.257 2	17.106 1	19.754 0	21.494 1	23.339 2	25.438 4
100	v, m^3/kg	0.003 70	0.004 27	0.004 87	0.008 90	0.012 85	0.016 83	0.022 84	0.032 94
	h, kJ/kg	504.52	653.87	814.00	1 882.29	2 917.30	3 946.88	5 492.26	8 072.99
	s, kJ/(kg K)	5.151 7	7.922 9	9.824 6	15.718 0	18.363 0	20.095 2	21.932 4	24.025 1

ARGON

Chemical formula: Ar
Molecular weight: 39.944
Melting point: 83.78 K
Normal boiling point: 87.29 K

Critical temperature: 150.86 K
Critical pressure: 4.898 MPa
Critical density: 536 kg/m^3
Normal vapor density: 1.78 kg/m^3 (@ 0 °C, 101.3 kPa)

Properties of the Liquid at Temperatures Below the Normal Boiling Point

	Temperature, °C												
	−200	-185.86	−160	−140	−120	−100	−50	0	20	50	100	150	200
	Temperature, K												
Property	73.15	87.29	113.15	133.15	153.15	173.15	223.15	273.15	293.15	323.15	373.15	423.15	473.15
Density, ρ_l (kg/m^3)	S	1 393	V	V	V	V	V	V	V	V	V	V	V
Specific heat capacity, $c_{p,l}$ (kJ/kg K)	S	1.083	V	V	V	V	V	V	V	V	V	V	V
Thermal conductivity, λ_l [(W/m^2)/(K/m)]	S	0.1232	V	V	V	V	V	V	V	V	V	V	V
Dynamic viscosity, η_l (10^{-5} Ns/m^2)	S	26.05	V	V	V	V	V	V	V	V	V	V	V

Properties of the Saturated Liquid and Vapor

T_{sat}, K	87.29	94.4	101.4	108.5	115.5	122.6	129.7	136.7	143.8	150.9
p_{sat}, kPa	101.3	201.6	362.2	601.5	938.2	1 393	1 987	2 738	3 702	4 898
ρ_ℓ, kg/m^3	1 393	1 348	1 301	1 251	1 197	1 137	1 068	986.7	877.6	535.6
ρ_g, kg/m^3	5.78	10.9	18.6	30.2	46.4	68.9	100.2	146.8	222.4	535.6
h_ℓ, kJ/kg	−116.1	−108.8	−101.1	−92.9	−84.2	−74.9	−64.5	−53.0	−40.2	−2.4
h_g, kJ/kg	43.5	45.8	47.6	48.7	49.0	48.2	46.0	41.7	33.3	−2.4
$\Delta h_{g,\ell}$, kJ/kg	159.6	154.6	148.9	141.6	133.2	123.1	110.6	94.7	73.5	
$c_{p,\ell}$, kJ/(kg K)	1.083	1.168	1.200	1.218	1.257	1.358	1.559	1.923	2.011	
$c_{p,g}$, kJ/(kg K)	0.548	0.569	0.626	0.665	0.745	0.866	1.067	1.509	2.951	
η_ℓ, μNs/m^2	260.5	211.9	176.6	150.8	131.1	116.1	101.7	84.2	63.9	27.9
η_g, μNs/m^2	7.43	8.04	8.69	9.39	10.2	10.5	12.1	13.6	15.8	27.9
λ_ℓ, (mW/m^2)/(K/m)	123.2	114.9	106.5	98.7	90.1	81.1	71.9	63.4	53.6	30
λ_g, (mW/m^2)/(K/m)	6.09	6.63	7.23	7.92	8.70	9.67	11.1	12.9	15.4	30
Pr_ℓ	2.29	2.15	1.99	1.86	1.83	1.94	2.21	2.55	2.40	
Pr_g	0.67	0.69	0.75	0.79	0.87	0.94	1.16	1.59	3.03	
σ, mN/m	14.50	12.77	11.28	9.35	7.73	6.18	4.71	3.34	2.61	1.75
$\beta_{e,\ell}$, kK^{-1}	4.58	5.01	5.50	6.41	7.80	9.61	12.6	16.8	23.7	41.8

Transport Properties of the Superheated Vapor

	Temperature, °C												
	− 150	− 100	− 50	0	25	100	200	300	400	500	1 000	1 500	2 000
Property (at low pressure)	Temperature, K												
	123.15	173.15	223.15	273.15	298.15	373.15	473.15	573.15	673.15	773.15	1 273.15	1 773.15	2 273.15
Specific heat capacity, $c_{p,g}$ (kJ/kg K)	0.519	0.519	0.519	0.519	0.519	0.519	0.519	0.519	0.519	0.519	0.519	0.519	0.519
Thermal conductivity, λ_g[(W/m^2)/(K/m)]	0.008	0.011	0.014	0.016	0.018	0.021	0.026	0.029	0.033	0.036	0.049	—	—
Dynamic viscosity, η_g(10^{-5} Ns/m^2)	1.02	1.40	1.76	2.10	2.26	2.68	3.21	3.67	4.10	4.49	6.15	7.42	—

NEON

Chemical formula: Ne
Molecular weight: 20.183
Melting point: 24.5 K
Normal boiling point: 27.09 K

Critical temperature: 44.4 K
Critical pressure: 2.654 MPa
Critical density: 483 kg/m^3
Normal vapor density: 0.90 kg/m^3
(@ 0 °C, 101.3 kPa)

Properties of the Saturated Liquid and Vapor

T_{sat}, K	27.1	29	31	33	35	37	39	41	43	44.4
p_{sat}, kPa	101.3	174	284	439	646	916	1 260	1 688	2 216	2 654
ρ_ℓ, kg/m^3	1 205	1 170	1 131	1 089	1 043	992	932	859	754	483
ρ_g, kg/m^3	9.57	15.7	25.0	38.0	56.0	80.8	115.3	164.5	243.5	483
h_ℓ, kJ/kg	5.01	9.00	13.38	17.96	22.74	27.70	32.93	38.67	46.32	
h_g, kJ/kg	91.11	92.23	93.00	93.27	92.94	91.85	89.83	86.58	81.13	
$\Delta h_{g,\ell}$, kJ/kg	86.1	83.2	79.6	75.3	70.2	64.2	56.9	47.9	34.8	
$c_{p,\ell}$, kJ/(kg K)	1.87	1.92	1.97	2.05	2.14	2.29	2.49	2.77	3.23	
$c_{p,g}$, kJ/(kg K)	1.31	1.42	1.61	1.84	2.02	2.42	3.06	4.45	8.08	
η_ℓ, μNs/m^2	127	105	92.5	77.9	65.9	56.6	47.3	38.6	27.8	16.7
η_g, μNs/m^2	4.63	5.04	5.43	5.83	6.12	6.82	7.43	8.47	11.9	16.7
λ_ℓ, (mW/m^2)/(K/m)	113	110	108	102	96.5	88.6	78.9	67.3	50.3	33
λ_g, (mW/m^2)/(K/m)	7.7	8.5	9.3	10.2	11.1	12.4	13.9	16.0	22.8	33
Pr_ℓ	2.10	1.83	1.69	1.57	1.46	1.46	1.49	1.59	1.69	
Pr_g	0.79	0.84	0.94	1.05	1.11	1.33	1.64	2.36	4.22	
σ, mN/m	4.78	4.15	3.50	2.87	2.26	1.69	1.15	0.65	0.20	
$\beta_{e,\ell}$, kK^{-1}	1.47	1.66	1.91	2.23	2.68	3.32	4.47	7.10	31.0	

Transport Properties of the Superheated Vapor

Property (at low pressure)	Temperature, °C												
	−150	−100	−50	0	25	100	200	300	400	500	1 000	1 500	2 000
	Temperature, K												
	123.15	173.15	223.15	273.15	298.15	373.15	473.15	573.15	673.15	773.15	1 273.15	1 773.15	2 273.15
Specific heat capacity, $c_{p,g}$ (kJ/kg K)	1.030	1.030	1.030	1.030	1.030	1.030	1.030	1.030	1.030	1.030	1.030	1.030	1.030
Thermal conductivity, λ_g[(W/m^2)/(K/m)]	0.027	0.034	0.041	0.046	0.049	0.057	0.067	0.077	0.087	0.097	0.132	0.154	0.180
Dynamic viscosity, η_g(10^{-5} Ns/m^2)	1.67	2.14	2.58	2.99	3.12	3.65	4.26	4.89	5.32	5.81	7.81	9.95	11.68

KRYPTON

Chemical formula: Kr
Molecular weight: 83.80
Melting point: 115.8 K
Boiling point: 119.8 K

Critical temperature: 209.4 K
Critical pressure: 5.502 MPa
Critical density: 1088 kg/m^3
Normal vapor density: 3.74 kg/m^3
(@ 0°C, 101.3 kPa)

Transport Properties of the Superheated Vapor

	Temperature, °C												
	−150	−100	−50	0	25	100	200	300	400	500	1 000	1 500	2 000
Property (at low pressure)	Temperature, K												
	123.15	173.15	223.15	273.15	298.15	373.15	473.15	573.15	673.15	773.15	1 273.15	1 773.15	2 273.15
Specific heat capacity, $c_{p,g}$ (kJ/kg K)	0.247	0.247	0.247	0.247	0.247	0.247	0.247	0.247	0.247	0.247	0.247	0.247	0.247
Thermal conductivity, λ_g[(W/m^2)/(K/m)]	0.004	0.006	0.007	0.009	0.010	0.012	0.014	0.016	0.018	0.021	0.030	0.035	0.041
Dynamic viscosity, η_g(10^{-5} Ns/m^2)	1.05	1.49	1.91	2.33	2.52	3.06	3.74	4.38	4.91	5.39	7.55	9.39	11.02

XENON

Chemical formula: Xe
Molecular weight: 131.30
Melting point: 161.3 K
Boiling point: 165.0 K

Critical temperature: 289.7 K
Critical pressure: 5.822 MPa
Critical density: 1113 kg/m^3
Normal vapor density: 5.86 kg/m^3
(@ 0°C, 101.3 kPa)

Transport Properties of the Superheated Vapor

	Temperature, °C												
	−150	−100	−50	0	25	100	200	300	400	500	1 000	1 500	2 000
Property (at low pressure)	Temperature, K												
	123.15	173.15	223.15	273.15	298.15	373.15	473.15	573.15	673.15	773.15	1 273.15	1 773.15	2 273.15
Specific heat capacity, $c_{p,g}$ (kJ/kg K)	S	0.159	0.159	0.159	0.159	0.159	0.159	0.159	0.159	0.159	0.159	0.159	0.159
Thermal conductivity, λ_g[(W/m^2)/(K/m)]	S	0.003	0.004	0.005	0.006	0.007	0.008	0.010	0.012	0.013	0.018	0.022	0.026
Dynamic viscosity, η_g(10^{-5} Ns/m^2)	S	1.39	1.78	2.11	2.29	2.83	3.50	4.15	4.73	5.24	7.38	9.22	1.084

HYDROGEN

Chemical formula: H_2
Molecular weight: 2.016 0
Melting point: 13.95 K
Normal boiling point: 20.38 K

Critical temperature: 33.23 K
Critical pressure: 1.316 MPa
Critical density: 31.6 kg/m^3
Normal vapor density: 0.09 kg/m^3 (@ 0°C, 101.3 kPa)

Properties of the Saturated Liquid and Vapor

T_{sat}, K	20.38	21	23	25	27	29	30	31	32	33.23
p_{sat}, kPa	101.3	121	204	321	479	685	808	946	1 100	1 316
ρ_ℓ, kg/m^3	71.1	70.4	67.9	65.0	61.6	57.4	54.9	51.7	47.5	31.6
ρ_g, kg/m^3	1.31	1.56	2.49	3.88	5.81	8.60	10.5	13.0	16.6	31.6
h_ℓ, kJ/kg	262	268	291	317	348	384	406	431	463	561
h_g, kJ/kg	718	721	729	732	730	719	710	694	671	561
$\Delta h_{g,\ell}$, kJ/kg	456	453	438	415	382	335	304	263	208	
$c_{p,\ell}$, kJ/(kg K)	9.74	10.2	11.8	13.7	16.4	21.1	25.5	33.8	55.8	
$c_{p,g}$, kJ/(kg K)	11.7	11.9	13.0	14.6	17.5	23.3	28.9	39.7	69.1	
η_ℓ, μNs/m^2	12.7	12.0	10.5	9.05	7.84	6.76	6.27	5.67	5.00	3.38
η_g, μNs/m^2	1.12	1.17	1.30	1.43	1.56	1.74	1.86	2.00	2.19	3.38
λ_ℓ, (mW/m^2)/(K/m)	119	121	126	127	122	112	106	100	91	60
λ_g, (mW/m^2)/(K/m)	16.3	16.9	19.2	22	25	29	31	35	40	60
Pr_ℓ	1.04	1.02	0.99	0.98	1.05	1.27	1.51	1.92	3.07	
Pr_g	0.80	0.82	0.88	0.95	1.09	1.40	1.73	2.27	3.78	
σ, mN/m	1.92	1.81	1.47	1.13	0.796	0.483	0.333	0.207	0.106	
$\beta_{e,\ell}$, kK^{-1}	15.9	16.8	20.6	26.3	35.1	52.0	68.4	95.3	172.9	

Transport Properties of the Superheated Vapor

	Temperature, °C												
	−150	−100	−50	0	25	100	200	300	400	500	1 000	1 500	2 000
Property (at low pressure)	Temperature, K												
	123.15	173.15	223.15	273.15	298.15	373.15	473.15	573.15	673.15	773.15	1 273.15	1 773.15	2 273.15
Specific heat capacity, $c_{p,g}$ (kJ/kg K)	11.85	13.00	13.50	14.05	14.34	14.41	14.41	14.41	14.41	14.55	15.51	16.95	17.36
Thermal conductivity, λ_g[(W/m^2)/(K/m)]	0.073	0.113	0.141	0.171	0.181	0.211	0.249	0.285	0.321	0.367	0.519	0.680	0.836
Dynamic viscosity, η_g(10^{-5} Ns/m^2)	0.488	0.618	0.734	0.841	0.892	1.04	1.22	1.39	1.54	1.69	2.38	2.86	3.28

Thermodynamic Properties of the Superheated Vapor

Pressure, MPa	Property	Value at different temperatures, K							
		100	200	300	400	600	800	1 200	1 500
0.101 325	v, m³/kg	4.070	8.147	12.22	16.29	24.43	32.57	48.85	61.06
(20.28)	h, kJ/kg	1 399.8	2 971.3	4 509.6	5 976.6	8 880.9	11 806.0	17 833.5	22 547.2
	s, kJ/(kg K)	42.689	53.475	59.729	63.952	69.840	74.046	80.146	83.650
0.20	v, m³/kg	2.061	4.130	6.194	8.257	12.38	16.51	24.75	30.94
(22.81)	h, kJ/kg	1 398.3	2 971.3	4 510.1	5 977.3	8 881.7	11 806.9	17 834.3	22 548.0
	s, kJ/(kg K)	39.869	50.667	56.924	61.147	67.035	71.242	77.342	80.846
0.50	v, m³/kg	0.824 1	1.656	2.482	3.307	4.957	6.607	9.906	12.38
(27.12)	h, kJ/kg	1 393.6	2 971.5	4 511.6	5 979.4	8 884.2	11 809.5	17 836.9	22 550.6
	s, kJ/(kg K)	36.046	46.880	53.143	57.367	63.257	67.463	73.563	77.067
1.0	v, m³/kg	0.411 9	0.830 8	1.245	1.658	2.483	3.307	4.957	6.194
(31.26)	h, kJ/kg	1 386.0	2 971.7	4 514.1	5 982.8	8 888.4	11 813.8	17 841.3	22 554.9
	s, kJ/(kg K)	33.114	44.008	50.280	54.507	60.398	64.605	70.705	74.209
2.0	v, m³/kg	0.205 9	0.418 4	0.626 1	0.832 8	1.245	1.658	2.482	3.101
	h, kJ/kg	1 371.4	2 972.3	4 519.1	5 989.7	8 896.7	11 822.6	17 850.1	22 563.5
	s, kJ/(kg K)	30.114	41.122	47.413	51.645	57.539	61.747	67.847	71.351
4.0	v, m³/kg	0.103 4	0.212 3	0.316 8	0.420 4	0.626 8	0.832 9	1.245	1.554
	h, kJ/kg	1 345.2	2 974.1	4 529.3	6 003.5	8 913.3	11 840.0	17 867.6	22 580.8
	s, kJ/(kg K)	26.993	38.213	44.538	48.781	54.681	58.890	64.990	68.494
7.0	v, m³/kg	0.060 19	0.124 0	0.184 3	0.243 6	0.361 6	0.479 3	0.714 6	0.891 1
	h, kJ/kg	1 313.9	2 978.3	4 544.9	6 024.2	8 938.1	11 866.0	17 893.8	22 606.6
	s, kJ/(kg K)	24.349	35.834	42.207	46.465	52.374	56.584	62.685	66.188
10	v, m³/kg	0.043 45	0.088 77	0.131 2	0.172 9	0.255 6	0.337 9	0.502 5	0.626 0
	h, kJ/kg	1 292.9	2 984.2	4 561.0	6 045.1	8 962.9	11 892.0	17 920.0	22 632.3
	s, kJ/(kg K)	22.615	34.299	40.715	44.987	50.904	55.116	61.217	64.720
20	v, m³/kg	0.025 28	0.047 95	0.069 45	0.090 43	0.131 8	0.172 9	0.255 1	0.316 7
	h, kJ/kg	1 281.7	3 015.9	4 617.9	6 115.4	9 045.0	11 978.2	18 007.0	22 718.0
	s, kJ/(kg K)	19.291	31.274	37.794	42.105	48.047	52.265	58.367	61.869
50	v, m³/kg	0.015 51	0.024 01	0.032 57	0.040 97	0.057 51	0.073 91	0.106 6	0.131 1
	h, kJ/kg	1 456.2	3 185.6	4 816.6	6 336.4	9 290.7	12 233.9	18 265.8	22 973.9
	s, kJ/(kg K)	15.362	27.268	33.906	38.282	44.275	48.508	54.614	58.113
100	v, m³/kg	0.011 92	0.016 08	0.020 31	0.024 48	0.032 72	0.040 86	0.057 08	0.069 23
	h, kJ/kg	1 850.1	3 553.2	5 192.3	6 725.3	9 700.6	12 655.0	18 692.0	23 396.7
	s, kJ/(kg K)	12.618	24.314	30.983	35.397	41.433	45.682	51.794	55.291

NITROGEN

Chemical formula: N_2
Molecular weight: 28.016
Melting point: 63.15 K
Normal boiling point: 77.35 K

Critical temperature: 126.25 K
Critical pressure: 3.396 MPa
Critical density: 304 kg/m^3
Normal vapor density: 1.25 kg/m^3
(@ 0 °C, 101.3 kPa)

Properties of the Liquid at Temperatures Below the Normal Boiling Point

	Temperature, °C												
	−200	−180	−160	−140	−120	−100	−50	0	20	50	100	150	200
	Temperature, K												
Property	73.15	93.15	113.15	133.15	153.15	173.15	223.15	273.15	293.15	323.15	373.15	423.15	473.15
Density, ρ_l (kg/m^3)	829	V	V	V	V	V	V	V	V	V	V	V	V
Specific heat capacity, $c_{p,l}$ (kJ/kg K)	1.99	V	V	V	V	V	V	V	V	V	V	V	V
Thermal conductivity, λ_l [(W/m^2)/(K/m)]	0.141	V	V	V	V	V	V	V	V	V	V	V	V
Dynamic viscosity, η_l (10^{-5} Ns/m^2)	19.7	V	V	V	V	V	V	V	V	V	V	V	V

Properties of the Saturated Liquid and Vapor

T_{sat}, K	77.35	85	90	95	100	105	110	115	120	126
p_{sat}, kPa	101.3	290	360	540	778	1 083	1 467	1 940	2 515	3 357
ρ_ℓ, kg/m^3	807.10	771.01	746.27	719.42	691.08	660.5	626.17	583.43	528.54	379.22
ρ_g, kg/m^3	4.621	9.833	15.087	22.286	31.989	44.984	62.578	87.184	124.517	237.925
h_ℓ, kJ/kg	−120.8	−105.7	−95.6	−85.2	−74.5	−63.8	−51.4	−38.1	−21.4	17.4
h_g, kJ/kg	76.8	82.3	85.0	86.8	87.7	87.4	85.6	81.8	74.3	49.5
$\Delta h_{g,\ell}$, kJ/kg	197.6	188.0	180.5	172.2	162.2	150.7	137.0	119.9	95.7	32.1
$c_{p,\ell}$, kJ/(kg K)	2.064	2.096	2.140	2.211	2.311	2.467	2.711	3.180	4.347	
$c_{p,g}$, kJ/(kg K)	1.123	1.192	1.258	1.350	1.474	1.666	1.975	2.586	4.136	
η_ℓ, μNs/m^2	163	127	110	97.2	86.9	78.5	70.8	59.9	48.4	19.1
η_g, μNs/m^2	5.41	5.60	6.36	6.80	7.28	7.82	8.42	9.25	10.68	19.1
λ_ℓ, (mW/m^2)/(K/m)	136.7	122.9	112.0	104.0	95.5	88.0	80.2	70.4	62.8	52.8
λ_g, (mW/m^2)/(K/m)	7.54	8.18	9.04	9.77	10.60	11.69	14.50	20.76	30.91	51.11
Pr_ℓ	2.46	2.17	2.10	2.07	2.10	2.20	2.39	2.71	3.35	
Pr_g	0.81	0.82	0.89	0.94	1.01	1.11	1.15	1.16	1.43	
σ, mN/m	8.85	7.20	6.16	4.59	3.67	2.79	1.98	1.18	0.52	0.01
$\beta_{e,\ell}$, kK^{-1}	5.65	6.46	7.26	8.47	9.69	12.1	15.7	23.0	37.5	

Transport Properties of the Superheated Vapor

	Temperature, °C												
	− 150	− 100	− 50	0	25	100	200	300	400	500	1 000	1 500	2 000
Property (at low pressure)	Temperature, K												
	123.15	173.15	223.15	273.15	298.15	373.15	473.15	573.15	673.15	773.15	1 273.15	1 773.15	2 273.15
Specific heat capacity, $c_{p,g}$ (kJ/kg K)	1.058	1.046	1.038	1.038	1.038	1.038	1.047	1.068	1.089	1.114	1.223	1.273	1.306
Thermal conductivity, λ_g[(W/m^2)/(K/m)]	0.012	0.017	0.021	0.024	0.026	0.031	0.037	0.042	0.047	0.052	0.072	0.104	0.125
Dynamic viscosity, η_g(10^{-5} Ns/m^2)	0.836	1.14	1.41	1.66	1.78	2.09	2.47	2.82	3.14	3.42	4.61	5.56	6.36

Thermodynamic Properties of the Superheated Vapor

Pressure, MPa	Property	Value at different temperatures, K							
		200	300	400	500	600	800	1 000	1 200
0.101 325	v, m^3/kg	0.584 5	0.878 6	1.172	1.465	1.758	2.344	2.930	3.516
(77.35)	h, kJ/kg	357.00	461.14	565.30	670.20	776.59	995.91	1 224.55	1 461.29
	s, kJ/(kg K)	3.982 9	4.405 1	4.704 8	4.938 8	5.132 7	5.447 9	5.702 8	5.918 5
0.50	v, m^3/kg	0.117 4	0.177 9	0.237 8	0.297 4	0.356 9	0.475 8	0.594 6	0.713 4
(93.98)	h, kJ/kg	355.05	460.25	564.89	670.03	776.59	996.10	1 224.84	1 461.63
	s, kJ/(kg K)	3.501 9	3.928 6	4.229 7	4.464 3	4.658 5	4.973 9	5.228 9	5.444 7
1.0	v, m^3/kg	0.058 10	0.088 90	0.119 1	0.149 0	0.178 8	0.238 4	0.297 8	0.357 2
(103.7)	h, kJ/kg	352.58	459.16	564.37	669.84	776.60	996.33	1 225.20	1.462.07
	s, kJ/(kg K)	3.287 0	3.719 5	4.022 2	4.257 5	4.452 1	4.767 9	5.023 1	5.238 9
2.0	v, m^3/kg	0.028 44	0.044 40	0.059 71	0.074 81	0.089 80	0.119 7	0.149 4	0.179 1
(115.6)	h, kJ/kg	347.60	457.00	563.37	669.47	776.62	996.81	1 225.92	1 462.94
	s, kJ/(kg K)	3.062 6	3.507 0	3.813 1	4.049 9	4.245 2	4.561 6	4.817 0	5.033 0
5.0	v, m^3/kg	0.010 71	0.017 75	0.024 13	0.030 31	0.036 40	0.048 43	0.060 38	0.072 30
	h, kJ/kg	332.45	450.83	560.56	668.48	776.77	998.29	1 228.12	1 465.57
	s, kJ/(kg K)	2.733 0	3.215 4	3.531 4	3.772 2	3.969 6	4.288 0	4.544 2	4.760 6
10	v, m^3/kg	0.005 02	0.008 95	0.012 32	0.015 51	0.018 61	0.024 70	0.030 71	0.036 69
	h, kJ/kg	308.56	441.78	556.63	667.31	777.34	1 000.90	1 231.86	1 470.01
	s, kJ/(kg K)	2.434 1	2.979 7	3.310 6	3.557 6	3.758 2	4.079 6	4.337 1	4.554 1
20	v, m^3/kg	0.002 69	0.004 70	0.006 49	0.008 15	0.009 75	0.012 85	0.015 88	0.018 89
	h, kJ/kg	280.28	428.93	551.48	666.60	779.54	1 006.65	1 239.64	1 479.06
	s, kJ/(kg K)	2.116 8	2.726 1	3.079 5	3.336 5	3.542 5	3.869 0	4.128 8	4.347 0
40	v, m^3/kg	0.001 88	0.002 78	0.003 68	0.004 53	0.005 35	0.006 94	0.008 48	0.009 99
	h, kJ/kg	272.71	421.05	550.24	670.61	787.47	1 019.85	1 256.13	1 497.70
	s, kJ/(kg K)	1.861 4	2.466 4	2.838 9	3.107 8	3.320 9	3.655 1	3.918 6	4.138 7
70	v, m^3/kg	0.001 56	0.002 04	0.002 54	0.003 02	0.003 50	0.004 42	0.005 31	0.006 18
	h, kJ/kg	287.71	431.56	562.23	685.58	805.40	1 042.66	1 282.44	1 526.54
	s, kJ/(kg K)	1.682 0	2.267 6	2.644 2	2.919 7	3.138 2	3.479 4	3.746 9	3.969 4
100	v, m^3/kg	0.001 42	0.001 75	0.002 09	0.002 43	0.002 76	0.003 41	0.004 04	0.004 66
	h, kJ/kg	309.67	451.53	582.06	706.35	827.54	1 067.65	1 309.89	1 555.93
	s, kJ/(kg K)	1.568 9	2.146 1	2.522 2	2.799 7	3.020 8	3.366 2	3.636 3	3.860 6
200	v, m^3/kg	0.001 21	0.001 38	0.001 55	0.001 72	0.001 89	0.002 22	0.002 54	0.002 86
	h, kJ/kg	393.11	534.36	664.90	790.06	912.85	1 157.26	1 404.11	1 654.43
	s, kJ/(kg K)	1.337 1	1.911 8	2.287 8	2.567 3	2.791 2	3.142 8	3.418 1	3.646 2

OXYGEN

Chemical formula: O_2
Molecular weight: 32.00
Melting point: 54.35 K
Normal boiling point: 90.18 K

Critical temperature: 154.77 K
Critical pressure: 5.090 MPa
Critical density: 405 kg/m^3
Normal vapor density: 1.43 kg/m^3
(@ 0°C, 101.3 kPa)

Properties of the Liquid at Temperatures Below the Normal Boiling Point

	Temperature, °C												
	−200	−180	−160	−140	−120	−100	−50	0	20	50	100	150	200
	Temperature, K												
Property	73.15	93.15	113.15	133.15	153.15	173.15	223.15	273.15	293.15	323.15	373.15	423.15	473.15
Density, ρ_l (kg/m^3)	1 226	V	V	V	V	V	V	V	V	V	V	V	V
Specific heat capacity, $c_{p,l}$ (kJ/kg K)	1.67	V	V	V	V	V	V	V	V	V	V	V	V
Thermal conductivity, λ_l [(W/m^2)/(K/m)]	0.179	V	V	V	V	V	V	V	V	V	V	V	V
Dynamic viscosity, η_l (10^{-5} Ns/m^2)	31.4	V	V	V	V	V	V	V	V	V	V	V	V

Properties of the Saturated Liquid and Vapor

T_{sat}, K	90.18	97	104	111	118	125	132	140	146	154
p_{sat}, kPa	101.3	196	352	583	908	1 348	1 924	2 782	3 591	3 939
ρ_ℓ, kg/m^3	1 135.72	1 102.05	1 065.07	1 025.64	982.32	934.58	880.28	808.41	737.56	557.10
ρ_g, kg/m^3	4.48	8.23	14.14	22.79	35.03	52.05	75.81	116.12	163.34	304.41
h_ℓ, kJ/kg	−133.4	−122.1	−110.3	−98.2	−85.4	−71.8	−57.8	−38.9	−23.2	10.6
h_g, kJ/kg	78.9	83.8	88.0	91.2	93.3	93.9	92.8	88.4	81.4	56.7
$\Delta h_{g,\ell}$, kJ/kg	212.3	205.9	198.3	189.4	178.7	165.7	150.1	127.3	104.6	46.1
$c_{p,\ell}$, kJ/(kg K)	1.63	1.66	1.70	1.76	1.86	2.00	2.22	2.63	3.28	
$c_{p,g}$, kJ/(kg K)	0.96	1.00	1.05	1.12	1.23	1.36	1.68	2.27	3.63	
η_ℓ, μNs/m^2	195.83	161.75	136.55	116.80	101.20	89.00	80.15	69.66	60.65	42.48
η_g, μNs/m^2	6.85	7.50	8.35	9.36	10.6	11.24	13.35	15.8	18.5	26.9
λ_ℓ, (mW/m^2)/(K/m)	148	139	130	121	111	102	92.5	82.0	71.2	
λ_g, (mW/m^2)/(K/m)	8.5	9.5	10.5	11.7	13.4	14.8	16.9	20.1	23.6	35.2
Pr_ℓ	2.16	1.93	1.79	1.70	1.70	1.75	1.92	2.23	2.79	
Pr_g	0.77	0.79	0.84	0.90	0.97	1.03	1.33	1.78	2.85	19.93
σ, mN/m	13.19	11.53	9.88	8.27	6.71	5.20	3.77	2.23	1.18	0.40
$\beta_{e,\ell}$, kK^{-1}	4.26	4.71	5.39	6.30	7.38	9.65	11.9	17.6	29.0	545.0

Transport Properties of the Superheated Vapor

Property (at low pressure)	−150	−100	−50	0	25	100	200	300	400	500	1 000	1 500	2 000
	Temperature, K												
	123.15	173.15	223.15	273.15	298.15	373.15	473.15	573.15	673.15	773.15	1 273.15	1 773.15	2 273.15
Specific heat capacity, $c_{p,g}$ (kJ/kg K)	0.927	0.909	0.903	0.909	0.913	0.934	0.963	0.992	1.026	1.051	1.122	1.164	1.197
Thermal conductivity, λ_g[(W/m^2)/(K/m)]	0.011	0.016	0.020	0.024	0.026	0.032	0.039	0.045	0.052	0.058	0.086	0.115	0.139
Dynamic viscosity, η_g(10^{-5} Ns/m^2)	0.955	1.31	1.63	1.92	2.03	2.43	2.88	3.29	3.67	4.03	5.59	6.92	8.20

(Column headings: Temperature, °C.)

Thermodynamic Properties of the Superheated Vapor

Pressure, MPa	Property	200	300	400	500	600	700	800	1 000
		Value at different temperatures, K							
0.050	v, m^3/kg	1.038	1.558	2.079	2.598	3.118	3.638	4.158	5.197
(83.94)	h, kJ/kg	374.65	466.10	559.00	654.63	753.40	855.13	959.43	1 174.09
	s, kJ/(kg K)	4.127 5	4.498 2	4.765 3	4.978 6	5.158 6	5.315 3	5.454 6	5.693 9
0.101 325	v, m^3/kg	0.511 3	0.768 8	1.026	1.282	1.539	1.795	2.052	2.565
(90.19)	h, kJ/kg	374.39	465.97	558.94	654.60	753.39	855.14	959.45	1 174.11
	s, kJ/(kg K)	3.943 1	4.314 3	4.581 6	4.795 0	4.975 0	5.131 8	5.271 0	5.510 4
0.20	v, m^3/kg	0.258 3	0.389 3	0.519 7	0.649 8	0.779 8	0.909 8	1.040	1.300
(97.24)	h, kJ/kg	373.89	465.73	558.82	654.54	753,38	855.15	959.48	1 174.17
	s, kJ/(kg K)	3.764 7	4.137 0	4.404 7	4.618 1	4.798 2	4.955 1	5.094 4	5.333 7
0.50	v, m^3/kg	0.102 4	0.155 4	0.207 9	0.260 1	0.312 2	0.364 2	0.416 2	0.520 1
(108.8)	h, kJ/kg	372.35	464.99	558.44	654.37	753.33	855.19	959.57	1 174.33
	s, kJ/(kg K)	3.521 2	3.396 9	4.165 6	4.379 6	4.559 9	4.716 9	4.856 2	5.095 7
1.0	v, m^3/kg	0.050 39	0.077 49	0.103 9	0.130 2	0.156 3	0.182 3	0.208 3	0.260 3
(119.6)	h, kJ/kg	369.73	463.76	557.81	654.09	753.25	855.25	959.73	1 174.59
	s, kJ/(kg K)	3.332 0	3.713 5	3.984 0	4.198 7	4.379 4	4.536 6	4.676 0	4.915 6
2.0	v, m^3/kg	0.024 39	0.038 53	0.051 98	0.065 20	0.078 32	0.091 37	0.104 4	0.130 4
(132.7)	h, kJ/kg	364.33	461.29	556.57	653.52	753.11	855.38	960.04	1 175.13
	s, kJ/(kg K)	3.132 8	3.526 6	3.300 7	4.016 9	4.198 4	4.356 0	4.495 8	4.735 6
5.0	v, m^3/kg	0.008 76	0.015 16	0.020 82	0.026 24	0.031 56	0.036 81	0.042 04	0.052 44
(154.4)	h, kJ/kg	346.74	453.93	552.90	651.86	752.69	855.78	961.00	1 176.75
	s, kJ/(kg K)	2.831 0	3.268 4	3.553 3	3.774 0	3.957 8	4.116 7	4.257 1	4.497 7
10	v, m^3/kg	0.003 61	0.007 43	0.010 46	0.013 27	0.015 98	0.018 64	0.021 26	0.026 47
	h, kJ/kg	313.19	442.21	547.14	649.30	752.12	856.53	962.66	1 179.50
	s, kJ/(kg K)	2.523 9	3.055 9	3.358 2	3.586 1	3.773 6	3.934 5	4.076 2	4.318 0
20	v, m^3/kg	0.001 73	0.003 70	0.005 34	0.006 82	0.008 22	0.009 57	0.010 89	0.013 50
	h, kJ/kg	268.46	422.72	537.55	645.22	751.58	858.40	966.22	1 185.16
	s, kJ/(kg K)	2.183 4	2.820 2	3.151 5	3.391 9	3.585 8	3.750 5	3.894 4	4.138 6

FLUORINE

Chemical formula: F_2
Molecular weight: 38.00
Melting point: 53 K
Normal boiling point: 85.2 K

Critical temperature: 144 K
Critical pressure: 5.320 MPa
Critical density: 535 kg/m^3
Normal vapor density: 1.70 kg/m^3
(@ 0°C, 101.3 kPa)

Properties of the Liquid at Temperatures Below the Normal Boiling Point

	Temperature, °C												
	−200	−180	−160	−140	−120	−100	−50	0	20	50	100	150	200
	Temperature, K												
Property	73.15	93.15	113.15	133.15	153.15	173.15	223.15	273.15	293.15	323.15	373.15	423.15	473.15
Density, ρ_l (kg/m^3)	1 140	V	V	V	V	V	V	V	V	V	V	V	V
Specific heat capacity, $c_{p,l}$ (kJ/kg K)	1.51	V	V	V	V	V	V	V	V	V	V	V	V
Thermal conductivity, λ_l [(W/m^2)/(K/m)]	(0.155)	V	V	V	V	V	V	V	V	V	V	V	V
Dynamic viscosity, η_l (10^{-5} Ns/m^2)	34.9	V	V	V	V	V	V	V	V	V	V	V	V

Properties of the Saturated Liquid and Vapor

T_{sat}, K	85.2	95	100	105	110	117.5	122.5	127.5	132.5	142.5
p_{sat}, kPa	101.3	278	428	634	903	1 439	1 903	2 470	3 159	4 987
ρ_ℓ, kg/m^3	1 524	1 435	1 393	1 349	1 303	1 229	1 174	1 112	1 040	749.0
ρ_g, kg/m^3	8.30	14.4	21.7	31.7	44.9	72.0	97.2	130.5	176.8	378.8
h_ℓ, kJ/kg	−359.8	−342.5	−333.8	−324.9	−316.1	−302.2	−292.4	−282.0	−270.9	−236.8
h_g, kJ/kg	−168.2	−164.2	−162.4	−161.1	−160.5	−160.9	−162.3	−165.0	−169.5	−192.4
$\Delta h_{g,\ell}$, kJ/kg	191.6	178.3	171.4	163.8	155.6	141.3	130.1	117.0	101.4	44.4
$c_{p,\ell}$, kJ/(kg K)	1.55	1.60	1.65	1.72	1.83	2.04	2.23	2.47	2.75	3.47
$c_{p,g}$, kJ/(kg K)	0.795	0.881	0.945	1.03	1.14	1.35	1.60	2.22	3.61	
η_ℓ, μNs/m^2	237	183	163	147	133	117	108	99.8	93.0	89.5
η_g, μNs/m^2	7.23	8.57	9.09	9.62	10.2	11.2	11.9	12.9	14.2	21.4
λ_ℓ, (mW/m^2)/(K/m)	158	143	135	128	121	114	102	94	85	53
λ_g, (mW/m^2)/(K/m)	7.0	8.3	8.9	9.5	10.5	11.8	13.2	15.0	17.0	26.5
Pr_ℓ	2.33	2.77	1.99	1.98	2.01	2.09	2.36	2.62	3.01	5.86
Pr_g	0.82	0.91	0.97	1.04	1.11	1.28	1.44	1.91	3.01	
σ, mN/m	14.25	12.13	11.03	9.92	8.79	7.06	5.87	3.91	3.38	0.56
$\beta_{e,\ell}$, kK^{-1}	6.30	6.59	6.88	7.55	8.52	10.6	12.5	16.2	23.2	

Transport Properties of the Superheated Vapor

	Temperature, °C												
	−150	−100	−50	0	25	100	200	300	400	500	1 000	1 500	2 000
Property	Temperature, K												
(at low pressure)	123.15	173.15	223.15	273.15	298.15	373.15	473.15	573.15	673.15	773.15	1 273.15	1 773.15	2 273.15
Specific heat capacity, $c_{p,g}$ (kJ/kg K)	0.766	0.755	0.795	0.816	0.825	0.862	0.904	0.921	0.938	0.950	0.988	1.001	1.009
Thermal conductivity, λ_g[(W/m^2)/(K/m)]	0.010	0.015	0.020	0.024	0.027	0.033	0.040	0.047	0.053	0.060	0.091	0.115	0.137
Dynamic viscosity, η_g(10^{-5} Ns/m^2)	0.890	1.25	1.56	2.09	2.42	2.79	3.30	3.90	4.37	4.81	7.67	10.3	12.5

CHLORINE

Chemical formula: Cl_2
Molecular weight: 70.914
Melting point: 172.65 K
Normal boiling point: 239.11 K

Critical temperature: 417.15 K
Critical pressure: 7.710 MPa
Critical density: 573 kg/m^3
Normal vapor density: 3.16 kg/m^3
(@ 0 °C, 101.3 kPa)

Properties of the Liquid at Temperatures Below the Normal Boiling Point

	Temperature, °C												
	−200	−180	−160	−140	−120	−100	−50	0	20	50	100	150	200
	Temperature, K												
Property	73.15	93.15	113.15	133.15	153.15	173.15	223.15	273.15	293.15	323.15	373.15	423.15	473.15
Density, ρ_l (kg/m^3)	S	S	S	S	S	1 717	1 598	V	V	V	V	V	V
Specific heat capacity, $c_{p,l}$ (kJ/kg K)	S	S	S	S	S	0.883	0.892	V	V	V	V	V	V
Thermal conductivity, λ_l [(W/m^2)/(K/m)]	S	S	S	S	S	0.198	0.186	V	V	V	V	V	V
Dynamic viscosity, η_l (10^{-5} Ns/m^2)	S	S	S	S	S	104.0	55.4	V	V	V	V	V	V

Properties of the Saturated Liquid and Vapor

T_{sat}, K	239	261	283	305	328	350	372	394	411	416
p_{sat}, kPa	101.3	241	501	928	1 575	2 501	3 769	5 452	7 036	7 634
ρ_ℓ, kg/m^3	1 563	1 503	1 439	1 370	1 295	1 213	1 116	992.1	838.9	700.8
ρ_g, kg/m^3	3.69	8.20	16.2	28.9	48.5	78.1	123.8	202.0	329.0	468.4
h_ℓ, kJ/kg	237.7	258.8	280.5	302.6	325.6	349.9	377.0	409.7	444.2	467.3
h_g, kJ/kg	525.4	533.9	541.7	548.1	552.8	554.8	553.0	543.9	523.4	499.8
$\Delta h_{g,\ell}$, kJ/kg	287.7	275.1	261.2	245.5	227.2	204.9	176.0	134.2	79.2	32.5
$c_{p,\ell}$, kJ/(kg K)	0.949	0.954	0.962	0.973	1.01	1.10	1.40	1.93		
$c_{p,g}$, kJ/(kg K)	0.497	0.528	0.579	0.644	0.748	0.985	1.40	1.98	4.40	
η_ℓ, μNs/m^2	496.5	428.6	372.6	331.3	285.5	270.5	248.5	230.0	217.6	214.2
η_g, μNs/m^2	10.9	12.5	13.6	14.8	16.0	17.5	19.5	22.3	28.0	35.9
λ_ℓ, (mW/m^2)/(K/m)	163	154	143	132	120	107	93.6	76.8	59.8	48.8
λ_g, (mW/m^2)/(K/m)	6.9	7.8	8.7	10.0	11.5	13.1	15.7	19.5	26.5	34.0
Pr_ℓ	2.89	2.66	2.52	2.45	2.41	2.79	3.71	5.76		
Pr_g	0.79	0.85	0.91	0.95	1.04	1.32	1.72	2.26	4.65	
σ, mN/m	26.90	23.58	19.96	16.53	13.14	9.64	6.34	3.13	0.76	0.12
$\beta_{e,\ell}$, kK^{-1}	1.69	1.95	2.27	2.70	3.38	4.16	6.08	10.5	26.0	

Transport Properties of the Superheated Vapor

	Temperature, °C												
	−150	−100	−50	0	25	100	200	300	400	500	1 000	1 500	2 000
Property (at low pressure)	Temperature, K												
	123.15	173.15	223.15	273.15	298.15	373.15	473.15	573.15	673.15	773.15	1 273.15	1 773.15	2 273.15
Specific heat capacity, $c_{p,g}$ (kJ/kg K)	S	L	L	0.473	0.477	0.494	0.507	0.515	0.523	0.528	0.536	0.544	0.548
Thermal conductivity, λ_g[(W/m^2)/(K/m)]	S	L	L	0.008	0.009	0.012	0.015	0.018	0.021	0.024	0.035	0.045	0.054
Dynamic viscosity, η_g(10^{-5} Ns/m^2)	S	L	L	1.23	1.34	1.68	2.10	2.50	2.86	3.22	4.68	5.90	6.99

BROMINE

Chemical formula: Br
Molecular weight: 159.81
Melting point: 264.9 K
Boiling point: 331.9 K

Critical temperature: 584 K
Critical pressure: 10.3 MPa
Critical density: 1258 kg/m^3
Normal vapor density: 7.13 kg/m^3
(@ 0 °C, 101.3 kPa)

Properties of the Liquid at Temperatures Below the Normal Boiling Point

	Temperature, °C												
	−200	−180	−160	−140	−120	−100	−50	0	20	50	100	150	200
	Temperature, K												
Property	73.15	93.15	113.15	133.15	153.15	173.15	223.15	273.15	293.15	323.15	373.15	423.15	473.15
Density, ρ_l (kg/m^3)	S	S	S	S	S	S	S	3 208	3 140	(3040)	V	V	V
Specific heat capacity, $c_{p,l}$ (kJ/kg K)	S	S	S	S	S	S	S	0.448	0.452	0.456	V	V	V
Thermal conductivity, λ_l [(W/m^2)/(K/m)]	S	S	S	S	S	S	S	(0.129)	0.124	0.117	V	V	V
Dynamic viscosity, η_l (10^{-5} Ns/m^2)	S	S	S	S	S	S	S	124	99.6	76.2	V	V	V

Transport Properties of the Superheated Vapor

	Temperature, °C												
	−150	−100	−50	0	25	100	200	300	400	500	1 000	1 500	2 000
Property	Temperature, K												
(at low pressure)	123.15	173.15	223.15	273.15	298.15	373.15	473.15	573.15	673.15	773.15	1 273.15	1 773.15	2 273.15
Specific heat capacity, $c_{p,g}$ (kJ/kg K)	S	S	S	L	L	0.227	0.229	0.230	0.231	0.232	0.234	0.235	0.237
Thermal conductivity, λ_g[(W/m^2)/(K/m)]	S	S	S	L	L	0.006	0.007	0.009	0.011	0.013	0.021	0.026	0.032
Dynamic viscosity, η_g(10^{-5} Ns/m^2)	S	S	S	L	L	1.88	2.37	2.92	3.40	3.87	5.98	7.73	9.25

IODINE

Chemical formula: I_2
Molecular weight: 253.81
Melting point: 386.8 K
Boiling point: 457.5 K

Critical temperature: 819 K
Critical pressure: 11.65 MPa
Critical density: 1637 kg/m^3
Normal vapor density: 11.33 kg/m^3
(@ 0 °C, 101.3 kPa)

Properties of the Liquid at Temperatures Below the Normal Boiling Point

	Temperature, °C												
	−200	−180	−160	−140	−120	−100	−50	0	20	50	100	150	200
	Temperature, K												
Property	73.15	93.15	113.15	133.15	153.15	173.15	223.15	273.15	293.15	323.15	373.15	423.15	473.15
Density, ρ_l (kg/m^3)	S	S	S	S	S	S	S	S	S	S	S	3 780	V
Specific heat capacity, $c_{p,l}$ (kJ/kg K)	S	S	S	S	S	S	S	S	S	S	S	0.318	V
Thermal conductivity, λ_l [(W/m^2)/(K/m)]	S	S	S	S	S	S	S	S	S	S	S	0.112	V
Dynamic viscosity, η_l (10^{-5} Ns/m^2)	S	S	S	S	S	S	S	S	S	S	S	179.8	V

Transport Properties of the Superheated Vapor

	Temperature, °C												
	−150	−100	−50	0	25	100	200	300	400	500	1 000	1 500	2 000
Property (at low pressure)	Temperature, K												
	123.15	173.15	223.15	273.15	298.15	373.15	473.15	573.15	673.15	773.15	1 273.15	1 773.15	2 273.15
Specific heat capacity, $c_{p,g}$ (kJ/kg K)	S	S	S	S	S	S	0.147	0.148	0.149	0.149	0.150	0.151	0.153
Thermal conductivity, λ_g[(W/m^2)/(K/m)]	S	S	S	S	S	S	0.005	0.006	0.007	0.008	0.011	0.015	0.018
Dynamic viscosity, η_g(10^{-5} Ns/m^2)	S	S	S	S	S	S	2.18	2.64	3.08	3.52	5.33	6.86	8.19

SULFUR

Chemical formula: S_2
Molecular weight: 64.12
Melting point: 385.95 K
Boiling point: 717.75 K

Critical temperature: 1314 K
Critical pressure: 20.7 MPa
Critical density: —
Normal vapor density: 2.86 kg/m^3
(@ 0 °C, 101.3 kPa)

Properties of the Liquid at Temperatures Below the Normal Boiling Point

	Temperature, °C												
	−150	−100	−75	−50	−25	0	20	50	100	150	200	250	300
	Temperature, K												
	123.15	173.15	198.15	223.15	248.15	273.15	293.15	323.15	373.15	423.15	473.15	523.15	573.15
Density, ρ_l (kg/m^3)	S	S	S	S	S	S	S	S	S	1 780	1 744	—	—
Specific heat capacity, $c_{p,l}$ (kJ/kg K)	S	S	S	S	S	S	S	S	S	0.570	0.570	0.570	0.570
Thermal conductivity, λ_l [(W/m^2)/(K/m)]	S	S	S	S	S	S	S	S	S	0.137	0.154	—	—
Dynamic viscosity, η_l (10^{-5} Ns/m^2)	S	S	S	S	S	S	S	S	S	705	79.1	—	—

Transport Properties of the Superheated Vapor

	Temperature, °C												
	−150	−100	−50	0	25	100	200	300	400	500	1 000	1 500	2 000
Property (at low pressure)	Temperature, K												
	123.15	173.15	223.15	273.15	298.15	373.15	473.15	573.15	673.15	773.15	1 273.15	1 773.15	2 273.15
Specific heat capacity, $c_{p,g}$ (kJ/kg K)	S	S	S	—	—	—	L	L	L	0.574	0.590	0.599	0.603
Thermal conductivity, λ_g[(W/m^2)/(K/m)]	S	S	S	S	S	S	L	L	L	0.014	0.023	0.032	0.039
Dynamic viscosity, η_g(10^{-5} Ns/m^2)	S	S	S	S	S	S	L	L	L	1.76	2.84	3.81	4.68

MERCURY

Chemical formula: Hg
Molecular weight: 200.51
Melting point: 234.32 K
Normal boiling point: 630.1 K

Critical temperature: 1763.2 K
Critical pressure: 151.0 MPa
Critical density: 5500 kg/m^3
Normal vapor density: 8.95 kg/m^3
(@ 0 °C, 101.3 kPa)

Properties of the Liquid at Temperatures Below the Normal Boiling Point

	Temperature, °C												
	−150	−100	−75	−50	−25	0	20	50	100	150	200	250	300
	Temperature, K												
Property	123.15	173.15	198.15	223.15	248.15	273.15	293.15	323.15	373.15	423.15	473.15	523.15	573.15
Density, ρ_l (kg/m^3)	S	S	S	S	(13660)	13595	13546	13473	13351	13231	13112	12993	12874
Specific heat capacity, $c_{p,l}$ (kJ/kg K)	S	S	S	S	(1.41)	0.1404	0.1396	0.1385	0.1371	0.1361	0.1355	0.1352	0.1353
Thermal conductivity, λ_l [(W/m^2)/(K/m)]	S	S	S	S	(7.85)	8.175	8.447	8.842	9.475	10.074	10.64	11.18	11.69
Dynamic viscosity, η_l (10^{-5} Ns/m^2)	S	S	S	S	(190)	168.7	155.6	140.0	124.1	112.0	103.9	97.5	92.6

Properties of the Saturated Liquid and Vapor

T_{sat}, K	630.1	650	700	750	800	850	900	950	1 000	1 050
p_{sat}, kPa	101.3	145	316	620	1 120	1 880	2 990	4 530	6 580	9 230
ρ_ℓ, kg/m^3	12 737	12 688	12 567	12 444	12 318	12 190	12 059	11 927	11 791	11 650
ρ_g, kg/m^3	3.91	5.37	10.9	20.1	34.2	54.6	82.7	119.9	167.7	227.3
h_ℓ, kJ/kg	91.8	94.5	101.3	108.2	115.2	122.3	129.5	136.9	144.4	153.8
h_g, kJ/kg	386.7	388.7	393.6	398.4	403.0	407.4	411.6	415.5	419.1	423.0
$\Delta h_{g,\ell}$, kJ/kg	294.9	294.2	292.3	290.2	287.8	285.1	282.1	278.6	274.7	269.2
$c_{p,\ell}$, kJ/(kg K)	0.136	0.136	0.137	0.138	0.140	0.142	0.144	0.146	0.149	0.153
$c_{p,g}$, kJ/(kg K)	0.104	0.104	0.105	0.106	0.107	0.108	0.109	0.111	0.113	0.116
η_ℓ, μNs/m^2	884	870	841	816	794	776	760	746	736	723
η_g, μNs/m^2	61.7	63.5	68.6	73.5	78.4	83.5	88.4	93.2	98.0	103.0
λ_ℓ, (mW/m^2)/(K/m)	121.9	123.6	128.0	131.9	135.1	137.8	141.8	144.5	146.9	147.9
λ_g, (mW/m^2)/(K/m)	10.4	10.8	11.7	12.6	13.5	14.4	15.3	16.2	17.2	18.1
Pr_ℓ	0.987	0.957	0.900	0.854	0.823	0.800	0.772	0.754	0.744	0.748
Pr_g	0.617	0.612	0.616	0.618	0.621	0.626	0.630	0.637	0.644	0.660
σ, mN/m										
$\beta_{e,\ell}$, kK^{-1}	0.194	0.193	0.195	0.203	0.212	0.221	0.230	0.241	0.253	0.269

Transport Properties of the Superheated Vapor

Property (at low pressure)	Temperature, °C: −150	−100	−50	0	25	100	200	300	400	500	1 000	1 500	2 000
	Temperature, K: 123.15	173.15	223.15	273.15	298.15	373.15	473.15	573.15	673.15	773.15	1 273.15	1 773.15	2 273.15
Specific heat capacity, $c_{p,g}$ (kJ/kg K)	S	S	S	L	L	L	L	L	0.103 6	0.103 6	0.103 6	0.103 6	0.103 6
Thermal conductivity, λ_g[(W/m^2)/(K/m)]	S	S	S	L	L	L	L	L	0.011	0.015	0.019	0.025	0.031
Dynamic viscosity, η_g(10^{-5} Ns/m^2)	S	S	S	L	L	L	L	L	6.62	8.62	12.26	16.27	19.80

Thermodynamic Properties of the Superheated Vapor

Pressure, MPa	Property	900	1 000	1 100	1 200	1 300	1 400	1 500	1 600
0.20	v, m^3/kg	0.186 1	0.206 9	0.227 7	0.248 4	0.269 2	0.289 9	0.310 7	0.331 4
(670.0)	h, kJ/kg	353.51	363.91	374.29	384.68	395.05	405.43	415.80	426.17
	s, kJ/(kg K)	0.537 5	0.548 4	0.558 3	0.567 4	0.575 7	0.583 4	0.590 5	0.597 2
0.50	v, m^3/kg	0.074 22	0.082 58	0.090 91	0.099 24	0.107 6	0.115 9	0.124 2	0.132 5
(733.2)	h, kJ/kg	353.20	363.65	374.07	384.48	394.89	405.28	415.67	426.06
	s, kJ/(kg K)	0.499 3	0.510 3	0.520 2	0.529 3	0.537 6	0.545 3	0.552 5	0.559 2
1.0	v, m^3/kg	0.036 92	0.041 13	0.045 32	0.049 51	0.053 68	0.057 85	0.062 02	0.066 18
(789.7)	h, kJ/kg	352.68	363.21	373.70	384.16	394.61	405.04	415.46	425.87
	s, kJ/(kg K)	0.470 2	0.481 3	0.491 3	0.500 4	0.508 7	0.516 5	0.523 7	0.530 4
2.0	v, m^3/kg	0.018 27	0.020 40	0.022 53	0.024 64	0.026 74	0.028 84	0.030 93	0.033 02
(855.8)	h, kJ/kg	351.62	362.32	372.95	383.52	394.05	404.55	415.02	425.48
	s, kJ/(kg K)	0.440 7	0.452 0	0.462 1	0.471 3	0.479 7	0.487 5	0.494 8	0.501 5
5.0	v, m^3/kg		0.007 96	0.008 85	0.009 72	0.010 58	0.011 43	0.012 28	0.013 13
(962.8)	h, kJ/kg		359.56	370.63	381.54	392.34	403.06	413.71	424.32
	s, kJ/(kg K)		0.412 2	0.422 8	0.432 3	0.440 9	0.448 9	0.456 2	0.463 0
10	v, m^3/kg			0.004 28	0.004 74	0.005 18	0.005 63	0.006 06	0.006 50
(1064)	h, kJ/kg			366.56	378.11	389.40	400.51	411.48	422.34
	s, kJ/(kg K)			0.391 6	0.401 7	0.410 7	0.418 9	0.426 5	0.433 5

DOWTHERM A

Chemical formula: $(C_6H_5)_2O$ (73.5%); $(C_6H_5)_2$ (26.5%)
Molecular weight: 166
Melting point: 285.15 K
Normal boiling point: 530.25 K

Critical temperature: 770.15 K
Critical pressure: 3.134 MPa
Critical density: 315.5 kg/m^3
Normal vapor density: 7.41 kg/m^3 (@ 0°C, 101.3 kPa)

Properties of the Liquid at Temperatures Below the Normal Boiling Point

	Temperature, °C												
	−150	−100	−75	−50	−25	0	20	50	100	150	200	250	300
	Temperature, K												
Property	123.15	173.15	198.15	223.15	248.15	273.15	293.15	323.15	373.15	423.15	473.15	523.15	573.15
Density, ρ_l (kg/m^3)	S	S	S	S	S	S	1 060	1 036	995	951	906	858	V
Specific heat capacity, $c_{p,l}$ (kJ/kg K)	S	S	S	S	S	S	1.574	1.660	1.800	1.947	2.087	2.219	V
Thermal conductivity, λ_l [(W/m^2)/(K/m)]	S	S	S	S	S	S	0.141	0.137	0.132	0.125	0.119	0.113	V
Dynamic viscosity, η_l (10^{-5} Ns/m^2)	S	S	S	S	S	S	380	215	100	58	39	28	

Properties of the Saturated Liquid and Vapor

T_{sat}, K	530.25	555	580	605	630	655	680	700	730	770.15
p_{sat}, kPa	101.3	170.4	270	411	600	848	1 170	1 470	2 040	3 134
ρ_ℓ, kg/m^3	851.9	826	799	770	740	706	670	637	573	315.5
ρ_g, kg/m^3	3.96	6.47	10.0	15.2	22.0	31.8	45.1	60.1	100	315.5
h_ℓ, kJ/kg	465	522	580	642	703	769	835	890	970	
h_g, kJ/kg	761	806	850	897	942	990	1 035	1 070	1 110	
$\Delta h_{g,\ell}$, kJ/kg	296	284	270	255	239	221	200	180	140	
$c_{p,\ell}$, kJ/(kg K)	2.24	2.32	2.40	2.47	2.53	2.59	2.69	2.83	3.26	
$c_{p,g}$, kJ/(kg K)	1.83	1.91	1.97	2.03	2.10	2.16	2.24	2.34	2.54	
η_ℓ, μNs/m^2	273	236	206	180	160	145	132	124	115	
η_g, μNs/m^2	10.1	10.6	11.2	11.7	12.2	12.7	13.3	13.9	14.5	
λ_ℓ, (mW/m^2)/(K/m)	112	109	106	103	100	97	94	91	88	
λ_g, (mW/m^2)/(K/m)	20.4	22.6	24.8	27.2	29.8	32.6	35.6	38.8	42.0	
Pr_ℓ	5.46	5.02	4.66	4.32	4.05	3.87	3.78	3.86	4.26	
Pr_g	0.91	0.90	0.89	0.87	0.86	0.84	0.84	0.84	0.88	
σ, mN/m										
$\beta_{e,\ell}$, kK^{-1}	1.20	1.33	1.50	1.70	2.02	2.41	2.93	3.73	6.85	

DOWTHERM J

Chemical formula: $C_{10}H_{14}$
Molecular weight: 134
Melting point: < 235.37 K
Normal boiling point: 454.26 K

Critical temperature: 656.15 K
Critical pressure: 2.837 MPa
Critical density: 273.82 kg/m^3
Normal vapor density: 5.98 kg/m^3
(@ 0°C, 101.3 kPa)

Properties of the Liquid at Temperatures Below the Normal Boiling Point

	Temperature, °C												
	−150	−100	−75	−50	−25	0	20	50	100	150	200	250	300
	Temperature, K												
Property	123.15	173.15	198.15	223.15	248.15	273.15	293.15	323.15	373.15	423.15	473.15	523.15	573.15
Density, ρ_l (kg/m^3)	S	S	S	917	897	888	872	842	801	754	V	V	V
Specific heat capacity, $c_{p,l}$ (kJ/kg K)	S	S	S	1.650	1.713	1.772	1.830	1.924	2.093	2.278	V	V	V
Thermal conductivity, λ_l [(W/m^2)/(K/m)]	S	S	S	0.137	0.135	0.134	0.133	0.130	0.126	0.122	V	V	V
Dynamic viscosity, η_l (10^{-5} Ns/m^2)	S	S	S	410	225	140	90	62	36	22	V	V	V

Properties of the Saturated Liquid and Vapor

T_{sat}, K	454.26	480	500	520	540	560	580	600	620	656.15
p_{sat}, kPa	101.3	187	279	351	536	804	1 070	1 460	1 870	2 837
ρ_ℓ, kg/m^3	729.3	705.3	683.5	660.9	636.6	609.5	580.8	545.4	505.3	273.8
ρ_g, kg/m^3	3.75	6.70	9.79	14.2	20.3	28.7	39.5	63.9	96.4	273.8
h_ℓ, kJ/kg	330	394	444	497	552	609	665	721	779	
h_g, kJ/kg	635	684	721	760	799	838	876	910	937	
$\Delta h_{g,\ell}$, kJ/kg	305	290	277	263	247	229	211	189	158	
$c_{p,\ell}$, kJ/(kg K)	2.40	2.51	2.58	2.66	2.75	2.84	2.95	3.08	3.25	
$c_{p,g}$, kJ/(kg K)	1.91	2.00	2.06	2.12	2.17	2.24	2.29	2.34	2.39	
η_ℓ, μNs/m^2	172	149	133	121	109	101	93	85	78	
η_g, μNs/m^2	8.60	9.05	9.40	9.75	10.10	10.45	10.80	11.15	11.40	
λ_ℓ, (mW/m^2)/(K/m)	118.5	117.0	115.6	114.2	111.8	110.4	109.0	107.6	106.2	
λ_g, (mW/m^2)/(K/m)										
Pr_ℓ	3.48	3.20	2.97	2.82	2.68	2.60	2.52	2.43	2.39	
Pr_g										
σ, mN/m										
$\beta_{e,\ell}$, kK^{-1}	1.20	1.46	1.64	1.73	2.20	2.76	3.54	4.69	7.52	

REFERENCES

1. Bondi, A., Estimation of the Heat Capacity of Liquids, *Ind. Eng. Chem. Fundam.*, vol. 5, pp. 442–449, 1966.
2. Borreson, R. W., Schorr, G. R., and Yaws, C. L., Correlation Constants for Chemical Compounds—Heat Capacities of Gases, *Chem. Eng.*, Aug. 16, pp. 79–81, 1976.
3. Brock, J. R., and Bird, R. B., Surface Tension and the Principle of Corresponding States, *AIChE J.*, vol. 1, pp. 174–177, 1955.
4. Chang, H.-Y., Thermal Conductivities of Gases at Atmosphere Pressure, *Chem. Eng.*, vol. 80, no. 9, pp. 122–123, 1973.
5. Chen, N. H., Generalized Correlation for Latent Heat of Vaporization, *J. Chem. Eng. Data,* vol. 10, pp. 207–210, 1965.
6. Chow, W. M., and Bright, J. A., Jr., Heat Capacities of Organic Liquids, *Chem. Eng. Prog.*, vol. 49, pp. 175–180, 1953.
7. Das, T. R., and Kuloor, N. R., Thermodynamic Properties of *n*-Butane, *Indian J. of Tech.*, vol. 5, pp. 33–39, 1967.
8. Dixon, J. A., and Schiesser, R. W., Viscosities of Benzene-d6 and Cyclohexane-d12, *J. Phys. Chem.*, vol. 58, pp. 430–432, 1954.
9. Dow Chemical Dowtherm, A Heat Transfer Fluid, Form Nos. 176–1337-78 and 176-1240-78R, Functional Products and Systems Department, Dow Chemical Co., Michigan.
10. Edmister, W. C., Applied Hydrocarbon Thermodynamics, p. 56, Gulf Publishing Co., Houston, Texas, 1961.
11. E.S.D.U. Physical Data, Chemical Engineering Sub. Series, vols. 1–7, Engineering Sciences Data Unit, London, 1987.
12. Fluid Properties Research, Inc., Oklahoma State University, Stillwater, Oklahoma, 1979.
13. Fugasi, P., and Rudi, C. E. J., Specific Heats of Organic Vapors, *Ind. Eng. Chem.*, vol. 30, p. 1029, 1938.
14. Gallant, R. W., *Physical Properties of Hydrocarbons,* vols. 1 and 2, Gulf Publishing Co., Houston, Texas, 1970.
15. Geist, J. M., and Cannon, M. R., Viscosities of Pure Hydrocarbons, *Ind. Eng. Chem. Anal. Ed.*, vol. 18, pp. 611–613, 1946.
16. Gomez-Nieto, M., and Thodos, G., Generalized Treatment for the Vapor Pressure Behavior of Polar and Hydrogen-Bonding Compounds, *Can. J. Chem. Eng.*, vol. 55, pp. 445–449, 1977.
17. Gomez-Nieto, M., and Thodos, G., Generalized Vapor Pressure Equation for Nonpolar Substances, *Ind. Eng. Chem. Fundam.*, vol. 17, pp. 45–51, 1978.
18. Gorin, C. E., and Yaws, C. L., Correlation Constants for Chemical Compounds-Heat of Vaporization, *Chem. Eng.*, vol. 83, pp. 85–87, 1976.
19. *GPSA Engineering Data Book,* Gas Processors Suppliers Association, Tulsa, oklahoma, 1977.
20. Gronier, W. S., and Thodos, G., Viscosity and Thermal Conductivity of Ammonia in the Gaseous and Liquid States, *J. Chem. Eng. Data,* vol. 6, pp. 240–244, 1961.
21. Gunn, R. D., and Yamada, T., A Corresponding States Correlation of Saturated Liquid Volumes, *AIChE J.*, vol. 17, pp. 1341–1345, 1971.
22. Horvath, A. L., *Physical Properties of Inorganic Compounds SI Units,* Crane, Russak & Company, New York, 1975.
23. IAPS, Eighth International Conference on The Properties of Steam, Giens, France, September 1974, Release on Thermal Conductivity of Water Substance, December 1977, International Association for the Properties of Steam, Brown Univ., Providence, Rhode Island.
24. International Critical Tables.
25. Jasper, J. J., The Surface Tension of Pure Liquid Compounds, *J. Phys. Chem. Ref. Data,* vol. 1, pp. 841–1009, 1972.
26. Jossi, J. A., Stiel, L. I., and Thodos, G., The Viscosity of Pure Substances in the Dense Gaseous and Liquid Phases, *AIChE J.*, vol. 8, pp. 59–63, 1962.
27. Kudchadker, A. P., Alani, G. H., and Zwolinski, B. J., Critical Constants of Organic Substances, vol. 68, pp. 729–735, 1968.
28. Lee, B. I., and Kesler, M. G., A Generalized Thermodynamic Correlation Based on Three-Parameter Corresponding States, *AIChE J.*, vol. 21, pp. 510–527, 1975.
29. Letsou, A., and Stiel, L. I., Viscosities of Saturated Nonpolar Liquids at Elevated Pressures, *AIChE J.*, vol. 19, pp. 409–411, 1973.
30. Liquide, L., *Gas Encyclopaedia,* Elsevier Scientific Publishing Co., Amsterdam, Netherlands, 1976.
31. Livingston, J., Morgan, R., and Owen, F. T., The Weight of a Falling Drop and the Laws of Tate, *J. Am. Chem. Soc.*, vol. 33, p. 1713, 1911.
32. Lydersen, A. L., Estimation of Critical Properties of Organic Compounds, University of Wisconsin College of Engineering, Eng. Exp. Stn. Rep. 3, Madison, April 1955.
33. Lyman, T. J., and Danner, R. P., Correlation of Liquid Heat Capacities with a Four-Parameter Corresponding States Method, *AIChE J.*, vol. 22, pp. 759–765, 1976.
34. Mathews, J. F., Critical Constants of Inorganic Substances, *Chem. Rev.*, vol. 72, no. 1, 1972.
35. Miller, J. W., Jr., Gordon, R. S., and Yaws, C. L., Correlation Constants for Liquids-Heat Capacities, *Chem. Eng.*, Oct. 25, pp. 129–131, 1976.
36. Miller, J. W., Jr., and Yaws, C. L., Correlation Constants for Liquids-Surface Tension, *Chem. Engr.*, vol. 83, no. 22, pp. 127–129, 1976.
37. Miller, J. W., Jr., McGinley, J. J., and Yaws, C. L., Correlation Constants for Liquids-Thermal Conductivities, *Chem. Eng.*, Oct. 25, pp. 133–135, 1976.
38. Miller, J. W., Jr., Gordon, R. S., and Yaws, C. L., Correlation Constants for Chemical Compounds-Liquid Viscosity, *Chem. Eng.*, vol. 86, no. 24, pp. 157–159, 1976.
39. Miller, J. W., Jr., Gordon, R. S., and Yaws, C. L., Correlation Constants for Chemical Compounds-Gas Viscosity, *Chem. Eng.*, vol. 86, no. 24, pp. 155–157, 1976.
40. Misic, D., and Thodos, G., The Thermal Conductivity of Hydrocarbon Gases at Normal Pressure, *AIChE J.*, vol. 7, pp. 264–267, 1961.
41. Patel, P. M., Schorr, G. R., Shah, P. N., and Yaws, C. L., Vapor Pressure, *Chem. Eng.*, pp. 159–161, Nov. 22, 1976.
42. Pennington, R. E., and Kobe, K. A., The Thermodynamic Properties of Acetone, *J. Am. Chem. Soc.*, vol. 79, pp. 300–305, 1957.
43. Perry, R. H., and Chilton, C. H., *Chemical Engineer's Handbook,* 5th Ed., McGraw-Hill, New York, 1973.
44. Raznjevic, K., *Handbook of Thermodynamic Tables and Charts,* 1st Ed., Hemisphere Publishing Corp., Washington, D.C., 1976.
45. Reid, R. C., Prausnitz, J. M., and Sherwood, T. K., *The Properties of Gases and Liquids,* 3d ed., McGraw-Hill, New York, 1977.
46. Reynolds, W. C., *Thermodynamic Properties in SI,* Department of Mechanical Engineering, Stanford University, California, 1979.
47. Rihani, D. N., and Doraiswamy, L. K., Estimation of Heat Capacity of Organic Compounds from Group Contributions, *Ind. Eng. Chem. Fundam.*, vol. 4, pp. 17–21, 1965.
48. Robbins, L. A., and Kingrea, C. L., Estimate Thermal Conductivity, *Hydrocarbon Proc. Pet. Ref.*, vol. 41, no. 5, pp. 133–136, 1962.
49. Sakiadis, B. C., and Coates, J., Studies in Thermal Conductivity of Liquids, *AIChE J.*, vol. 1, pp. 275–188, 1955.
50. Shah, P. N., and Yaws, C. L., Densities of Liquids, *Chem. Eng.*, pp. 131–133, Oct. 25, 1976.

51. Stiel, L. I., and Thodos, G., The Viscosity of Polar Substances in the Dense Gaseous and Liquid Regions, *AIChE J.*, vol. 10, pp. 275–277, 1964.
52. Stiel, L. I., and Thodos, G., The Thermal Conductivities of Nonpolar Substances in the Dense Gaseous and Liquid Regions, *AIChE J.*, vol. 10, pp. 26–29, 1964.
53. Stiel, L. I., and Thodos, G., The Viscosities of Polar Gases at Normal Pressures, *AIChE J.*, vol. 8, pp. 229–232, 1962.
54. Stiel, L. I., and Thodos G., The Viscosities of Nonpolar Gases at Normal Pressures, *AIChE J.*, vol. 7, pp. 611–615, 1961.
55. Svehla, R. A., Estimated Viscosities and Thermal Conductivities of Gases at High Temperatures, Tech. Rept. R132, NASA, Lewis Research Center, Cleveland, Ohio, 1962.
56. *Technical Data Book-Petroleum Refining,* American Petroleum Institute, Division of Refining, D.C., 1970.
57. T.R.C., Selected Values of Properties of Hydrocarbons and Related Compounds, looseleaf data collection, Texas A & M University, College Station, Tex., Extant 1984.
58. Thinh, T. P., Duran, J. L., Ramalho, R. S., and Kaliaguine, S., Equations Improve C_p^0 Predictions, *Hyd. Proc.*, vol. 50, pp. 98–104, January, 1971.
59. Timmermans, J., *Physico-Chemical Constants of Pure Organic Compounds,* Elsevier Publishing Co., pp. 303–325, New York, 1950.
60. Touloukian, Y. S., Powell, R. W., Ho, C. Y., and Klemens, P. G., *Thermophysical Properties of Matter,* vol. 1, IFI/Plenum, New York, 1970.
61. Touloukian, Y. S., Liley, P. E., and Saxena, S. C., *Thermophysical Properties of Matter,* vol. 3, IFI/Plenum, New York, 1970.
62. Toloukian, Y. S., and Makitu, T., *Thermophysical Properties of Matter,* vol. 6, IFI/Plenum, New York, 1970.
63. Touloukian, Y. S., Saxena, S. C., and Hestermanns, P., *Thermophysical Properties of Matter,* vol. 11, IFI/Plenum, New York, 1975.
64. Van Velzen, D., Cardozo, R. L., and Langenkamp, H., A Liquid Viscosity-Temperature-Chemical Constitution Relation for Organic Compounds, *Ind. Eng. Chem. Fundam.*, vol. 11, pp. 20–25, 1972.
65. Vargaftik, N. B., *Tables on the Thermophysical Properties of Liquids and Gases,* 2d ed., Hemisphere Publishing Corp., Washington, D.C., 1975.
66. Vines, R. G., and Bennett, L. A., The Thermal Conductivity of Organic Vapors. The Relationship between Thermal Conductivity and Viscosity, and the Significance of the Euken Factor. *J. Chem. Phys.*, vol. 22, pp. 360–366, 1954.
67. *Wärmeatlas,* VDI-Verlag GmGH, Düsseldorf, 1984.
68. Watson, K. M., Thermodynamics of the Liquid State, *Ind. Eng. Chem.*, vol. 35, pp. 398–400, 1943.
69. Weast, R. C., ed., *Handbook of Chemistry and Physics,* 54th ed., Chemical Rubber Co., Cleveland, 1974.
70. Yoor, P., and Thodos, G., Viscosity of Nonpolar Gaseous Mixtures at Normal Pressures, *AIChE J.*, vol. 16, pp. 300–304, 1970.
71. Yuan, T. F., and Stiel, L. I., Heat Capacity of Saturated Nonpolar and Polar Liquids, *Ind. Eng. Chem. Fundam.*, vol. 9, pp. 393–400, 1970.

CHAPTER 3
Properties of Water and Steam

3.1. Steam Tables

The tables in this section are reprinted with permission from NBS/NRC Steam Tables by Haar, L., Gallagher, J. S., and Kell, G. S. (Hemisphere, Washington, D.C., 1984). The symbols and nomenclature for this section are as given in the original text.

SYMBOLS AND NOMENCLATURE FOR THE TABLES

Symbol	Property	Units
h	specific enthalpy	kJ/kg
P	pressure	bar = 0.1 MPa
Pr	Prandtl number ($= \eta c_p/\lambda$)	dimensionless
r	specific enthalpy of vaporization	kJ/kg
s	specific entropy	kJ/(kg K)
t	Celsius temperature	°C
t_s	temperature at saturation	
u	specific internal energy	kJ/kg
v	specific volume	m^3/kg
ϵ	static dielectric constant	dimensionless
η	viscosity	10^{-6} kg/(s m) = MPa s
λ	thermal conductivity	mW/(K m)
ρ	density	kg/m^3
σ	surface tension	kg/s^2 = N/m
ϕ	specific entropy of vaporization	kJ/(kg K)

Subscripts

g	denotes a saturated vapor state
l	denotes a saturated liquid state

The reference state for all property values is the liquid at the triple point, for which state the specific internal energy and the specific entropy have been set to zero.

Table 1 Saturation (temperature)

t (°C)	P	ρ_l	ρ_g	h_l	h_g	r	s_l	s_g	ϕ	v_l ($\times 10^3$)	v_g ($\times 10^3$)
0.01	0.0061173	999.78	0.004855	0.00	2500.5	2500.5	0.00000	9.1541	9.1541	1.00022	205990
	0.0065716	999.85	0.005196	4.18	2502.4	2498.2	0.01528	9.1277	9.1124	1.00015	192440
	0.0070605	999.90	0.005563	8.40	2504.2	2495.8	0.03064	9.1013	9.0707	1.00010	179760
	0.0075813	999.93	0.005952	12.61	2506.0	2493.4	0.04592	9.0752	9.0292	1.00007	168020
	0.0081359	999.95	0.006364	16.82	2507.9	2491.1	0.06112	9.0492	8.9881	1.00005	157130
5	0.0087260	999.94	0.006802	21.02	2509.7	2488.7	0.07626	9.0236	8.9473	1.00006	147020
	0.0093537	999.92	0.007265	25.22	2511.5	2486.3	0.09133	8.9981	8.9068	1.00008	137650
	0.0100209	999.89	0.007756	29.42	2513.4	2484.0	0.10633	8.9729	8.8666	1.00011	128940
	0.0107297	999.84	0.008275	33.61	2515.2	2481.6	0.12127	8.9479	8.8266	1.00016	120850
	0.0114825	999.77	0.008824	37.80	2517.1	2479.3	0.13615	8.9232	8.7870	1.00023	113320
10	0.012281	999.69	0.009405	41.99	2518.9	2476.9	0.15097	8.8986	8.7477	1.00031	106320
	0.013129	999.60	0.010019	46.18	2520.7	2474.5	0.16573	8.8743	8.7086	1.00040	99810
	0.014027	999.49	0.010668	50.36	2522.6	2472.2	0.18044	8.8502	8.6698	1.00051	93740
	0.014979	999.37	0.011353	54.55	2524.4	2469.8	0.19509	8.8263	8.6313	1.00063	88090
	0.015988	999.24	0.012075	58.73	2526.2	2467.5	0.20969	8.8027	8.5930	1.00076	82810
15	0.017056	999.09	0.012837	62.92	2528.0	2465.1	0.22424	8.7792	8.5550	1.00091	77900
	0.018185	998.93	0.013641	67.10	2529.9	2462.8	0.23873	8.7560	8.5173	1.00107	73310
	0.019380	998.76	0.014488	71.28	2531.7	2460.4	0.25317	8.7330	8.4798	1.00124	69020
	0.020644	998.58	0.015380	75.47	2533.5	2458.1	0.26757	8.7101	8.4426	1.00142	65020
	0.021979	998.39	0.016319	79.65	2535.3	2455.7	0.28191	8.6875	8.4056	1.00161	61280
20	0.023388	998.19	0.017308	83.84	2537.2	2453.3	0.29621	8.6651	8.3689	1.00182	57778
	0.024877	997.97	0.018347	88.02	2539.0	2451.0	0.31045	8.6428	8.3324	1.00203	54503
	0.026447	997.75	0.019441	92.20	2540.8	2448.6	0.32465	8.6208	8.2962	1.00226	51438
	0.028104	997.52	0.020590	96.39	2542.6	2446.2	0.33880	8.5990	8.2602	1.00249	48568
	0.029850	997.27	0.021797	100.57	2544.5	2443.9	0.35290	8.5773	8.2244	1.00274	45878
25	0.031691	997.02	0.023065	104.75	2546.3	2441.5	0.36696	8.5558	8.1889	1.00299	43357
	0.033629	996.75	0.024395	108.94	2548.1	2439.2	0.38096	8.5346	8.1536	1.00326	40992
	0.035670	996.48	0.025791	113.12	2549.9	2436.8	0.39492	8.5135	8.1185	1.00353	38773
	0.037818	996.20	0.027255	117.30	2551.7	2434.4	0.40884	8.4926	8.0837	1.00381	36690
	0.040078	995.91	0.028791	121.49	2553.5	2432.0	0.42271	8.4718	8.0491	1.00411	34734
30	0.042455	995.61	0.030399	125.67	2555.3	2429.7	0.43653	8.4513	8.0147	1.00441	32896
	0.044953	995.30	0.032084	129.85	2557.1	2427.3	0.45031	8.4309	7.9806	1.00472	31168
	0.047578	994.99	0.033849	134.04	2559.0	2424.9	0.46404	8.4107	7.9466	1.00504	29543
	0.050335	994.66	0.035696	138.22	2560.8	2422.5	0.47772	8.3906	7.9129	1.00537	28014
	0.053229	994.33	0.037629	142.40	2562.6	2420.2	0.49137	8.3708	7.8794	1.00570	26575
35	0.056267	993.99	0.039650	146.59	2564.4	2417.8	0.50496	8.3511	7.8461	1.00605	25220
	0.059454	993.64	0.041764	150.77	2566.2	2415.4	0.51851	8.3315	7.8130	1.00640	23944
	0.062795	993.28	0.043973	154.95	2568.0	2413.0	0.53202	8.3122	7.7802	1.00676	22741
	0.066298	992.92	0.046281	159.14	2569.8	2410.6	0.54549	8.2930	7.7475	1.00713	21607
	0.069969	992.55	0.048691	163.32	2571.6	2408.2	0.55891	8.2739	7.7150	1.00751	20538
40	0.073814	992.17	0.05121	167.50	2573.4	2405.9	0.57228	8.2550	7.6828	1.00789	19528
	0.077840	991.78	0.05383	171.69	2575.2	2403.5	0.58562	8.2363	7.6507	1.00829	18576
	0.082054	991.39	0.05657	175.87	2576.9	2401.1	0.59891	8.2177	7.6188	1.00869	17676
	0.086464	990.99	0.05943	180.05	2578.7	2398.7	0.61216	8.1993	7.5872	1.00909	16826
	0.091076	990.58	0.06241	184.23	2580.5	2396.3	0.62537	8.1810	7.5557	1.00951	16023
45	0.095898	990.17	0.06552	188.42	2582.3	2393.9	0.63853	8.1629	7.5244	1.00993	15263
	0.100938	989.74	0.06875	192.60	2584.1	2391.5	0.65166	8.1450	7.4933	1.01036	14545
	0.106205	989.32	0.07212	196.78	2585.9	2389.1	0.66474	8.1271	7.4624	1.01080	13866
	0.111706	988.88	0.07563	200.96	2587.6	2386.7	0.67778	8.1094	7.4317	1.01124	13222
	0.117449	988.44	0.07928	205.14	2589.4	2384.3	0.69078	8.0919	7.4011	1.01170	12614
50	0.12344	987.99	0.08308	209.33	2591.2	2381.9	0.70374	8.0745	7.3708	1.01215	12037
	0.12970	987.54	0.08703	213.51	2593.0	2379.5	0.71666	8.0573	7.3406	1.01262	11490
	0.13623	987.08	0.09114	217.69	2594.7	2377.0	0.72954	8.0401	7.3106	1.01309	10972
	0.14303	986.61	0.09541	221.87	2596.5	2374.6	0.74238	8.0232	7.2808	1.01357	10481
	0.15012	986.13	0.09985	226.06	2598.3	2372.2	0.75518	8.0063	7.2511	1.01406	10015

Table 1 Saturation (temperature) (*Continued*)

t (°C)	P	ρ_l	ρ_g	h_l	h_g	r	s_l	s_g	ϕ	v_l (×10³)	v_g (×10³)
55	0.15752	985.65	0.10446	230.24	2600.0	2369.8	0.76795	7.9896	7.2216	1.01455	9573.
	0.16522	985.17	0.10925	234.42	2601.8	2367.4	0.78067	7.9730	7.1923	1.01505	9153.
	0.17324	984.68	0.11423	238.60	2603.5	2364.9	0.79336	7.9566	7.1632	1.01556	8754.
	0.18159	984.18	0.11939	242.79	2605.3	2362.5	0.80600	7.9402	7.1342	1.01608	8376.
	0.19028	983.67	0.12475	246.97	2607.0	2360.1	0.81862	7.9240	7.1054	1.01660	8016.
60	0.19932	983.16	0.13030	251.15	2608.8	2357.6	0.83119	7.9080	7.0768	1.01712	7674.
	0.20873	982.65	0.13607	255.34	2610.5	2355.2	0.84373	7.8920	7.0483	1.01766	7349.
	0.21851	982.13	0.14204	259.52	2612.3	2352.8	0.85622	7.8762	7.0200	1.01820	7040.
	0.22868	981.60	0.14824	263.71	2614.0	2350.3	0.86869	7.8605	6.9918	1.01875	6746.
	0.23925	981.07	0.15465	267.89	2615.8	2347.9	0.88112	7.8450	6.9638	1.01930	6466.
65	0.25022	980.53	0.16130	272.08	2617.5	2345.4	0.89351	7.8295	6.9360	1.01986	6200.
	0.26163	979.98	0.16819	276.26	2619.2	2343.0	0.90586	7.8142	6.9083	1.02043	5946.
	0.27347	979.43	0.17532	280.45	2620.9	2340.5	0.91819	7.7989	6.8808	1.02100	5704.
	0.28576	978.88	0.18269	284.63	2622.7	2338.0	0.93047	7.7838	6.8534	1.02158	5474.
	0.29852	978.32	0.19033	288.82	2624.4	2335.6	0.94272	7.7689	6.8261	1.02216	5254.
70	0.31176	977.75	0.19823	293.01	2626.1	2333.1	0.95494	7.7540	6.7990	1.02276	5044.6
	0.32549	977.18	0.20640	297.20	2627.8	2330.6	0.96713	7.7392	6.7721	1.02336	4844.9
	0.33972	976.60	0.21485	301.39	2629.5	2328.1	0.97928	7.7246	6.7453	1.02396	4654.4
	0.35448	976.02	0.22358	305.58	2631.2	2325.7	0.99139	7.7100	6.7186	1.02457	4472.6
	0.36978	975.43	0.23261	309.77	2632.9	2323.2	1.00348	7.6956	6.6921	1.02519	4299.0
75	0.38563	974.84	0.24194	313.96	2634.6	2320.7	1.01553	7.6813	6.6657	1.02581	4133.3
	0.40205	974.24	0.25158	318.15	2636.3	2318.2	1.02754	7.6670	6.6395	1.02644	3975.0
	0.41905	973.64	0.26153	322.34	2638.0	2315.7	1.03953	7.6529	6.6134	1.02708	3823.7
	0.43665	973.03	0.27180	326.54	2639.7	2313.2	1.05149	7.6389	6.5874	1.02772	3679.1
	0.45487	972.41	0.28241	330.73	2641.4	2310.7	1.06341	7.6250	6.5616	1.02837	3541.0
80	0.47373	971.79	0.29336	334.93	2643.1	2308.1	1.07530	7.6112	6.5359	1.02902	3408.8
	0.49324	971.17	0.30465	339.12	2644.7	2305.6	1.08716	7.5975	6.5103	1.02969	3282.4
	0.51342	970.54	0.31631	343.32	2646.4	2303.1	1.09899	7.5838	6.4849	1.03035	3161.5
	0.53428	969.91	0.32832	347.52	2648.1	2300.6	1.11079	7.5703	6.4595	1.03103	3045.8
	0.55585	969.27	0.34072	351.72	2649.7	2298.0	1.12255	7.5569	6.4344	1.03171	2935.0
85	0.57815	968.62	0.35349	355.92	2651.4	2295.5	1.13429	7.5436	6.4093	1.03239	2828.9
	0.60119	967.98	0.36666	360.12	2653.1	2292.9	1.14600	7.5304	6.3844	1.03308	2727.3
	0.62499	967.32	0.38023	364.32	2654.7	2290.4	1.15768	7.5172	6.3595	1.03378	2630.0
	0.64958	966.66	0.39420	368.52	2656.4	2287.8	1.16932	7.5042	6.3349	1.03449	2536.8
	0.67496	966.00	0.40860	372.73	2658.0	2285.3	1.18094	7.4912	6.3103	1.03520	2447.4
90	0.70117	965.33	0.42343	376.93	2659.6	2282.7	1.19253	7.4784	6.2858	1.03591	2361.7
	0.72823	964.66	0.43870	381.14	2661.3	2280.1	1.20409	7.4656	6.2615	1.03664	2279.5
	0.75614	963.98	0.45441	385.35	2662.9	2277.5	1.21563	7.4529	6.2373	1.03736	2200.7
	0.78495	963.30	0.47058	389.56	2664.5	2275.0	1.22713	7.4403	6.2132	1.03810	2125.0
	0.81465	962.61	0.48723	393.77	2666.1	2272.4	1.23861	7.4278	6.1892	1.03884	2052.4
95	0.84529	961.92	0.5043	397.98	2667.7	2269.8	1.25006	7.4154	6.1653	1.03959	1982.8
	0.87688	961.22	0.5220	402.20	2669.4	2267.2	1.26148	7.4030	6.1416	1.04034	1915.9
	0.90945	960.52	0.5401	406.41	2671.0	2264.5	1.27287	7.3908	6.1179	1.04110	1851.6
	0.94301	959.82	0.5587	410.63	2672.5	2261.9	1.28424	7.3786	6.0944	1.04186	1789.9
	0.97759	959.11	0.5778	414.84	2674.1	2259.3	1.29557	7.3665	6.0709	1.04264	1730.6
100	1.0132	958.39	0.5975	419.06	2675.7	2256.7	1.30689	7.3545	6.0476	1.04341	1673.6
	1.0499	957.67	0.6177	423.28	2677.3	2254.0	1.31817	7.3426	6.0244	1.04420	1618.9
	1.0877	956.95	0.6385	427.51	2678.9	2251.4	1.32943	7.3307	6.0013	1.04499	1566.2
	1.1266	956.22	0.6598	431.73	2680.5	2248.7	1.34066	7.3189	5.9783	1.04578	1515.5
	1.1667	955.49	0.6817	435.95	2682.0	2246.1	1.35187	7.3072	5.9553	1.04659	1466.8
105	1.2079	954.75	0.7042	440.18	2683.6	2243.4	1.36305	7.2956	5.9325	1.04739	1420.0
	1.2503	954.01	0.7273	444.41	2685.1	2240.7	1.37420	7.2840	5.9098	1.04821	1374.9
	1.2939	953.26	0.7511	448.64	2686.7	2238.0	1.38533	7.2726	5.8872	1.04903	1331.4
	1.3388	952.51	0.7754	452.87	2688.2	2235.3	1.39644	7.2612	5.8647	1.04986	1289.6
	1.3850	951.75	0.8004	457.10	2689.7	2232.6	1.40751	7.2498	5.8423	1.05069	1249.4

Table 1 Saturation (temperature) (*Continued*)

t (°C)	P	ρ_l	ρ_g	h_l	h_g	r	s_l	s_g	ϕ	v_l (×10³)	v_g (×10³)
110	1.4324	951.00	0.8260	461.34	2691.3	2229.9	1.41857	7.2386	5.8200	1.05153	1210.6
	1.4812	950.23	0.8523	465.57	2692.8	2227.2	1.42960	7.2274	5.7978	1.05238	1173.3
	1.5313	949.46	0.8793	469.81	2694.3	2224.5	1.44060	7.2163	5.7757	1.05323	1137.3
	1.5829	948.69	0.9069	474.05	2695.8	2221.8	1.45158	7.2052	5.7536	1.05409	1102.6
	1.6358	947.91	0.9353	478.29	2697.3	2219.0	1.46253	7.1942	5.7317	1.05495	1069.2
115	1.6902	947.13	0.9643	482.54	2698.8	2216.3	1.47347	7.1833	5.7099	1.05582	1037.0
	1.7461	946.34	0.9941	486.78	2700.3	2213.5	1.48437	7.1725	5.6881	1.05670	1005.9
	1.8034	945.55	1.0247	491.03	2701.8	2210.8	1.49526	7.1617	5.6664	1.05758	975.9
	1.8623	944.76	1.0559	495.28	2703.3	2208.0	1.50612	7.1510	5.6449	1.05847	947.0
	1.9228	943.96	1.0880	499.53	2704.7	2205.2	1.51695	7.1403	5.6234	1.05937	919.1
120	1.9848	943.16	1.1208	503.78	2706.2	2202.4	1.52776	7.1297	5.6020	1.06027	892.2
	2.0485	942.35	1.1545	508.03	2707.6	2199.6	1.53855	7.1192	5.5807	1.06118	866.2
	2.1139	941.54	1.1889	512.29	2709.1	2196.8	1.54932	7.1087	5.5594	1.06210	841.1
	2.1809	940.72	1.2242	516.55	2710.5	2194.0	1.56006	7.0983	5.5383	1.06302	816.9
	2.2496	939.90	1.2603	520.81	2712.0	2191.2	1.57078	7.0880	5.5172	1.06395	793.5
125	2.3201	939.07	1.2972	525.07	2713.4	2188.3	1.58148	7.0777	5.4962	1.06488	770.9
	2.3924	938.24	1.3351	529.33	2714.8	2185.5	1.59216	7.0675	5.4753	1.06582	749.0
	2.4666	937.41	1.3738	533.60	2716.2	2182.6	1.60281	7.0573	5.4545	1.06677	727.9
	2.5425	936.57	1.4134	537.86	2717.6	2179.8	1.61344	7.0472	5.4338	1.06772	707.5
	2.6204	935.73	1.4539	542.13	2719.0	2176.9	1.62405	7.0372	5.4131	1.06869	687.8
130	2.7002	934.88	1.4954	546.41	2720.4	2174.0	1.63464	7.0272	5.3925	1.06965	668.7
	2.7820	934.03	1.5378	550.68	2721.8	2171.1	1.64521	7.0172	5.3720	1.07063	650.3
	2.8657	933.18	1.5811	554.96	2723.2	2168.2	1.65575	7.0074	5.3516	1.07161	632.5
	2.9515	932.32	1.6255	559.23	2724.5	2165.3	1.66628	6.9975	5.3313	1.07260	615.2
	3.0393	931.45	1.6708	563.52	2725.9	2162.4	1.67678	6.9878	5.3110	1.07359	598.5
135	3.1293	930.59	1.7172	567.80	2727.2	2159.4	1.68726	6.9780	5.2908	1.07459	582.4
	3.2214	929.71	1.7646	572.08	2728.6	2156.5	1.69772	6.9684	5.2706	1.07560	566.7
	3.3157	928.84	1.8130	576.37	2729.9	2153.5	1.70816	6.9587	5.2506	1.07661	551.6
	3.4122	927.96	1.8625	580.66	2731.2	2150.6	1.71858	6.9492	5.2306	1.07764	536.9
	3.5109	927.07	1.9130	584.95	2732.5	2147.6	1.72898	6.9397	5.2107	1.07866	522.7
140	3.6119	926.18	1.9647	589.24	2733.8	2144.6	1.73936	6.9302	5.1908	1.07970	508.99
	3.7153	925.29	2.0174	593.54	2735.1	2141.6	1.74972	6.9208	5.1711	1.08074	495.68
	3.8211	924.39	2.0713	597.84	2736.4	2138.6	1.76006	6.9114	5.1513	1.08179	482.78
	3.9292	923.49	2.1264	602.14	2737.7	2135.6	1.77038	6.9021	5.1317	1.08285	470.28
	4.0398	922.58	2.1826	606.44	2739.0	2132.5	1.78068	6.8928	5.1121	1.08391	458.17
145	4.1529	921.67	2.2400	610.75	2740.2	2129.5	1.79096	6.8836	5.0926	1.08498	446.43
	4.2685	920.76	2.2986	615.06	2741.5	2126.4	1.80122	6.8744	5.0732	1.08606	435.05
	4.3867	919.84	2.3584	619.37	2742.7	2123.3	1.81146	6.8652	5.0538	1.08715	424.01
	4.5075	918.92	2.4195	623.68	2743.9	2120.3	1.82169	6.8562	5.0345	1.08824	413.31
	4.6310	917.99	2.4818	628.00	2745.2	2117.2	1.83189	6.8471	5.0152	1.08934	402.93
150	4.7572	917.06	2.5454	632.32	2746.4	2114.1	1.84208	6.8381	4.9960	1.09044	392.86
	4.8861	916.12	2.6104	636.64	2747.6	2110.9	1.85224	6.8291	4.9769	1.09156	383.09
	5.0178	915.18	2.6766	640.96	2748.8	2107.8	1.86239	6.8202	4.9578	1.09268	373.61
	5.1523	914.24	2.7442	645.29	2750.0	2104.7	1.87252	6.8113	4.9388	1.09381	364.41
	5.2896	913.29	2.8131	649.62	2751.1	2101.5	1.88263	6.8025	4.9198	1.09495	355.48
155	5.4299	912.33	2.8834	653.95	2752.3	2098.3	1.89273	6.7937	4.9010	1.09609	346.81
	5.5732	911.38	2.9551	658.28	2753.4	2095.2	1.90280	6.7849	4.8821	1.09724	338.40
	5.7194	910.41	3.0282	662.62	2754.6	2092.0	1.91286	6.7762	4.8633	1.09840	330.23
	5.8687	909.45	3.1028	666.96	2755.7	2088.8	1.92290	6.7675	4.8446	1.09957	322.29
	6.0211	908.48	3.1788	671.30	2756.8	2085.5	1.93292	6.7589	4.8260	1.10074	314.58
160	6.1766	907.50	3.2564	675.65	2758.0	2082.3	1.94293	6.7503	4.8073	1.10193	307.09
	6.3353	906.52	3.3354	680.00	2759.1	2079.1	1.95292	6.7417	4.7888	1.10312	299.82
	6.4973	905.54	3.4159	684.35	2760.1	2075.8	1.96289	6.7332	4.7703	1.10432	292.75
	6.6625	904.55	3.4980	688.71	2761.2	2072.5	1.97284	6.7247	4.7518	1.10552	285.87
	6.8310	903.56	3.5817	693.07	2762.3	2069.2	1.98278	6.7162	4.7334	1.10674	279.19

Table 1 Saturation (temperature) (*Continued*)

t (°C)	P	ρ_l	ρ_g	h_l	h_g	r	s_l	s_g	ϕ	v_l ($\times 10^3$)	v_g ($\times 10^3$)
165	7.0029	902.56	3.6670	697.43	2763.3	2065.9	1.99271	6.7078	4.7151	1.10796	272.70
	7.1783	901.56	3.7539	701.79	2764.4	2062.6	2.00261	6.6994	4.6968	1.10919	266.39
	7.3570	900.55	3.8424	706.16	2765.4	2059.3	2.01250	6.6910	4.6785	1.11043	260.25
	7.5394	899.54	3.9326	710.53	2766.4	2055.9	2.02237	6.6827	4.6603	1.11168	254.28
	7.7252	898.53	4.0245	714.90	2767.5	2052.5	2.03223	6.6744	4.6422	1.11293	248.48
170	7.9147	897.51	4.1181	719.28	2768.5	2049.2	2.04207	6.6662	4.6241	1.11420	242.83
	8.1078	896.48	4.2135	723.66	2769.4	2045.8	2.05190	6.6579	4.6060	1.11547	237.33
	8.3047	895.46	4.3106	728.05	2770.4	2042.4	2.06171	6.6498	4.5880	1.11675	231.99
	8.5053	894.42	4.4095	732.43	2771.4	2038.9	2.07150	6.6416	4.5701	1.11804	226.78
	8.7098	893.38	4.5102	736.83	2772.3	2035.5	2.08128	6.6335	4.5522	1.11934	221.72
175	8.9180	892.34	4.6127	741.22	2773.3	2032.0	2.09105	6.6254	4.5343	1.12065	216.79
	9.1303	891.30	4.7172	745.62	2774.2	2028.6	2.10080	6.6173	4.5165	1.12196	211.99
	9.3464	890.24	4.8235	750.02	2775.1	2025.1	2.11054	6.6092	4.4987	1.12329	207.32
	9.5666	889.19	4.9317	754.43	2776.0	2021.6	2.12026	6.6012	4.4810	1.12462	202.77
	9.7909	888.13	5.0418	758.84	2776.9	2018.1	2.12996	6.5932	4.4633	1.12596	198.34
180	10.019	887.06	5.154	763.25	2777.8	2014.5	2.13966	6.5853	4.4456	1.12732	194.03
	10.252	885.99	5.268	767.67	2778.6	2011.0	2.14934	6.5774	4.4280	1.12868	189.82
	10.489	884.92	5.384	772.09	2779.5	2007.4	2.15900	6.5694	4.4104	1.13005	185.73
	10.730	883.84	5.502	776.51	2780.3	2003.8	2.16865	6.5616	4.3929	1.13143	181.74
	10.975	882.75	5.623	780.94	2781.2	2000.2	2.17829	6.5537	4.3754	1.13282	177.85
185	11.225	881.67	5.745	785.37	2782.0	1996.6	2.18791	6.5459	4.3580	1.13422	174.06
	11.479	880.57	5.870	789.81	2782.8	1993.0	2.19752	6.5381	4.3406	1.13563	170.37
	11.738	879.47	5.996	794.25	2783.6	1989.3	2.20712	6.5303	4.3232	1.13704	166.77
	12.001	878.37	6.125	798.69	2784.3	1985.6	2.21670	6.5226	4.3059	1.13847	163.26
	12.269	877.26	6.256	803.14	2785.1	1982.0	2.22628	6.5148	4.2886	1.13991	159.84
190	12.542	876.15	6.390	807.60	2785.8	1978.2	2.23583	6.5071	4.2713	1.14136	156.50
	12.819	875.03	6.525	812.06	2786.6	1974.5	2.24538	6.4994	4.2541	1.14282	153.25
	13.101	873.91	6.663	816.52	2787.3	1970.8	2.25491	6.4918	4.2369	1.14429	150.08
	13.388	872.78	6.804	820.98	2788.0	1967.0	2.26444	6.4841	4.2197	1.14576	146.98
	13.680	871.65	6.946	825.46	2788.7	1963.2	2.27395	6.4765	4.2026	1.14725	143.96
195	13.976	870.51	7.091	829.93	2789.4	1959.4	2.28344	6.4689	4.1855	1.14875	141.02
	14.278	869.37	7.239	834.41	2790.0	1955.6	2.29293	6.4613	4.1684	1.15026	138.14
	14.585	868.22	7.389	838.90	2790.7	1951.8	2.30241	6.4538	4.1514	1.15178	135.34
	14.897	867.07	7.541	843.39	2791.3	1947.9	2.31187	6.4463	4.1344	1.15332	132.60
	15.214	865.91	7.697	847.88	2791.9	1944.0	2.32132	6.4387	4.1174	1.15486	129.93
200	15.537	864.74	7.854	852.38	2792.5	1940.1	2.33076	6.4312	4.1005	1.15641	127.32
	15.864	863.57	8.014	856.89	2793.1	1936.2	2.34019	6.4238	4.0836	1.15798	124.77
	16.197	862.40	8.177	861.40	2793.7	1932.3	2.34961	6.4163	4.0667	1.15955	122.29
	16.536	861.22	8.343	865.91	2794.2	1928.3	2.35902	6.4089	4.0498	1.16114	119.86
	16.880	860.04	8.511	870.43	2794.8	1924.4	2.36842	6.4014	4.0330	1.16274	117.49
205	17.229	858.85	8.682	874.96	2795.3	1920.4	2.37781	6.3940	4.0162	1.16435	115.17
	17.584	857.65	8.856	879.49	2795.8	1916.3	2.38719	6.3866	3.9994	1.16597	112.91
	17.945	856.45	9.033	884.02	2796.3	1912.3	2.39656	6.3793	3.9827	1.16761	110.70
	18.311	855.25	9.213	888.56	2796.8	1908.2	2.40591	6.3719	3.9660	1.16925	108.55
	18.684	854.03	9.395	893.11	2797.3	1904.1	2.41526	6.3646	3.9493	1.17091	106.44
210	19.062	852.82	9.581	897.66	2797.7	1900.0	2.42460	6.3572	3.9326	1.17258	104.38
	19.446	851.59	9.769	902.22	2798.1	1895.9	2.43393	6.3499	3.9160	1.17427	102.36
	19.836	850.37	9.961	906.78	2798.6	1891.8	2.44326	6.3426	3.8993	1.17596	100.40
	20.232	849.13	10.155	911.35	2798.9	1887.6	2.45257	6.3353	3.8827	1.17767	98.47
	20.634	847.89	10.353	915.93	2799.3	1883.4	2.46187	6.3280	3.8662	1.17939	96.59
215	21.042	846.65	10.554	920.51	2799.7	1879.2	2.47117	6.3208	3.8496	1.18113	94.75
	21.457	845.40	10.758	925.10	2800.0	1874.9	2.48046	6.3135	3.8331	1.18288	92.96
	21.878	844.14	10.965	929.69	2800.4	1870.7	2.48974	6.3063	3.8166	1.18464	91.20
	22.305	842.88	11.176	934.29	2800.7	1866.4	2.49901	6.2991	3.8001	1.18641	89.48
	22.738	841.61	11.389	938.90	2801.0	1862.1	2.50827	6.2919	3.7836	1.18820	87.80

Table 1 Saturation (temperature) (*Continued*)

t (°C)	P	ρ_l	ρ_g	h_l	h_g	r	s_l	s_g	ϕ	v_l ($\times 10^3$)	v_g ($\times 10^3$)
220	23.178	840.34	11.607	943.51	2801.3	1857.8	2.51753	6.2847	3.7671	1.19000	86.16
	23.625	839.06	11.827	948.13	2801.5	1853.4	2.52678	6.2775	3.7507	1.19182	84.55
	24.078	837.77	12.052	952.75	2801.8	1849.0	2.53602	6.2703	3.7343	1.19365	82.98
	24.538	836.48	12.279	957.38	2802.0	1844.6	2.54525	6.2631	3.7179	1.19549	81.44
	25.005	835.18	12.511	962.02	2802.2	1840.2	2.55448	6.2559	3.7015	1.19735	79.93
225	25.479	833.87	12.745	966.67	2802.4	1835.7	2.56370	6.2488	3.6851	1.19922	78.46
	25.959	832.56	12.984	971.32	2802.6	1831.2	2.57292	6.2416	3.6687	1.20111	77.02
	26.446	831.25	13.226	975.98	2802.7	1826.7	2.58212	6.2345	3.6524	1.20301	75.61
	26.941	829.92	13.472	980.65	2802.9	1822.2	2.59133	6.2274	3.6361	1.20493	74.23
	27.442	828.59	13.722	985.32	2803.0	1817.7	2.60052	6.2203	3.6197	1.20687	72.88
230	27.951	827.25	13.976	990.00	2803.1	1813.1	2.60971	6.2131	3.6034	1.20882	71.55
	28.467	825.91	14.233	994.69	2803.1	1808.5	2.61890	6.2060	3.5871	1.21078	70.26
	28.990	824.56	14.495	999.39	2803.2	1803.8	2.62808	6.1989	3.5709	1.21276	68.99
	29.521	823.21	14.761	1004.09	2803.2	1799.2	2.63725	6.1918	3.5546	1.21476	67.75
	30.059	821.84	15.031	1008.80	2803.3	1794.5	2.64642	6.1847	3.5383	1.21678	66.53
235	30.604	820.47	15.304	1013.52	2803.3	1789.7	2.65559	6.1777	3.5221	1.21881	65.34
	31.157	819.10	15.583	1018.25	2803.2	1785.0	2.66475	6.1706	3.5058	1.22086	64.17
	31.718	817.71	15.865	1022.98	2803.2	1780.2	2.67390	6.1635	3.4896	1.22292	63.03
	32.286	816.32	16.152	1027.72	2803.1	1775.4	2.68306	6.1564	3.4734	1.22500	61.91
	32.863	814.93	16.443	1032.48	2803.1	1770.6	2.69220	6.1494	3.4572	1.22710	60.82
240	33.447	813.52	16.739	1037.24	2803.0	1765.7	2.70135	6.1423	3.4409	1.22922	59.74
	34.039	812.11	17.039	1042.00	2802.8	1760.8	2.71049	6.1352	3.4247	1.23136	58.69
	34.639	810.69	17.344	1046.78	2802.7	1755.9	2.71963	6.1282	3.4085	1.23351	57.66
	35.247	809.27	17.653	1051.57	2802.5	1751.0	2.72876	6.1211	3.3923	1.23569	56.65
	35.863	807.83	17.967	1056.36	2802.3	1746.0	2.73789	6.1140	3.3761	1.23788	55.66
245	36.488	806.39	18.286	1061.16	2802.1	1741.0	2.74702	6.1070	3.3600	1.24009	54.69
	37.121	804.94	18.610	1065.98	2801.9	1735.9	2.75615	6.0999	3.3438	1.24232	53.73
	37.762	803.49	18.939	1070.80	2801.6	1730.8	2.76528	6.0929	3.3276	1.24458	52.80
	38.412	802.02	19.273	1075.63	2801.4	1725.7	2.77440	6.0858	3.3114	1.24685	51.89
	39.070	800.55	19.612	1080.47	2801.1	1720.6	2.78352	6.0787	3.2952	1.24914	50.99
250	39.737	799.07	19.956	1085.32	2800.7	1715.4	2.79264	6.0717	3.2790	1.25145	50.111
	40.412	797.58	20.305	1090.18	2800.4	1710.2	2.80176	6.0646	3.2629	1.25379	49.248
	41.096	796.09	20.660	1095.05	2800.0	1705.0	2.81088	6.0575	3.2467	1.25614	48.403
	41.789	794.59	21.020	1099.93	2799.6	1699.7	2.82000	6.0505	3.2305	1.25852	47.573
	42.491	793.07	21.386	1104.82	2799.2	1694.4	2.82911	6.0434	3.2143	1.26092	46.760
255	43.202	791.55	21.757	1109.72	2798.8	1689.1	2.83823	6.0363	3.1981	1.26334	45.962
	43.922	790.03	22.134	1114.63	2798.3	1683.7	2.84735	6.0292	3.1819	1.26578	45.180
	44.651	788.49	22.517	1119.55	2797.8	1678.3	2.85646	6.0222	3.1657	1.26825	44.412
	45.390	786.94	22.905	1124.48	2797.3	1672.8	2.86558	6.0151	3.1495	1.27074	43.658
	46.137	785.39	23.300	1129.43	2796.8	1667.4	2.87470	6.0080	3.1333	1.27325	42.919
260	46.895	783.83	23.700	1134.38	2796.2	1661.9	2.88382	6.0009	3.1170	1.27579	42.194
	47.661	782.25	24.107	1139.34	2795.6	1656.3	2.89294	5.9938	3.1008	1.27836	41.482
	48.437	780.67	24.520	1144.32	2795.0	1650.7	2.90206	5.9866	3.0846	1.28095	40.783
	49.223	779.08	24.939	1149.31	2794.4	1645.1	2.91119	5.9795	3.0683	1.28356	40.098
	50.018	777.48	25.365	1154.31	2793.7	1639.4	2.92031	5.9724	3.0521	1.28620	39.424
265	50.823	775.87	25.797	1159.32	2793.0	1633.7	2.92944	5.9652	3.0358	1.28887	38.764
	51.638	774.25	26.236	1164.35	2792.3	1628.0	2.93858	5.9581	3.0195	1.29156	38.115
	52.463	772.63	26.682	1169.38	2791.6	1622.2	2.94771	5.9509	3.0032	1.29429	37.478
	53.298	770.99	27.135	1174.43	2790.8	1616.3	2.95685	5.9437	2.9869	1.29704	36.853
	54.143	769.34	27.595	1179.49	2790.0	1610.5	2.96599	5.9365	2.9705	1.29981	36.239
270	54.999	767.68	28.061	1184.57	2789.1	1604.6	2.97514	5.9293	2.9542	1.30262	35.636
	55.864	766.01	28.536	1189.66	2788.3	1598.6	2.98429	5.9221	2.9378	1.30546	35.044
	56.740	764.34	29.017	1194.76	2787.4	1592.6	2.99345	5.9149	2.9215	1.30833	34.462
	57.627	762.65	29.506	1199.87	2786.5	1586.6	3.00261	5.9077	2.9051	1.31122	33.891
	58.524	760.95	30.003	1205.00	2785.5	1580.5	3.01178	5.9004	2.8886	1.31415	33.330

Table 1 Saturation (temperature) (*Continued*)

t (°C)	P	ρ_l	ρ_g	h_l	h_g	r	s_l	s_g	ϕ	v_l ($\times 10^3$)	v_g ($\times 10^3$)
275	59.431	759.24	30.507	1210.15	2784.5	1574.4	3.02095	5.8931	2.8722	1.31711	32.779
	60.350	757.52	31.020	1215.30	2783.5	1568.2	3.03013	5.8859	2.8557	1.32011	32.237
	61.279	755.78	31.541	1220.47	2782.5	1562.0	3.03931	5.8786	2.8392	1.32313	31.705
	62.219	754.04	32.069	1225.66	2781.4	1555.8	3.04850	5.8712	2.8227	1.32619	31.182
	63.170	752.28	32.607	1230.86	2780.3	1549.4	3.05770	5.8639	2.8062	1.32929	30.669
280	64.132	750.52	33.152	1236.08	2779.2	1543.1	3.06691	5.8565	2.7896	1.33242	30.164
	65.105	748.74	33.707	1241.31	2778.0	1536.7	3.07613	5.8492	2.7730	1.33558	29.668
	66.089	746.95	34.270	1246.56	2776.8	1530.2	3.08535	5.8418	2.7564	1.33878	29.180
	67.085	745.14	34.843	1251.82	2775.5	1523.7	3.09458	5.8344	2.7398	1.34202	28.701
	68.092	743.33	35.424	1257.10	2774.3	1517.2	3.10382	5.8269	2.7231	1.34530	28.229
285	69.111	741.50	36.015	1262.40	2773.0	1510.6	3.11308	5.8195	2.7064	1.34862	27.766
	70.141	739.66	36.616	1267.71	2771.6	1503.9	3.12234	5.8120	2.6896	1.35197	27.310
	71.183	737.81	37.226	1273.04	2770.2	1497.2	3.13161	5.8045	2.6729	1.35537	26.863
	72.237	735.94	37.847	1278.39	2768.8	1490.4	3.14089	5.7969	2.6560	1.35881	26.422
	73.303	734.06	38.478	1283.75	2767.4	1483.6	3.15019	5.7894	2.6392	1.36229	25.989
290	74.380	732.16	39.119	1289.14	2765.9	1476.7	3.15950	5.7818	2.6223	1.36581	25.563
	75.470	730.26	39.770	1294.54	2764.3	1469.8	3.16882	5.7742	2.6054	1.36938	25.144
	76.572	728.33	40.433	1299.96	2762.8	1462.8	3.17815	5.7665	2.5884	1.37300	24.732
	77.686	726.40	41.106	1305.40	2761.2	1455.8	3.18750	5.7589	2.5714	1.37666	24.327
	78.813	724.45	41.791	1310.86	2759.5	1448.7	3.19686	5.7511	2.5543	1.38037	23.928
295	79.952	722.48	42.488	1316.34	2757.8	1441.5	3.20623	5.7434	2.5372	1.38412	23.536
	81.103	720.50	43.196	1321.84	2756.1	1434.3	3.21563	5.7356	2.5200	1.38793	23.150
	82.268	718.50	43.917	1327.36	2754.3	1427.0	3.22503	5.7278	2.5028	1.39179	22.770
	83.445	716.49	44.650	1332.90	2752.5	1419.6	3.23446	5.7200	2.4855	1.39570	22.397
	84.635	714.46	45.395	1338.47	2750.7	1412.2	3.24390	5.7121	2.4682	1.39967	22.029
300	85.838	712.41	46.154	1344.05	2748.7	1404.7	3.25336	5.7042	2.4508	1.40369	21.667
	87.054	710.35	46.926	1349.66	2746.8	1397.1	3.26284	5.6962	2.4334	1.40777	21.310
	88.283	708.27	47.711	1355.29	2744.8	1389.5	3.27233	5.6882	2.4159	1.41190	20.960
	89.526	706.17	48.510	1360.95	2742.8	1381.8	3.28185	5.6802	2.3983	1.41610	20.614
	90.782	704.05	49.324	1366.63	2740.7	1374.0	3.29139	5.6721	2.3807	1.42035	20.274
305	92.051	701.92	50.15	1372.33	2738.5	1366.2	3.30095	5.6640	2.3630	1.42467	19.940
	93.334	699.76	51.00	1378.06	2736.3	1358.3	3.31053	5.6558	2.3453	1.42906	19.610
	94.631	697.59	51.85	1383.81	2734.1	1350.3	3.32014	5.6476	2.3275	1.43351	19.285
	95.942	695.40	52.73	1389.59	2731.8	1342.2	3.32977	5.6393	2.3096	1.43803	18.966
	97.267	693.18	53.62	1395.40	2729.4	1334.0	3.33943	5.6310	2.2916	1.44262	18.651
310	98.605	690.95	54.52	1401.23	2727.0	1325.8	3.34911	5.6226	2.2735	1.44728	18.340
	99.958	688.70	55.45	1407.10	2724.6	1317.5	3.35882	5.6142	2.2554	1.45202	18.035
	101.326	686.42	56.39	1412.99	2722.1	1309.1	3.36856	5.6057	2.2372	1.45683	17.734
	102.707	684.12	57.35	1418.91	2719.5	1300.6	3.37832	5.5972	2.2189	1.46173	17.437
	104.104	681.80	58.33	1424.86	2716.9	1292.0	3.38812	5.5886	2.2005	1.46670	17.145
315	105.51	679.46	59.32	1430.84	2714.2	1283.3	3.39795	5.5799	2.1820	1.47176	16.856
	106.94	677.09	60.34	1436.86	2711.4	1274.6	3.40781	5.5712	2.1634	1.47691	16.572
	108.38	674.70	61.38	1442.90	2708.6	1265.7	3.41770	5.5624	2.1447	1.48215	16.293
	109.84	672.28	62.44	1448.99	2705.7	1256.7	3.42763	5.5535	2.1259	1.48748	16.017
	111.31	669.83	63.51	1455.10	2702.8	1247.6	3.43760	5.5446	2.1070	1.49291	15.745
320	112.79	667.36	64.62	1461.25	2699.7	1238.5	3.44760	5.5356	2.0880	1.49843	15.476
	114.29	664.87	65.74	1467.44	2696.6	1229.2	3.45765	5.5265	2.0688	1.50406	15.212
	115.81	662.34	66.89	1473.67	2693.5	1219.8	3.46773	5.5173	2.0496	1.50980	14.951
	117.34	659.78	68.06	1479.93	2690.2	1210.3	3.47786	5.5081	2.0302	1.51565	14.693
	118.89	657.20	69.26	1486.24	2686.9	1200.7	3.48803	5.4987	2.0107	1.52161	14.439
325	120.46	654.58	70.48	1492.58	2683.5	1190.9	3.49825	5.4893	1.9911	1.52769	14.189
	122.04	651.93	71.73	1498.97	2680.1	1181.1	3.50852	5.4798	1.9713	1.53390	13.942
	123.64	649.25	73.00	1505.40	2676.5	1171.1	3.51884	5.4702	1.9513	1.54024	13.698
	125.25	646.53	74.31	1511.88	2672.9	1161.0	3.52921	5.4605	1.9313	1.54671	13.457
	126.88	643.78	75.65	1518.41	2669.1	1150.7	3.53963	5.4506	1.9110	1.55332	13.219

Table 1 Saturation (temperature) (*Continued*)

t (°C)	P	ρ_l	ρ_g	h_l	h_g	r	s_l	s_g	ϕ	v_l ($\times 10^3$)	v_g ($\times 10^3$)
330	128.52	641.0	77.01	1525.0	2665.3	1140.3	3.5501	5.4407	1.8906	1.5601	12.985
	130.19	638.2	78.41	1531.6	2661.4	1129.8	3.5607	5.4307	1.8700	1.5670	12.753
	131.87	635.3	79.84	1538.3	2657.4	1119.1	3.5713	5.4205	1.8493	1.5740	12.524
	133.57	632.4	81.31	1545.0	2653.3	1108.3	3.5819	5.4103	1.8283	1.5813	12.298
	135.28	629.5	82.82	1551.8	2649.0	1097.2	3.5927	5.3999	1.8072	1.5887	12.075
335	137.01	626.5	84.36	1558.6	2644.7	1086.1	3.6035	5.3894	1.7859	1.5963	11.854
	138.76	623.4	85.94	1565.5	2640.3	1074.7	3.6144	5.3787	1.7643	1.6040	11.636
	140.53	620.3	87.56	1572.5	2635.7	1063.2	3.6253	5.3679	1.7426	1.6120	11.421
	142.32	617.2	89.22	1579.5	2631.1	1051.5	3.6364	5.3569	1.7205	1.6202	11.208
	144.12	614.0	90.93	1586.7	2626.3	1039.6	3.6475	5.3458	1.6983	1.6286	10.997
340	145.94	610.8	92.69	1593.8	2621.3	1027.5	3.6587	5.3345	1.6758	1.6373	10.788
	147.78	607.5	94.50	1601.1	2616.3	1015.2	3.6701	5.3231	1.6530	1.6462	10.582
	149.64	604.1	96.36	1608.4	2611.1	1002.7	3.6815	5.3114	1.6299	1.6553	10.378
	151.52	600.7	98.27	1615.8	2605.7	989.9	3.6930	5.2996	1.6066	1.6647	10.176
	153.42	597.2	100.24	1623.3	2600.2	976.9	3.7047	5.2876	1.5829	1.6745	9.976
345	155.33	593.7	102.27	1630.9	2594.5	963.6	3.7164	5.2753	1.5589	1.6845	9.778
	157.27	590.0	104.37	1638.6	2588.7	950.1	3.7283	5.2629	1.5345	1.6948	9.581
	159.22	586.3	106.53	1646.4	2582.7	936.3	3.7404	5.2502	1.5098	1.7056	9.387
	161.20	582.5	108.77	1654.3	2576.5	922.2	3.7526	5.2372	1.4847	1.7166	9.194
	163.20	578.7	111.08	1662.3	2570.1	907.8	3.7649	5.2240	1.4591	1.7281	9.002
350	165.21	574.7	113.48	1670.4	2563.5	893.0	3.7774	5.2105	1.4331	1.7401	8.812
	167.25	570.6	115.96	1678.7	2556.6	877.9	3.7901	5.1967	1.4066	1.7525	8.623
	169.31	566.4	118.54	1687.1	2549.6	862.4	3.8030	5.1825	1.3796	1.7654	8.436
	171.38	562.2	121.22	1695.7	2542.2	846.6	3.8161	5.1681	1.3520	1.7788	8.249
	173.48	557.8	124.01	1704.4	2534.6	830.2	3.8294	5.1532	1.3238	1.7929	8.064
355	175.61	553.2	126.92	1713.3	2526.7	813.5	3.8429	5.1379	1.2950	1.8076	7.879
	177.75	548.5	129.95	1722.4	2518.5	796.2	3.8568	5.1222	1.2655	1.8230	7.695
	179.92	543.7	133.13	1731.7	2510.0	778.3	3.8709	5.1060	1.2352	1.8392	7.512
	182.11	538.7	136.46	1741.2	2501.1	759.9	3.8853	5.0893	1.2040	1.8563	7.328
	184.32	533.5	139.96	1750.9	2491.8	740.8	3.9001	5.0721	1.1719	1.8744	7.145
360	186.55	528.1	143.65	1761.0	2482.0	721.1	3.9153	5.0542	1.1388	1.8936	6.962
	188.81	522.5	147.54	1771.3	2471.8	700.5	3.9310	5.0355	1.1046	1.9140	6.778
	191.10	516.6	151.68	1782.0	2461.0	679.0	3.9471	5.0161	1.0690	1.9358	6.593
	193.40	510.4	156.08	1793.1	2449.6	656.5	3.9638	4.9958	1.0320	1.9592	6.407
	195.74	503.9	160.80	1804.6	2437.5	632.9	3.9812	4.9745	0.9933	1.9845	6.219
365.0	198.09	497.0	165.88	1816.7	2424.6	607.9	3.9994	4.9520	0.9526	2.0120	6.028
	199.28	493.4	168.58	1822.9	2417.8	594.8	4.0088	4.9402	0.9314	2.0268	5.932
	200.48	489.7	171.39	1829.3	2410.7	581.3	4.0185	4.9280	0.9096	2.0422	5.835
	201.68	485.8	174.33	1835.9	2403.3	567.4	4.0284	4.9155	0.8870	2.0585	5.736
	202.89	481.8	177.42	1842.7	2395.6	552.9	4.0387	4.9024	0.8637	2.0757	5.636
367.5	204.11	477.6	180.67	1849.8	2387.6	537.8	4.0493	4.8888	0.8395	2.0939	5.535
	205.33	473.2	184.11	1857.1	2379.2	522.1	4.0602	4.8746	0.8143	2.1133	5.432
	206.56	468.6	187.75	1864.7	2370.3	505.6	4.0717	4.8597	0.7880	2.1340	5.326
	207.80	463.8	191.63	1872.6	2360.9	488.3	4.0836	4.8440	0.7604	2.1563	5.218
	209.05	458.6	195.79	1880.9	2350.9	470.0	4.0962	4.8274	0.7313	2.1804	5.107
370.0	210.30	453.1	200.29	1889.7	2340.2	450.4	4.1094	4.8098	0.7003	2.2068	4.993
	211.56	447.2	205.21	1899.1	2328.5	429.4	4.1236	4.7907	0.6671	2.2361	4.873
	212.83	440.7	210.64	1909.3	2315.8	406.5	4.1389	4.7700	0.6311	2.2689	4.747
	214.11	433.5	216.74	1920.5	2301.6	381.2	4.1558	4.7471	0.5913	2.3067	4.614
	215.39	425.3	223.74	1933.0	2285.5	352.5	4.1748	4.7212	0.5464	2.3515	4.469
372.5	216.69	415.4	232.1	1947.7	2266.6	318.9	4.1971	4.6910	0.4939	2.4074	4.309
	217.99	402.4	242.7	1966.6	2243.0	276.4	4.2258	4.6536	0.4277	2.4850	4.121
	219.30	385.0	259.0	1991.6	2207.3	215.7	4.2640	4.5977	0.3337	2.5974	3.861
373.976	220.55	322		2086		0	4.409		0	3.106	

Table 2 Saturation (pressure)

P (bar)	t (°C)	ρ_l	ρ_g	h_l	h_g	r	s_l	s_g	ϕ	v_l ($\times 10^3$)	v_g ($\times 10^3$)
0.0061173	0.010	999.78	0.004855	0.00	2500.5	2500.5	0	9.1541	9.1541	1.00022	205990.
0.010	6.970	999.89	0.007740	29.27	2513.3	2484.1	0.10581	8.9737	8.8678	1.00011	129190
0.025	21.080	997.96	0.018433	88.36	2539.1	2450.8	0.31160	8.6411	8.3295	1.00205	54249.
0.050	32.881	994.70	0.035472	137.67	2560.5	2422.9	0.47594	8.3930	7.9171	1.00533	28191
0.075	40.299	992.05	0.051982	168.74	2573.9	2405.1	0.57625	8.2494	7.6732	1.00801	19237.
0.100	45.817	989.82	0.06815	191.83	2583.8	2391.9	0.64926	8.1482	7.4990	1.01028	14674.
0.15	53.983	986.14	0.09977	225.95	2598.2	2372.3	0.75486	8.0066	7.2517	1.01405	10023.
0.20	60.073	983.13	0.13072	251.46	2608.9	2357.5	0.83211	7.9068	7.0747	1.01716	7650.
0.25	64.980	980.54	0.16117	271.99	2617.4	2345.5	0.89326	7.8298	6.9366	1.01985	6204.8
0.50	81.339	970.96	0.30856	340.54	2645.3	2304.8	1.09117	7.5928	6.5017	1.02991	3240.9
0.75	91.783	964.13	0.45095	384.43	2662.5	2278.1	1.21309	7.4557	6.2426	1.03721	2217.5
1.0	99.632	958.66	0.5902	417.51	2675.1	2257.6	1.30273	7.3589	6.0562	1.04313	1694.3
1.5	111.378	949.94	0.8624	467.18	2693.4	2226.2	1.43376	7.2232	5.7894	1.05270	1159.5
2.0	120.241	942.96	1.1289	504.80	2706.5	2201.7	1.53035	7.1272	5.5968	1.06049	885.9
2.5	127.443	937.04	1.3912	535.49	2716.8	2181.4	1.60753	7.0528	5.4453	1.06719	718.8
3.0	133.555	931.84	1.6505	561.61	2725.3	2163.7	1.67211	6.9921	5.3200	1.07315	605.9
3.5	138.891	927.17	1.9074	584.48	2732.4	2147.9	1.72785	6.9407	5.2129	1.07855	524.27
4.0	143.643	922.91	2.1624	604.90	2738.5	2133.6	1.77700	6.8961	5.1191	1.08353	462.46
5.0	151.866	915.31	2.6677	640.38	2748.6	2108.2	1.86104	6.8214	4.9604	1.09253	374.86
6.0	158.863	908.61	3.1683	670.71	2756.7	2086.0	1.93155	6.7601	4.8285	1.10058	315.63
7.0	164.983	902.58	3.6655	697.35	2763.3	2066.0	1.99254	6.7079	4.7154	1.10794	272.81
8.0	170.444	897.05	4.1603	721.23	2768.9	2047.7	2.04644	6.6625	4.6161	1.11476	240.37
9.0	175.388	891.94	4.6531	742.93	2773.6	2030.7	2.09484	6.6222	4.5274	1.12116	214.91
10.0	179.916	887.15	5.144	762.88	2777.7	2014.8	2.13885	6.5859	4.4471	1.12720	194.38
12.5	189.848	876.32	6.369	806.92	2785.7	1978.8	2.23439	6.5083	4.2739	1.14114	157.01
15.0	198.327	866.69	7.592	844.86	2791.5	1946.6	2.31496	6.4438	4.1288	1.15382	131.72
17.5	205.764	857.93	8.815	878.42	2795.7	1917.3	2.38498	6.3884	4.0034	1.16559	113.44
20.0	212.417	849.85	10.041	908.69	2798.7	1890.0	2.44714	6.3396	3.8924	1.17667	99.59
22.5	218.452	842.30	11.272	936.38	2800.8	1864.4	2.50320	6.2958	3.7926	1.18722	88.72
25.0	223.989	835.19	12.508	961.98	2802.2	1840.2	2.55439	6.2560	3.7016	1.19733	79.95
27.5	229.114	828.44	13.751	985.86	2803.0	1817.1	2.60158	6.2195	3.6179	1.20709	72.72
30.0	233.892	821.99	15.001	1008.30	2803.3	1795.0	2.64544	6.1855	3.5401	1.21656	66.66
35.0	242.595	809.84	17.527	1049.64	2802.6	1752.9	2.72508	6.1240	3.3989	1.23481	57.05
40.0	250.392	798.49	20.092	1087.24	2800.6	1713.4	2.79623	6.0689	3.2727	1.25237	49.771
45.0	257.474	787.75	22.700	1121.90	2797.6	1675.7	2.86080	6.0188	3.1580	1.26943	44.053
50.0	263.977	777.51	25.355	1154.22	2793.7	1639.5	2.92013	5.9725	3.0524	1.28615	39.440
55.0	270.001	767.68	28.062	1184.60	2789.1	1604.5	2.97518	5.9294	2.9542	1.30263	35.636
60.0	275.621	758.16	30.824	1213.37	2783.9	1570.5	3.02667	5.8887	2.8620	1.31898	32.442
65.0	280.893	748.92	33.646	1240.78	2778.1	1537.3	3.07517	5.8500	2.7748	1.33525	29.721
70.0	285.864	739.90	36.533	1267.02	2771.8	1504.8	3.12111	5.8130	2.6919	1.35153	27.373
75.0	290.570	731.07	39.488	1292.25	2765.0	1472.8	3.16485	5.7775	2.6126	1.36786	25.324
80.0	295.042	722.38	42.516	1316.61	2757.8	1441.2	3.20668	5.7431	2.5365	1.38430	23.520
85.0	299.305	713.82	45.623	1340.21	2750.1	1409.9	3.24683	5.7098	2.4629	1.40091	21.919
90.0	303.379	705.35	48.814	1363.15	2742.0	1378.8	3.28553	5.6772	2.3917	1.41773	20.486
95.0	307.282	696.96	52.10	1385.49	2733.5	1348.0	3.32292	5.6453	2.3224	1.43481	19.195
100.0	311.031	688.61	55.47	1407.33	2724.5	1317.2	3.35918	5.6140	2.2548	1,45220	18.026
105.0	314.637	680.29	58.96	1428.72	2715.2	1286.4	3.39444	5.5831	2.1887	1.46995	16.962
110.0	318.112	671.99	62.55	1449.73	2705.4	1255.7	3.42882	5.5526	2.1238	1.48812	15.987
115.0	321.466	663.67	66.27	1470.40	2695.2	1224.8	3.46242	5.5223	2.0599	1.50677	15.091
120.0	324.709	655.33	70.11	1490.79	2684.6	1193.8	3.49534	5.4922	1.9968	1.52596	14.263
125.0	327.847	646.93	74.10	1510.95	2673.4	1162.5	3.52769	5.4621	1.9344	1.54576	13.495

Table 2 Saturation (pressure) (*Continued*)

P (bar)	t (°C)	ρ_l	ρ_g	h_l	h_g	r	s_l	s_g	ϕ	v_l ($\times 10^3$)	v_g ($\times 10^3$)
130.0	330.888	638.5	78.25	1530.9	2661.9	1130.9	3.5595	5.4319	1.8724	1.5663	12.780
135.0	333.837	629.9	82.56	1550.7	2649.8	1099.0	3.5910	5.4017	1.8107	1.5875	12.112
140.0	336.701	621.2	87.06	1570.5	2637.1	1066.7	3.6221	5.3712	1.7491	1.6097	11.486
145.0	339.485	612.4	91.77	1590.2	2623.9	1033.8	3.6530	5.3404	1.6874	1.6328	10.896
150.0	342.192	603.4	96.71	1609.9	2610.1	1000.2	3.6837	5.3093	1.6255	1.6571	10.340
155.0	344.827	594.3	101.91	1629.6	2595.6	965.9	3.7144	5.2775	1.5631	1.6828	9.812
160.0	347.394	584.8	107.40	1649.5	2580.3	930.8	3.7452	5.2451	1.5000	1.7099	9.311
165.0	349.896	575.1	113.23	1669.6	2564.2	894.6	3.7761	5.2119	1.4358	1.7388	8.832
170.0	352.335	565.0	119.43	1689.9	2547.1	857.2	3.8073	5.1777	1.3704	1.7698	8.373
175.0	354.715	554.5	126.09	1710.7	2529.0	818.3	3.8390	5.1422	1.3032	1.8033	7.931
180.0	357.038	543.6	133.27	1731.9	2509.6	777.6	3.8713	5.1053	1.2339	1.8398	7.504
185.0	359.306	531.9	141.10	1753.9	2488.7	734.8	3.9046	5.0664	1.1618	1.8799	7.087
190.0	361.522	519.5	149.72	1776.7	2466.0	689.3	3.9391	5.0252	1.0860	1.9248	6.679
195.0	363.686	506.1	159.35	1800.8	2441.1	640.3	3.9754	4.9809	1.0054	1.9759	6.275
200.0	365.800	491.3	170.36	1826.5	2413.2	586.7	4.0142	4.9323	0.9181	2.0353	5.870
205.0	367.865	474.6	183.32	1854.8	2381.0	526.2	4.0568	4.8776	0.8208	2.1070	5.455
210.0	369.881	454.8	199.46	1887.1	2342.0	454.8	4.1055	4.8128	0.7072	2.1988	5.014
212.5	370.871	442.9	209.55	1906.0	2318.2	412.2	4.1340	4.7739	0.6399	2.2578	4.772
215.0	371.848	428.5	222.08	1928.2	2289.1	361.0	4.1675	4.7270	0.5595	2.3335	4.503
217.5	372.813	409.2	239.6	1957.0	2249.4	292.3	4.2113	4.6637	0.4525	2.444	4.173
220.55	373.976	322		2086		0	4.409		0	3.106	

Table 3 Compressed water and superheated steam

t (°C)	0.1 bar (t_s = 45.817 °C) v (×10³)	ρ	h	u	s
t_l	1.01028	989.82	191.83	191.82	0.64926
t_g	14674	0.06815	2583.8	2437.0	8.1482
0	1.00022	999.78	-0.03	-0.04	-0.00015
5	1.00005	999.95	21.03	21.02	0.07626
10	1.00030	999.70	42.00	41.99	0.15097
15	1.00091	999.10	62.92	62.91	0.22423
20	1.00181	998.19	83.84	83.83	0.29621
25	1.00299	997.02	104.76	104.75	0.36695
30	1.00441	995.61	125.68	125.67	0.43653
35	1.00605	993.99	146.59	146.58	0.50496
40	1.00789	992.17	167.51	167.50	0.57228
45	1.00993	990.17	188.42	188.41	0.63853
50	14869	0.06725	2591.8	2443.1	8.1731
55	15103	0.06621	2601.3	2450.3	8.2024
60	15336	0.06521	2610.8	2457.5	8.2313
65	15569	0.06423	2620.4	2464.7	8.2596
70	15802	0.06328	2629.9	2471.8	8.2876
75	16035	0.06236	2639.4	2479.0	8.3151
80	16268	0.06147	2648.9	2486.2	8.3422
85	16500	0.06061	2658.4	2493.4	8.3690
90	16732	0.05976	2667.9	2500.6	8.3954
95	16964	0.05895	2677.4	2507.8	8.4214
100	17196	0.05815	2687.0	2515.0	8.4471
105	17428	0.05738	2696.5	2522.2	8.4724
110	17660	0.05662	2706.0	2529.4	8.4974
115	17892	0.05589	2715.5	2536.6	8.5222
120	18124	0.05518	2725.1	2543.8	8.5466
125	18356	0.05448	2734.6	2551.1	8.5707
130	18587	0.05380	2744.2	2558.3	8.5946
135	18819	0.05314	2753.7	2565.6	8.6181
140	19050	0.05249	2763.3	2572.8	8.6415
145	19282	0.05186	2772.9	2580.1	8.6645
150	19513	0.051248	2782.5	2587.4	8.6873
155	19744	0.050648	2792.1	2594.7	8.7099
160	19976	0.050061	2801.7	2602.0	8.7322
165	20207	0.049488	2811.3	2609.3	8.7543
170	20438	0.048928	2821.0	2616.6	8.7762
175	20670	0.048379	2830.6	2623.9	8.7978
180	20901	0.047845	2840.3	2631.3	8.8193
185	21132	0.047321	2850.0	2638.6	8.8405
190	21363	0.046809	2859.6	2646.0	8.8615
195	21594	0.046308	2869.3	2653.4	8.8823
200	21826	0.045818	2879.0	2660.8	8.9030
205	22057	0.045338	2888.8	2668.2	8.9234
210	22288	0.044868	2898.5	2675.6	8.9437
215	22519	0.044407	2908.3	2683.1	8.9637
220	22750	0.043956	2918.0	2690.5	8.9836
225	22981	0.043514	2927.8	2698.0	9.0034
230	23212	0.043081	2937.6	2705.5	9.0229
235	23443	0.042656	2947.4	2713.0	9.0423
240	23674	0.042240	2957.2	2720.5	9.0615
245	23905	0.041832	2967.1	2728.0	9.0806
250	24136	0.041432	2976.9	2735.5	9.0995
255	24367	0.041039	2986.8	2743.1	9.1183
260	24598	0.040654	2996.6	2750.7	9.1369
265	24829	0.040275	3006.5	2758.2	9.1554
270	25060	0.039904	3016.4	2765.8	9.1737
275	25291	0.039540	3026.4	2773.4	9.1919
280	25522	0.039182	3036.3	2781.1	9.2099
285	25753	0.038831	3046.2	2788.7	9.2278
290	25984	0.038485	3056.2	2796.4	9.2456
295	26215	0.038146	3066.2	2804.0	9.2633

t (°C)	0.1 bar v (×10³)	ρ	h	u	s
300	26446	0.037813	3076.2	2811.7	9.2808
305	26677	0.037486	3086.2	2819.4	9.2982
310	26908	0.037164	3096.2	2827.2	9.3154
315	27138	0.036848	3106.3	2834.9	9.3326
320	27369	0.036537	3116.3	2842.6	9.3496
325	27600	0.036232	3126.4	2850.4	9.3665
330	27831	0.035931	3136.5	2858.2	9.3833
335	28062	0.035635	3146.6	2866.0	9.4000
340	28293	0.035345	3156.7	2873.8	9.4166
345	28524	0.035058	3166.9	2881.6	9.4330
350	28755	0.034777	3177.0	2889.5	9.4494
355	28986	0.034500	3187.2	2897.3	9.4657
360	29216	0.034227	3197.4	2905.2	9.4818
365	29447	0.033959	3207.6	2913.1	9.4978
370	29678	0.033695	3217.8	2921.0	9.5138
375	29909	0.033435	3228.0	2928.9	9.5296
380	30140	0.033179	3238.3	2936.9	9.5454
385	30371	0.032926	3248.5	2944.8	9.5610
390	30602	0.032678	3258.8	2952.8	9.5766
395	30832	0.032433	3269.1	2960.8	9.5921
400	31063	0.032192	3279.4	2968.8	9.6075
410	31525	0.031721	3300.1	2984.8	9.6379
420	31987	0.031263	3320.8	3001.0	9.6681
430	32448	0.030818	3341.6	3017.1	9.6979
440	32910	0.030386	3362.5	3033.4	9.7274
450	33372	0.029966	3383.4	3049.7	9.7565
460	33833	0.029557	3404.5	3066.1	9.7854
470	34295	0.029159	3425.5	3082.6	9.8139
480	34757	0.028772	3446.7	3099.1	9.8422
490	35218	0.028394	3467.9	3115.7	9.8702
500	35680	0.028027	3489.2	3132.4	9.8979
520	36603	0.027320	3531.9	3165.9	9.9525
540	37526	0.026648	3575.0	3199.7	10.0061
560	38450	0.026008	3618.3	3233.8	10.0587
580	39373	0.025398	3661.9	3268.1	10.1104
600	40296	0.024816	3705.7	3302.8	10.1612
620	41219	0.024261	3749.9	3337.7	10.2112
640	42142	0.023729	3794.3	3372.9	10.2604
660	43065	0.023220	3839.1	3408.4	10.3089
680	43989	0.022733	3884.1	3444.2	10.3566
700	44912	0.022266	3929.4	3480.2	10.4036
720	45835	0.021817	3974.9	3516.6	10.4500
740	46758	0.021387	4020.8	3553.2	10.4957
760	47681	0.020973	4066.9	3590.1	10.5408
780	48604	0.020574	4113.4	3627.3	10.5853
800	49527	0.020191	4160.1	3664.8	10.6292
850	51840	0.019292	4278.1	3759.7	10.7367
900	54140	0.018470	4397.8	3856.4	10.8410
950	56450	0.017715	4519.2	3954.7	10.9423
1000	58760	0.017019	4642.3	4054.7	11.0410
1100	63370	0.015779	4893.2	4259.5	11.2307
1200	67990	0.014708	5150.2	4470.3	11.4113
1300	72600	0.013773	5412.9	4686.8	11.5838
1400	77220	0.012950	5680.8	4908.6	11.7489
1500	81840	0.012220	5953.5	5135.1	11.9071
1600	86450	0.011567	6230.6	5366.1	12.0592
1700	91070	0.010981	6511.8	5601.1	12.2054
1800	95680	0.010451	6796.7	5839.9	12.3463
1900	100300	0.009970	7085.1	6082.2	12.4821
2000	104910	0.009532	7376.7	6327.6	12.6133

Table 3 Compressed water and superheated steam (*Continued*)

t	0.5 bar (t_s = 81.339 °C)				
(°C)	v (× 10^3)	ρ	h	u	s
t_f	1.02991	970.96	340.54	340.49	1.09117
t_g	3240.9	0.30856	2645.3	2483.3	7.5928
0	1.00020	999.80	0.01	-0.04	-0.00015
5	1.00003	999.97	21.07	21.02	0.07626
10	1.00028	999.72	42.04	41.99	0.15096
15	1.00089	999.11	62.96	62.91	0.22423
20	1.00179	998.21	83.88	83.83	0.29620
25	1.00297	997.04	104.80	104.75	0.36694
30	1.00439	995.63	125.71	125.66	0.43652
35	1.00603	994.01	146.63	146.58	0.50495
40	1.00787	992.19	167.54	167.49	0.57227
45	1.00991	990.18	188.45	188.40	0.63852
50	1.01214	988.01	209.36	209.31	0.70372
55	1.01454	985.67	230.27	230.22	0.76793
60	1.01711	983.18	251.18	251.13	0.83117
65	1.01985	980.54	272.10	272.05	0.89349
70	1.02275	977.76	293.02	292.97	0.95493
75	1.02581	974.84	313.97	313.92	1.01552
80	1.02902	971.80	334.93	334.88	1.07530
85	3275.9	0.30526	2652.6	2488.8	7.6132
90	3323.6	0.30088	2662.5	2496.3	7.6406
95	3371.3	0.29663	2672.3	2503.7	7.6676
100	3418.8	0.29250	2682.1	2511.2	7.6941
105	3466.2	0.28850	2691.9	2518.6	7.7202
110	3513.5	0.28462	2701.7	2526.0	7.7459
115	3560.7	0.28084	2711.5	2533.5	7.7712
120	3607.8	0.27717	2721.2	2540.9	7.7962
125	3654.9	0.27360	2731.0	2548.2	7.8208
130	3701.9	0.27013	2740.7	2555.6	7.8451
135	3748.9	0.26674	2750.5	2563.0	7.8691
140	3795.8	0.26345	2760.2	2570.4	7.8928
145	3842.7	0.26023	2769.9	2577.8	7.9162
150	3889.5	0.25710	2779.7	2585.2	7.9394
155	3936.3	0.25405	2789.4	2592.6	7.9622
160	3983.0	0.25106	2799.1	2600.0	7.9848
165	4029.8	0.24815	2808.9	2607.4	8.0072
170	4076.4	0.24531	2818.6	2614.8	8.0293
175	4123.1	0.24254	2828.3	2622.2	8.0511
180	4169.7	0.23982	2838.1	2629.6	8.0728
185	4216.3	0.23717	2847.9	2637.0	8.0942
190	4262.9	0.23458	2857.6	2644.5	8.1154
195	4309.5	0.23205	2867.4	2651.9	8.1364
200	4356.0	0.22957	2877.2	2659.4	8.1572
205	4402.5	0.22714	2887.0	2666.8	8.1778
210	4449.0	0.22477	2896.8	2674.3	8.1982
215	4495.5	0.22244	2906.6	2681.8	8.2184
220	4542.0	0.22017	2916.4	2689.3	8.2384
225	4588.4	0.21794	2926.2	2696.8	8.2582
230	4634.9	0.21576	2936.1	2704.3	8.2779
235	4681.3	0.21362	2945.9	2711.9	8.2974
240	4727.7	0.21152	2955.8	2719.4	8.3167
245	4774.1	0.20946	2965.7	2727.0	8.3358
250	4820.5	0.20745	2975.6	2734.5	8.3548
255	4866.9	0.20547	2985.5	2742.1	8.3737
260	4913.3	0.20353	2995.4	2749.7	8.3924
265	4959.7	0.20163	3005.3	2757.3	8.4109
270	5006.	0.19976	3015.3	2765.0	8.4293
275	5052.	0.19793	3025.2	2772.6	8.4475
280	5099.	0.19613	3035.2	2780.2	8.4656
285	5145.	0.19436	3045.2	2787.9	8.4836
290	5191.	0.19263	3055.2	2795.6	8.5014
295	5238.	0.19092	3065.2	2803.3	8.5191

t	0.5 bar				
(°C)	v (× 10^3)	ρ	h	u	s
300	5284	0.18925	3075.2	2811.0	8.5367
305	5330	0.18761	3085.2	2818.7	8.5541
310	5377	0.18599	3095.3	2826.4	8.5714
315	5423	0.18440	3105.3	2834.2	8.5886
320	5469	0.18284	3115.4	2842.0	8.6057
325	5516	0.18131	3125.5	2849.7	8.6226
330	5562	0.17980	3135.6	2857.5	8.6395
335	5608	0.17831	3145.8	2865.4	8.6562
340	5654	0.17685	3155.9	2873.2	8.6728
345	5701	0.17542	3166.1	2881.0	8.6893
350	5747	0.17401	3176.2	2888.9	8.7057
355	5793	0.17262	3186.4	2896.8	8.7220
360	5839	0.17125	3196.6	2904.6	8.7381
365	5886	0.16990	3206.8	2912.5	8.7542
370	5932	0.16858	3217.1	2920.5	8.7702
375	5978	0.16727	3227.3	2928.4	8.7861
380	6024	0.16599	3237.6	2936.4	8.8018
385	6071	0.16473	3247.9	2944.3	8.8175
390	6117	0.16348	3258.1	2952.3	8.8331
395	6163	0.16225	3268.5	2960.3	8.8486
400	6209	0.16105	3278.8	2968.3	8.8640
410	6302	0.15868	3299.5	2984.4	8.8945
420	6394	0.15639	3320.3	3000.5	8.9247
430	6487	0.15416	3341.1	3016.7	8.9545
440	6579	0.15199	3362.0	3033.0	8.9840
450	6672	0.14989	3382.9	3049.3	9.0132
460	6764	0.14784	3404.0	3065.8	9.0421
470	6857	0.14585	3425.1	3082.2	9.0707
480	6949	0.14391	3446.2	3098.8	9.0989
490	7041	0.14202	3467.4	3115.4	9.1269
500	7134	0.14018	3488.7	3132.0	9.1547
520	7319	0.13664	3531.5	3165.6	9.2093
540	7503	0.13327	3574.6	3199.4	9.2629
560	7688	0.13007	3617.9	3233.5	9.3156
580	7873	0.12702	3661.5	3267.9	9.3673
600	8058	0.12411	3705.4	3302.5	9.4182
620	8242	0.12132	3749.6	3337.5	9.4682
640	8427	0.11866	3794.1	3372.7	9.5174
660	8612	0.11612	3838.8	3408.2	9.5659
680	8797	0.11368	3883.8	3444.0	9.6136
700	8981	0.11134	3929.1	3480.1	9.6606
720	9166	0.10910	3974.7	3516.4	9.7070
740	9351	0.10694	4020.6	3553.1	9.7527
760	9535	0.10487	4066.7	3590.0	9.7979
780	9720	0.10288	4113.2	3627.2	9.8424
800	9905	0.10096	4159.9	3664.7	9.8863
850	10366	0.09647	4277.9	3759.6	9.9938
900	10828	0.09235	4397.7	3856.3	10.0981
950	11290	0.08858	4519.1	3954.6	10.1995
1000	11751	0.08510	4642.2	4054.6	10.2981
1100	12675	0.07890	4893.1	4259.4	10.4878
1200	13598	0.07354	5150.1	4470.2	10.6684
1300	14521	0.06887	5412.8	4686.8	10.8409
1400	15444	0.06475	5680.7	4908.5	11.0060
1500	16367	0.06110	5953.4	5135.0	11.1643
1600	17290	0.05784	6230.5	5366.0	11.3164
1700	18213	0.05490	6511.7	5601.1	11.4626
1800	19137	0.05226	6796.7	5839.9	11.6035
1900	20060	0.04985	7085.1	6082.1	11.7393
2000	20983	0.04766	7376.7	6327.6	11.8705

Table 3 Compressed water and superheated steam (*Continued*)

t	1.0 bar	(t_s = 99.632 °C)			
(°C)	v (× 10³)	ρ	h	u	s
t_f	1.04313	958.66	417.51	417.41	1.30273
t_g	1694.3	0.5902	2675.1	2505.7	7.3589
0	1.00017	999.83	0.06	−0.04	−0.00015
5	1.00001	999.99	21.12	21.02	0.07626
10	1.00026	999.74	42.08	41.98	0.15096
15	1.00086	999.14	63.01	62.91	0.22422
20	1.00177	998.23	83.93	83.83	0.29619
25	1.00295	997.06	104.84	104.74	0.36693
30	1.00437	995.65	125.76	125.66	0.43650
35	1.00601	994.03	146.67	146.57	0.50493
40	1.00785	992.21	167.59	167.48	0.57225
45	1.00989	990.21	188.49	188.39	0.63849
50	1.01212	988.03	209.40	209.30	0.70370
55	1.01452	985.69	230.31	230.21	0.76790
60	1.01709	983.20	251.22	251.12	0.83115
65	1.01983	980.56	272.14	272.04	0.89347
70	1.02272	977.78	293.07	292.96	0.95490
75	1.02578	974.86	314.01	313.90	1.01549
80	1.02900	971.82	334.97	334.86	1.07526
85	1.03237	968.64	355.95	355.85	1.13426
90	1.03590	965.35	376.96	376.85	1.19251
95	1.03958	961.93	397.99	397.89	1.25004
100	1696.1	0.5896	2675.9	2506.3	7.3609
105	1720.5	0.5812	2686.1	2514.0	7.3880
110	1744.8	0.5731	2696.2	2521.7	7.4146
115	1769.0	0.5653	2706.3	2529.4	7.4408
120	1793.1	0.5577	2716.3	2537.0	7.4665
125	1817.1	0.5503	2726.3	2544.6	7.4918
130	1841.1	0.5432	2736.3	2552.2	7.5167
135	1865.0	0.5362	2746.3	2559.8	7.5413
140	1888.9	0.5294	2756.2	2567.3	7.5655
145	1912.7	0.5228	2766.1	2574.9	7.5893
150	1936.4	0.51641	2776.1	2582.4	7.6129
155	1960.2	0.51016	2785.9	2589.9	7.6361
160	1983.8	0.50407	2795.8	2597.5	7.6591
165	2007.5	0.49813	2805.7	2605.0	7.6818
170	2031.1	0.49234	2815.6	2612.5	7.7042
175	2054.7	0.48669	2825.5	2620.0	7.7263
180	2078.3	0.48117	2835.3	2627.5	7.7482
185	2101.8	0.47579	2845.2	2635.0	7.7699
190	2125.3	0.47052	2855.1	2642.5	7.7913
195	2148.8	0.46538	2864.9	2650.1	7.8125
200	2172.3	0.46035	2874.8	2657.6	7.8335
205	2195.7	0.45544	2884.7	2665.1	7.8543
210	2219.1	0.45063	2894.6	2672.7	7.8748
215	2242.5	0.44592	2904.5	2680.2	7.8952
220	2265.9	0.44132	2914.4	2687.8	7.9153
225	2289.3	0.43681	2924.3	2695.3	7.9353
230	2312.7	0.43240	2934.2	2702.9	7.9551
235	2336.1	0.42807	2944.1	2710.5	7.9747
240	2359.4	0.42384	2954.0	2718.1	7.9942
245	2382.7	0.41969	2963.9	2725.7	8.0134
250	2406.1	0.41562	2973.9	2733.3	8.0325
255	2429.4	0.41163	2983.8	2740.9	8.0515
260	2452.7	0.40772	2993.8	2748.5	8.0702
265	2476.0	0.40388	3003.8	2756.2	8.0889
270	2499.2	0.40012	3013.8	2763.8	8.1073
275	2522.5	0.39643	3023.8	2771.5	8.1257
280	2545.8	0.39281	3033.8	2779.2	8.1438
285	2569.1	0.38925	3043.8	2786.9	8.1619
290	2592.3	0.38576	3053.8	2794.6	8.1798
295	2615.6	0.38233	3063.9	2802.3	8.1975

t	1.0 bar				
(°C)	v (× 10³)	ρ	h	u	s
300	2638.8	0.37896	3073.9	2810.1	8.2152
305	2662.0	0.37565	3084.0	2817.8	8.2327
310	2685.3	0.37240	3094.1	2825.6	8.2500
315	2708.5	0.36921	3104.2	2833.3	8.2673
320	2731.7	0.36607	3114.3	2841.1	8.2844
325	2754.9	0.36299	3124.4	2848.9	8.3014
330	2778.1	0.35995	3134.6	2856.7	8.3183
335	2801.3	0.35697	3144.7	2864.6	8.3350
340	2824.5	0.35404	3154.9	2872.4	8.3517
345	2847.7	0.35116	3165.1	2880.3	8.3682
350	2870.9	0.34832	3175.3	2888.2	8.3846
355	2894.1	0.34553	3185.5	2896.1	8.4009
360	2917.3	0.34278	3195.7	2904.0	8.4172
365	2940.5	0.34008	3205.9	2911.9	8.4333
370	2963.7	0.33742	3216.2	2919.8	8.4493
375	2986.9	0.33480	3226.4	2927.8	8.4652
380	3010.0	0.33222	3236.7	2935.7	8.4810
385	3033.2	0.32968	3247.0	2943.7	8.4967
390	3056.4	0.32719	3257.3	2951.7	8.5123
395	3079.5	0.32472	3267.7	2959.7	8.5278
400	3102.7	0.32230	3278.0	2967.7	8.5432
410	3149.0	0.31756	3298.7	2983.8	8.5738
420	3195.3	0.31296	3319.5	3000.0	8.6040
430	3241.6	0.30849	3340.4	3016.2	8.6339
440	3287.9	0.30414	3361.3	3032.5	8.6634
450	3334.2	0.29992	3382.3	3048.9	8.6927
460	3380.5	0.29582	3403.3	3065.3	8.7216
470	3426.8	0.29182	3424.5	3081.8	8.7502
480	3473.0	0.28793	3445.6	3098.3	8.7785
490	3519.3	0.28415	3466.9	3115.0	8.8065
500	3565.5	0.28046	3488.2	3131.6	8.8342
520	3658.0	0.27337	3531.0	3165.2	8.8889
540	3750.5	0.26663	3574.1	3199.1	8.9426
560	3843.0	0.26021	3617.5	3233.2	8.9953
580	3935.5	0.25410	3661.1	3267.6	9.0470
600	4027.9	0.24827	3705.0	3302.3	9.0979
620	4120.3	0.24270	3749.2	3337.2	9.1480
640	4212.8	0.23737	3793.7	3372.4	9.1972
660	4305.2	0.23228	3838.5	3408.0	9.2457
680	4397.6	0.22740	3883.5	3443.8	9.2935
700	4490.0	0.22272	3928.8	3479.8	9.3405
720	4582.4	0.21823	3974.4	3516.2	9.3869
740	4674.8	0.21391	4020.3	3552.9	9.4326
760	4767.1	0.20977	4066.5	3589.8	9.4778
780	4859.5	0.20578	4112.9	3627.0	9.5223
800	4951.9	0.20194	4159.7	3664.5	9.5662
850	5183.	0.19295	4277.7	3759.4	9.6738
900	5414.	0.18472	4397.5	3856.1	9.7781
950	5645.	0.17716	4518.9	3954.5	9.8794
1000	5876.	0.17020	4642.0	4054.5	9.9781
1100	6337.	0.15780	4893.0	4259.3	10.1678
1200	6799.	0.14708	5150.0	4470.1	10.3485
1300	7260.	0.13773	5412.7	4686.7	10.5210
1400	7722.	0.12950	5680.6	4908.4	10.6861
1500	8184.	0.12219	5953.3	5135.0	10.8444
1600	8645.	0.11567	6230.5	5366.0	10.9964
1700	9107.	0.10981	6511.7	5601.0	11.1427
1800	9568.	0.10451	6796.6	5839.8	11.2835
1900	10030.	0.09970	7085.1	6082.1	11.4194
2000	10492.	0.09531	7376.7	6327.5	11.5506

Table 3 Compressed water and superheated steam (*Continued*)

t	2.0 bar (t_s = 120.241 °C)				
(°C)	v (× 10³)	ρ	h	u	s
t_f	1.06049	942.96	504.80	504.59	1.53036
t_g	885.9	1.1289	2706.5	2529.4	7.1272
0	1.00012	999.88	0.16	−0.04	−0.00014
5	0.99996	1000.04	21.22	21.02	0.07626
10	1.00021	999.79	42.18	41.98	0.15095
15	1.00082	999.18	63.11	62.91	0.22421
20	1.00173	998.28	84.02	83.82	0.29617
25	1.00290	997.11	104.93	104.73	0.36690
30	1.00432	995.70	125.85	125.65	0.43647
35	1.00596	994.07	146.76	146.56	0.50489
40	1.00781	992.25	167.67	167.47	0.57221
45	1.00985	990.25	188.58	188.38	0.63845
50	1.01207	988.07	209.49	209.29	0.70365
55	1.01447	985.73	230.40	230.19	0.76785
60	1.01704	983.24	251.31	251.10	0.83109
65	1.01978	980.60	272.22	272.02	0.89341
70	1.02268	977.82	293.15	292.94	0.95484
75	1.02574	974.91	314.09	313.88	1.01543
80	1.02895	971.86	335.05	334.84	1.07520
85	1.03232	968.69	356.03	355.82	1.13419
90	1.03585	965.39	377.03	376.83	1.19244
95	1.03953	961.97	398.07	397.86	1.24997
100	1.04336	958.44	419.14	418.93	1.30681
105	1.04735	954.79	440.24	440.03	1.36298
110	1.05150	951.02	461.38	461.17	1.41852
115	1.05580	947.14	482.56	482.35	1.47344
120	1.06027	943.16	503.78	503.57	1.52776
125	897.8	1.1138	2716.6	2537.1	7.1527
130	910.3	1.0985	2727.1	2545.1	7.1789
135	922.7	1.0837	2737.6	2553.0	7.2047
140	935.1	1.0694	2748.0	2561.0	7.2300
145	947.4	1.0555	2758.3	2568.8	7.2549
150	959.7	1.0420	2768.6	2576.7	7.2793
155	971.9	1.0289	2778.9	2584.5	7.3034
160	984.1	1.0162	2789.1	2592.3	7.3271
165	996.2	1.0038	2799.3	2600.0	7.3505
170	1008.3	0.9918	2809.4	2607.8	7.3736
175	1020.4	0.9800	2819.6	2615.5	7.3963
180	1032.4	0.9686	2829.7	2623.2	7.4188
185	1044.4	0.9575	2839.8	2630.9	7.4409
190	1056.4	0.9466	2849.9	2638.6	7.4628
195	1068.4	0.9360	2859.9	2646.3	7.4845
200	1080.3	0.9257	2870.0	2653.9	7.5059
205	1092.2	0.9156	2880.1	2661.6	7.5270
210	1104.1	0.9057	2890.1	2669.3	7.5479
215	1116.0	0.8961	2900.2	2677.0	7.5686
220	1127.9	0.8866	2910.2	2684.6	7.5891
225	1139.7	0.8774	2920.3	2692.3	7.6094
230	1151.6	0.8684	2930.3	2700.0	7.6294
235	1163.4	0.8596	2940.4	2707.7	7.6493
240	1175.2	0.8509	2950.4	2715.4	7.6690
245	1187.0	0.8425	2960.5	2723.1	7.6885
250	1198.8	0.8342	2970.5	2730.8	7.7078
255	1210.6	0.8261	2980.6	2738.5	7.7269
260	1222.3	0.8181	2990.6	2746.2	7.7459
265	1234.1	0.8103	3000.7	2753.9	7.7647
270	1245.8	0.8027	3010.8	2761.6	7.7833
275	1257.6	0.7952	3020.9	2769.4	7.8018
280	1269.3	0.7878	3031.0	2777.1	7.8201
285	1281.0	0.7806	3041.1	2784.9	7.8383
290	1292.8	0.7735	3051.2	2792.6	7.8563
295	1304.5	0.7666	3061.3	2800.4	7.8742

t	2.0 bar				
(°C)	v (× 10³)	ρ	h	u	s
300	1316.2	0.7598	3071.4	2808.2	7.8920
305	1327.9	0.7531	3081.6	2816.0	7.9096
310	1339.6	0.7465	3091.7	2823.8	7.9271
315	1351.3	0.7400	3101.9	2831.6	7.9444
320	1362.9	0.7337	3112.0	2839.4	7.9616
325	1374.6	0.7275	3122.2	2847.3	7.9787
330	1386.3	0.7214	3132.4	2855.2	7.9957
335	1398.0	0.7153	3142.6	2863.0	8.0125
340	1409.6	0.7094	3152.8	2870.9	8.0293
345	1421.3	0.7036	3163.1	2878.8	8.0459
350	1432.9	0.6979	3173.3	2886.7	8.0624
355	1444.6	0.6922	3183.6	2894.6	8.0788
360	1456.2	0.6867	3193.8	2902.6	8.0951
365	1467.9	0.6812	3204.1	2910.5	8.1112
370	1479.5	0.6759	3214.4	2918.5	8.1273
375	1491.2	0.6706	3224.7	2926.5	8.1433
380	1502.8	0.6654	3235.0	2934.5	8.1591
385	1514.4	0.6603	3245.4	2942.5	8.1749
390	1526.1	0.6553	3255.7	2950.5	8.1906
395	1537.7	0.6503	3266.1	2958.5	8.2061
400	1549.3	0.6454	3276.4	2966.6	8.2216
410	1572.6	0.6359	3297.2	2982.7	8.2523
420	1595.8	0.6266	3318.1	2998.9	8.2826
430	1619.0	0.6177	3339.0	3015.2	8.3125
440	1642.2	0.6089	3360.0	3031.5	8.3421
450	1665.5	0.6004	3381.0	3047.9	8.3714
460	1688.7	0.5922	3402.1	3064.4	8.4004
470	1711.9	0.5842	3423.3	3080.9	8.4291
480	1735.0	0.5764	3444.5	3097.5	8.4574
490	1758.2	0.5688	3465.8	3114.1	8.4855
500	1781.4	0.5614	3487.1	3130.8	8.5133
520	1827.8	0.54712	3530.0	3164.5	8.5681
540	1874.1	0.53359	3573.2	3198.4	8.6218
560	1920.4	0.52072	3616.6	3232.5	8.6746
580	1966.7	0.50846	3660.3	3267.0	8.7264
600	2013.0	0.49677	3704.3	3301.7	8.7773
620	2059.3	0.48560	3748.5	3336.7	8.8274
640	2105.6	0.47493	3793.0	3371.9	8.8767
660	2151.8	0.46472	3837.8	3407.5	8.9253
680	2198.1	0.45494	3882.9	3443.3	8.9731
700	2244.3	0.44557	3928.3	3479.4	9.0201
720	2290.6	0.43657	3973.9	3515.8	9.0666
740	2336.8	0.42793	4019.8	3552.4	9.1123
760	2383.0	0.41963	4066.0	3589.4	9.1575
780	2429.3	0.41165	4112.5	3626.6	9.2020
800	2475.5	0.40396	4159.2	3664.1	9.2460
850	2591.0	0.38595	4277.3	3759.1	9.3536
900	2706.6	0.36947	4397.1	3855.8	9.4579
950	2822.1	0.35435	4518.6	3954.2	9.5593
1000	2937.5	0.34042	4641.7	4054.2	9.6580
1100	3168.5	0.31561	4892.8	4259.1	9.8478
1200	3399.4	0.29417	5149.8	4470.0	10.0284
1300	3630.3	0.27546	5412.5	4686.5	10.2010
1400	3861.1	0.25899	5680.5	4908.3	10.3661
1500	4092.0	0.24438	5953.2	5134.8	10.5244
1600	4322.8	0.23133	6230.4	5365.8	10.6764
1700	4553.6	0.21960	6511.6	5600.9	10.8227
1800	4784.5	0.20901	6796.6	5839.7	10.9636
1900	5015.3	0.19939	7085.0	6082.0	11.0994
2000	5246.1	0.19062	7376.7	6327.4	11.2306

Table 3 Compressed water and superheated steam (*Continued*)

t (°C)	5.0 bar (t_s = 151.866 °C) v (× 10^3)	ρ	h	u	s
t_l	1.09253	915.31	640.38	639.84	1.86104
t_g	374.86	2.6677	2748.6	2561.2	6.8214
0	0.99997	1000.03	0.47	−0.03	−0.00012
5	0.99981	1000.19	21.52	21.02	0.07625
10	1.00007	999.93	42.47	41.97	0.15092
15	1.00068	999.32	63.39	62.89	0.22416
20	1.00159	998.41	84.30	83.80	0.29610
25	1.00277	997.24	105.21	104.71	0.36683
30	1.00419	995.83	126.12	125.62	0.43638
35	1.00583	994.21	147.03	146.53	0.50479
40	1.00767	992.39	167.94	167.44	0.57209
45	1.00971	990.38	188.84	188.34	0.63832
50	1.01194	988.20	209.75	209.24	0.70351
55	1.01434	985.87	230.65	230.14	0.76770
60	1.01691	983.37	251.56	251.05	0.83093
65	1.01964	980.74	272.47	271.96	0.89324
70	1.02254	977.96	293.39	292.88	0.95466
75	1.02560	975.04	314.33	313.82	1.01524
80	1.02881	972.00	335.29	334.77	1.07500
85	1.03218	968.82	356.26	355.75	1.13399
90	1.03570	965.53	377.27	376.75	1.19222
95	1.03938	962.11	398.30	397.78	1.24974
100	1.04321	958.58	419.36	418.84	1.30657
105	1.04720	954.93	440.46	439.94	1.36274
110	1.05134	951.17	461.60	461.07	1.41827
115	1.05564	947.29	482.77	482.24	1.47318
120	1.06010	943.31	503.99	503.46	1.52749
125	1.06473	939.21	525.25	524.72	1.58123
130	1.06952	935.00	546.56	546.03	1.63442
135	1.07448	930.69	567.92	567.38	1.68707
140	1.07961	926.26	589.33	588.79	1.73922
145	1.08493	921.72	610.80	610.26	1.79087
150	1.09043	917.07	632.33	631.79	1.84205
155	378.24	2.6438	2755.8	2566.7	6.8383
160	383.58	2.6070	2767.2	2575.4	6.8648
165	388.87	2.5715	2778.5	2584.1	6.8907
170	394.12	2.5373	2789.7	2592.6	6.9160
175	399.33	2.5042	2800.7	2601.1	6.9408
180	404.50	2.4722	2811.7	2609.5	6.9652
185	409.64	2.4412	2822.6	2617.8	6.9891
190	414.74	2.4111	2833.5	2626.1	7.0126
195	419.82	2.3820	2844.2	2634.3	7.0358
200	424.87	2.3537	2854.9	2642.5	7.0585
205	429.90	2.3261	2865.6	2650.7	7.0810
210	434.90	2.2994	2876.2	2658.8	7.1031
215	439.89	2.2733	2886.8	2666.9	7.1249
220	444.85	2.2479	2897.4	2674.9	7.1463
225	449.80	2.2232	2907.9	2683.0	7.1676
230	454.73	2.1991	2918.4	2691.0	7.1885
235	459.65	2.1756	2928.8	2699.0	7.2092
240	464.55	2.1526	2939.3	2707.0	7.2297
245	469.44	2.1302	2949.7	2715.0	7.2499
250	474.32	2.1083	2960.1	2723.0	7.2699
255	479.18	2.0869	2970.5	2730.9	7.2897
260	484.04	2.0660	2980.9	2738.9	7.3092
265	488.88	2.0455	2991.3	2746.8	7.3286
270	493.72	2.0255	3001.6	2754.8	7.3478
275	498.54	2.0059	3012.0	2762.7	7.3668
280	503.36	1.9867	3022.4	2770.7	7.3856
285	508.17	1.9679	3032.7	2778.6	7.4042
290	512.97	1.9494	3043.1	2786.6	7.4227
295	517.76	1.9314	3053.4	2794.5	7.4410

t (°C)	5.0 bar v (× 10^3)	ρ	h	u	s
300	522.5	1.9137	3063.7	2802.5	7.4591
305	527.3	1.8963	3074.1	2810.4	7.4771
310	532.1	1.8793	3084.4	2818.4	7.4949
315	536.9	1.8626	3094.8	2826.4	7.5126
320	541.6	1.8463	3105.2	2834.3	7.5301
325	546.4	1.8302	3115.5	2842.3	7.5475
330	551.1	1.8144	3125.9	2850.3	7.5647
335	555.9	1.7989	3136.2	2858.3	7.5819
340	560.6	1.7837	3146.6	2866.3	7.5989
345	565.4	1.7687	3157.0	2874.3	7.6157
350	570.1	1.7540	3167.4	2882.3	7.6325
355	574.8	1.7396	3177.8	2890.4	7.6491
360	579.6	1.7254	3188.2	2898.4	7.6656
365	584.3	1.7114	3198.6	2906.4	7.6819
370	589.0	1.6977	3209.0	2914.5	7.6982
375	593.7	1.6842	3219.4	2922.6	7.7144
380	598.5	1.6710	3229.9	2930.7	7.7304
385	603.2	1.6579	3240.3	2938.7	7.7463
390	607.9	1.6451	3250.8	2946.8	7.7622
395	612.6	1.6324	3261.3	2955.0	7.7779
400	617.3	1.6200	3271.7	2963.1	7.7935
410	626.7	1.5957	3292.7	2979.4	7.8245
420	636.1	1.5721	3313.8	2995.7	7.8550
430	645.5	1.5493	3334.8	3012.1	7.8852
440	654.8	1.5271	3356.0	3028.6	7.9151
450	664.2	1.5056	3377.2	3045.1	7.9446
460	673.6	1.4847	3398.4	3061.6	7.9738
470	682.9	1.4643	3419.7	3078.2	8.0026
480	692.3	1.4445	3441.1	3094.9	8.0312
490	701.6	1.4253	3462.5	3111.7	8.0594
500	710.9	1.4066	3483.9	3128.5	8.0873
520	729.6	1.3706	3527.0	3162.2	8.1424
540	748.2	1.3365	3570.4	3196.3	8.1964
560	766.9	1.3040	3614.0	3230.6	8.2493
580	785.5	1.2731	3657.8	3265.1	8.3013
600	804.1	1.2437	3701.9	3299.9	8.3524
620	822.7	1.2156	3746.3	3335.0	8.4027
640	841.2	1.1887	3791.0	3370.3	8.4521
660	859.8	1.1630	3835.9	3406.0	8.5008
680	878.4	1.1385	3881.0	3441.9	8.5487
700	896.9	1.1149	3926.5	3478.0	8.5959
720	915.5	1.0923	3972.2	3514.5	8.6424
740	934.0	1.0706	4018.2	3551.2	8.6882
760	952.6	1.0498	4064.5	3588.2	8.7334
780	971.1	1.0297	4111.0	3625.5	8.7781
800	989.6	1.0105	4157.8	3663.0	8.8221
850	1036.0	0.9653	4276.1	3758.1	8.9298
900	1082.2	0.9240	4396.0	3854.9	9.0342
950	1128.5	0.8861	4517.6	3953.3	9.1357
1000	1174.8	0.8512	4640.8	4053.4	9.2345
1100	1267.3	0.7891	4892.0	4258.4	9.4244
1200	1359.7	0.7354	5149.2	4469.4	9.6052
1300	1452.1	0.6886	5412.0	4686.0	9.7777
1400	1544.6	0.6474	5680.1	4907.8	9.9429
1500	1636.9	0.6109	5952.9	5134.4	10.1013
1600	1729.3	0.57826	6230.1	5365.4	10.2534
1700	1821.7	0.54894	6511.4	5600.5	10.3996
1800	1914.1	0.52245	6796.4	5839.4	10.5405
1900	2006.4	0.49840	7084.9	6081.7	10.6764
2000	2098.8	0.47647	7376.5	6327.2	10.8076

Table 3 Compressed water and superheated steam (*Continued*)

t	10.0 bar (t_s = 179.916 °C)				
(°C)	v (× 10³)	ρ	h	u	s
t_f	1.12720	887.15	762.88	761.75	2.13885
t_g	194.38	5.1445	2777.7	2583.3	6.5859
0	0.99971	1000.29	0.98	−0.02	−0.00008
5	0.99957	1000.43	22.02	21.02	0.07625
10	0.99983	1000.17	42.96	41.96	0.15088
15	1.00044	999.56	63.87	62.87	0.22408
20	1.00136	998.64	84.77	83.77	0.29600
25	1.00254	997.47	105.67	104.67	0.36670
30	1.00396	996.05	126.58	125.57	0.43622
35	1.00560	994.43	147.48	146.48	0.50461
40	1.00745	992.60	168.38	167.37	0.57190
45	1.00949	990.60	189.28	188.27	0.63811
50	1.01171	988.42	210.18	209.17	0.70328
55	1.01411	986.08	231.08	230.06	0.76746
60	1.01668	983.59	251.98	250.96	0.83067
65	1.01941	980.96	272.88	271.86	0.89296
70	1.02231	978.18	293.80	292.78	0.95436
75	1.02536	975.26	314.73	313.71	1.01492
80	1.02857	972.22	335.68	334.66	1.07467
85	1.03194	969.05	356.66	355.62	1.13364
90	1.03546	965.76	377.65	376.62	1.19186
95	1.03913	962.35	398.68	397.64	1.24937
100	1.04295	958.81	419.74	418.70	1.30618
105	1.04694	955.17	440.83	439.78	1.36233
110	1.05107	951.41	461.96	460.91	1.41784
115	1.05537	947.54	483.13	482.08	1.47274
120	1.05982	943.56	504.34	503.28	1.52704
125	1.06444	939.46	525.60	524.53	1.58076
130	1.06922	935.26	546.90	545.83	1.63393
135	1.07417	930.95	568.25	567.18	1.68657
140	1.07930	926.53	589.66	588.58	1.73870
145	1.08460	922.00	611.12	610.03	1.79033
150	1.09009	917.36	632.64	631.55	1.84149
155	1.09577	912.60	654.22	653.13	1.89220
160	1.10165	907.73	675.87	674.77	1.94247
165	1.10773	902.74	697.60	696.49	1.99233
170	1.11403	897.64	719.40	718.28	2.04181
175	1.12056	892.41	741.28	740.16	2.09091
180	194.43	5.1432	2777.9	2583.5	6.5864
185	197.36	5.0669	2790.6	2593.2	6.6142
190	200.25	4.9938	2803.0	2602.8	6.6412
195	203.09	4.9239	2815.3	2612.2	6.6675
200	205.90	4.8566	2827.4	2621.5	6.6932
205	208.68	4.7920	2839.3	2630.6	6.7183
210	211.43	4.7296	2851.1	2639.7	6.7428
215	214.16	4.6695	2862.8	2648.6	6.7668
220	216.86	4.6114	2874.3	2657.5	6.7904
225	219.53	4.5551	2885.8	2666.2	6.8135
230	222.19	4.5006	2897.1	2674.9	6.8362
235	224.83	4.4478	2908.4	2683.6	6.8586
240	227.45	4.3966	2919.6	2692.2	6.8805
245	230.05	4.3468	2930.8	2700.7	6.9021
250	232.64	4.2984	2941.9	2709.2	6.9235
255	235.22	4.2513	2952.9	2717.7	6.9444
260	237.79	4.2055	2963.9	2726.1	6.9652
265	240.34	4.1608	2974.9	2734.5	6.9856
270	242.88	4.1173	2985.8	2742.9	7.0058
275	245.41	4.0748	2996.6	2751.2	7.0257
280	247.93	4.0334	3007.5	2759.6	7.0454
285	250.44	3.9930	3018.3	2767.9	7.0648
290	252.94	3.9534	3029.1	2776.1	7.0841
295	255.44	3.9148	3039.8	2784.4	7.1031

t	10.0 bar				
(°C)	v (× 10³)	ρ	h	u	s
300	257.93	3.8771	3050.6	2792.7	7.1219
305	260.41	3.8401	3061.3	2800.9	7.1406
310	262.88	3.8040	3072.0	2809.1	7.1590
315	265.35	3.7686	3082.7	2817.4	7.1773
320	267.81	3.7340	3093.4	2825.6	7.1954
325	270.27	3.7001	3104.1	2833.8	7.2133
330	272.72	3.6668	3114.7	2842.0	7.2311
335	275.16	3.6342	3125.4	2850.2	7.2486
340	277.60	3.6023	3136.1	2858.5	7.2661
345	280.04	3.5710	3146.7	2866.7	7.2834
350	282.47	3.5402	3157.3	2874.9	7.3005
355	284.90	3.5101	3168.0	2883.1	7.3175
360	287.32	3.4804	3178.6	2891.3	7.3344
365	289.74	3.4514	3189.3	2899.5	7.3511
370	292.16	3.4228	3199.9	2907.7	7.3678
375	294.57	3.3948	3210.5	2916.0	7.3842
380	296.98	3.3673	3221.2	2924.2	7.4006
385	299.38	3.3402	3231.8	2932.4	7.4168
390	301.79	3.3136	3242.5	2940.7	7.4329
395	304.19	3.2875	3253.1	2948.9	7.4489
400	306.58	3.2617	3263.8	2957.2	7.4648
410	311.37	3.2116	3285.1	2973.7	7.4963
420	316.15	3.1631	3306.5	2990.3	7.5273
430	320.92	3.1161	3327.8	3006.9	7.5579
440	325.68	3.0705	3349.3	3023.6	7.5882
450	330.43	3.0263	3370.7	3040.3	7.6180
460	335.18	2.9835	3392.2	3057.0	7.6475
470	339.92	2.9419	3413.7	3073.8	7.6767
480	344.65	2.9015	3435.3	3090.6	7.7055
490	349.38	2.8622	3456.9	3107.5	7.7340
500	354.10	2.8241	3478.6	3124.5	7.7622
520	363.53	2.7508	3522.0	3158.5	7.8177
540	372.94	2.6814	3565.7	3192.8	7.8721
560	382.34	2.6155	3609.6	3227.2	7.9254
580	391.72	2.5528	3653.7	3262.0	7.9778
600	401.09	2.4932	3698.1	3297.0	8.0292
620	410.45	2.4363	3742.7	3332.2	8.0797
640	419.81	2.3821	3787.5	3367.7	8.1293
660	429.15	2.3302	3832.6	3403.4	8.1782
680	438.48	2.2806	3877.9	3439.5	8.2263
700	447.81	2.2331	3923.6	3475.7	8.2736
720	457.13	2.1876	3969.4	3512.3	8.3203
740	466.45	2.1439	4015.6	3549.1	8.3663
760	475.76	2.1019	4062.0	3586.2	8.4116
780	485.06	2.0616	4108.6	3623.6	8.4563
800	494.36	2.0228	4155.5	3661.2	8.5005
850	517.6	1.9320	4274.0	3756.4	8.6084
900	540.8	1.8491	4394.2	3853.4	8.7130
950	564.0	1.7730	4516.0	3951.9	8.8147
1000	587.2	1.7030	4639.3	4052.1	8.9136
1100	633.5	1.5785	4890.8	4257.3	9.1037
1200	679.8	1.4710	5148.2	4468.4	9.2846
1300	726.1	1.3772	5411.2	4685.1	9.4573
1400	772.4	1.2947	5679.4	4907.0	9.6225
1500	818.6	1.2216	5952.3	5133.7	9.7810
1600	864.8	1.1563	6229.6	5364.8	9.9331
1700	911.0	1.0976	6511.0	5599.9	10.0794
1800	957.2	1.0447	6796.1	5838.8	10.2204
1900	1003.4	0.9966	7084.6	6081.2	10.3563
2000	1049.6	0.9527	7376.3	6326.7	10.4875

Table 3 Compressed water and superheated steam (*Continued*)

t	20 bar	(t_s = 212.417 °C)			
(°C)	v (× 10³)	ρ	h	u	s
t_l	1.17667	849.85	908.69	906.33	2.44714
t_g	99.59	10.041	2798.7	2599.5	6.3396
0	0.99921	1000.79	2.00	0.00	0.00000
5	0.99908	1000.92	23.01	21.01	0.07623
10	0.99935	1000.65	43.94	41.94	0.15079
15	0.99998	1000.02	64.83	62.83	0.22393
20	1.00090	999.10	85.71	83.71	0.29579
25	1.00209	997.92	106.60	104.60	0.36643
30	1.00351	996.50	127.49	125.48	0.43592
35	1.00516	994.87	148.38	146.37	0.50427
40	1.00701	993.04	169.27	167.25	0.57151
45	1.00905	991.04	190.16	188.14	0.63768
50	1.01127	988.86	211.04	209.02	0.70282
55	1.01366	986.52	231.93	229.90	0.76696
60	1.01623	984.03	252.82	250.78	0.83014
65	1.01896	981.39	273.71	271.68	0.89240
70	1.02185	978.62	294.62	292.58	0.95377
75	1.02490	975.71	315.54	313.49	1.01430
80	1.02810	972.67	336.48	334.42	1.07401
85	1.03146	969.50	357.44	355.38	1.13295
90	1.03497	966.22	378.43	376.36	1.19115
95	1.03863	962.81	399.44	397.37	1.24862
100	1.04244	959.28	420.49	418.41	1.30540
105	1.04641	955.64	441.57	439.48	1.36152
110	1.05054	951.89	462.69	460.59	1.41700
115	1.05482	948.03	483.85	481.74	1.47186
120	1.05926	944.06	505.05	502.93	1.52613
125	1.06386	939.97	526.29	524.16	1.57982
130	1.06862	935.78	547.58	545.44	1.63296
135	1.07356	931.48	568.92	566.77	1.68556
140	1.07866	927.07	590.31	588.15	1.73766
145	1.08395	922.55	611.75	609.59	1.78925
150	1.08942	917.92	633.26	631.08	1.84037
155	1.09507	913.18	654.82	652.63	1.89104
160	1.10092	908.33	676.46	674.26	1.94128
165	1.10698	903.36	698.16	695.95	1.99110
170	1.11325	898.27	719.94	717.72	2.04053
175	1.11974	893.06	741.80	739.56	2.08958
180	1.12647	887.73	763.75	761.50	2.13829
185	1.13344	882.27	785.80	783.53	2.18666
190	1.14067	876.68	807.94	805.66	2.23473
195	1.14817	870.95	830.19	827.89	2.28252
200	1.15596	865.08	852.56	850.25	2.33005
205	1.16405	859.07	875.06	872.73	2.37735
210	1.17248	852.89	897.69	895.35	2.42444
215	100.46	9.954	2806.4	2605.4	6.3553
220	102.12	9.793	2820.9	2616.6	6.3848
225	103.74	9.640	2835.0	2627.5	6.4133
230	105.32	9.495	2848.8	2638.2	6.4409
235	106.88	9.356	2862.4	2648.6	6.4677
240	108.41	9.224	2875.6	2658.8	6.4937
245	109.92	9.097	2888.7	2668.9	6.5191
250	111.41	8.976	2901.6	2678.8	6.5438
255	112.88	8.859	2914.3	2688.5	6.5679
260	114.33	8.747	2926.8	2698.1	6.5915
265	115.77	8.638	2939.2	2707.6	6.6146
270	117.19	8.534	2951.4	2717.0	6.6373
275	118.59	8.432	2963.5	2726.4	6.6595
280	119.98	8.334	2975.6	2735.6	6.6814
285	121.37	8.240	2987.5	2744.8	6.7028
290	122.74	8.147	2999.3	2753.8	6.7239
295	124.10	8.058	3011.1	2762.9	6.7447

t	20 bar				
(°C)	v (× 10³)	ρ	h	u	s
300	125.45	7.971	3022.7	2771.8	6.7651
305	126.79	7.887	3034.3	2780.8	6.7853
310	128.13	7.805	3045.9	2789.6	6.8052
315	129.45	7.725	3057.4	2798.5	6.8248
320	130.77	7.647	3068.8	2807.3	6.8441
325	132.09	7.571	3080.2	2816.0	6.8633
330	133.39	7.497	3091.5	2824.8	6.8822
335	134.69	7.424	3102.8	2833.5	6.9008
340	135.99	7.354	3114.1	2842.1	6.9193
345	137.28	7.284	3125.4	2850.8	6.9375
350	138.56	7.217	3136.6	2859.4	6.9556
355	139.84	7.151	3147.7	2868.0	6.9735
360	141.12	7.086	3158.9	2876.7	6.9911
365	142.39	7.023	3170.0	2885.2	7.0087
370	143.66	6.961	3181.1	2893.8	7.0260
375	144.92	6.900	3192.2	2902.4	7.0432
380	146.18	6.841	3203.3	2911.0	7.0602
385	147.44	6.782	3214.4	2919.5	7.0771
390	148.69	6.725	3225.4	2928.0	7.0938
395	149.94	6.669	3236.5	2936.6	7.1104
400	151.19	6.614	3247.5	2945.1	7.1269
410	153.68	6.507	3269.5	2962.2	7.1594
420	156.15	6.404	3291.6	2979.3	7.1914
430	158.62	6.304	3313.6	2996.3	7.2229
440	161.07	6.208	3335.6	3013.4	7.2539
450	163.52	6.115	3357.5	3030.5	7.2845
460	165.97	6.025	3379.5	3047.6	7.3148
470	168.40	5.938	3401.5	3064.7	7.3446
480	170.83	5.854	3423.6	3081.9	7.3740
490	173.25	5.772	3445.6	3099.1	7.4031
500	175.67	5.693	3467.7	3116.3	7.4318
520	180.49	5.541	3511.9	3150.9	7.4882
540	185.29	5.397	3556.2	3185.6	7.5435
560	190.07	5.261	3600.7	3220.6	7.5975
580	194.84	5.132	3645.4	3255.7	7.6505
600	199.60	5.010	3690.2	3291.0	7.7024
620	204.35	4.8937	3735.3	3326.6	7.7535
640	209.08	4.7828	3780.5	3362.4	7.8036
660	213.81	4.6770	3826.0	3398.4	7.8528
680	218.53	4.5760	3871.7	3434.6	7.9013
700	223.25	4.4794	3917.6	3471.1	7.9490
720	227.95	4.3869	3963.8	3507.9	7.9959
740	232.65	4.2982	4010.2	3544.9	8.0422
760	237.35	4.2132	4056.9	3582.2	8.0878
780	242.04	4.1316	4103.8	3619.7	8.1328
800	246.73	4.0531	4150.9	3657.5	8.1771
850	258.42	3.8696	4269.9	3753.1	8.2855
900	270.10	3.7023	4390.5	3850.3	8.3905
950	281.76	3.5491	4512.7	3949.1	8.4925
1000	293.41	3.4082	4636.4	4049.5	8.5916
1100	316.67	3.1579	4888.4	4255.0	8.7821
1200	339.89	2.9421	5146.2	4466.4	8.9633
1300	363.09	2.7542	5409.5	4683.3	9.1363
1400	386.27	2.5889	5678.0	4905.4	9.3017
1500	409.43	2.4424	5951.1	5132.3	9.4603
1600	432.58	2.3117	6228.7	5363.5	9.6125
1700	455.72	2.1943	6510.2	5598.8	9.7589
1800	478.85	2.0883	6795.4	5837.7	9.8999
1900	501.97	1.9921	7084.1	6080.2	10.0359
2000	525.10	1.9044	7376.0	6325.8	10.1672

Table 3 Compressed water and superheated steam (*Continued*)

t (°C)	50 bar (t_s = 263.977 °C) v (× 10^3)	ρ	h	u	s
t_f	1.28614	777.52	1154.20	1147.77	2.92011
t_g	39.440	25.355	2793.7	2596.5	5.9725
0	0.99769	1002.31	5.05	0.06	0.00020
5	0.99762	1002.39	25.99	21.00	0.07616
10	0.99794	1002.07	46.85	41.86	0.15050
15	0.99859	1001.41	67.69	62.70	0.22345
20	0.99954	1000.46	88.52	83.53	0.29514
25	1.00074	999.26	109.37	104.36	0.36564
30	1.00218	997.83	130.22	125.21	0.43499
35	1.00383	996.18	151.07	146.05	0.50321
40	1.00568	994.35	171.92	166.89	0.57034
45	1.00772	992.34	192.77	187.73	0.63640
50	1.00994	990.16	213.63	208.58	0.70144
55	1.01233	987.82	234.48	229.42	0.76547
60	1.01489	985.33	255.34	250.26	0.82855
65	1.01760	982.70	276.20	271.11	0.89071
70	1.02048	979.93	297.07	291.97	0.95199
75	1.02351	977.03	317.96	312.85	1.01243
80	1.02669	974.00	338.87	333.74	1.07205
85	1.03002	970.85	359.80	354.65	1.13090
90	1.03351	967.58	380.75	375.59	1.18900
95	1.03714	964.19	401.73	396.55	1.24639
100	1.04093	960.68	422.75	417.54	1.30308
105	1.04487	957.06	443.79	438.57	1.35911
110	1.04895	953.33	464.88	459.63	1.41449
115	1.05320	949.49	486.00	480.73	1.46926
120	1.05759	945.54	507.16	501.87	1.52343
125	1.06215	941.49	528.36	523.05	1.57703
130	1.06686	937.33	549.61	544.28	1.63007
135	1.07174	933.06	570.91	565.55	1.68257
140	1.07679	928.68	592.26	586.88	1.73456
145	1.08202	924.20	613.66	608.25	1.78605
150	1.08742	919.61	635.12	629.68	1.83706
155	1.09300	914.91	656.64	651.17	1.88761
160	1.09878	910.10	678.22	672.73	1.93773
165	1.10475	905.18	699.87	694.35	1.98743
170	1.11093	900.14	721.59	716.04	2.03673
175	1.11733	894.99	743.40	737.81	2.08565
180	1.12395	889.72	765.28	759.66	2.13421
185	1.13081	884.32	787.26	781.60	2.18244
190	1.13792	878.80	809.33	803.64	2.23035
195	1.14529	873.14	831.50	825.78	2.27798
200	1.15293	867.35	853.79	848.03	2.32533
205	1.16088	861.42	876.20	870.40	2.37245
210	1.16913	855.34	898.74	892.89	2.41934
215	1.17772	849.10	921.42	915.53	2.46604
220	1.18666	842.70	944.25	938.32	2.51258
225	1.19599	836.13	967.25	961.27	2.55898
230	1.20573	829.37	990.43	984.40	2.60528
235	1.21592	822.42	1013.81	1007.73	2.65151
240	1.22659	815.27	1037.40	1031.26	2.69770
245	1.23780	807.89	1061.22	1055.03	2.74390
250	1.24958	800.27	1085.30	1079.05	2.79014
255	1.26200	792.39	1109.66	1103.35	2.83649
260	1.27513	784.23	1134.33	1127.95	2.88298
265	39.631	25.233	2798.0	2599.9	5.9805
270	40.533	24.671	2818.2	2615.6	6.0179
275	41.399	24.155	2837.5	2630.5	6.0532
280	42.230	23.680	2855.9	2644.8	6.0867
285	43.033	23.238	2873.7	2658.5	6.1186
290	43.812	22.825	2890.8	2671.7	6.1491
295	44.567	22.438	2907.4	2684.5	6.1785

t (°C)	50 bar v (× 10^3)	ρ	h	u	s
300	45.304	22.073	2923.5	2697.0	6.2067
305	46.021	21.729	2939.2	2709.1	6.2340
310	46.725	21.402	2954.5	2720.9	6.2604
315	47.414	21.091	2969.5	2732.5	6.2860
320	48.091	20.794	2984.3	2743.8	6.3109
325	48.754	20.511	2998.7	2754.9	6.3352
330	49.410	20.239	3012.9	2765.9	6.3588
335	50.053	19.979	3026.9	2776.6	6.3819
340	50.689	19.728	3040.7	2787.2	6.4045
345	51.316	19.487	3054.3	2797.7	6.4266
350	51.93	19.255	3067.7	2808.0	6.4482
355	52.55	19.031	3081.0	2818.2	6.4695
360	53.15	18.814	3094.1	2828.4	6.4903
365	53.75	18.604	3107.2	2838.4	6.5108
370	54.35	18.401	3120.1	2848.3	6.5309
375	54.93	18.203	3132.9	2858.2	6.5507
380	55.52	18.012	3145.5	2868.0	6.5703
385	56.10	17.826	3158.2	2877.7	6.5895
390	56.67	17.646	3170.7	2887.3	6.6084
395	57.24	17.470	3183.1	2896.9	6.6271
400	57.81	17.299	3195.5	2906.5	6.6456
410	58.93	16.969	3220.1	2925.4	6.6818
420	60.04	16.656	3244.4	2944.2	6.7172
430	61.14	16.357	3268.5	2962.8	6.7517
440	62.22	16.072	3292.5	2981.4	6.7856
450	63.30	15.798	3316.3	2999.8	6.8187
460	64.37	15.536	3340.0	3018.2	6.8513
470	65.42	15.285	3363.6	3036.5	6.8833
480	66.48	15.043	3387.1	3054.7	6.9147
490	67.52	14.810	3410.5	3072.9	6.9456
500	68.56	14.586	3433.9	3091.1	6.9760
520	70.62	14.160	3480.5	3127.4	7.0355
540	72.66	13.762	3527.0	3163.7	7.0934
560	74.68	13.390	3573.4	3200.0	7.1498
580	76.69	13.039	3619.8	3236.4	7.2048
600	78.69	12.709	3666.2	3272.8	7.2586
620	80.67	12.396	3712.7	3309.4	7.3112
640	82.64	12.101	3759.3	3346.1	7.3628
660	84.60	11.820	3805.9	3382.9	7.4133
680	86.56	11.553	3852.7	3419.9	7.4630
700	88.50	11.299	3899.7	3457.1	7.5117
720	90.44	11.056	3946.8	3494.5	7.5596
740	92.38	10.825	3994.0	3532.1	7.6067
760	94.31	10.604	4041.5	3570.0	7.6531
780	96.23	10.392	4089.1	3608.0	7.6988
800	98.15	10.189	4137.0	3646.3	7.7438
850	102.93	9.716	4257.5	3742.9	7.8536
900	107.68	9.287	4379.4	3841.0	7.9598
950	112.42	8.895	4502.7	3940.6	8.0627
1000	117.15	8.536	4627.4	4041.7	8.1626
1100	126.56	7.901	4881.1	4248.3	8.3543
1200	135.94	7.356	5140.2	4460.5	8.5365
1300	145.29	6.883	5404.5	4678.1	8.7101
1400	154.62	6.468	5673.8	4900.8	8.8760
1500	163.93	6.100	5947.7	5128.1	9.0350
1600	173.24	5.7725	6225.9	5359.7	9.1876
1700	182.53	5.4786	6507.9	5595.3	9.3343
1800	191.81	5.2134	6793.6	5834.5	9.4755
1900	201.09	4.9728	7082.6	6077.2	9.6117
2000	210.37	4.7536	7374.8	6323.0	9.7431

Table 3 Compressed water and superheated steam (*Continued*)

t (°C)	100 bar (t_s = 311.031 °C) v (× 10^3)	ρ	h	u	s
t_l	1.45216	688.63	1407.28	1392.75	3.35912
t_g	18.025	55.48	2724.5	2544.3	5.6139
0	0.99521	1004.81	10.10	0.15	0.00045
5	0.99522	1004.80	30.92	20.97	0.07599
10	0.99561	1004.41	51.69	41.73	0.14998
15	0.99631	1003.70	72.44	62.47	0.22262
20	0.99730	1002.71	93.20	83.22	0.29405
25	0.99853	1001.48	113.97	103.98	0.36431
30	0.99998	1000.02	134.75	124.75	0.43344
35	1.00165	998.36	155.54	145.52	0.50146
40	1.00350	996.51	176.33	166.30	0.56839
45	1.00554	994.49	197.13	187.07	0.63428
50	1.00775	992.31	217.93	207.85	0.69914
55	1.01013	989.97	238.73	228.62	0.76301
60	1.01267	987.48	259.53	249.40	0.82592
65	1.01537	984.86	280.34	270.19	0.88793
70	1.01822	982.10	301.16	290.98	0.94905
75	1.02122	979.22	322.00	311.79	1.00934
80	1.02437	976.21	342.85	332.61	1.06881
85	1.02767	973.07	363.73	353.45	1.12751
90	1.03112	969.82	384.63	374.32	1.18546
95	1.03471	966.46	405.56	395.21	1.24270
100	1.03844	962.98	426.52	416.13	1.29924
105	1.04233	959.39	447.51	437.08	1.35512
110	1.04636	955.70	468.53	458.07	1.41036
115	1.05053	951.90	489.59	479.09	1.46498
120	1.05486	947.99	510.70	500.15	1.51899
125	1.05935	943.98	531.84	521.24	1.57243
130	1.06398	939.87	553.02	542.38	1.62531
135	1.06878	935.65	574.26	563.57	1.67765
140	1.07374	931.33	595.53	584.80	1.72947
145	1.07886	926.90	616.86	606.08	1.78079
150	1.08416	922.38	638.25	627.41	1.83162
155	1.08963	917.74	659.69	648.79	1.88199
160	1.09528	913.00	681.19	670.24	1.93192
165	1.10113	908.16	702.75	691.74	1.98142
170	1.10717	903.20	724.39	713.31	2.03051
175	1.11342	898.14	746.09	734.96	2.07922
180	1.11988	892.96	767.88	756.68	2.12756
185	1.12656	887.66	789.74	778.48	2.17555
190	1.13348	882.24	811.70	800.37	2.22322
195	1.14064	876.70	833.76	822.35	2.27058
200	1.14806	871.03	855.91	844.43	2.31766
205	1.15576	865.23	878.18	866.63	2.36448
210	1.16375	859.29	900.57	888.94	2.41106
215	1.17205	853.21	923.09	911.37	2.45743
220	1.18068	846.97	945.75	933.95	2.50361
225	1.18966	840.58	968.56	956.67	2.54964
230	1.19902	834.02	991.54	979.55	2.59553
235	1.20878	827.28	1014.69	1002.60	2.64131
240	1.21898	820.36	1038.03	1025.84	2.68702
245	1.22966	813.23	1061.58	1049.29	2.73270
250	1.24085	805.90	1085.36	1072.96	2.77837
255	1.25261	798.34	1109.39	1096.87	2.82408
260	1.26498	790.52	1133.69	1121.04	2.86988
265	1.27804	782.45	1158.29	1145.51	2.91580
270	1.29187	774.07	1183.22	1170.31	2.96192
275	1.30654	765.38	1208.52	1195.46	3.00828
280	1.32217	756.33	1234.23	1221.01	3.05497
285	1.33889	746.89	1260.40	1247.01	3.10206
290	1.35687	736.99	1287.09	1273.52	3.14967
295	1.37630	726.58	1314.39	1300.62	3.19792

t (°C)	100 bar v (× 10^3)	ρ	h	u	s
300	1.39746	715.58	1342.38	1328.40	3.24697
305	1.42070	703.88	1371.19	1356.98	3.29702
310	1.44648	691.34	1400.99	1386.53	3.34835
315	18.592	53.786	2750.6	2564.7	5.6585
320	19.248	51.952	2780.6	2588.2	5.7093
325	19.855	50.366	2808.1	2609.6	5.7555
330	20.421	48.969	2833.6	2629.4	5.7979
335	20.956	47.719	2857.5	2648.0	5.8374
340	21.464	46.590	2880.1	2665.5	5.8745
345	21.950	45.559	2901.6	2682.2	5.9094
350	22.416	44.612	2922.2	2698.1	5.9425
355	22.865	43.734	2942.0	2713.3	5.9741
360	23.300	42.918	2961.0	2728.0	6.0043
365	23.722	42.155	2979.4	2742.2	6.0333
370	24.132	41.438	2997.3	2756.0	6.0612
375	24.532	40.763	3014.7	2769.4	6.0882
380	24.923	40.124	3031.7	2782.5	6.1143
385	25.305	39.518	3048.3	2795.2	6.1395
390	25.679	38.942	3064.5	2807.7	6.1641
395	26.047	38.392	3080.5	2820.0	6.1880
400	26.408	37.867	3096.1	2832.0	6.2114
410	27.113	36.883	3126.6	2855.5	6.2563
420	27.797	35.976	3156.2	2878.3	6.2994
430	28.463	35.133	3185.1	2900.5	6.3408
440	29.114	34.347	3213.4	2922.3	6.3807
450	29.752	33.611	3241.1	2943.6	6.4194
460	30.378	32.919	3268.4	2964.6	6.4568
470	30.993	32.266	3295.3	2985.4	6.4932
480	31.598	31.647	3321.8	3005.8	6.5287
490	32.195	31.061	3348.0	3026.1	6.5633
500	32.784	30.503	3374.0	3046.2	6.5971
520	33.940	29.463	3425.3	3085.9	6.6625
540	35.073	28.512	3475.8	3125.1	6.7255
560	36.186	27.635	3525.8	3164.0	6.7862
580	37.281	26.824	3575.4	3202.6	6.8451
600	38.361	26.068	3624.7	3241.1	6.9022
620	39.427	25.363	3673.8	3279.5	6.9577
640	40.482	24.702	3722.7	3317.9	7.0119
660	41.527	24.081	3771.5	3356.2	7.0648
680	42.562	23.495	3820.3	3394.6	7.1165
700	43.590	22.941	3869.0	3433.1	7.1671
720	44.610	22.417	3917.7	3471.6	7.2167
740	45.623	21.919	3966.5	3510.3	7.2653
760	46.631	21.445	4015.4	3549.1	7.3131
780	47.633	20.994	4064.4	3588.1	7.3600
800	48.630	20.564	4113.5	3627.2	7.4062
850	51.10	19.568	4236.7	3725.7	7.5184
900	53.55	18.673	4360.9	3825.3	7.6266
950	55.99	17.861	4486.1	3926.3	7.7311
1000	58.40	17.122	4612.5	4028.5	7.8324
1100	63.20	15.822	4869.0	4236.9	8.0263
1200	67.96	14.714	5130.3	4450.6	8.2100
1300	72.70	13.755	5396.4	4669.4	8.3847
1400	77.41	12.918	5667.1	4893.0	8.5516
1500	82.11	12.178	5942.2	5121.1	8.7112
1600	86.80	11.521	6221.3	5353.3	8.8644
1700	91.48	10.932	6504.2	5589.5	9.0115
1800	96.15	10.401	6790.6	5829.2	9.1531
1900	100.81	9.920	7080.3	6072.3	9.2895
2000	105.46	9.482	7373.1	6318.4	9.4212

Table 3 Compressed water and superheated steam (*Continued*)

t (°C)	200 bar (t_s = 365.800 °C) v (× 10³)	ρ	h	u	s
t_f	2.0360	491.2	1826.7	1786.0	4.0146
t_g	5.874	170.25	2413.6	2296.1	4.9330
0	0.99037	1009.73	20.08	0.28	0.00066
5	0.99054	1009.55	40.69	20.88	0.07543
10	0.99105	1009.03	61.27	41.45	0.14876
15	0.99185	1008.21	81.86	62.03	0.22084
20	0.99291	1007.14	102.48	82.62	0.29176
25	0.99420	1005.84	123.11	103.23	0.36157
30	0.99569	1004.33	143.77	123.86	0.43028
35	0.99738	1002.63	164.44	144.50	0.49792
40	0.99924	1000.76	185.13	165.14	0.56449
45	1.00128	998.72	205.81	185.79	0.63003
50	1.00349	996.53	226.50	206.43	0.69456
55	1.00585	994.19	247.20	227.08	0.75811
60	1.00836	991.71	267.90	247.73	0.82072
65	1.01102	989.10	288.61	268.39	0.88242
70	1.01383	986.36	309.33	289.05	0.94325
75	1.01677	983.50	330.07	309.73	1.00324
80	1.01986	980.52	350.82	330.42	1.06243
85	1.02309	977.43	371.59	351.13	1.12084
90	1.02646	974.22	392.39	371.86	1.17851
95	1.02997	970.90	413.22	392.62	1.23546
100	1.03361	967.48	434.07	413.40	1.29172
105	1.03740	963.95	454.95	434.20	1.34732
110	1.04132	960.32	475.87	455.04	1.40227
115	1.04538	956.59	496.82	475.91	1.45659
120	1.04958	952.76	517.81	496.81	1.51032
125	1.05393	948.83	538.83	517.75	1.56346
130	1.05842	944.81	559.90	538.73	1.61603
135	1.06305	940.69	581.00	559.74	1.66806
140	1.06784	936.47	602.15	580.79	1.71957
145	1.07279	932.15	623.35	601.89	1.77056
150	1.0779	927.7	644.6	623.0	1.8211
155	1.0832	923.2	665.9	644.2	1.8711
160	1.0886	918.6	687.2	665.5	1.9207
165	1.0942	913.9	708.6	686.7	1.9698
170	1.1000	909.1	730.1	708.1	2.0185
175	1.1059	904.2	751.6	729.5	2.0668
180	1.1121	899.2	773.2	751.0	2.1147
185	1.1185	894.1	794.9	772.5	2.1623
190	1.1251	888.8	816.6	794.1	2.2095
195	1.1318	883.5	838.5	815.8	2.2564
200	1.1389	878.1	860.4	837.6	2.3030
205	1.1461	872.5	882.4	859.5	2.3493
210	1.1537	866.8	904.5	881.5	2.3953
215	1.1615	861.0	926.8	903.5	2.4411
220	1.1695	855.0	949.1	925.7	2.4866
225	1.1779	849.0	971.6	948.0	2.5320
230	1.1866	842.7	994.2	970.5	2.5771
235	1.1956	836.4	1017.0	993.1	2.6221
240	1.2051	829.8	1039.9	1015.8	2.6670
245	1.2148	823.1	1063.0	1038.7	2.7118
250	1.2251	816.3	1086.3	1061.8	2.7565
255	1.2357	809.2	1109.7	1085.0	2.8012
260	1.2469	802.0	1133.4	1108.5	2.8458
265	1.2586	794.5	1157.3	1132.2	2.8905
270	1.2709	786.9	1181.5	1156.1	2.9352
275	1.2838	779.0	1206.0	1180.3	2.9800
280	1.2974	770.8	1230.7	1204.8	3.0250
285	1.3118	762.3	1255.8	1229.6	3.0701
290	1.3270	753.6	1281.3	1254.7	3.1155
295	1.3432	744.5	1307.1	1280.3	3.1612

t (°C)	200 bar v (× 10³)	ρ	h	u	s
300	1.3605	735.0	1333.4	1306.2	3.2073
305	1.3790	725.1	1360.3	1332.7	3.2539
310	1.3990	714.8	1387.7	1359.7	3.3011
315	1.4206	703.9	1415.7	1387.3	3.3490
320	1.4442	692.4	1444.5	1415.6	3.3978
325	1.4702	680.2	1474.2	1444.8	3.4476
330	1.4990	667.1	1505.0	1475.0	3.4988
335	1.5314	653.0	1537.0	1506.4	3.5518
340	1.5685	637.5	1570.7	1539.4	3.6070
345	1.6120	620.4	1606.6	1574.3	3.6652
350	1.6645	600.8	1645.4	1612.1	3.7277
355	1.7314	577.6	1688.6	1654.0	3.7968
360	1.8248	548.0	1739.7	1703.2	3.8778
365	1.9900	502.5	1810.4	1770.6	3.9890
370	6.905	144.82	2523.8	2385.7	5.1050
375	7.668	130.42	2601.0	2447.6	5.2246
380	8.256	121.13	2658.6	2493.4	5.3131
385	8.752	114.27	2706.0	2531.0	5.3855
390	9.188	108.83	2747.0	2563.3	5.4476
395	9.583	104.35	2783.6	2592.0	5.5025
400	9.946	100.54	2816.9	2617.9	5.5521
405	10.284	97.24	2847.5	2641.8	5.5975
410	10.602	94.32	2876.1	2664.1	5.6395
415	10.903	91.72	2903.0	2684.9	5.6787
420	11.190	89.37	2928.5	2704.7	5.7156
425	11.464	87.23	2952.7	2723.4	5.7505
430	11.728	85.26	2975.9	2741.4	5.7836
435	11.983	83.45	2998.3	2758.6	5.8152
440	12.230	81.77	3019.8	2775.2	5.8455
445	12.469	80.20	3040.6	2791.2	5.8746
450	12.701	78.73	3060.8	2806.8	5.9026
460	13.149	76.05	3099.6	2836.6	5.9559
470	13.577	73.65	3136.6	2865.0	6.0060
480	13.988	71.49	3172.0	2892.3	6.0534
490	14.384	69.52	3206.3	2918.6	6.0986
500	14.769	67.71	3239.4	2944.1	6.1417
520	15.506	64.49	3303.2	2993.1	6.2232
540	16.208	61.70	3364.2	3040.1	6.2992
560	16.883	59.23	3423.2	3085.5	6.3708
580	17.536	57.03	3480.6	3129.8	6.4389
600	18.169	55.04	3536.7	3173.3	6.5039
620	18.786	53.23	3591.7	3216.0	6.5663
640	19.390	51.57	3646.0	3258.2	6.6264
660	19.981	50.05	3699.7	3300.1	6.6845
680	20.562	48.63	3752.8	3341.6	6.7408
700	21.133	47.32	3805.5	3382.8	6.7955
720	21.696	46.09	3857.9	3424.0	6.8488
740	22.252	44.94	3910.0	3465.0	6.9008
760	22.802	43.86	3961.9	3505.9	6.9515
780	23.345	42.84	4013.7	3546.8	7.0012
800	23.883	41.87	4065.4	3587.8	7.0498
850	25.207	39.67	4194.4	3690.3	7.1673
900	26.508	37.72	4323.5	3793.3	7.2797
950	27.785	35.99	4452.9	3897.1	7.3877
1000	29.053	34.42	4582.8	4001.8	7.4919
1200	34.00	29.41	5111.	4431.	7.877
1400	38.83	25.75	5654.	4878.	8.223
1600	43.59	22.94	6213.	5341.	8.538
1800	48.32	20.69	6785.	5819.	8.828
2000	53.02	18.86	7370.	6310.	9.098

Table 3 Compressed water and superheated steam (*Continued*)

t	500 bar				
(°C)	$v\ (\times 10^3)$	ρ	h	u	s
0	0.97674	1023.82	49.20	0.36	-0.00076
5	0.97734	1023.19	69.30	20.43	0.07216
10	0.97818	1022.31	89.43	40.52	0.14388
15	0.97924	1021.20	109.61	60.65	0.21454
20	0.98049	1019.90	129.85	80.83	0.28419
25	0.98192	1018.41	150.15	101.05	0.35283
30	0.98352	1016.75	170.48	121.31	0.42046
35	0.98528	1014.94	190.84	141.58	0.48709
40	0.98718	1012.98	211.23	161.87	0.55270
45	0.98923	1010.89	231.62	182.16	0.61732
50	0.99141	1008.66	252.03	202.46	0.68097
55	0.99373	1006.31	272.45	222.76	0.74367
60	0.99617	1003.84	292.88	243.07	0.80544
65	0.99874	1001.26	313.31	263.38	0.86633
70	1.00143	998.57	333.76	283.69	0.92636
75	1.00425	995.77	354.23	304.01	0.98557
80	1.00719	992.86	374.71	324.35	1.04398
85	1.01024	989.86	395.21	344.69	1.10162
90	1.01342	986.76	415.73	365.06	1.15852
95	1.01671	983.56	436.27	385.44	1.21471
100	1.02012	980.27	456.84	405.84	1.27021
105	1.02365	976.90	477.44	426.26	1.32504
110	1.02729	973.43	498.06	446.70	1.37922
115	1.03106	969.88	518.72	467.16	1.43277
120	1.03494	966.24	539.40	487.65	1.48571
125	1.03894	962.52	560.11	508.16	1.53807
130	1.04306	958.72	580.85	528.70	1.58984
135	1.04730	954.83	601.63	549.27	1.64106
140	1.05167	950.87	622.44	569.86	1.69173
145	1.05617	946.82	643.28	590.48	1.74188
150	1.0608	942.7	664.2	611.1	1.7915
160	1.0704	934.2	706.0	652.5	1.8893
170	1.0806	925.4	748.1	694.0	1.9853
180	1.0914	916.3	790.3	735.7	2.0795
190	1.1028	906.8	832.7	777.6	2.1720
200	1.1148	897.0	875.3	819.6	2.2631
210	1.1276	886.9	918.2	861.9	2.3529
220	1.1411	876.4	961.4	904.4	2.4413
230	1.1554	865.5	1005.0	947.2	2.5287
240	1.1706	854.2	1048.9	990.3	2.6151
250	1.1869	842.5	1093.2	1033.8	2.7006
260	1.2042	830.4	1137.9	1077.7	2.7854
270	1.2228	817.8	1183.3	1122.1	2.8696
280	1.2428	804.6	1229.2	1167.0	2.9534
290	1.2643	790.9	1275.8	1212.5	3.0368
300	1.2876	776.6	1323.1	1258.7	3.1202
310	1.3129	761.7	1371.3	1305.6	3.2035
320	1.3406	745.9	1420.5	1353.4	3.2871
330	1.3710	729.4	1470.7	1402.2	3.3712
340	1.4046	711.9	1522.3	1452.1	3.4559
350	1.4422	693.4	1575.3	1503.2	3.5417
355	1.4627	683.7	1602.5	1529.3	3.5851
360	1.4845	673.6	1630.1	1555.8	3.6289
365	1.5077	663.3	1658.2	1582.8	3.6731
370	1.5326	652.5	1686.9	1610.2	3.7179
375	1.5593	641.3	1716.1	1638.2	3.7632
380	1.5880	629.7	1746.1	1666.7	3.8093
385	1.6191	617.6	1776.8	1695.8	3.8561
390	1.6529	605.0	1808.3	1725.6	3.9037
395	1.6898	591.8	1840.7	1756.2	3.9524
400	1.7301	578.0	1874.1	1787.6	4.0022
405	1.7746	563.5	1908.6	1819.9	4.0533
410	1.8237	548.3	1944.4	1853.2	4.1058
415	1.8783	532.4	1981.5	1887.5	4.1600
420	1.9392	515.7	2020.0	1923.1	4.2158
425	2.0074	498.2	2060.2	1959.8	4.2736
430	2.0838	479.9	2102.0	1997.8	4.3332
435	2.1692	461.0	2145.5	2037.1	4.3949
440	2.2646	441.6	2190.6	2077.4	4.4584
445	2.3701	421.9	2237.1	2118.6	4.5233
450	2.4858	402.3	2284.7	2160.4	4.5894
455	2.6109	383.0	2332.9	2202.4	4.6558
460	2.7440	364.4	2381.2	2244.0	4.7219
465	2.8835	346.8	2429.1	2284.9	4.7870
470	3.0273	330.3	2476.0	2324.7	4.8503
475	3.1734	315.1	2521.7	2363.0	4.9115
480	3.320	301.2	2565.7	2399.7	4.9702
485	3.466	288.5	2608.0	2434.7	5.0262
490	3.611	276.9	2648.5	2467.9	5.0794
495	3.753	266.5	2687.2	2499.6	5.1300
500	3.892	257.0	2724.2	2529.6	5.1780
505	4.028	248.28	2759.6	2558.3	5.2237
510	4.160	240.37	2793.5	2585.5	5.2671
515	4.290	233.12	2826.1	2611.6	5.3085
520	4.416	226.47	2857.3	2636.5	5.3480
525	4.539	220.33	2887.4	2660.4	5.3858
530	4.659	214.66	2916.4	2683.4	5.4220
535	4.776	209.40	2944.4	2705.6	5.4568
540	4.890	204.50	2971.5	2727.0	5.4902
545	5.002	199.93	2997.7	2747.7	5.5224
550	5.111	195.65	3023.3	2767.7	5.5535
560	5.323	187.85	3072.3	2806.1	5.6127
570	5.527	180.92	3118.9	2842.5	5.6684
580	5.724	174.71	3163.5	2877.3	5.7209
590	5.914	169.09	3206.3	2910.6	5.7709
600	6.098	163.99	3247.7	2942.7	5.8184
620	6.451	155.01	3326.5	3003.9	5.9077
640	6.787	147.35	3401.1	3061.8	5.9903
660	7.107	140.70	3472.5	3117.1	6.0677
680	7.416	134.85	3541.2	3170.4	6.1406
700	7.713	129.64	3607.8	3222.1	6.2097
720	8.002	124.97	3672.6	3272.6	6.2757
740	8.282	120.74	3736.0	3321.9	6.3388
760	8.555	116.89	3798.2	3370.4	6.3996
780	8.822	113.35	3859.3	3418.1	6.4582
800	9.083	110.09	3919.5	3465.3	6.5148
850	9.715	102.93	4067.1	3581.3	6.6492
900	10.322	96.88	4211.5	3695.4	6.7751
950	10.908	91.67	4353.9	3808.5	6.8940
1000	11.479	87.12	4495.0	3921.1	7.0070
1200	13.65	73.28	5055.	4372.	7.415
1400	15.70	63.69	5618.	4833.	7.774
1600	17.69	56.52	6189.	5305.	8.096
1800	19.64	50.91	6771.	5789.	8.391
2000	21.57	46.36	7363.	6284.	8.664

Table 3 Compressed water and superheated steam (*Continued*)

t (°C)	1000 bar $v(\times 10^3)$	ρ	h	u	s
0	0.95666	1045.31	95.40	−0.27	−0.00854
5	0.95777	1044.09	114.96	19.18	0.06243
10	0.95902	1042.73	134.59	38.69	0.13238
15	0.96041	1041.23	154.31	58.27	0.20141
20	0.96191	1039.59	174.11	77.92	0.26954
25	0.96354	1037.84	193.98	97.63	0.33676
30	0.96529	1035.96	213.91	117.38	0.40304
35	0.96714	1033.98	233.88	137.17	0.46838
40	0.96910	1031.89	253.88	156.97	0.53276
45	0.97116	1029.69	273.90	176.78	0.59618
50	0.97333	1027.40	293.93	196.60	0.65866
55	0.97559	1025.02	313.98	216.42	0.72021
60	0.97796	1022.54	334.03	236.24	0.78087
65	0.98042	1019.97	354.10	256.06	0.84066
70	0.98298	1017.32	374.18	275.88	0.89960
75	0.98563	1014.58	394.27	295.71	0.95773
80	0.98838	1011.76	414.38	315.54	1.01508
85	0.99122	1008.86	434.51	335.38	1.07167
90	0.99415	1005.88	454.65	355.23	1.12752
95	0.99718	1002.83	474.81	375.10	1.18267
100	1.00030	999.70	495.00	394.97	1.23713
105	1.00350	996.51	515.21	414.86	1.29092
110	1.00681	993.24	535.43	434.75	1.34407
115	1.01020	989.91	555.69	454.67	1.39658
120	1.01368	986.50	575.96	474.59	1.44848
125	1.01726	983.04	596.26	494.53	1.49978
130	1.02092	979.50	616.58	514.48	1.55050
135	1.02469	975.91	636.92	534.45	1.60064
140	1.02854	972.25	657.28	554.43	1.65023
145	1.03249	968.53	677.67	574.42	1.69928
150	1.0365	964.8	698.1	594.4	1.7478
160	1.0449	957.0	739.0	634.5	1.8433
170	1.0537	949.0	780.0	674.6	1.9369
180	1.0629	940.8	821.1	714.8	2.0286
190	1.0725	932.4	862.3	755.0	2.1185
200	1.0826	923.7	903.6	795.3	2.2068
210	1.0931	914.8	945.1	835.8	2.2936
220	1.1042	905.6	986.7	876.3	2.3789
230	1.1158	896.2	1028.5	917.0	2.4628
240	1.1279	886.6	1070.6	957.8	2.5455
250	1.1406	876.7	1112.8	998.7	2.6270
260	1.1540	866.6	1155.3	1039.9	2.7075
270	1.1680	856.1	1198.0	1081.2	2.7870
280	1.1828	845.4	1241.1	1122.8	2.8655
290	1.1984	834.5	1284.5	1164.6	2.9432
300	1.2148	823.2	1328.2	1206.7	3.0202
310	1.2321	811.6	1372.3	1249.1	3.0965
320	1.2504	799.8	1416.8	1291.8	3.1722
330	1.2697	787.6	1461.8	1334.8	3.2473
340	1.2902	775.1	1507.2	1378.2	3.3220
350	1.3120	762.2	1553.1	1421.9	3.3962
360	1.3351	749.0	1599.5	1466.0	3.4701
370	1.3597	735.4	1646.4	1510.5	3.5437
380	1.3859	721.5	1694.0	1555.4	3.6170
390	1.4139	707.2	1742.1	1600.7	3.6902
400	1.4439	692.6	1790.9	1646.5	3.7632

t (°C)	1000 bar $v(\times 10^3)$	ρ	h	u	s
410	1.4759	677.5	1840.4	1692.8	3.8361
420	1.5103	662.1	1890.5	1739.5	3.9089
430	1.5471	646.4	1941.3	1786.6	3.9817
440	1.5867	630.2	1992.8	1834.2	4.0545
450	1.6292	613.8	2045.1	1882.2	4.1273
460	1.6749	597.1	2098.1	1930.6	4.2000
470	1.7239	580.1	2151.7	1979.3	4.2727
480	1.7766	562.9	2206.0	2028.3	4.3453
490	1.8329	545.6	2260.9	2077.6	4.4176
500	1.8932	528.2	2316.2	2126.9	4.4897
510	1.9573	510.9	2372.0	2176.2	4.5613
520	2.0253	493.8	2427.9	2225.4	4.6324
530	2.0969	476.9	2484.0	2274.3	4.7026
540	2.1721	460.4	2539.9	2322.7	4.7718
550	2.2504	444.4	2595.5	2370.5	4.8398
560	2.3315	428.9	2650.7	2417.5	4.9064
570	2.4148	414.1	2705.2	2463.7	4.9714
580	2.5001	400.0	2758.8	2508.8	5.0347
590	2.5867	386.6	2811.6	2552.9	5.0961
600	2.6743	373.9	2863.4	2595.9	5.1558
610	2.762	362.0	2914.0	2637.8	5.2135
620	2.851	350.8	2963.6	2678.5	5.2693
630	2.939	340.2	3012.1	2718.2	5.3233
640	3.027	330.3	3059.5	2756.8	5.3755
650	3.115	321.0	3105.8	2794.4	5.4259
660	3.202	312.3	3151.1	2831.0	5.4748
670	3.288	304.1	3195.5	2866.7	5.5220
680	3.373	296.4	3238.9	2901.5	5.5678
690	3.458	289.2	3281.4	2935.6	5.6121
700	3.542	282.4	3323.1	2968.9	5.6552
710	3.624	275.9	3364.0	3001.5	5.6970
720	3.706	269.8	3404.1	3033.5	5.7377
730	3.787	264.1	3443.6	3064.9	5.7772
740	3.867	258.6	3482.5	3095.8	5.8158
750	3.946	253.4	3520.7	3126.1	5.8534
760	4.024	248.50	3558.4	3156.0	5.8900
770	4.101	243.81	3595.6	3185.5	5.9258
780	4.178	239.35	3632.3	3214.5	5.9608
790	4.254	235.09	3668.5	3243.2	5.9951
800	4.328	231.03	3704.3	3271.5	6.0286
820	4.476	223.42	3774.8	3327.2	6.0937
840	4.621	216.42	3843.9	3381.8	6.1563
860	4.762	209.97	3911.7	3435.5	6.2167
880	4.902	204.00	3978.5	3488.3	6.2751
900	5.039	198.45	4044.3	3540.4	6.3317
920	5.174	193.28	4109.3	3591.9	6.3866
940	5.307	188.44	4173.5	3642.8	6.4400
960	5.438	183.90	4237.1	3693.3	6.4920
980	5.567	179.63	4300.1	3743.4	6.5427
1000	5.694	175.61	4362.6	3793.2	6.5921
1200	6.896	145.01	4970.	4280.	7.035
1400	8.007	124.89	5563.	4762.	7.413
1600	9.064	110.33	6155.	5248.	7.747
1800	10.085	99.16	6751.	5743.	8.049
2000	11.083	90.23	7354.	6245.	8.327

Table 4 Specific heat capacity at constant pressure

t (°C)	*P* (bar) 0	1	5	10	20	50	100	200	500	1000
0	1.859	4.228	4.226	4.223	4.218	4.202	4.177	4.130	4.021	3.909
20	1.863	4.183	4.182	4.180	4.177	4.168	4.153	4.125	4.054	3.968
40	1.868	4.182	4.181	4.180	4.178	4.170	4.159	4.137	4.078	4.002
60	1.875	4.183	4.182	4.181	4.178	4.172	4.161	4.141	4.086	4.012
80	1.882	4.194	4.193	4.192	4.190	4.183	4.173	4.153	4.098	4.023
100	1.890	2.042	4.216	4.215	4.213	4.206	4.195	4.174	4.117	4.039
120	1.899	2.005	4.248	4.247	4.244	4.237	4.224	4.201	4.140	4.057
140	1.908	1.986	4.288	4.286	4.284	4.275	4.261	4.234	4.165	4.075
160	1.918	1.977	2.267	4.337	4.334	4.323	4.306	4.275	4.195	4.094
180	1.929	1.974	2.188	2.556	4.399	4.386	4.365	4.327	4.231	4.115
200	1.940	1.975	2.138	2.400	4.486	4.469	4.442	4.394	4.277	4.141
220	1.951	1.980	2.106	2.301	2.861	4.583	4.547	4.482	4.335	4.173
240	1.963	1.986	2.087	2.236	2.635	4.740	4.689	4.601	4.409	4.213
260	1.975	1.994	2.076	2.194	2.490	4.967	4.889	4.761	4.504	4.262
280	1.987	2.003	2.071	2.165	2.394	3.614	5.186	4.983	4.623	4.321
300	2.000	2.013	2.069	2.147	2.328	3.181	5.675	5.311	4.775	4.391
310	2.006	2.018	2.070	2.141	2.303	3.033	6.073	5.541	4.866	4.430
320	2.012	2.023	2.071	2.136	2.282	2.914	5.726	5.846	4.970	4.472
330	2.018	2.029	2.073	2.132	2.265	2.817	4.932	6.273	5.088	4.517
340	2.025	2.035	2.075	2.130	2.250	2.738	4.404	6.933	5.225	4.564
350	2.031	2.040	2.078	2.128	2.239	2.672	4.027	8.138	5.384	4.615
360	2.037	2.046	2.081	2.128	2.229	2.616	3.746	11.461	5.571	4.668
370	2.044	2.052	2.085	2.128	2.221	2.570	3.528	18.863	5.794	4.725
380	2.051	2.058	2.088	2.129	2.214	2.530	3.355	10.329	6.061	4.784
390	2.057	2.064	2.093	2.130	2.209	2.497	3.215	7.714	6.388	4.846
400	2.064	2.070	2.097	2.132	2.205	2.468	3.100	6.371	6.789	4.911
420	2.077	2.083	2.106	2.137	2.201	2.423	2.924	4.966	7.871	5.047
440	2.090	2.095	2.116	2.143	2.199	2.389	2.799	4.232	9.169	5.189
460	2.104	2.108	2.127	2.150	2.200	2.365	2.706	3.782	9.635	5.331
480	2.117	2.121	2.138	2.159	2.203	2.347	2.637	3.482	8.636	5.460
500	2.131	2.135	2.150	2.168	2.208	2.335	2.584	3.269	7.239	5.557
520	2.145	2.148	2.161	2.178	2.213	2.327	2.544	3.113	6.127	5.604
540	2.158	2.162	2.174	2.189	2.221	2.322	2.513	2.996	5.336	5.581
560	2.173	2.175	2.186	2.200	2.229	2.320	2.489	2.905	4.775	5.484
580	2.187	2.189	2.199	2.212	2.238	2.320	2.472	2.834	4.367	5.324
600	2.201	2.203	2.212	2.224	2.247	2.322	2.458	2.778	4.062	5.123
650	2.236	2.238	2.245	2.255	2.274	2.333	2.440	2.682	3.567	4.581
700	2.272	2.273	2.279	2.287	2.303	2.351	2.437	2.627	3.283	4.129
750	2.307	2.308	2.313	2.320	2.333	2.373	2.444	2.597	3.106	3.797
800	2.342	2.343	2.348	2.353	2.364	2.398	2.456	2.583	2.992	3.561
850	2.377	2.378	2.382	2.386	2.396	2.424	2.474	2.579	2.916	3.392
900	2.411	2.412	2.415	2.419	2.427	2.452	2.494	2.583	2.866	3.269
950	2.445	2.446	2.448	2.452	2.459	2.480	2.516	2.593	2.833	3.179
1000	2.478	2.478	2.481	2.484	2.490	2.508	2.540	2.606	2.812	3.113
1100	2.540	2.541	2.543	2.545	2.550	2.564	2.59	2.64	2.80	3.03
1200	2.599	2.599	2.601	2.603	2.606	2.618	2.64	2.68	2.80	2.98
1300	2.653	2.654	2.655	2.656	2.660	2.669	2.68	2.72	2.81	2.96
1400	2.704	2.704	2.705	2.706	2.709	2.717	2.73	2.76	2.83	2.96
1500	2.750	2.750	2.751	2.752	2.754	2.761	2.77	2.79	2.86	2.96
1600	2.792	2.792	2.793	2.794	2.796	2.801	2.81	2.83	2.88	2.97
1700	2.831	2.831	2.832	2.833	2.834	2.839	2.85	2.86	2.91	2.98
1800	2.867	2.867	2.868	2.868	2.870	2.874	2.88	2.89	2.93	3.00
1900	2.901	2.901	2.901	2.902	2.903	2.907	2.91	2.92	2.96	3.01
2000	2.931	2.931	2.932	2.933	2.934	2.937	2.94	2.95	2.98	3.03

Table 5 Viscosity

P (bar)	t (°C) 0	25	50	75	100	150	200	250	300	350	375
1	1792	890.8	547.1	378.4	12.28	14.19	16.18	18.22	20.29	22.37	23.41
5	1791	890.7	547.1	378.5	282.4	182.0	16.07	18.15	20.25	22.35	23.39
10	1790	890.6	547.2	378.6	282.6	182.1	15.93	18.07	20.20	22.32	23.37
25	1786	890.3	547.5	379.0	283.0	182.5	133.9	17.83	20.06	22.24	23.32
50	1780	889.8	547.9	379.6	283.6	183.2	134.5	106.1	19.86	22.15	23.27
75	1775	889.3	548.3	380.2	284.3	183.8	135.1	106.8	19.74	22.13	23.28
100	1769	888.9	548.7	380.9	284.9	184.4	135.7	107.5	86.42	22.18	23.35
125	1764	888.5	549.1	381.5	285.6	185.1	136.3	108.2	87.40	22.39	23.52
150	1759	888.1	549.5	382.1	286.3	185.7	136.9	108.8	88.32	22.91	23.84
175	1754	887.7	550.0	382.7	286.9	186.3	137.5	109.5	89.21	66.85	24.45
200	1749	887.4	550.4	383.4	287.6	186.9	138.1	110.1	90.06	69.21	25.79
225	1744	887.1	550.9	384.0	288.2	187.6	138.7	110.7	90.88	71.10	47.65
250	1739	886.8	551.3	384.6	288.9	188.2	139.3	111.4	91.67	72.71	58.09
275	1735	886.6	551.8	385.2	289.5	188.8	139.9	112.0	92.43	74.14	61.87
300	1731	886.4	552.3	385.9	290.2	189.4	140.5	112.6	93.18	75.43	64.49
350	1722	886.0	553.3	387.2	291.5	190.6	141.6	113.8	94.61	77.71	68.31
400	1714	885.8	554.3	388.4	292.8	191.8	142.8	114.9	95.98	79.72	71.21
450	1707	885.6	555.3	389.7	294.2	193.1	143.9	116.1	97.28	81.52	73.61
500	1700	885.5	556.4	391.0	295.5	194.3	145.0	117.2	98.55	83.19	75.70
550	1694	885.6	557.5	392.3	296.8	195.5	146.1	118.3	99.76	84.73	77.57
600	1687	885.7	558.6	393.6	298.1	196.7	147.2	119.4	100.9	86.19	79.27
650	1682	885.9	559.7	395.0	299.4	197.9	148.3	120.4	102.1	87.57	80.85
700	1676	886.2	560.9	396.3	300.8	199.0	149.3	121.5	103.2	88.88	82.33
800	1667	887.1	563.3	399.0	303.4	201.4	151.5	123.5	105.4	91.35	85.05
900	1659	888.3	565.8	401.7	306.1	203.8	153.6	125.5	107.4	93.65	87.54
1000	1653	889.9	568.4	404.4	308.7	206.1	155.6	127.5	109.4	95.82	89.84

P (bar)	400	425	450	475	500	550	600	650	700	750	800
1	24.45	25.49	26.52	27.55	28.57	30.61	32.61	34.60	36.55	38.48	40.38
5	24.44	25.48	26.52	27.55	28.58	30.62	32.63	34.61	36.57	38.50	40.39
10	24.42	25.47	26.52	27.55	28.58	30.63	32.64	34.63	36.59	38.52	40.42
25	24.39	25.46	26.52	27.57	28.61	30.67	32.70	34.70	36.66	38.59	40.50
50	24.38	25.47	26.55	27.62	28.67	30.76	32.81	34.82	36.79	38.73	40.63
75	24.40	25.52	26.61	27.70	28.77	30.87	32.93	34.95	36.93	38.88	40.78
100	24.49	25.62	26.72	27.82	28.89	31.01	33.09	35.11	37.09	39.04	40.94
125	24.65	25.77	26.88	27.98	29.06	31.18	33.26	35.28	37.27	39.21	41.11
150	24.91	26.01	27.10	28.19	29.27	31.38	33.45	35.48	37.46	39.39	41.29
175	25.32	26.34	27.39	28.46	29.52	31.62	33.68	35.69	37.66	39.59	41.48
200	25.96	26.80	27.77	28.79	29.82	31.89	33.92	35.93	37.88	39.80	41.68
225	27.03	27.44	28.26	29.20	30.18	32.19	34.20	36.18	38.12	40.03	41.89
250	29.00	28.36	28.89	29.70	30.61	32.54	34.50	36.45	38.38	40.26	42.11
275	33.73	29.70	29.71	30.32	31.12	32.93	34.84	36.75	38.64	40.51	42.35
300	43.83	31.73	30.78	31.06	31.71	33.37	35.20	37.07	38.93	40.77	42.59
350	55.78	39.35	33.97	33.06	33.19	34.40	36.02	37.77	39.55	41.33	43.10
400	61.29	48.69	39.05	35.92	35.16	35.65	36.98	38.56	40.24	41.94	43.65
450	65.01	55.07	45.22	39.72	37.68	37.15	38.07	39.44	40.98	42.60	44.24
500	67.89	59.44	50.71	44.08	40.70	38.88	39.30	40.41	41.79	43.30	44.85
550	70.30	62.76	55.06	48.36	44.02	40.84	40.65	41.45	42.65	44.03	45.50
600	72.40	65.46	58.52	52.16	47.37	42.96	42.12	42.57	43.57	44.81	46.17
650	74.28	67.76	61.36	55.40	50.53	45.18	43.67	43.75	44.52	45.61	46.87
700	75.98	69.79	63.79	58.18	53.38	47.41	45.28	44.98	45.51	46.44	47.58
800	79.04	73.28	67.81	62.72	58.20	51.70	48.55	47.52	47.55	48.15	49.04
900	81.75	76.27	71.11	66.35	62.09	55.51	51.73	50.06	49.62	49.88	50.52
1000	84.22	78.92	73.97	69.42	65.32	58.80	54.66	52.50	51.65	51.58	51.98

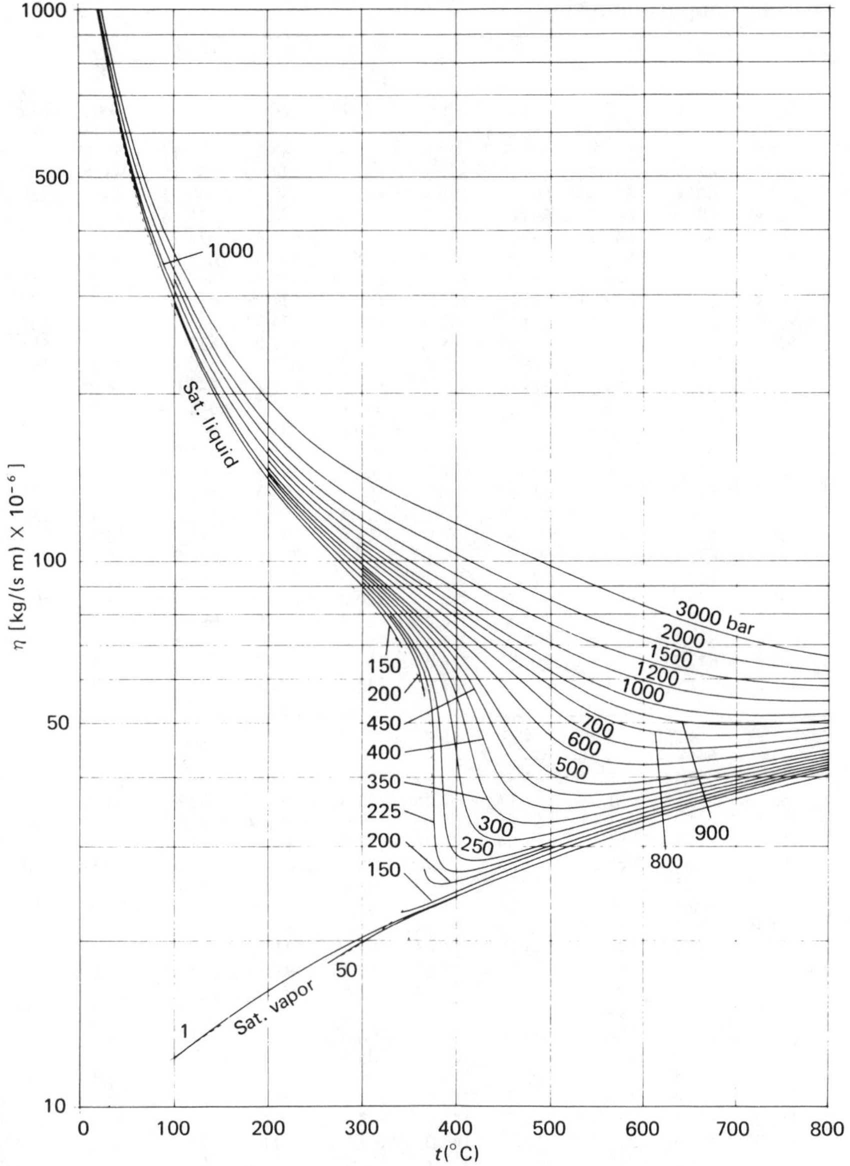

Figure 1 Viscosity.

Table 6 Thermal conductivity

P (bar)	t (°C) 0	25	50	75	100	150	200	250	300	350	375
1	561.0	607.2	643.6	666.8	25.08	28.85	33.28	38.17	43.42	48.96	51.83
5	561.3	607.4	643.7	667.0	679.3	682.1	34.93	39.18	44.09	49.44	52.25
10	561.5	607.6	644.0	667.2	679.6	682.4	37.21	40.51	44.95	50.06	52.79
25	562.4	608.3	644.7	668.0	680.4	683.4	664.2	45.16	47.82	52.06	54.53
50	563.7	609.4	645.8	669.2	681.8	685.1	666.4	622.7	53.86	55.99	57.87
75	565.1	610.5	647.0	670.5	683.2	686.8	668.6	625.9	63.11	61.06	62.00
100	566.5	611.7	648.2	671.7	684.5	688.5	670.7	629.0	550.9	68.10	67.35
125	567.9	612.8	649.3	673.0	685.9	690.2	672.8	632.0	556.5	79.15	74.68
150	569.3	613.9	650.5	674.2	687.2	691.8	674.9	635.0	561.8	100.9	85.54
175	570.6	615.1	651.6	675.5	688.6	693.5	677.0	637.9	566.8	452.5	103.7
200	572.0	616.2	652.8	676.7	690.0	695.1	679.1	640.8	571.6	463.3	142.3
225	573.4	617.3	654.0	678.0	691.3	696.8	681.2	643.6	576.2	472.8	441.5
250	574.8	618.5	655.1	679.2	692.7	698.4	683.2	646.3	580.7	481.4	411.4
275	576.1	619.6	656.3	680.4	694.0	700.1	685.3	649.1	585.0	489.1	425.8
300	577.5	620.8	657.4	681.7	695.3	701.7	687.3	651.8	589.1	496.3	438.0
350	580.2	623.0	659.8	684.1	698.0	704.9	691.3	657.0	597.1	509.3	457.5
400	582.9	625.3	662.1	686.6	700.7	708.2	695.3	662.2	604.6	521.0	473.2
450	585.5	627.5	664.4	689.1	703.3	711.4	699.3	667.2	611.7	531.8	486.6
500	588.1	629.8	666.7	691.5	706.0	714.6	703.2	672.1	618.5	541.7	498.5
550	590.7	632.0	668.9	693.9	708.6	717.7	707.0	676.9	625.1	551.0	509.4
600	593.3	634.2	671.2	696.3	711.2	720.9	710.9	681.6	631.3	559.7	519.4
650	595.8	636.4	673.5	698.7	713.8	724.0	714.7	686.3	637.4	568.0	528.8
700	598.3	638.6	675.7	701.1	716.4	727.2	718.5	690.8	643.2	575.9	537.7
800	603.1	642.9	680.2	705.9	721.5	733.4	726.0	699.8	654.5	590.6	554.1
900	607.8	647.2	684.6	710.5	726.6	739.5	733.4	708.6	665.1	604.2	569.1
1000	612.2	651.3	688.9	715.2	731.6	745.6	740.7	717.2	675.4	616.8	583.0

P (bar)	400	425	450	475	500	550	600	650	700	750	800
1	54.76	57.74	60.77	63.85	66.97	73.35	79.89	86.57	93.37	100.3	107.3
5	55.13	58.08	61.08	64.14	67.25	73.61	80.13	86.80	93.59	100.5	107.5
10	55.61	58.51	61.48	64.51	67.60	73.93	80.44	87.09	93.87	100.8	107.8
25	57.15	59.89	62.75	65.69	68.71	74.94	81.39	88.01	94.75	101.6	108.5
50	60.06	62.49	65.10	67.86	70.74	76.79	83.13	89.67	96.34	103.1	109.9
75	63.56	65.54	67.82	70.33	73.03	78.84	85.04	91.49	98.08	104.8	111.5
100	67.89	69.19	70.99	73.16	75.61	81.11	87.14	93.47	99.97	106.5	113.2
125	73.40	73.63	74.73	76.43	78.53	83.62	89.43	95.63	102.0	108.5	115.0
150	80.69	79.13	79.19	80.20	81.85	86.39	91.92	97.96	104.2	110.6	116.9
175	90.76	86.10	84.54	84.58	85.61	89.45	94.63	100.47	106.6	112.8	119.0
200	105.5	95.12	91.04	89.70	89.89	92.81	97.57	103.2	109.1	115.2	121.2
225	128.6	107.1	99.01	95.70	94.75	96.51	100.7	106.0	111.8	117.7	123.5
250	169.3	123.2	108.8	102.7	100.3	100.6	104.1	109.1	114.6	120.3	126.0
275	249.1	145.5	121.0	111.0	106.6	105.0	107.8	112.4	117.6	123.1	128.6
300	330.1	176.3	136.0	120.6	113.7	109.8	111.7	115.8	120.7	126.0	131.3
350	384.5	259.4	176.5	144.9	130.7	120.8	120.3	123.3	127.5	132.2	137.0
400	414.0	323.3	227.6	175.8	151.6	133.5	130.0	131.5	134.8	138.8	143.1
450	435.0	363.4	276.3	211.3	176.0	147.9	140.6	140.3	142.6	145.9	149.6
500	451.6	391.5	315.6	247.0	202.7	163.7	152.1	149.8	150.9	153.4	156.4
550	465.5	412.8	346.5	279.6	229.7	180.6	164.3	159.8	159.6	161.2	163.4
600	477.7	430.0	371.2	308.0	255.6	198.0	177.0	170.1	168.5	169.1	170.6
650	488.6	444.5	391.4	332.5	279.6	215.4	189.9	180.6	177.6	177.2	177.9
700	498.7	457.1	408.5	353.6	301.5	232.4	202.7	191.0	186.7	185.3	185.2
800	516.8	478.5	436.0	388.0	339.1	264.5	227.5	211.5	204.4	201.1	199.6
900	533.1	496.6	457.9	414.9	369.8	293.5	250.6	230.6	221.1	216.1	213.1
1000	548.0	512.7	476.3	436.9	395.1	319.3	271.8	248.0	236.2	229.7	225.5

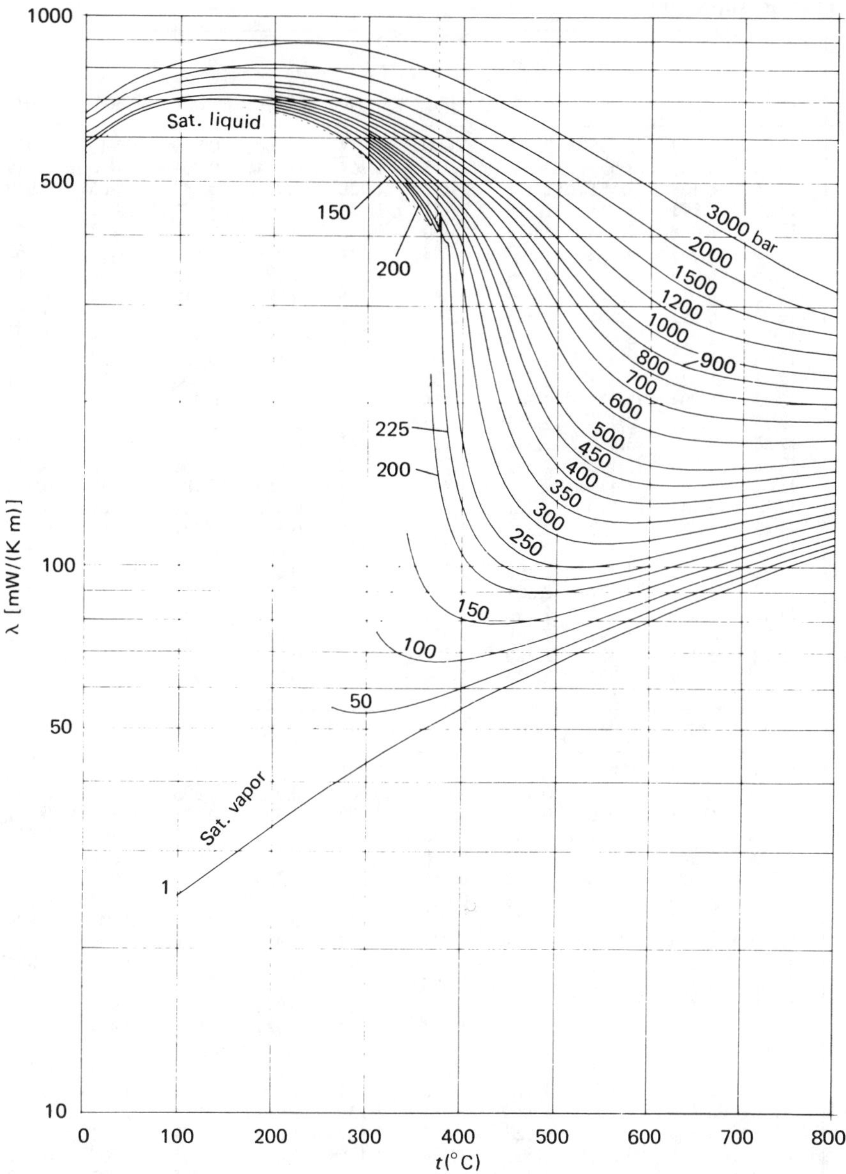

Figure 2 Thermal conductivity.

Table 7 Prandtl number

P (bar)	t (°C) 0	25	50	75	100	150	200	250	300	350	375
1	13.50	6.137	3.555	2.378	1.000	0.974	0.960	0.950	0.941	0.932	0.928
5	13.48	6.133	3.553	2.377	1.753	1.151	0.984	0.964	0.950	0.939	0.934
10	13.46	6.128	3.551	2.377	1.752	1.150	1.028	0.987	0.965	0.949	0.942
25	13.39	6.113	3.546	2.374	1.751	1.150	0.903	1.096	1.021	0.982	0.969
50	13.27	6.088	3.538	2.371	1.750	1.149	0.902	0.825	1.173	1.057	1.025
75	13.15	6.063	3.529	2.367	1.748	1.148	0.900	0.821	1.466	1.162	1.098
100	13.04	6.039	3.521	2.364	1.746	1.147	0.899	0.817	0.890	1.312	1.191
125	12.93	6.015	3.513	2.360	1.744	1.146	0.897	0.813	0.874	1.545	1.314
150	12.83	5.992	3.505	2.357	1.743	1.145	0.896	0.810	0.860	2.006	1.484
175	12.73	5.970	3.497	2.354	1.741	1.145	0.895	0.806	0.848	1.379	1.753
200	12.63	5.947	3.490	2.350	1.740	1.144	0.894	0.803	0.837	1.216	2.353
225	12.53	5.926	3.482	2.347	1.738	1.143	0.892	0.800	0.827	1.121	8.125
250	12.43	5.904	3.475	2.344	1.736	1.142	0.891	0.798	0.818	1.057	1.937
275	12.34	5.883	3.467	2.341	1.735	1.142	0.890	0.795	0.810	1.009	1.485
300	12.25	5.863	3.460	2.338	1.733	1.141	0.889	0.793	0.803	0.973	1.291
350	12.08	5.823	3.446	2.332	1.731	1.140	0.887	0.788	0.790	0.919	1.103
400	11.92	5.785	3.433	2.326	1.728	1.138	0.885	0.784	0.779	0.880	1.005
450	11.77	5.748	3.420	2.321	1.725	1.137	0.883	0.780	0.769	0.850	0.943
500	11.62	5.713	3.407	2.315	1.723	1.136	0.882	0.777	0.761	0.827	0.899
550	11.48	5.680	3.395	2.310	1.721	1.135	0.880	0.773	0.753	0.807	0.866
600	11.36	5.648	3.384	2.305	1.718	1.134	0.879	0.770	0.747	0.791	0.840
650	11.23	5.617	3.372	2.301	1.716	1.134	0.878	0.768	0.741	0.777	0.818
700	11.12	5.588	3.362	2.296	1.714	1.133	0.876	0.765	0.735	0.765	0.800
800	10.91	5.533	3.342	2.288	1.711	1.131	0.874	0.761	0.726	0.745	0.772
900	10.72	5.483	3.324	2.280	1.707	1.130	0.872	0.757	0.718	0.730	0.750
1000	10.55	5.439	3.307	2.273	1.704	1.129	0.870	0.753	0.712	0.717	0.733

P (bar)	400	425	450	475	500	550	600	650	700	750	800
1	0.924	0.921	0.917	0.914	0.911	0.905	0.899	0.894	0.890	0.886	0.882
5	0.929	0.925	0.921	0.917	0.913	0.907	0.901	0.895	0.891	0.886	0.882
10	0.936	0.931	0.926	0.921	0.917	0.909	0.902	0.897	0.891	0.887	0.883
25	0.958	0.949	0.941	0.934	0.928	0.917	0.908	0.900	0.894	0.889	0.884
50	1.002	0.983	0.969	0.957	0.946	0.929	0.916	0.906	0.898	0.891	0.886
75	1.055	1.024	1.000	0.982	0.966	0.943	0.925	0.911	0.901	0.894	0.888
100	1.118	1.070	1.035	1.008	0.988	0.956	0.933	0.917	0.904	0.895	0.889
125	1.195	1.123	1.073	1.038	1.010	0.970	0.942	0.921	0.907	0.896	0.889
150	1.290	1.183	1.116	1.069	1.034	0.984	0.950	0.926	0.909	0.897	0.889
175	1.408	1.253	1.163	1.102	1.059	0.998	0.958	0.930	0.911	0.898	0.889
200	1.568	1.336	1.215	1.138	1.085	1.013	0.966	0.934	0.912	0.897	0.888
225	1.812	1.436	1.273	1.177	1.112	1.027	0.974	0.937	0.913	0.897	0.887
250	2.273	1.564	1.339	1.219	1.141	1.042	0.981	0.941	0.914	0.896	0.885
275	3.363	1.737	1.416	1.264	1.171	1.057	0.988	0.944	0.914	0.895	0.883
300	3.329	1.979	1.506	1.314	1.203	1.073	0.996	0.946	0.914	0.894	0.881
350	1.693	2.415	1.729	1.427	1.272	1.104	1.010	0.951	0.914	0.890	0.876
400	1.290	1.922	1.900	1.548	1.345	1.135	1.023	0.955	0.912	0.886	0.871
450	1.118	1.479	1.791	1.627	1.413	1.167	1.037	0.959	0.911	0.882	0.865
500	1.021	1.245	1.542	1.601	1.454	1.195	1.049	0.962	0.909	0.877	0.858
550	0.957	1.109	1.335	1.488	1.447	1.218	1.061	0.965	0.907	0.872	0.852
600	0.911	1.022	1.190	1.354	1.396	1.231	1.071	0.968	0.905	0.868	0.846
650	0.876	0.962	1.089	1.235	1.320	1.231	1.078	0.971	0.904	0.864	0.841
700	0.849	0.917	1.016	1.140	1.238	1.219	1.083	0.974	0.903	0.860	0.836
800	0.808	0.855	0.919	1.005	1.096	1.164	1.079	0.977	0.903	0.855	0.828
900	0.778	0.813	0.859	0.920	0.992	1.091	1.060	0.976	0.903	0.853	0.823
1000	0.755	0.782	0.817	0.863	0.919	1.020	1.030	0.970	0.903	0.853	0.821

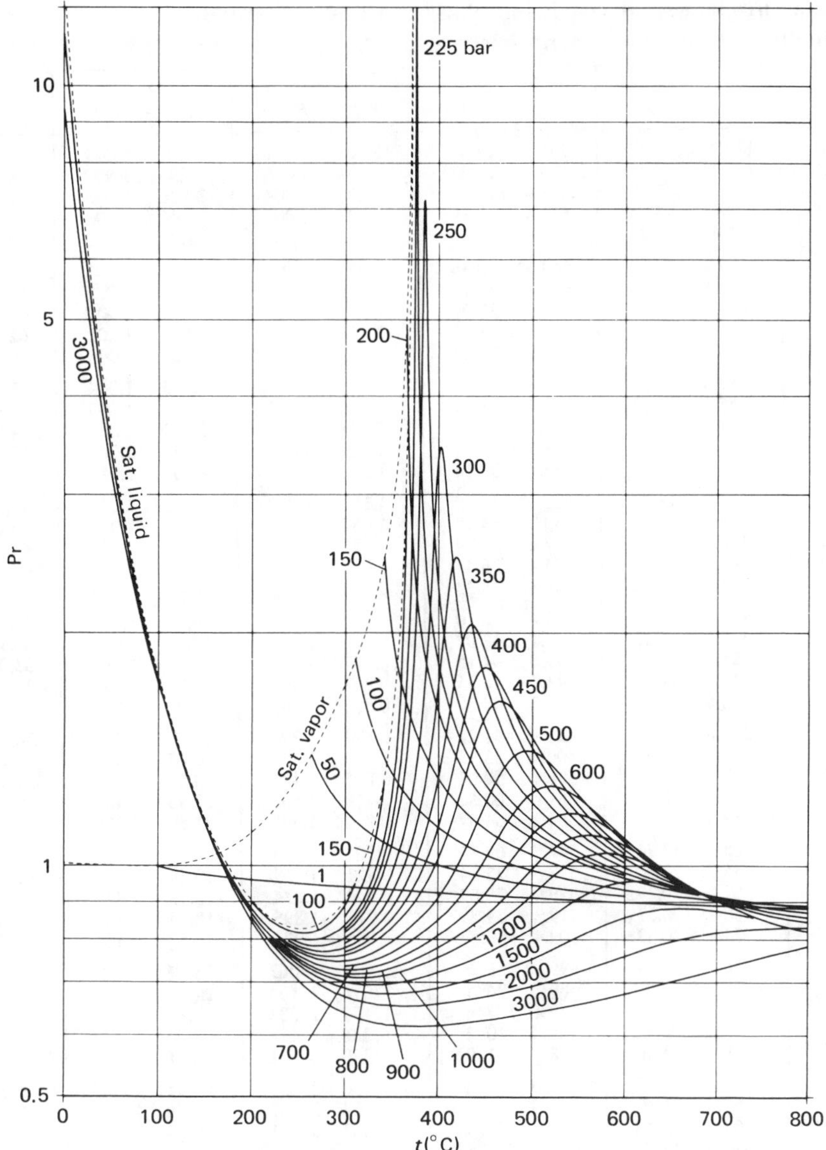

Figure 3 Prandtl number.

Table 8 Properties for coexisting phases: viscosity, thermal, conductivity, Prandtl number, dielectric constant, surface tension

t (°C)	η_l	η_g	λ_l	λ_g	Pr_l	Pr_g	ϵ_l	ϵ_g	σ
0.01	1791.	9.22	561.0	17.07	13.50	1.008	87.81	1.000	75.65
10	1308.	9.46	580.0	17.62	9.444	1.006	83.98	1.000	74.22
20	1003.	9.73	598.4	18.23	7.010	1.004	80.26	1.000	72.74
30	798.	10.01	615.4	18.89	5.423	1.003	76.66	1.000	71.20
40	653.	10.31	630.5	19.60	4.332	1.002	73.21	1.000	69.60
50	547.1	10.62	643.5	20.36	3.555	1.001	69.90	1.001	67.95
60	466.8	10.94	654.3	21.19	2.984	1.000	66.73	1.001	66.24
70	404.5	11.26	663.1	22.07	2.554	0.999	63.71	1.001	64.49
80	355.0	11.60	670.0	23.01	2.223	0.999	60.82	1.002	62.68
90	315.1	11.93	675.3	24.02	1.962	0.999	58.06	1.003	60.82
100	282.3	12.28	679.1	25.09	1.753	1.000	55.43	1.004	58.92
110	255.1	12.62	681.7	26.24	1.584	1.001	52.92	1.005	56.97
120	232.2	12.97	683.2	27.46	1.444	1.004	50.53	1.007	54.97
130	212.8	13.32	683.7	28.76	1.328	1.007	48.24	1.009	52.94
140	196.3	13.67	683.3	30.14	1.232	1.013	46.05	1.011	50.86
150	182.0	14.02	682.1	31.59	1.151	1.020	43.96	1.014	48.75
160	169.6	14.37	680.0	33.12	1.082	1.030	41.96	1.018	46.60
170	158.9	14.72	677.1	34.74	1.025	1.042	40.05	1.022	44.41
180	149.4	15.07	673.4	36.44	0.977	1.058	38.21	1.028	42.20
190	141.0	15.42	668.8	38.23	0.937	1.077	36.44	1.034	39.95
200	133.6	15.78	663.4	40.10	0.904	1.101	34.75	1.041	37.68
210	127.0	16.13	657.1	42.07	0.878	1.128	33.11	1.050	35.39
220	121.0	16.49	649.8	44.15	0.857	1.161	31.53	1.061	33.08
230	115.5	16.85	641.4	46.35	0.842	1.200	30.00	1.073	30.75
240	110.5	17.22	632.0	48.70	0.832	1.244	28.52	1.088	28.40
250	105.8	17.59	621.4	51.22	0.827	1.296	27.08	1.105	26.05
260	101.5	17.98	609.4	53.98	0.828	1.355	25.67	1.127	23.70
270	97.4	18.38	596.1	57.04	0.835	1.423	24.29	1.152	21.35
280	93.4	18.80	581.4	60.51	0.848	1.502	22.93	1.183	19.00
290	89.6	19.25	565.2	64.57	0.870	1.593	21.59	1.220	16.67
300	85.8	19.74	547.7	69.47	0.901	1.699	20.26	1.267	14.37
305	84.0	20.00	538.5	72.34	0.921	1.759	19.59	1.295	13.23
310	82.1	20.28	529.0	75.59	0.944	1.824	18.92	1.326	12.10
315	80.2	20.57	519.3	79.29	0.973	1.895	18.25	1.361	10.98
320	78.4	20.89	509.4	83.57	1.006	1.974	17.57	1.402	9.87
325	76.4	21.24	499.4	88.6	1.047	2.063	16.88	1.449	8.78
330	74.4	21.62	489.2	94.5	1.096	2.164	16.18	1.503	7.71
335	72.4	22.04	479.0	101.6	1.157	2.282	15.47	1.567	6.66
340	70.3	22.52	468.7	110.2	1.236	2.424	14.73	1.643	5.64
345	68.1	23.07	458.2	121.0	1.340	2.601	13.96	1.735	4.64
350	65.7	23.72	447.8	134.8	1.486	2.835	13.15	1.849	3.67
355	63.1	24.52	437.3	153.0	1.704	3.167	12.29	1.996	2.75
360	60.2	25.54	427.5	178.5	2.068	3.691	11.34	2.193	1.89
365	56.7	26.94	420.3	217.8	2.796	4.676	10.24	2.481	1.09
370	52.0	29.26	428.8	301.2	5.08	7.66	8.82	2.98	0.40
371	50.7	30.00	439.3	337.2	6.41	9.30	8.44	3.15	0.28
372	49.1	30.95	462.8	397.4	9.11	12.43	7.99	3.37	0.17
373	46.7	32.38	545.4	538.3	18.69	21.19	7.35	3.71	0.07
373.976	39.0		∞		∞		5.34		0.

Tables 9 and 10 are reprinted from *VDI—Wärmeatlas* [2].

Table 9 Thermal expansion coefficient $\beta = (1/v)(\partial v/\partial T)_p$ of liquid water as a function of pressure and temperature[a]

	t (°C)								
P (bar)	0	20	50	100	150	200	250	300	350
1	−0.085 2	0.206 7	0.462 3						
5	−0.083 8	0.207 2	0.462 2	0.753 9	1.024				
10	−0.082 0	0.207 9	0.462 0	0.753 0	1.022				
50	−0.067 8	0.213 3	0.460 5	0.745 5	1.007	1.347	1.936		
100	−0.049 9	0.220 1	0.458 9	0.736 6	0.990 2	1.312	1.848	3.189	
150	−0.032 0	0.227 2	0.457 4	0.728 1	0.974 0	1.281	1.772	2.883	
200	−0.014 2	0.234 3	0.456 2	0.720 0	0.958 7	1.251	1.704	2.648	6.923
250	+0.003 3	0.241 6	0.455 1	0.712 2	0.944 2	1.224	1.643	2.460	5.162
300	0.020 5	0.248 9	0.454 2	0.704 7	0.930 3	1.198	1.589	2.306	4.276
350	0.037 3	0.256 2	0.453 4	0.697 5	0.917 2	1.175	1.539	2.176	3.718
400	0.053 5	0.263 6	0.452 8	0.690 7	0.904 6	1.152	1.494	2.065	3.324
450	0.069 0	0.270 9	0.452 3	0.684 1	0.892 6	1.131	1.453	1.968	3.027
500	0.083 6	0.278 2	0.452 0	0.677 7	0.881 1	1.111	1.415	1.884	2.791
600	0.110 0	0.292 6	0.451 7	0.665 7	0.859 6	1.075	1.348	1.742	2.439
700	0.131 7	0.306 5	0.451 8	0.654 5	0.839 7	1.042	1.290	1.626	2.186
800	0.147 5	0.319 6	0.452 3	0.644 1	0.821 3	1.012	1.238	1.530	1.994
900	0.156 5	0.331 7	0.453 0	0.634 3	0.804 2	0.984 4	1.193	1.448	1.843
1 000	0.157 6	0.342 6	0.454 0	0.625 2	0.788 2	0.959 4	1.152	1.377	1.720

[a] β in 10^{-3}/K.

Table 10 Thermal diffusivity κ of liquid water as a function of pressure and temperature[a]

	t (°C)								
P (bar)	0	20	50	100	150	200	250	300	350
1	0.135	0.144	0.156						
10	0.135	0.144	0.156	0.168	0.173				
50	0.137	0.146	0.157	0.170	0.175	0.173	0.159		
100	0.138	0.147	0.158	0.171	0.175	0.174	0.162	0.134	
150	0.140	0.148	0.159	0.171	0.176	0.175	0.166	0.140	
200	0.140	0.149	0.160	0.172	0.176	0.176	0.167	0.147	0.093
250	0.142	0.150	0.161	0.173	0.178	0.177	0.169	0.151	0.108
300	0.145	0.151	0.162	0.173	0.180	0.179	0.171	0.155	0.120
350	0.145	0.152	0.163	0.175	0.180	0.181	0.174	0.159	0.128
400	0.147	0.153	0.163	0.175	0.181	0.181	0.178	0.162	0.135
450	0.147	0.154	0.164	0.176	0.181	0.182	0.178	0.166	0.141
500	0.147	0.155	0.164	0.177	0.182	0.183	0.178	0.169	0.147

[a] κ in 10^{-6} m²/s.

REFERENCES

1. Haar, L., Gallagher, J. S., and Kell, G. S., *Thermodynamic and Transport Properties and Computer Programs for Vapor and Liquid States of Water in S.I. Units*, NBS/NRC, Hemisphere, Washington, D.C., 1984.
2. *VDI—Wärmeatlas*, 2d ed., Verein Deutsches Ingenieure, Düsseldorf, 1974.

3.2. Properties of Liquid Heavy Water

Critical properties: p_c = 216.59 bar; T_c = 370.66 °C; T_c = 643.81 K; ρ_c = 350 kg/m^3; molecular weight: $\tilde{M}$ = 20.028 kg/kmol; triple point: T_t = 3.82 °C; critical temperature: 370.66 °C; critical pressure: 21.659 MPa; critical density: 350 kg/m^3.

Table 1 Properties of liquid heavy water at p = 1 bar[a,b]

T, °C	ρ, kg/m³	c_p, kJ/kg K	β, 10^{-3}/K	λ, 10^{-3} W/m K	η, 10^{-6} kg/m s	ν, 10^{-6} m²/s	κ, 10^{-6} m²/s	Pr
3.82	1 105.5	4.210	–0.132	564	2 058	1.861	0.121	15.35
10.00	1 106.0	4.231	–0.027	570	1 689	1.527	0.122	12.53
20.00	1 105.4	4.242	0.121	580	1 253	1.133	0.124	9.17
30.00	1 103.3	4.240	0.242	589	972	0.881	0.126	7.00
40.00	1 100.0	4.232	0.343	597	785	0.713	0.128	5.56
50.00	1 095.7	4.220	0.428	605	648	0.591	0.131	4.52
60.00	1 090.6	4.207	0.502	611	545	0.500	0.133	3.75
70.00	1 084.7	4.194	0.569	617	468	0.432	0.136	3.18
80.00	1 078.2	4.181	0.631	621	409	0.379	0.138	2.75
90.00	1 071.0	4.170	0.689	625	363	0.339	0.140	2.42

[a] T, temperature; ρ, density; c_p, specific heat capacity; β, thermal expansion coefficient; λ, thermal conductivity; η, dynamic viscosity; υ, kinematic viscosity; κ, thermal diffusivity; Pr, Prandtl number.

[b] Properties according to [6, 9].

Table 2 Density ρ of liquid heavy water as a function of pressure and temperature[a]

Pressure, bar	Temperature, °C								
	3.82	20	50	100	150	200	250	300	350
1	1 105.5	1 105.4	1 095.7	1 063.4					
10	1 106.0	1 105.9	1 096.1	1 063.8	1 017.6				
50	1 108.3	1 108.0	1 098.1	1 065.9	1 020.1	961.2	885.1		
100	1 111.1	1 110.6	1 100.5	1 068.5	1 023.2	965.3	891.5	788.3	
150	1 113.9	1 113.1	1 102.9	1 071.0	1 026.3	969.4	897.5	800.3	
170	1 115.0	1 114.1	1 103.9	1 072.0	1 027.5	970.9	899.9	804.7	625.8

[a] ρ in kg/m³.

Table 3 Constant pressure specific heat c_p of liquid heavy water as a function of pressure and temperature[a]

Pressure, bar	Temperature, °C								
	3.82	20	50	100	150	200	250	300	350
1	4.210	4.242	4.220	4.161					
10	4.207	4.239	4.218	4.159	4.172				
50	4.189	4.225	4.209	4.151	4.160	4.295	4.630		
100	4.167	4.208	4.197	4.140	4.145	4.269	4.571	5.424	
150	4.145	4.192	4.186	4.130	4.131	4.246	4.518	5.218	
170	4.136	4.186	4.182	4.126	4.126	4.237	4.499	5.151	9.933

[a]c_p in kJ/kg K.

Table 4 Thermal expansion coefficient $\beta = (1/v)(\partial v/\partial T)_p$ of liquid heavy water as a function of pressure and temperature[a]

Pressure, bar	Temperature, °C								
	3.82	20	50	100	150	200	250	300	350
1	−0.132 1	0.121 2	0.428 0	0.745 4					
10	−0.130 0	0.122 9	0.428 1	0.743 8	1.032 9				
50	−0.119 8	0.130 5	0.428 5	0.737 0	1.018 1	1.375 8	1.977 5		
100	−0.105 5	0.140 1	0.429 1	0.728 9	1.000 6	1.340 3	1.886 4	3.288 5	
150	−0.089 8	0.149 6	0.429 6	0.721 1	0.984 0	1.307 4	1.807 0	2.956 2	
170	−0.083 2	0.153 4	0.429 9	0.718 1	0.977 6	1.294 9	1.778 0	2.848 5	11.0377

[a]β in 10^{-3}/K.

Table 5 Thermal conductivity λ of liquid heavy water as a function of pressure and temperature[a]

Pressure, bar	Temperature, °C								
	3.82	20	50	100	150	200	250	300	350
1	564	580	605	628					
10	565	580	605	628	620				
50	567	583	608	630	624	591	536		
100	571	586	610	632	628	594	541	471	
150	574	589	613	635	632	597	546	478	
170	575	591	614	635	634	598	548	481	368

[a]λ in 10^{-3} W/m K.

Table 6 Dynamic viscosity η of liquid heavy water as a function of pressure and temperature[a]

Pressure, bar	Temperature, °C								
	3.82	20	50	100	150	200	250	300	350
1	2 057.9	1 252.9	647.8	325.4					
10	2 056.9	1 252.9	648.1	325.6	207.6				
50	2 052.5	1 252.5	649.1	326.6	209.3	150.1	116.7		
100	2 046.6	1 251.9	650.4	327.9	211.5	151.2	118.0	94.2	
150	2 040.4	1 251.2	651.7	329.3	213.7	152.3	119.3	95.7	
170	2 037.8	1 250.9	652.2	329.8	214.6	152.8	119.9	96.4	68.3

[a]η in 10^{-6} kg/m s.

Table 7 Kinematic viscosity ν of liquid heavy water as a function of pressure and temperature[a]

Pressure, bar	Temperature, °C								
	3.82	20	50	100	150	200	250	300	350
1	1.861	1.133	0.591	0.306					
10	1.860	1.133	0.591	0.306	0.204				
50	1.852	1.130	0.591	0.306	0.205	0.156	0.132		
100	1.842	1.127	0.591	0.307	0.207	0.157	0.132	0.119	
150	1.832	1.124	0.591	0.307	0.208	0.157	0.133	0.120	
170	1.828	1.123	0.591	0.308	0.209	0.157	0.133	0.120	0.109

[a] ν in 10^{-6} m²/s.

Table 8 Prandtl number of liquid heavy water as a function of pressure and temperature

Pressure, bar	Temperature, °C								
	3.82	20	50	100	150	200	250	300	350
1	15.35	9.17	4.52	2.16					
10	15.32	9.15	4.52	2.15	1.40				
50	15.15	9.08	4.50	2.15	1.40	1.09	1.01		
100	14.94	8.99	4.47	2.15	1.40	1.09	1.00	1.08	
150	14.74	8.90	4.45	2.14	1.40	1.08	0.99	1.05	
170	14.66	8.87	4.44	2.14	1.40	1.08	0.98	1.03	1.85

Table 9 Thermal diffusivity κ of liquid heavy water as a function of pressure and temperature[a]

Pressure, bar	Temperature, °C								
	3.82	20	50	100	150	200	250	300	350
1	0.121	0.124	0.131	0.142					
10	0.121	0.124	0.131	0.142	0.146				
50	0.122	0.125	0.131	0.142	0.147	0.143	0.131		
100	0.123	0.125	0.132	0.143	0.148	0.144	0.133	0.110	
150	0.124	0.126	0.133	0.143	0.149	0.145	0.135	0.114	
170	0.125	0.127	0.133	0.144	0.149	0.145	0.135	0.116	0.059

[a] κ in 10^{-6} m²/s.

REFERENCES

1. Kazavchinskii, Y. Z., et al., *Heavy Water Thermophysical Properties.* Moskau, Leningrad, 1963.
2. Blank, G., Neue Bestimmung des kritischen Punktes von leichtem und schwerem Wasser, *Wärme Stoffübertrag.*, vol. 2, pp. S.53–59, 1969.
3. Straub, J., Scheffler, K., Albrecht, F., and Wittl, H., Programm zur Berechnung der Zustands- und Transportgrocen von schwerem Wasser. TUM-MW 708-8008, Technische Universität München (private communication), 1980.
4. Vargaftik, N. B., Volyak, L. D., and Volkov, B. M. Investigating the Surface Tension of H_2O and D_2O at Near-Critical Temperatures, *Teploenergetica,* vol. 20, no. 9, pp. S.80–82, 1973.
5. Tanashita, I., Watanabe, K., Uematsu, M., and Eguchi, K., Evaluation and Correlations of Saturation Pressure of Light and Heavy Water, *Proc. 8th Int. Conf. on Prop. of Steam, Griens,* vol. 1, p. 560, 1974.
6. Nagashima, A., and Matsunaga, N., Transport Properties of D_2O: A Compilation and Correlation of Available Data, *Proc. 9th Int. Conf. on Prop. of Steam, München,* 1979.
7. Hill, P. G., and McMillan, R. D. C., Development of a Fundamental Equation of State for Heavy Water, *Proc. 9th Int. Conf. on Prop. of Steam, München,* 1979.

8. Hill, P. G., and McMillan, R. D. C., Thermodynamic Properties of D_2O in Liquid-Vapour Equilibrium, *Proc. 9th Int. Conf. on Prop. of Steam, München*, 1979.
9. Hill, P. G., McMillan, R. D. C., and Lee, V., Tables of Thermodynamic Properties of Heavy Water in S.I. Units, AECL 7531, Sheridan Park, Ontario, Canada, 1981.
10. Ulrych, G., D_2O-Dampftafel: Die 1967-IFC-Formulation für Industriegebrauch als Grundlage einer Dampftafel für schweres Wasser (D_2O) und Zusammenstellung der Transportgrößen, *Fortschr. Ber. VDI*, vol. 6, no. 90, 1981.
11. Hill, P. G., McMillan, R. D. C., and Lee, V., A Fundamental Equation of State for Heavy Water., *J. Phys. Chem. Ref. Data*, vol. 11, no. 1, 1982.

3.3. Properties of Sea Water

Values of thermal conductivity, dynamic viscosity, heat capacity, density, and Prandtl number for seawater and its concentrates are given. The data on which these recommended values are based are the most reliable available and cover a wide range of both temperature and concentration.

The concentration of seawater samples is normally defined by a single term, either salinity or chlorinity. This is possible since for all practical purposes the relative composition of seawater is constant. This does not imply that all samples have the same composition but merely that all ions are present in the same ratios and that the only variation is the amount of pure water present. Thus, if the concentration of any one ion is measured, the amount of all the other ions can be accurately found by calculation.

Salinity was originally defined as the weight in grams of the dissolved inorganic matter in 1 kg of seawater after all bromide and iodide have been replaced by an equivalent amount of chloride and all carbonate has been converted to oxide. Chlorinity is in effect the halide concentration of seawater. The two are related by

$$S‰ = 1.806\,55\ Cl‰ \qquad (1)$$

(The symbol ‰ denotes parts per thousand of seawater.) Chlorinity and salinity are now more accurately found by measuring the ratio of the electrical conductivity of the unknown sample to that of a "standard seawater" sample.

Table 1 Composition of seawater

Ion	Concentration in g/kg of seawater (salinity, 35 g/kg)
Chloride	19.344
Sodium	10.773
Sulfate	2.712
Magnesium	1.294
Calcium	0.412
Potassium	0.399
Bicarbonate	1.142
Bromide	0.067
Strontium	0.008
Boron	0.004
Fluoride	0.001 3

Table 1 gives the major components of seawater at its "normal" salinity of 35‰, though samples from different seas may vary by more than the range 32–40‰. Samples reported at other salinities may be considered merely as having more or less pure water, but at the highest concentrations reported some salts (notably calcium sulfates) will have exceeded their solubilities and precipitated from solutions.

Tables 2–9 appear on the following pages. References for these tables and equations representing the change in value with temperature and salinity can be found in Data Item Number 77024 published by the Engineering Sciences Data Unit, 251-259 Regent Street, London W1R 7AD.

Table 2 Thermal conductivity of seawater and its concentrates (mW/m^2)/(K/m)

t, °C	Salinity, g/kg															
	10	20	30	35[a]	40	50	60	70	80	90	100	110	120	130	140	150
0	570	569	567	566	565	563	562	560	558	556	554	552	550	548	546	544
10	587	586	584	584	583	581	580	578	577	575	573	571	570	568	566	564
20	603	602	600	600	599	598	597	595	594	592	591	589	588	586	585	583
30	617	616	615	614	614	613	612	611	609	608	607	606	604	603	602	600
40	629	629	628	628	627	626	626	625	624	623	622	621	620	618	617	616
50	641	640	640	639	639	639	638	637	637	636	635	634	633	632	631	630
60	651	650	650	650	650	649	649	649	648	648	647	646	646	645	644	644
70	659	659	659	659	659	659	659	658	658	658	658	657	657	656	656	655
80	666	667	667	667	667	667	667	667	667	667	667	667	666	666	666	666
90	672	673	673	673	674	674	674	674	675	675	675	675	675	675	675	675
100	677	678	678	679	679	680	680	681	681	681	682	682	682	682	682	683
110	681	682	683	683	683	684	685	685	686	687	687	688	688	688	689	689
120	683	684	685	686	686	687	688	689	690	691	691	692	693	693	694	694
130	685	686	687	688	688	690	691	692	693	694	695	695	696	697	698	699
140	685	687	688	689	689	691	692	693	694	696	697	698	699	700	701	702
150	684	686	688	688	689	691	692	694	695	696	698	699	700	701	702	703
160	683	684	686	687	688	690	691	693	694	696	697	699	700	701	703	704
170	680	682	684	685	686	687	689	691	693	694	696	698	699	701	702	704
180	676	678	680	681	682	684	686	686	690	692	694	695	697	699	700	702

[a] "Normal" seawater.

Table 3 Dynamic viscosity of seawater and its concentrates (10^{-3} Ns/m^2)

t, °C	Salinity, g/kg														
	10	20	30	40	50	60	70	80	90	100	110	120	130	140	150
0	1.802	1.831	1.861	1.893	1.928	1.965	2.005	2.049	2.096	2.147	2.202	2.261	2.326	2.395	2.470
10	1.327	1.350	1.375	1.401	1.429	1.459	1.491	1.526	1.563	1.603	1.646	1.693	1.743	1.797	1.855
20	1.021	1.041	1.061	1.083	1.106	1.131	1.157	1.185	1.216	1.248	1.283	1.321	1.361	1.404	1.451
30	0.814	0.830	0.848	0.866	0.886	0.906	0.929	0.952	0.977	1.004	1.033	1.064	1.098	1.133	1.171
40	0.667	0.681	0.696	0.712	0.729	0.747	0.765	0.786	0.807	0.830	0.854	0.880	0.908	0.938	0.970
50	0.559	0.571	0.585	0.599	0.613	0.629	0.645	0.662	0.681	0.700	0.721	0.744	0.768	0.793	0.821
60	0.477	0.488	0.500	0.512	0.525	0.539	0.553	0.568	0.584	0.602	0.620	0.639	0.660	0.682	0.706
70	0.414	0.424	0.434	0.445	0.457	0.469	0.481	0.495	0.509	0.524	0.540	0.558	0.576	0.595	0.616
80	0.364	0.373	0.382	0.392	0.402	0.413	0.424	0.436	0.449	0.463	0.477	0.492	0.508	0.525	0.544
90	0.323	0.331	0.340	0.349	0.358	0.368	0.378	0.389	0.400	0.412	0.425	0.439	0.453	0.469	0.485
100	0.290	0.297	0.305	0.313	0.322	0.331	0.340	0.350	0.360	0.371	0.383	0.395	0.408	0.422	0.436
110	0.262	0.269	0.276	0.284	0.291	0.300	0.308	0.317	0.326	0.336	0.347	0.358	0.370	0.382	0.395
120	0.239	0.245	0.252	0.259	0.266	0.273	0.281	0.289	0.298	0.307	0.317	0.327	0.337	0.349	0.361
130	0.219	0.225	0.231	0.237	0.244	0.251	0.258	0.266	0.273	0.282	0.291	0.300	0.310	0.320	0.331
140	0.201	0.207	0.213	0.219	0.225	0.231	0.238	0.245	0.252	0.260	0.268	0.277	0.286	0.295	0.305
150	0.187	0.192	0.197	0.203	0.208	0.214	0.221	0.227	0.234	0.241	0.249	0.256	0.265	0.273	0.283
160	0.173	0.178	0.183	0.189	0.194	0.200	0.205	0.211	0.218	0.224	0.231	0.239	0.246	0.254	0.263
170	0.162	0.167	0.171	0.176	0.181	0.186	0.192	0.198	0.203	0.210	0.216	0.223	0.230	0.237	0.245
180	0.152	0.156	0.161	0.165	0.170	0.175	0.180	0.185	0.191	0.196	0.202	0.209	0.215	0.222	0.230

For "normal" seawater see Table 4.

Table 4 Dynamic viscosity of seawater (10^{-3} Ns/m^2)

t, °C	Salinity, g/kg										
	30	31	32	33	34	35[a]	36	37	38	39	40
0	1.861	1.864	1.867	1.871	1.874	1.877	1.880	1.883	1.887	1.890	1.893
10	1.375	1.377	1.380	1.382	1.365	1.388	1.390	1.393	1.396	1.398	1.401
20	1.061	1.063	1.065	1.068	1.070	1.072	1.074	1.076	1.078	1.081	1.083
30	0.848	0.850	0.851	0.853	0.855	0.857	0.859	0.861	0.862	0.864	0.866
40	0.696	0.698	0.699	0.701	0.702	0.704	0.706	0.707	0.709	0.710	0.712
50	0.585	0.586	0.587	0.589	0.590	0.592	0.593	0.594	0.596	0.597	0.599
60	0.500	0.501	0.503	0.504	0.505	0.506	0.507	0.509	0.510	0.511	0.512
70	0.434	0.435	0.437	0.438	0.439	0.440	0.441	0.442	0.443	0.444	0.445
80	0.382	0.383	0.384	0.385	0.386	0.387	0.388	0.389	0.390	0.391	0.392
90	0.340	0.341	0.342	0.343	0.343	0.344	0.345	0.346	0.347	0.348	0.349
100	0.305	0.306	0.307	0.308	0.308	0.309	0.310	0.311	0.312	0.312	0.313
110	0.276	0.277	0.278	0.278	0.279	0.280	0.281	0.281	0.282	0.283	0.284
120	0.252	0.252	0.253	0.254	0.254	0.255	0.256	0.257	0.257	0.258	0.259
130	0.231	0.231	0.232	0.233	0.233	0.234	0.235	0.235	0.236	0.237	0.237
140	0.213	0.213	0.214	0.215	0.215	0.216	0.216	0.217	0.218	0.218	0.219
150	0.197	0.198	0.198	0.199	0.199	0.200	0.200	0.201	0.202	0.202	0.203
160	0.183	0.184	0.184	0.185	0.186	0.186	0.187	0.187	0.188	0.188	0.189
170	0.171	0.172	0.172	0.173	0.173	0.174	0.174	0.175	0.175	0.176	0.176
180	0.161	0.161	0.161	0.162	0.162	0.163	0.163	0.164	0.164	0.165	0.165

[a]"Normal" seawater.

Table 5 Heat capacity of seawater and its concentrates (kJ/kg K)

t, °C	Salinity, g/kg 10	20	30	40	50	60	70	80	90	100	110	120	130	140	150
0	4.143	4.081	4.021	3.964	3.910	3.858	3.809	3.763	3.720	3.679	3.641	3.606	3.573	3.543	3.516
10	4.136	4.077	4.020	3.965	3.913	3.863	3.815	3.770	3.727	3.686	3.648	3.612	3.579	3.547	3.518
20	4.131	4.074	4.020	3.967	3.917	3.868	3.822	3.777	3.735	3.694	3.656	3.619	3.584	3.552	3.521
30	4.128	4.074	4.021	3.971	3.922	3.874	3.829	3.785	3.743	3.702	3.663	3.626	3.591	3.557	3.525
40	4.127	4.075	4.024	3.975	3.927	3.881	3.836	3.793	3.751	3.710	3.671	3.633	3.597	3.562	3.529
50	4.128	4.078	4.029	3.981	3.934	3.888	3.844	3.801	3.759	3.719	3.679	3.641	3.604	3.568	3.533
60	4.131	4.082	4.034	3.987	3.941	3.896	3.853	3.810	3.768	3.727	3.687	3.649	3.611	3.574	3.538
70	4.137	4.088	4.041	3.995	3.950	3.905	3.861	3.819	3.777	3.736	3.696	3.657	3.618	3.581	3.544
80	4.144	4.096	4.050	4.004	3.959	3.914	3.871	3.828	3.786	3.745	3.704	3.665	3.626	3.588	3.551
90	4.154	4.106	4.059	4.014	3.968	3.924	3.880	3.837	3.795	3.754	3.713	3.673	3.634	3.595	3.558
100	4.165	4.118	4.071	4.025	3.979	3.934	3.891	3.847	3.805	3.763	3.722	3.682	3.642	3.603	3.565
110	4.179	4.131	4.083	4.037	3.991	3.946	3.901	3.857	3.815	3.772	3.731	3.690	3.651	3.612	3.573
120	4.195	4.146	4.097	4.050	4.003	3.957	3.912	3.868	3.825	3.782	3.740	3.700	3.659	3.620	3.582
130	4.213	4.162	4.113	4.064	4.016	3.970	3.924	3.879	3.835	3.792	3.750	3.709	3.669	3.629	3.591
140	4.233	4.181	4.129	4.079	4.030	3.982	3.936	3.890	3.845	3.802	3.760	3.718	3.678	3.639	3.601
150	4.255	4.201	4.148	4.096	4.045	3.996	3.948	3.902	3.856	3.812	3.769	3.728	3.688	3.649	3.611
160	4.279	4.222	4.167	4.113	4.061	4.010	3.961	3.913	3.867	3.823	3.780	3.738	3.698	3.659	3.622
170	4.306	4.246	4.188	4.132	4.078	4.025	3.974	3.926	3.878	3.833	3.790	3.748	3.708	3.670	3.634
180	4.334	4.271	4.210	4.152	4.095	4.041	3.988	3.938	3.890	3.844	3.800	3.758	3.719	3.681	3.646

For "normal" seawater see Table 6.

Table 6 Heat capacity of seawater (kJ/kg K)

t, °C	Salinity, g/kg										
	30	31	32	33	34	35[a]	36	37	38	39	40
0	4.021	4.015	4.010	4.004	3.998	3.992	3.987	3.981	3.975	3.970	3.964
10	4.020	4.014	4.009	4.003	3.998	3.992	3.987	3.981	3.976	3.971	3.965
20	4.020	4.015	4.009	4.004	3.999	3.993	3.988	3.983	3.978	3.973	3.967
30	4.021	4.016	4.011	4.006	4.001	3.996	3.991	3.986	3.981	3.976	3.971
40	4.024	4.019	4.014	4.009	4.004	4.000	3.995	3.990	3.985	3.980	3.975
50	4.029	4.024	4.019	4.014	4.009	4.004	4.000	3.995	3.990	3.985	3.981
60	4.034	4.029	4.025	4.020	4.015	4.011	4.006	4.001	3.997	3.992	3.987
70	4.041	4.037	4.032	4.027	4.023	4.018	4.013	4.009	4.004	4.000	3.995
80	4.050	4.045	4.040	4.036	4.031	4.027	4.022	4.017	4.013	4.008	4.004
90	4.059	4.055	4.050	4.046	4.041	4.036	4.032	4.027	4.023	4.018	4.014
100	4.071	4.066	4.061	4.057	4.052	4.048	4.043	4.038	4.034	4.029	4.025
110	4.083	4.079	4.074	4.069	4.065	4.060	4.055	4.051	4.046	4.041	4.037
120	4.097	4.092	4.088	4.083	4.078	4.073	4.069	4.064	4.059	4.054	4.050
130	4.113	4.108	4.103	4.098	4.093	4.088	4.083	4.078	4.074	4.069	4.064
140	4.129	4.124	4.119	4.114	4.109	4.104	4.099	4.094	4.089	4.084	4.079
150	4.148	4.142	4.137	4.132	4.127	4.121	4.116	4.111	4.106	4.101	4.096
160	4.167	4.162	4.156	4.151	4.145	4.140	4.135	4.129	4.124	4.119	4.113
170	4.188	4.182	4.177	4.171	4.165	4.160	4.154	4.149	4.143	4.137	4.132
180	4.120	4.204	4.198	4.192	4.187	4.181	4.175	4.169	4.163	4.157	4.152

[a] "Normal" seawater.

Table 7 Density of seawater and its concentrates (kg/m^3)

t, °C	Salinity, g/kg 10	20	30	40	50	60	70	80	90	100	110	120	130	140	150
0	1 008.1	1 016.2	1 024.2	1 032.0	1 039.8	1 047.6	1 055.5	1 063.5	1 071.6	1 079.7	1 088.0	1 096.2	1 104.4	1 112.5	1 120.4
10	1 007.7	1 015.5	1 023.2	1 030.2	1 038.4	1 046.0	1 053.8	1 061.6	1 669.6	1 077.6	1 085.7	1 093.9	1 102.0	1 110.1	1 118.0
20	1 005.8	1 013.3	1 020.8	1 028.3	1 035.9	1 043.5	1 051.2	1 058.9	1 066.7	1 074.5	1 082.4	1 090.3	1 098.2	1 106.2	1 114.2
30	1 002.8	1 010.2	1 017.6	1 025.1	1 032.6	1 040.2	1 047.8	1 055.4	1 063.1	1 070.8	1 078.5	1 086.3	1 094.1	1 102.0	1 109.9
40	999.2	1 006.6	1 013.9	1 021.4	1 028.8	1 036.3	1 043.8	1 051.4	1 059.0	1 066.6	1 074.2	1 081.9	1 089.6	1 097.4	1 105.2
50	995.0	1 002.3	1 009.7	1 017.1	1 024.5	1 031.9	1 039.4	1 046.9	1 054.4	1 062.0	1 069.5	1 077.1	1 084.8	1 092.4	1 100.1
60	990.2	997.5	1 004.9	1 012.2	1 019.6	1 027.0	1 034.5	1 041.9	1 049.4	1 056.9	1 064.4	1 072.0	1 079.5	1 087.1	1 094.8
70	984.9	992.2	999.5	1 006.9	1 014.3	1 021.7	1 029.1	1 036.5	1 043.9	1 051.4	1 058.9	1 066.4	1 074.0	1 081.5	1 089.1
80	979.0	986.4	993.7	1 001.1	1 008.4	1 015.8	1 023.2	1 030.6	1 038.1	1 045.5	1 053.0	1 060.5	1 068.0	1 075.6	1 083.1
90	972.7	980.0	987.4	994.7	1 002.1	1 009.5	1 017.0	1 024.4	1 031.8	1 039.3	1 046.8	1 054.3	1 061.8	1 069.3	1 076.8
100	965.8	973.2	980.6	988.0	995.4	1 002.8	1 010.3	1 017.7	1 025.2	1 032.7	1 040.2	1 047.7	1 055.2	1 062.7	1 070.3
110	958.5	965.9	973.3	980.8	988.3	995.7	1 003.2	1 010.7	1 018.2	1 025.7	1 033.2	1 040.8	1 048.3	1 055.9	1 063.4
120	950.7	958.2	965.7	973.2	980.7	988.2	995.8	1 003.3	1 010.9	1 018.4	1 026.0	1 033.6	1 041.2	1 048.7	1 056.3
130	942.4	950.0	957.6	965.2	972.8	980.4	988.0	995.6	1 003.2	1 010.8	1 018.5	1 026.1	1 033.7	1 041.3	1 049.0
140	933.8	941.4	949.1	956.8	964.5	972.2	979.9	987.6	995.2	1 002.9	1 010.6	1 018.3	1 026.0	1 033.7	1 041.4
150	924.7	932.5	940.3	948.1	955.9	963.7	971.4	979.2	987.0	994.8	1 002.5	1 010.3	1 018.0	1 025.8	1 033.6
160	915.2	923.2	931.1	939.0	946.9	954.8	962.7	970.6	978.5	986.3	994.2	1 002.0	1 009.9	1 017.7	1 025.5
170	905.4	913.5	921.6	929.6	937.7	945.7	953.7	961.7	969.7	977.6	985.6	993.5	1 001.4	1 009.3	1 017.2
180	895.3	903.5	911.7	919.9	928.1	936.3	944.4	952.6	960.7	968.7	976.8	984.8	992.8	1 000.8	1 008.7

For "normal" seawater see Table 8.

Table 8 Density of seawater (kg/m³)

t, °C	Salinity, g/kg										
	30	31	32	33	34	35[a]	36	37	38	39	40
0	1 024.2	1 024.9	1 025.7	1 026.5	1 027.3	1 028.1	1 028.9	1 029.6	1 030.4	1 031.2	1 032.0
10	1 023.2	1 023.9	1 024.7	1 025.4	1 026.2	1 027.0	1 027.7	1 028.5	1 029.3	1 030.0	1 030.8
20	1 020.8	1 021.5	1 022.3	1 023.0	1 023.8	1 024.5	1 025.3	1 026.0	1 026.8	1 027.5	1 028.3
30	1 017.6	1 018.4	1 019.1	1 019.9	1 020.6	1 021.4	1 022.1	1 022.9	1 023.6	1 024.4	1 025.1
40	1 013.9	1 014.7	1 015.4	1 016.2	1 016.9	1 017.7	1 018.4	1 019.1	1 019.9	1 020.6	1 021.4
50	1 009.7	1 010.4	1 011.2	1 011.9	1 012.6	1 013.4	1 014.1	1 014.8	1 015.6	1 016.3	1 017.1
60	1 004.9	1 005.6	1 006.3	1 007.1	1 007.8	1 008.6	1 009.3	1 010.0	1 010.8	1 011.5	1 012.2
70	999.5	1 000.3	1 001.0	1 001.7	1 002.5	1 003.2	1 003.9	1 004.7	1 005.4	1 006.2	1 006.9
80	993.7	994.4	995.2	995.9	996.6	997.4	998.1	998.8	999.6	1 000.3	1 001.1
90	987.4	988.1	988.8	989.6	990.3	991.1	991.8	992.5	993.3	994.0	994.7
100	980.6	981.3	982.1	982.8	983.5	984.3	985.0	985.8	986.5	987.2	988.0
110	973.3	974.1	974.8	975.6	976.3	977.1	977.8	978.6	979.3	980.0	980.8
120	965.7	966.4	967.2	967.9	968.7	969.4	970.2	970.9	971.7	972.4	973.2
130	957.6	958.4	959.1	959.9	960.6	961.4	962.1	962.9	963.7	964.4	965.2
140	949.1	949.9	950.7	951.4	952.2	953.0	953.7	954.5	955.3	956.0	956.8
150	940.3	941.1	941.8	942.6	943.4	944.2	945.0	945.7	946.5	947.3	948.1
160	931.1	931.9	932.7	933.5	934.3	935.1	935.8	936.6	937.4	938.2	939.0
170	921.6	922.4	923.2	924.0	924.8	925.6	926.4	927.2	928.0	928.8	929.6
180	911.7	912.6	913.4	914.2	915.0	915.8	916.7	917.5	918.3	919.1	919.9

[a] "Normal" seawater.

Table 9 Prandtl number of seawater and its concentrates

t, °C	Salinity, g/kg															
	10	20	30	35[a]	40	50	60	70	80	90	100	110	120	130	140	150
0	13.1	13.1	13.2	13.2	13.3	13.4	13.5	13.6	13.8	14.0	14.3	14.5	14.8	15.2	15.5	16.0
10	9.35	9.39	9.46	9.49	9.53	9.62	9.72	9.84	9.97	10.1	10.3	10.5	10.7	11.0	11.2	11.6
20	6.99	7.04	7.11	7.13	7.17	7.24	7.33	7.43	7.53	7.67	7.80	7.96	8.13	8.32	8.52	8.76
30	5.45	5.49	5.54	5.58	5.60	5.67	5.74	5.82	5.92	6.01	6.12	6.24	6.39	6.54	6.69	6.88
40	4.38	4.41	4.46	4.48	4.51	4.57	4.63	4.70	4.78	4.86	4.95	5.05	5.16	5.28	5.42	5.56
50	3.60	3.64	3.68	3.71	3.73	3.77	3.83	3.89	3.95	4.02	4.10	4.18	4.28	4.38	4.48	4.60
60	3.03	3.06	3.10	3.12	3.14	3.19	3.24	3.28	3.34	3.40	3.47	3.54	3.61	3.69	3.78	3.88
70	2.60	2.63	2.66	2.68	2.70	2.74	2.78	2.82	2.87	2.92	2.98	3.04	3.11	3.18	3.25	3.33
80	2.26	2.29	2.32	2.34	2.35	2.39	2.42	2.46	2.50	2.55	2.60	2.65	2.71	2.77	2.83	2.90
90	2.00	2.02	2.05	2.06	2.08	2.11	2.14	2.18	2.21	2.25	2.29	2.34	2.39	2.44	2.50	2.56
100	1.78	1.80	1.83	1.84	1.86	1.88	1.92	1.94	1.98	2.01	2.05	2.09	2.13	2.18	2.23	2.28
110	1.61	1.63	1.65	1.66	1.68	1.70	1.73	1.75	1.78	1.81	1.84	1.88	1.92	1.96	2.00	2.05
120	1.47	1.49	1.51	1.51	1.53	1.55	1.57	1.60	1.62	1.65	1.68	1.71	1.75	1.78	1.82	1.86
130	1.35	1.37	1.38	1.39	1.40	1.42	1.44	1.46	1.49	1.51	1.54	1.57	1.60	1.63	1.66	1.70
140	1.24	1.26	1.28	1.29	1.30	1.31	1.33	1.35	1.37	1.39	1.42	1.44	1.47	1.50	1.53	1.56
150	1.16	1.18	1.19	1.20	1.21	1.22	1.24	1.26	1.27	1.30	1.32	1.34	1.36	1.39	1.42	1.45
160	1.08	1.10	1.11	1.12	1.13	1.14	1.16	1.17	1.19	1.21	1.23	1.25	1.28	1.30	1.32	1.35
170	1.03	1.04	1.05	1.06	1.06	1.07	1.09	1.10	1.12	1.13	1.16	1.17	1.20	1.22	1.24	1.26
180	0.975	0.983	0.997	1.00	1.00	1.02	1.03	1.04	1.06	1.07	1.09	1.10	1.13	1.14	1.17	1.19

[a] "Normal" seawater.

CHAPTER 4
Properties of Fluid Mixtures

4.1. Phase Behavior of Mixtures

A. Introduction

Liquid mixtures can take on a wide variety of characteristics. They can range from a simple, ideal, two-component mixture such as that formed by isobutane and normal butane, through a binary mixture with strongly nonideal behavior such as ethyl alcohol and water or acetone and water, to a boiling range cut of a crude oil that has a composition so complex that it is not normally identified in terms of mole fractions of pure components. Because of these widely diverse characteristics, liquid mixtures seldom are amenable to simple rules in prediction of their characteristics. In addition, the rules that work for one kind of mixture probably will not work well for another. In many of the cases that follow there will be different estimating techniques recommended for different kinds of mixtures.

In the world of design today there is heavy dependence on the digital computer for developing property information and for carrying out many of the design calculations. Techniques for carrying out these calculations will be discussed briefly. Attention will also be directed to calculations that can be carried out with a hand calculator or slide rule and give reasonably compatible results with the more-detailed computer calculations.

Gibbs [1] was the first to propose criteria for establishing the equilibrium state. His phase rule is:

$$F + P = C + 2 \qquad (1)$$

where F = number of degrees of freedom for the system
P = number of phases present
C = number of components or chemical constituents present,
2 = constant needed when considering both temperature and pressure as variables

The principles of the phase rule apply equally well to single components and to mixtures. In addition, it can be used [2-4] to describe steady-state systems that are not really in thermodynamic equilibrium.

In engineering work many different kinds of systems are encountered. In spite of the fact that there may be liquid-liquid, liquid-liquid-gas, and solid-liquid-gas systems, the combination of overriding importance in most processing applications is vapor-liquid equilibrium. In working with vapor-liquid equilibrium one immediately encounters two broad categories: hydrocarbons and nonhydrocarbons. The engineer working with hydrocarbon systems, almost by the nature of his systems, is working with multicomponent mixtures. Rare, indeed, is the hydrocarbon design situation where either a pure component or binary mixture is of prime importance.

On the other hand, the engineer involved with chemicals, even petrochemicals, frequently has only two components and rarely encounters more than three with major composition constituency. For this reason, the fundamental approaches utilized in the two sectors tend to be different. For this reason also, much of the following discussion will be divided between hydrocarbon and chemical systems.

While Eq. (1) enables the engineer to specify the necessary number of variables to make certain that the system is in equilibrium, it does not provide much additional information. Gibbs also formulated the free energy and suggested that at thermodynamic equilibrium the free energy for a given system was at a minimum. This concept has calculational value and has been used for both physical operations and chemically reacting systems. Because of the sheer volume of calculations involved, it is suitable only for digital-computer solution and in no way can be considered to be a tool for hand calculation. However, in complex systems such as encountered in a Claus unit, which produces sulfur from hydrogen sulfide [5], the concept of minimum free energy for the system provides a means for estimating compositions at the equilibrium state. It has been used also for systems in physical equilibrium such as in vapor–liquid flash calculations.

B. Vapor-liquid equilibrium

In vapor-liquid equilibrium, the overall composition of the mixture is generally known. The problem is to relate the instantaneous vapor and liquid compositions to the overall composition in such a way that both a material balance and the criteria for thermodynamic equilibrium are satisfied.

Frequently, the vapor and liquid will not actually be in equilibrium. However, in most cases, evaluation of properties at the equilibrium state will be sufficiently close for engineering needs. The criterion for equilibrium in a vapor-liquid mixture is

$$f_{i_v} = f_{i_l} \tag{2}$$

where f = fugacity ($\bar{F}_i = RT \ln f_i$). For an ideal gas, $f_i = cP$. For a real gas,

$$\lim_{P \to 0} \frac{f_i}{P} = 1.0$$

$\bar{F}$ = partial molal free energy
c = constant
v = vapor
l = liquid
i = any component

Equation (2) applies individually to each component in the mixture, whether there be two or twenty. The problem of predicting phase compositions resolves into relating the fugacities for each component to mixture compositions. This discussion will concentrate on primary variables–temperature, pressure, composition, etc. Such things as nuclear interference, gravitational or magnetic effects, etc., will be neglected. These things come up only in rare instances and are beyond the scope of this treatment.

According to the law of ideal solutions, the fugacity for a component in the vapor can be expressed as

$$f_{i_v} = f_{i_l}^0 \tilde{y}_i \tag{3}$$

For the same component in the liquid phase:

$$f_{i_l} = f_{i_l}^0 \tilde{x}_i \tag{4}$$

where f_{i_l} = fugacity of component i in the liquid mixture
f_{i_v} = fugacity of component i in the vapor mixture
$f_{i_l}^0$ = fugacity of component i in the pure liquid state
$f_{i_v}^0$ = fugacity of component i in the pure vapor state

Study of Eqs. (3) and (4) immediately discloses a potential problem: not all components in the mixture may exist in the pure liquid state at the conditions of the mixture. Indeed, this does present a serious problem in many hydrocarbon cases where the lightest component in the mixture may be well above its critical temperature.

Because the fugacity is an "unbounded" variable, different relationships are ordinarily used for correlation purposes. The fugacity coefficient is such a different variable, a bounded variable, always falling between 0 and 1.0.

An alternative form of Eq. (3) for the vapor phase is developed using the fugacity coefficient:

$$\gamma_{g_i} = \frac{f_{i_v}}{\tilde{y}_i \pi} \tag{5}$$

where $\tilde{y}$ = composition of the vapor phase
π = total pressure on the system
γ_{g_i} = fugacity coefficient

By definition, $\gamma \to 1.0$ as $\pi \to 0$. This must be true for all components in the mixture. The actual value of π when γ_{g_i} becomes close enough to 1.0 to be considered unity depends upon the nature of the system. For most hydrocarbon mixtures, at temperatures near the normal boiling point, this pressure is less than 100–150 psi. For mixtures containing associating materials (acetic acid-water), there may be significant departure of the fugacity coefficient at pressures much less than 1 atm.

For a component in the liquid phase an alternative is to use the activity coefficient which is defined as

$$\gamma_{l_i} = \frac{a_i}{\tilde{x}_i} \tag{6}$$

where a_i is the activity of component i in the liquid phase. The activity coefficient can also be related to the standard state fugacity by

$$\gamma_{l_i} = \frac{f_{i_l}}{\tilde{x}_i} f_{i_l}^0 \tag{7}$$

The standard state fugacity $f_{i_l}^0$ is defined at an arbitrarily chosen pressure and composition. Unfortunately, many workers do not clearly define their standard state for $f_{i_l}^0$ and this tends to minimize the contribution of their work.

Most equilibrium constant evaluations from the fugacity-activity coefficient basis are developed from one of the following expressions:

$$K = \frac{f_{i_l}/\tilde{x}_i}{f_{i_v}/y_i} \frac{1}{\pi} \quad (8)$$

$$K = \frac{f_{i_l}/P^0\tilde{x}_i}{f_{i_v}/\pi\tilde{y}_i} \frac{P^0}{\pi} \quad (9)$$

$$K = \frac{f_{i_l}/f_{i_l}^0 x_i f_{i_l}^0}{f_{i_v}/\pi\tilde{y}_i} \quad (10)$$

$$K = \frac{\gamma_{l_i} v_i}{\gamma_{g_i}} \quad (11)$$

where π = total system absolute pressure
P^0 = pure component vapor pressure
γ_{l_i} = activity coefficient of any component in the liquid mixture
f_{i_l} = fugacity of a pure liquid at system conditions
$f_{i_l}^0$ = fugacity of any component as a pure liquid
γ_{g_i} = fugacity coefficient of any component in the vapor mixture

As a matter of interest, Chao and Seader [6] used Eq. (11) for prediction of equilibrium constants and enthalpies. Their correlation was the first equilibrium constant correlation developed with the express intent of being used on a digital computer for phase equilibrium generation. They used the Pitzer acentric factor [7] as a third correlating parameter in their corresponding states-type approach.

C. Hydrocarbon phase behavior

In estimating mixture properties there are a number of characteristics of the mixture that can be used. Such characteristics as critical temperature, critical pressure, acentric factor, etc. are frequently referred to as correlating parameters. Estimating the correlating parameters for nearly ideal mixtures made up of light hydrocarbons is exceedingly difficult. Most often the critical temperature and critical pressure for a mixture are estimated as the Kay's rule or mole fraction average values; also referred to as "pseudocritical" temperature or pressure:

Critical temperature

$$T_{c_m} = \Sigma\, \tilde{x}_i T_{c_i} \quad (12)$$

Critical pressure

$$P_{c_m} = \Sigma\, \tilde{x}_i P_{c_i} \quad (13)$$

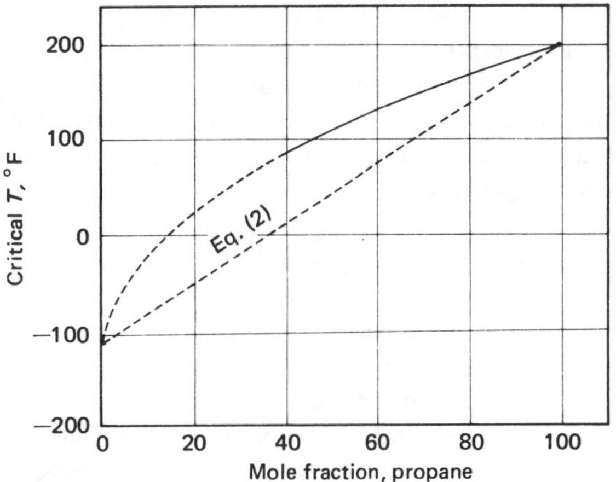

Figure 1 Composition-critical temperature diagram for methane-propane binary.

Equations (12) and (13) assume a linear relationship between composition and the critical property in question. Actual behavior is demonstrated by Figs. 1 and 2 for the methane-propane binary mixture. Data for preparation of the two figures were taken from Katz et al. [8]. Deviation of the critical temperature as shown in Fig. 1 does not appear to be particularly significant. However, as shown in Fig. 2 the influence of this deviation on the critical pressure causes significant deviations from the mole fraction average. The values used to calculate the correlating parameters are so different from the true values that there is good reason for significant error to show up in values calculated through correlations.

The acentric factor was defined by Pitzer et al. [9] as

$$\omega = -(\log P_r^0 - 1)_{T_r = 0.7} \quad (14)$$

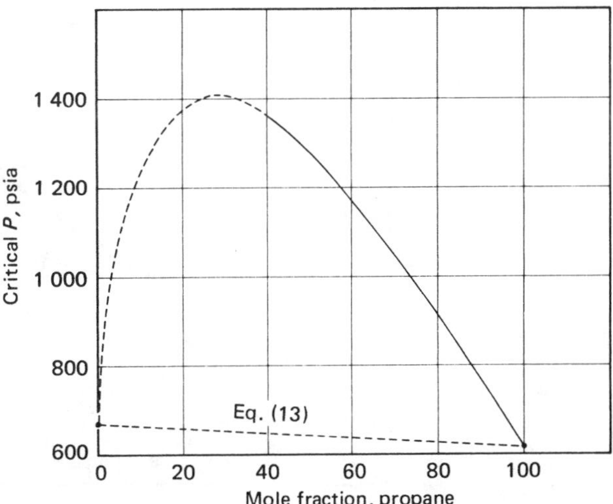

Figure 2 Composition-critical pressure for methane-propane binary.

where P_r^0 = reduced vapor pressure $= P^0/P_c$
ω = acentric factor
P^0 = vapor pressure in pascals

Edmister [10] showed that, when vapor pressure data are not available, the acentric factor can satisfactorily be approximated by

$$\omega = \frac{3}{7}\,\frac{\log\,[P_c/(1.013\,25 \times 10^5)]}{T_c/T_b - 1.0} - 1.0 \qquad (15)$$

where P_c = critical pressure in pascals
T_c = critical temperature in K
T_b = normal boiling point in K

There are many instances in which the vapor-liquid behavior of mixtures is of importance in heat transfer calculations. While the boiling point of a pure component is a fixed temperature at given pressure, this is not true for mixtures. The range of temperatures at which boiling/condensation can occur at a given pressure is dependent on the composition. Figure 3 shows a pressure-temperature diagram for a typical mixture of light hydrocarbons. The overall composition of the mixture whose behavior is depicted in Fig. 3 is constant. The composition of the vapor and liquid phases will change from point to point if, indeed, they do exist. Examination of Fig. 3 immediately discloses some significant differences between mixture and pure component behavior. The critical temperature for a pure component is defined as that temperature above which the material cannot be liquefied. Obviously this definition does not apply to the mixture whose behavior is depicted in Fig. 3. There is quite a range of temperatures higher than the critical temperature at which some liquid can exist in equilibrium with the vapor. For a pure component the critical pressure represents the maximum vapor pressure that the component can exert. The maximum pressure of the two-phase envelope on Fig. 3 exceeds the critical pressure by an amount that depends on the shape of the curve.

Defining the critical temperature for a mixture requires dependence on observations and experimental evidence. On pressure-temperature diagrams such as shown in Fig. 3, lines of constant percentage liquid tend to converge to the point thought to represent the critical temperature and pressure. The curves generated by calculating around the *P-T* envelope tend to extrapolate to the point thought to represent the critical of the mixture. Based on these observations the critical temperature of a mixture can be defined as that temperature above which all of the mixture cannot be liquefied.

Another significant difference in mixture behavior is the existence of so-called "retrograde" regions in the two-phase envelope. This is better illustrated by Fig. 4. Consider a constant pressure heating path beginning at point 1 and extending to point 2. At point 1 the mixture would be a subcooled liquid. Heating to point A at that pressure would result in the mixture entering the two-phase region. Until a point approximately at B is reached, additional heating would cause additional liquid formation. From point B to point C there would be vaporization of the liquid that had been formed. At point 2 the mixture would be a single-phase dense gas. The same kind of thing would occur on a constant temperature-pressure decrease path beginning at point 3 and extending to point 4. This retrograde behavior has been observed for many years in hydrocarbon reservoirs at high temperatures and pressures. Recently, however, Maddox and Erbar [11] have shown that this retrograde behavior can be exhibited at pressures below 3.5 MPa and temperatures below 200 K.

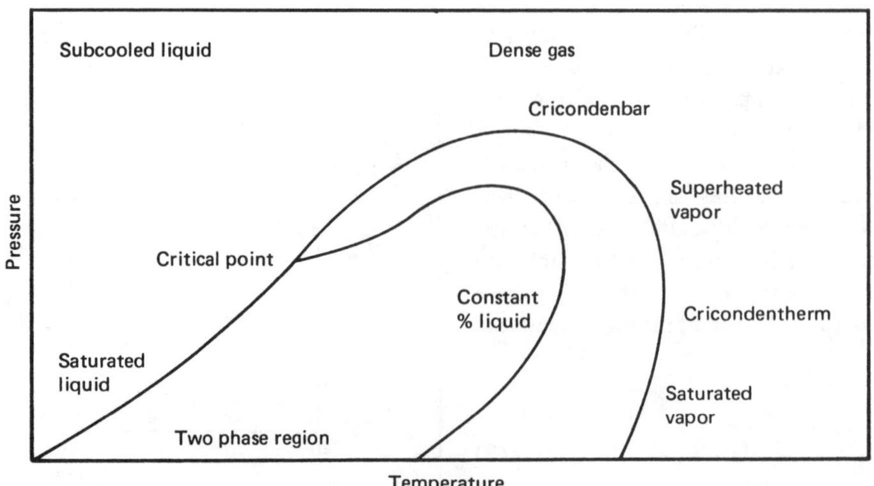

Figure 3 Phase envelope for a typical natural gas mixture.

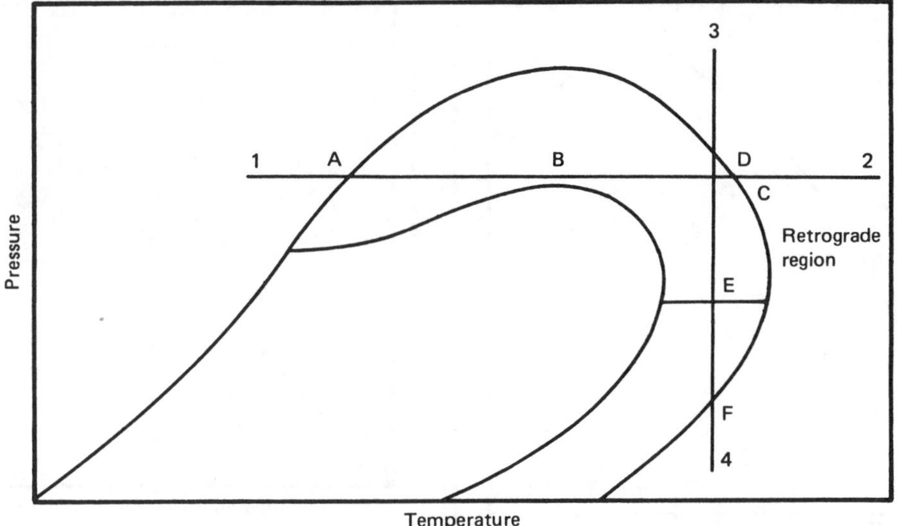

Figure 4 Retrograde behavior.

D. Nonhydrocarbon phase behavior

The vapor-liquid behavior of hydrocarbon mixtures has been investigated over a wider range of temperatures and pressures, particularly for the light hydrocarbon defined components, than most individual components in nonhydrocarbon mixtures. Hydrocarbons do illustrate some liquid-phase nonidealities. However, in most cases these are small when compared with the liquid-phase nonidealities that can exist in mixtures containing one or more nonhydrocarbons. Hydrocarbons, on the other hand, tend to be processed more frequently at pressures where pressure effects can cause serious shifts from ideal behavior. For these reasons, most attention in nonhydrocarbon mixtures tends to be directed to liquid-phase nonidealities.

Figure 5 shows a $\tilde{y}$-$\tilde{x}$, or equilibrium, diagram for a binary mixture that exhibits ideal behavior. The curve may be used to replace traditional bubble point–dew point calculations using Henry's Law:

$$\tilde{y}_i = K_i\tilde{x}_i \tag{16}$$

If experimentally determined equilibrium constants for the mixture are not available, Raoult's law may be used in place of Henry's law:

$$\pi\tilde{y}_i = P_i^0\,\tilde{x}_i \tag{17}$$

Nonideal vapor-liquid phase behavior is illustrated in Figs. 6 and 7. In Fig. 6 the nonideality is sufficient only to cause a point of inflection in the vapor-liquid equilibrium behavior. In Fig. 7 the $\tilde{y}$-$\tilde{x}$ curve actually crosses the $\tilde{y}$=$\tilde{x}$ line. This represents an azeotrope or constant boiling mixture that, from a vapor-liquid equilibrium standpoint, behaves as a pure component. In the nonideal system the K values, or equilibrium constants, may be expressed as

$$K = \gamma K^0 \tag{18}$$

Predicting the occurrence of the kind of behavior demonstrated in Figs. 6 or 7 is all but impossible. Experimental data on the particular system must be obtained and evaluated. Group-contribution methods such as those discussed in Sec. 4.2 do show some promise for prediction in this area.

E. Phase behavior of mixtures of undefined components

Mixtures of undefined composition can occur in two ways: when they constitute the entire mixture, such as in crude oil fractions, or when they constitute only a portion of the total mixture, such as in most natural gases. In either case, some experimental data are necessary before the vapor-liquid phase behavior of the mixture can be satisfactorily predicted.

In most cases where a crude oil or a fraction of a crude oil constitutes the entire mixture, vapor-liquid equilibrium behavior is desired at low pressures. In this case the main problem becomes one of evaluating vaporization characteristics. Edmister [10] details a procedure including sample calculations for making these predictions. The integral technique treats the mixture of undefined components as a continuum (an excellent approximation for crude oil mixtures). It utilizes the density, boiling range, and true boiling point distillation to evaluate vaporization characteristics. It is applicable to flash vaporization, continuous and batch distillation.

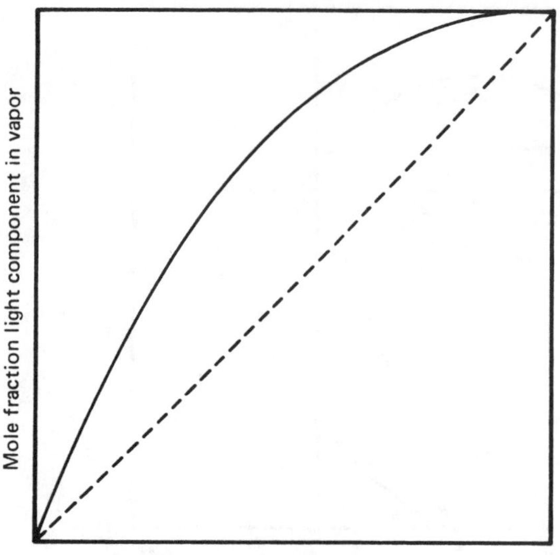

Figure 5 Equilibrium curve for ideal binary mixture.

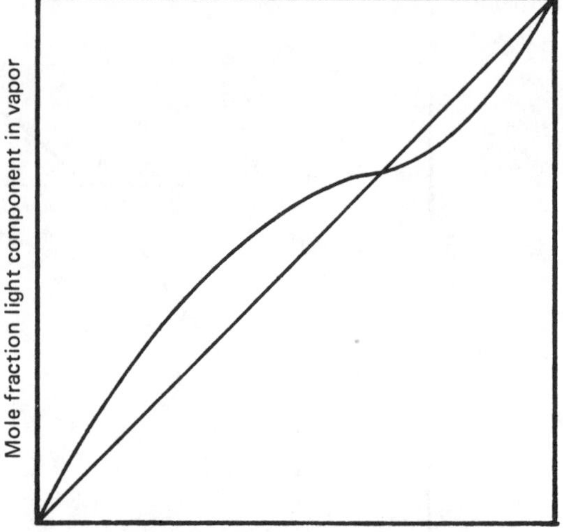

Figure 7 Equilibrium curve for a binary mixture with azeotrope.

For cases where the undefined component constitutes a small fraction of the total stream, and where an equation of state is to be used to predict vapor-liquid behavior, such as typical natural gas streams, more-detailed procedures are required. Every equation of state capable of handling mixtures of undefined components requires as input to the computer program information concerning certain "correlating parameters." These correlating parameters will include such characteristics as critical temperature, critical pressure, acentric factor, true boiling point curve, density, UOP characterization factor, etc. Wilson et al. [12] discuss the importance of characterizing undefined mixtures and describe a procedure for estimating the correlating parameters in such a way as to match experimentally determined bubble point-dew point-percent vaporization data. Maddox and Erbar [13] provide discussion in greater detail along with example problems worked out in detail.

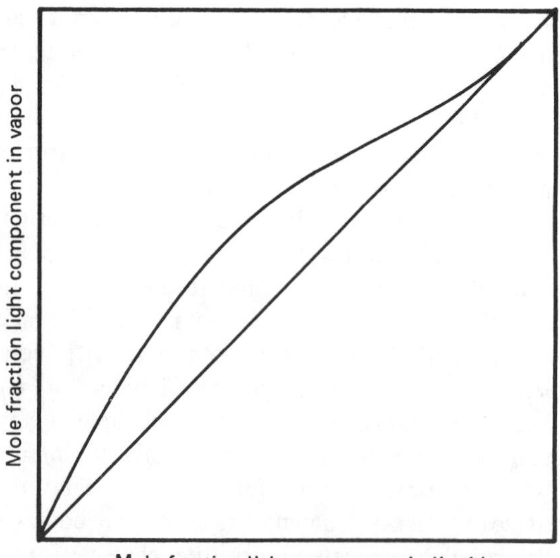

Figure 6 Equilibrium curve for nonideal binary mixture.

For a mixture of undefined composition, the most generally available information is the density (specific gravity) and the boiling range from either a simple or true boiling point distillation. From this information, other characteristics of the mixture can be predicted. Kesler and Lee [14] have presented equations that use density and boiling point to predict the following properties:

Critical temperature

$$T_c = 189.8 + 450.6\text{SG} + (0.424\,4 + 0.117\,4\text{SG})T_b + (0.144\,1 - 1.006\,9\text{SG})10^5/T_b \qquad (19)$$

Critical pressure

$$\ln P_c = 17.201\,9 - \frac{0.056\,6}{\text{SG}} - \left(0.436\,4 + \frac{4.121\,6}{\text{SG}} + \frac{0.213\,43}{\text{SG}^2}\right)10^{-3}T_b + \left(4.757\,9 + \frac{11.819}{\text{SG}} + \frac{1.530\,2}{\text{SG}^2}\right)10^{-7}T_b^2 - \left(2.450\,5 + \frac{9.900\,9}{\text{SG}^2}\right) \times 10^{-10}T_b^3 \qquad (20)$$

Acentric factor

$$\omega = -7.904 + 0.135\,2K - 0.007\,465K^2 + 8.359T_{br} + \frac{1.408 - 0.010\,63K}{T_{br}} \quad \text{for } T_{br} > 0.8 \qquad (21)$$

Molecular weight

$$MW = -12\,272.6 + 9\,486.4SG + (8.374\,1 - 5.991\,2SG)T_b + (0.555\,56 - 0.428\,24SG - 0.011\,433SG^2)\left(1.343\,7 - \frac{400.43}{T_b}\right)\frac{10^7}{T_b} + (0.171\,5 - 0.138\,7SG + 0.003\,317SG^2) \times \left(1.882\,8 - \frac{101.1}{T_b}\right)\frac{10^{12}}{T_b^3} \quad (22)$$

where T_b = normal boiling point (taken as the mean of the boiling temperature range), K

SG = specific gravity (i.e., the ratio of the fluid density to the density of water at the same temperature, normally taken at 15.56°C)

T_{br} = reduced normal boiling point = T_b/T_c

K = Watson characterization factor = 1.216 $(CABP)^{1/3}/SG$

CABP = cubic average boiling point (i.e., the cube root of the mean of the cubes of the maximum and minimum temperatures in the boiling range. Note that CABP ≈ T_b if the boiling range is small)

T_c = critical temperature, K

P_c = critical pressure, Pa

Equation (21) is intended for heavy mixtures having reduced normal boiling points greater than 0.8. For lighter mixtures, Eq. (23) should be used to predict the acentric factor.

$$\omega = (\ln P_{br}^s - 5.927\,14 + \frac{6.069\,48}{T_{br}} + 1.288\,62 \ln T_{br} - 0.169\,347\, T_{br}^6)/(15.251\,8 - \frac{15.687\,5}{T_{br}} - 13.472\,1 \ln T_{br} + 0.435\,77\, T_{br}^6) \quad (23)$$

where P_{br}^s = reduced vapor pressure at T_b.

Example

Calculate the properties of a 260°-315.6°C (533.15-588.75 K) boiling point range cut of a mid-continent crude oil which has a specific gravity of 0.842 3.

Solution

The first step in the solution is evaluation of the Watson characterization factor, K. This value is evaluated at the average of the boiling range, the mid-boiling point T_b (i.e., 560.95 K). The error introduced by this assumption will normally be small.

$$K = \frac{1.216\sqrt[3]{560.95}}{0.842\,3} = 11.90 \quad (24)$$

The critical temperature can be evaluated from Eq. 19,

$$T_c = 189.8 + 450.6(0.842\,3) + [0.424\,4 + 0.117\,4(0.842\,3)] \times 560.95 + [0.144\,1 - 1.006\,9(0.842\,3)]\,\frac{10^5}{T_b}$$

$$= 189.8 + 379.5 + 293.5 - 125.5 = 737.3 \text{ K}$$

$$T_{br} = \frac{560.95}{737.3} = 0.761$$

The critical pressure can be estimated by use of Eq. (20):

$$\ln P_c = 17.201\,9 - \frac{0.056\,6}{0.842\,3} - \left(0.436\,4 + \frac{4.121\,6}{SG} + \frac{0.213\,43}{SG^2}\right)10^{-3}T_b + \left(4.757\,9 + \frac{11.819}{SG} + \frac{1.530\,2}{SG^2}\right)10^{-7}T_b^2 - \left(2.450\,5 + \frac{9.900\,9}{SG^2}\right) \times 10^{-10}T_b^3 = 17.201\,9 - 0.067\,2 - 3.158\,4 + 0.659\,1 - 0.289\,6$$

$$\ln P_c = 14.345\,8$$

$$P_c = 1.699 \times 10^6 \text{ Pa}$$

$$P_{br} = \frac{1.013 \times 10^5}{1.699 \times 10^6} = 0.059\,62$$

Since the reduced boiling point is less than 0.8, Eq. (23) will be used to evaluate the acentric factor:

$$\omega = \{\ln(0.059\,2) - 5.927\,141 + 6.069\,48/0.761 + (1.288\,62 \ln 0.761) - [0.169\,347(0.761)^6]\}/ \times [15.251\,8 - (15.687\,5/0.761) - (13.472\,1 \ln 0.761) + 0.435\,79(0.761)^6]$$

$$= (-2.826\,83 - 5.927\,14 + 8.011\,4 - 0.351\,95 - 0.032\,89)/(15.251\,8 - 20.614\,32 + 3.679\,53 + 0.084\,64)$$

$$\omega = \frac{-1.127\,41}{-1.598\,35} = 0.782\,3$$

The molecular weight for the fraction is

$$MW = -12\,272.6 + 9\,486.4(0.842\,3) + [8.374\,1 - 5.991\,2(0.842\,3)]\,(560.95) + [0.555\,56 - 0.428\,24(0.842\,3) - 0.011\,433(0.842\,3)^2] \times \left(1.343\,7 - \frac{400.43}{560.95}\right)\frac{10^7}{T_b} + [0.171\,5 - 0.138\,7(0.842\,3) + 0.003\,817(0.842\,3)^2 \times \left(1.882\,8 - \frac{101.1}{T_b}\right)\frac{10^{12}}{T_b^3} = -12\,272.6 + 7\,990.4 + 1\,866.7 + 2\,096.9 + 553.3$$

$$MW = 234.6$$

The true critical temperature of mixtures of natural gas rich in methane can be estimated through use of Fig. 8. (Figure 8 is taken after the chart in the API Technical Data Book–Refining [16]). The method requires estimation of the weight average API gravity. The API gravity is related to specific gravity by the formula: API gravity = (141.5/SG) − 131.5. For most components, values of density (and, hence, specific gravity) from Chapter 2 can be used. However, the following values must be used for the indicated components:

Component	Effective API gravity	Effective specific gravity
Methane	440	0.247
Ethane	213	0.410
Carbon dioxide	42	0.816
Nitrogen	43.6	0.808

Proper use of Fig. 8 is illustrated in the following example.

Estimate the true critical temperature of the natural gas consisting of methane (0.914 mole fraction), ethane (0.061 mole fraction), propane (0.011 mole fraction) and carbon dioxide (0.014 mole fraction).

Solution

1. Obtain the molecular weight for each component from Chapter 2 and calculate the mole fraction average molecular weight by summing the product of the mole fraction and molecular weight for each component:

Component	Mole fraction	Mole weight, kg/kmole	Mole fraction × molecular weight
Methane	0.914	16.042	14.66
Ethane	0.061	30.068	1.83
Propane	0.011	44.094	0.49
Carbon dioxide	0.014	44.011	0.62
Mole fraction average molecular weight			17.60

2. Calculate the weight fraction of each component by dividing the product of the mole fraction and the molecular weight for each component by the mole fraction average molecular weight (e.g., for methane, weight fraction = 14.66/17.60 = 0.833).

3. Calculate the weight average API gravity.

Component	API gravity	Weight fraction	Weight fraction × API gravity
Methane	440	0.833	366.5
Ethane	213	0.104	22.2
Propane	147.2	0.028	4.1
Carbon dioxide	42.0	0.035	1.5
Weight fraction average API gravity =			394.3

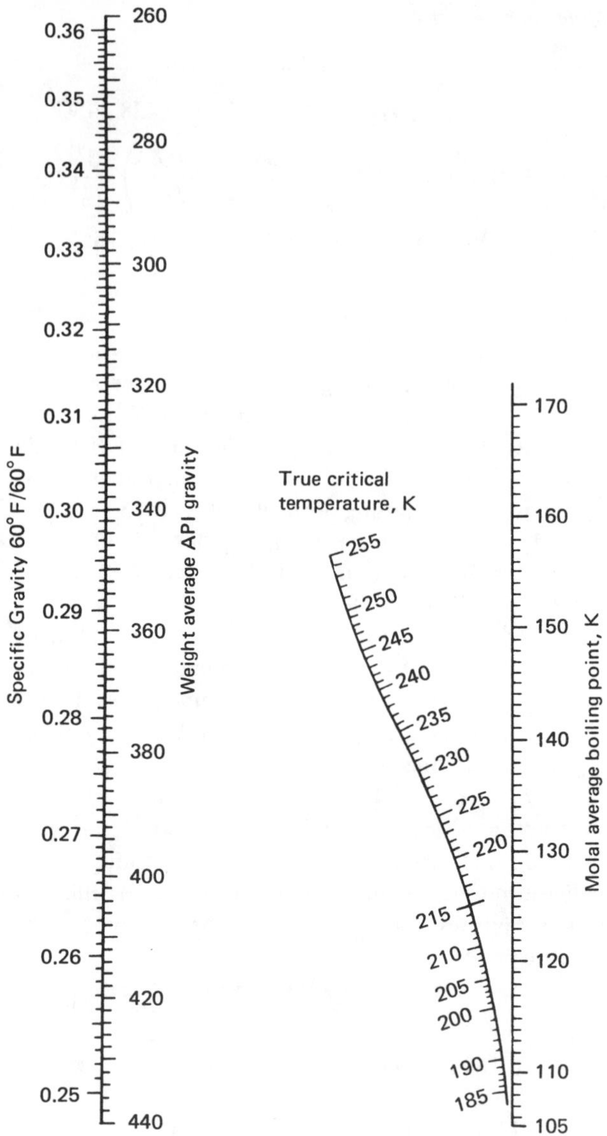

Figure 8 Critical temperature of natural gas mixtures.

4. Calculate the molal average boiling point.

Component	Boiling point, K	Boiling point × mole fraction
Methane	111.42	101.8
Ethane	184.52	11.3
Propane	231.1	2.5
Carbon dioxide	194.65	2.7
Mole fraction average boiling point =		118.3

5. Read the mixture true critical temperature from Fig. 8. Draw a line between 394.3 on the left hand (weight, average API gravity) scale and 118.3 on the right hand (mole fraction average boiling point) scale and read off the critical temperature (206 K) from the curved scale.

4.2. Thermodynamic Properties

A. Light hydrocarbons

There are reasonably reliable charts available from which equilibrium constants for light hydrocarbons can be obtained. Perhaps the best, and most widely used, are those published by the Gas Processors Association [17]. Values from the charts, if properly used, will agree closely with the values obtained from a good equation of state. In addition, they are easily and quickly used and ideal for hand calculations.

If more accurate equilibrium constants are desired, or if equilibrium conditions are being predicted by computer, an equation of state is used for estimating the equilibrium constant values. The most widely used equations of state for light hydrocarbons are the Soave version of the Redlich-Kwong (SRK) [18-20], the Starling version of the Benedict-Webb-Rubin (BWR) [21], and the Peng-Robinson (PR) [22]. Each has advantages and disadvantages, proponents and opponents. More importantly, different computer solutions for each equation may yield equilibrium constants that are significantly different. Erbar and Maddox [23] have made comparison calculations for simple equilibrium flash calculations that differed by as much as 20%. The same equation of state was used for all calculations, the programming of the solution was different.

Equilibrium coefficients are calculated from the equation of state using one of the Eqs. 4.1.(8)–4.1.(11). All thermodynamic quantities may be related to each other by mathematical manipulation. The fugacity coefficient is related to pressure, volume, temperature, and compressibility factor by

$$\ln \gamma_{g_i} = \frac{1}{\tilde{R}T} \int_v^{\infty} \left[\left(\frac{\partial P}{\partial n_i} \right)_{T,V,n_j} - \frac{\tilde{R}T}{V} \right] dV - \ln Z \tag{1}$$

where $\tilde{R}$ is the universal gas constant, Z is the compressibility factor $= PV/RT$, and n_i is the number of moles of component i in the mixture.

To develop the equations for calculating the fugacity coefficient, the basic equation of state is differentiated and substituted into Eq. (1). A direct analytical solution for the fugacity coefficient results once the constants have been evaluated, if the volumetric behavior of the system is known.

Prediction of enthalpies and entropies is handled differently. Ideal gas values for these variables are known rather accurately. The calculation is broken into two steps: an isobaric influence of temperature and an isothermal influence of pressure. Mixture values for ideal gas behavior are taken as the composition weighted average.

The equations necessary for use in the SRK equation are shown in Table 1. The SRK is a two-constant, or two-parameter, equation. The constants A and B can be determined from the correlating parameters for each component in the mixture.

The equations for the Peng-Robinson equation are shown in Table 2. It also is a two-constant equation of state.

The BWR equations are shown in Table 3. They have eleven constants that must be evaluated for each component. More PVT data are required for determination of the eleven constants than would be required for the two-constant equations.

The k_{ij} terms that appear in the equations are the

Table 1 Equations for Soave-Redlich-Kwong equation of state

$$K_i = \frac{\phi_i^L}{\phi_i^v}$$

$$\ln \phi_i = \frac{b_i}{b}(Z-1) - \ln(Z-B) - \frac{A}{B}\left[\frac{2(ac\alpha)_i}{ac\alpha} - \frac{b_i}{b}\right]\ln\left(1+\frac{B}{Z}\right)$$

$$Z^3 - Z^2 + (A - B - B^2)Z - AB = 0.0$$

$$A = \frac{ac\alpha P}{R^2T^2}; \quad B = b\frac{P}{T}; \quad b = \Sigma y_i b_i$$

$$b_i = \Omega_b \frac{RT_{ci}}{P_{ci}}; \quad a_{ci} = 0.427\ 47\frac{R^2T_{ci}^2}{P_{ci}}$$

$$\Omega_b = 0.086\ 67; \quad \alpha_{ci}^{0.5} = 1 + m_i(1 - T_{ri}^{0.5})$$

$$ac\alpha = \sum_i \sum_j x_i x_j a_{ci}^{0.5} a_{cj}^{0.5} \alpha_i^{0.5} \alpha_j^{0.5}(1-k_{ij})$$

$$(ac\alpha)_i = \sum_{j=1}^{n} x_j a_{ci}^{0.5} a_{cj}^{0.5} \alpha_i^{0.5} \alpha_j^{0.5}(1-k_{ij})$$

Table 2 Equations for Peng-Robinson equation of state

$$P = \frac{RT}{v-b} - \frac{a}{v(v+b)+b(v-b)}$$

$$Z^3 - (1-B)Z^2 + (A - B^2 - 2B)Z - (AB - B^2 - B^3) = 0.0$$

$$\ln \phi_i = \frac{b_i}{b}(Z-1) - \ln(Z-B) - \frac{A}{2\sqrt{2}B}\left(\frac{2\sum_i x_i a_{ik}}{a} - \frac{b_i}{b}\right)\ln\left(\frac{Z+2.414B}{Z-0.414B}\right)$$

$$\frac{\Delta H}{RT} = Z - 1 + \frac{T\,da/dT - a}{2\sqrt{2}b}\ln\left(\frac{Z+2.414B}{Z-0.414B}\right)$$

$$a = \sum_j^n \sum_i^n x_i x_j (a_{ci} a_{cj} \alpha_i \alpha_j)^{1/2}(1-k_{ij}) \qquad A = \frac{aP}{R^2T^2}$$

$$b = \Sigma x_i b_i \qquad B = \frac{bP}{PT}$$

$$a_{c_i} = 0.457\ 24\frac{R^2T_{ci}^2}{P_{c_i}}$$

$$b_i = 0.077\ 80\frac{RT_c}{P_{c_i}}$$

Table 3 Equations for Benedict-Webb-Rubin equation of state

$$p = \rho RT + \left(B_0RT - A_0 - \frac{C_0}{T^2} + \frac{D_0}{T^3} - \frac{E_0}{T^4}\right)\rho^2 + \left(bRT - a - \frac{d}{T}\right)\rho^3 + \alpha\left(a + \frac{d}{T}\right)\rho^6 + \frac{c\rho^3}{T^2}(1 + \gamma\rho^2)\exp(-\gamma\rho^2)$$

$$B_0 = \sum_i x_i B_{0i} \qquad b = \sum_i\sum_j\sum_k x_i x_j x_k b_i^{1/3} b_j^{1/3} b_k^{1/3}$$

$$A_0 = \sum_i\sum_j x_i x_j A_{0i}^{1/2} A_{0j}^{1/2}(1 - k_{ij}) \quad a = \sum_i\sum_j\sum_k x_i x_j x_k a_i^{1/3} a_j^{1/3} a_k^{1/3}$$

$$C_0 = \sum_i\sum_j x_i x_j C_{0i}^{1/2} C_{0j}^{1/2}(1 - k_{ij})^3 \quad \alpha = \sum_i\sum_j\sum_k x_i x_j x_k \alpha_i^{1/3} \alpha_j^{1/3} \alpha_k^{1/3}$$

$$\gamma = \sum_i\sum_j x_i x_j \gamma_i^{1/2} \gamma_j^{1/2} \qquad c = \sum_i\sum_j\sum_k x_i x_j x_k c_i^{1/3} c_j^{1/3} c_k^{1/3}$$

$$D_0 = \sum_i\sum_j x_i x_j D_{0i}^{1/2} D_{0j}^{1/2}(1 - k_{ij})^4 \qquad d = \sum_i\sum_j\sum_k x_i x_j x_k d_i^{1/3} d_j^{1/3} d_k^{1/3}$$

$$E_0 = \sum_i\sum_j x_i x_j E_{0i}^{1/2} E_{0j}^{1/2}(1 - k_{ij})^5$$

$$H - H^0 = \rho\left(B_0RT - A_0 - \frac{4C_0}{T^2} + \frac{5D_0}{T^3} - \frac{6E_0}{T^4}\right) + \frac{1}{2}\rho^2\left(2bRT - 3a - \frac{4d}{T}\right) + \frac{1}{5}\rho^5\alpha\left(6a + \frac{7d}{T}\right) + \frac{c}{\gamma T^2}\left[3 - \left(3 + \frac{1}{2}\gamma\rho^2 - \gamma^2\rho^4\right)\exp(-\gamma\rho^2)\right]$$

$$S - S^0 = -\sum_i x_i R \ln(RT\rho x_i) - \rho\left(B_0R + \frac{2C_0}{T^3} - \frac{3D_0}{T^4} + \frac{4E_0}{T^5}\right) - \frac{1}{2}\rho^2\left(bR + \frac{d}{T^2}\right) + \frac{1}{5}\rho^5\frac{\alpha d}{T^2} + \frac{2c}{\gamma T^3}\left[1 - \left(1 + \frac{1}{2}\rho^2\gamma\right)\exp(-\gamma\rho^2)\right]$$

$$RT\ln f_i = RT\ln(\rho RTx_i) + \rho(B_0 + B_{0i})RT + 2\rho\sum_j x_j\left[-(A_0A_{0i})^{1/2}(1 - k_{ij}) - \frac{(C_0C_{0i})^{1/2}}{T^2}(1 - k_{ij})^3 + \frac{(D_0D_{0i})^{1/2}}{T^3}(1 - k_{ij})^4 - \frac{(E_0E_{0i})^{1/2}}{T^4}(1 - k_{ij})^5\right] + \frac{\rho^2}{2}\left[3(b^2b_i)^{1/3}RT - 3(a^2a_i)^{1/3} - \frac{3(d^2d_i)^{1/3}}{T}\right] + \frac{\alpha\rho^5}{5}\left(3(a^2a_i)^{1/3} + \frac{3(d^2d_i)^{1/3}}{T}\right) + \frac{3\rho^5}{5}\left(a + \frac{d}{T}\right)(\alpha^2\alpha_i)^{1/3} + \frac{3(c^2c_i)^{1/3}\rho^2}{T^2}\left[\frac{1 - \exp(-\gamma\rho^2)}{\gamma\rho^2} - \frac{\exp(-\gamma\rho^2)}{2}\right] - \frac{2c}{\gamma T^2}\left(\frac{\gamma_i}{\gamma}\right)^{1/2}\left[1 - \exp(-\gamma\rho^2)\left(1 + \gamma\rho^2 + \frac{1}{2}\gamma^2\rho^4\right)\right]$$

binary interaction parameters used to help "force fit" data on recalcitrant mixtures.

B. Nonhydrocarbons

Techniques used to predict vapor-liquid equilibrium behavior of nonhydrocarbons must be capable of describing a variety of nonideal behaviors. For this reason, they tend to be different from those used for the relatively well-behaved hydrocarbon systems. The currently used techniques for developing the activity coefficients needed for vapor-liquid equilibrium determinations for the mixtures involve the use of group contributions. In this approach, the individual molecule is considered to be made up of specific functional groups, The behavior of the molecule can be estimated by summing the characteristics of the functional groups. There are two techniques for allocating the group characteristics, similar in principle but differing in detail.

(a) The Analytical Solution of Groups (ASOG)

This procedure [21–24] was actually developed in two steps. The first step was to express the activity coefficient at infinite dilution ($\gamma_{\ell i}^0$) as

$$\ln \gamma_{\ell i}^0 = f(n_a, n_b) \qquad (2)$$

where $n_a = n_b$ are the groups making up molecules in the binary mixture.

The second step in the development of ASOG, and

the currently most widely used, considers the activity coefficient to be made up of two contributions, one based on molecular size and one based on molecular interactions. The activity coefficient is calculated as the algebraic sum of these two contributions:

$$\log \gamma_{\ell_i} = \log \gamma_{\ell_i}^G + \log \gamma_{\ell_i}^S \tag{3}$$

While this procedure sounds simple, the application is a little more complicated, involving the following steps:

1. Determine the number of size groups ($-CH_2-$, $-CH_3$, $-CH-$, $-CO-$, $-OH$) in the molecular structure for each component in the mixture.
2. The size term R_i is calculated from the number of size groups for each component and the composition of each component in the mixture:

$$R_i = \frac{S_i}{\sum_j S_j \tilde{x}_j} \tag{4}$$

where S_i is the number of size groups in molecule i, S_j is the number of size groups in molecule j, and $\tilde{x}_j$ is the mole fraction of component j in the mixture.

3. Calculate the size contribution to the activity coefficient for each component:

$$\log \gamma_{\ell i}^S = \log R_i - 0.434(1 - \log R_i) \tag{5}$$

4. From the composition of each pure component in the mixture and the number of each type of interaction group, calculate the composition of each interaction group in the mixture by

$$X_K = \frac{\Sigma \tilde{x}_i \nu_{Ki}}{\sum_K \sum_i \tilde{x}_j \nu_{Ki}} \tag{6}$$

where X_K is the composition of interaction group K, ν_K is the number of interaction groups of type K.

5. Obtain the interaction parameter A_k for each function group in the mixture. Some interaction parameters are tabulated in Table 1. If not listed there, the interaction parameters can be obtained by regression analysis of experimental data.
6. Calculate the group activity coefficients from

$$\ln \Gamma_K = -\ln \sum_\ell x_\ell A_{K\ell} + 1 - \sum_\ell \frac{\chi_\ell A_{K\ell}}{\sum_m \chi_m A_{\ell m}} \tag{7}$$

7. Calculate the standard state group activity coefficients (Γ_K^*) using Eq. (7) for a solution containing each component of the mixture. If a component has only one kind of interaction group, such as water or benzene, the standard state activity coefficient will be 1.0. Other components will have a standard state activity coefficient for each interaction group.
8. Calculate the group interaction to the activity coefficient by

$$\ln \gamma_{\ell i}^G = \Sigma \nu_{K_i} \ln \Gamma_K - \sum_K \nu_{K_i} \ln \Gamma_K^* \tag{8}$$

9. Calculate the activity coefficient from Eq. (3). The important point of the ASOG approach is that the group parameters depend on the group, not the arrangement of the group in the molecule.

Example

Using the ASOG procedure calculate the composition of the vapor in equilibrium with a 20-mole % solution of *n*-butanol in water. The temperature at which a solution of this concentration would boil is 92.25°C for 1-atm pressure.

Since the ASOG procedure considers the CH_3 and CH_2 groups to be equivalent, *n*-butanol contains four CH_2 groups and one OH group. In the ASOG procedure, water is treated as one size group but as 1.4 interaction groups.

For a 20-mole % solution of *n*-butanol in water, the size terms are (let 1 be *n*-butanol and 2 be water)

$$R_1 = \frac{5}{5 \times 0.2 + 0.8 \times 1} = 2.777\,8$$

$$R_2 = \frac{1}{1.8} = 0.555\,6$$

The size contributions to the activity coefficient can now be calaculated:

$$\ln \gamma_1^S = 1 - 2.777\,8 + n(2.777\,8) = -0.756\,1$$

$$\ln \gamma_2^S = 1 - 0.555\,6 + n(0.555\,6) = -0.143\,3$$

Calculate the mole fraction of each interaction group based on the specified 20-mole % *n*-butanol solution.

$$X_{CH_2} = \frac{4 \times 0.2}{4 \times 0.2 + 1 \times 0.2 + 0.8 \times 1.4} = 0.377\,4$$

$$X_{OH} = 0.622\,6$$

The interaction parameters can now be determined from Table 1.

$$A_{11} = 1, \quad A_{12} = 0.305$$

$$A_{21} = 0.014\,7, \quad A_{22} = 1.0$$

The group activity coefficients can now be calculated.

$$\ln \Gamma_1 = -\ln(0.377\,4 + 0.622\,6 \times 0.305) + \left(1 - \frac{0.377\,4}{0.377\,4 + 0.622\,6 \times 0.305} - \frac{0.622\,6 \times 0.014\,7}{0.622\,6 + 0.377\,4 \times 0.014\,7}\right)$$

$$= 0.566\,9 + 1 - 0.665\,3 - 0.014\,6 = 0.887\,0$$

$$\ln \Gamma_2 = -\ln(0.622\,6 + 0.377\,4 \times 0.014\,7) + \left(1 - \frac{0.622\,6}{0.622\,6 + 0.377\,4 \times 0.014\,7} - \frac{0.377\,4 \times 0.305}{0.377\,4 + 0.622\,6 \times 0.305}\right)$$

$$= 0.465\,0 + 1 - 0.991\,2 - 0.202\,9 = 0.270\,9$$

For pure n-butanol the group mole fractions are

$$X_{CH_2} = \frac{4 \times 1.0}{4 \times 1.0 + 1 \times 1.0} = 0.8$$

$$X_{OH} = 0.2$$

The standard state group activity coefficients in n-butanol are

$$\ln \Gamma_1^* = -\ln(0.8 + 0.2 \times 0.305) + \left(1 - \frac{0.8}{0.8 + 0.2 \times 0.305} - \frac{0.2 \times 0.014\,7}{0.2 + 0.8 \times 0.014\,7}\right)$$

$$= 0.149\,7 + 1 - 0.929\,2 - 0.013\,9 = 0.206\,6$$

$$\ln \Gamma_2^* = -\ln(0.2 + 0.8 \times 0.014\,7) + \left(1 - \frac{0.2}{0.2 + 0.8 \times 0.014\,7} - \frac{0.305}{0.8 + 0.2 \times 0.305}\right)$$

$$= 1.552\,3 + 1.0 - 0.944\,5 - 0.283\,4 = 1.324\,4$$

For pure water the mole fraction of CH_2 is 0 and the mole fraction of OH is 1.0. Using these concentrations the standard state activity coefficients are

$$\ln \Gamma_1^* = -\ln(0 + 1.0 \times 0.305) + \left(1 - 0 + \frac{1.0 \times 0.014\,7}{1 + 0}\right)$$

$$= 1.187\,4 + 1 - 0.014\,7 = 2.172\,7$$

$$\ln \Gamma_2^* = 0.0$$

The group interaction contribution to the activity coefficient is

$$\ln \Gamma_1^G = (4 \times 0.887 + 1 \times 0.270\,9) - (4 \times 0.206\,6 + 1 \times 1.324\,6)$$

$$= 3.818\,9 - 0.826\,4 - 1.324\,4 = 1.668\,1$$

$$\ln \Gamma_2^G = (0 \times 0.887 + 1.4 \times 0.270\,9) - (0 \times 2.172\,7 + 1.4 \times 0) = 0.379\,3$$

The activity coefficients for n-butanol and water in 20-mole % solution are

$$\gamma_1 = \exp(1.668\,1 - 0.756\,1) = 2.489\,3$$

$$\gamma_2 = \exp(0.379\,3 - 0.143\,3) = 1.266\,3$$

The activity coefficient may be related to vapor and liquid composition by

$$\gamma = \frac{y\pi}{xP^0}$$

Palmer [27] gives equations for the vapor pressure of n-butanol and water:

$$nC_4OH \quad \log P^0 = 9.067\,429 - \frac{2\,295.061\,1}{254.29 + t}$$

$$H_2O \quad \log P^0 = 7.966\,81 - \frac{1\,668.21}{228.0 + t}$$

Perry et al. [30] give the constant boiling azeotropic temperature for n-butanol and water as 92.25°C. At this temperature the vapor composition in equilibrium with the 20% liquid solution would be

$$2.489\,3 + \frac{y_1}{0.2 \times 278.38} \qquad y_1 = 0.192\,9$$

$$1.266\,3 = \frac{y_2}{0.8 \times 572.43} \qquad y_2 = 0.807\,1$$

The calculated vapor composition is very nearly the same as the liquid composition. This actually represents a region of partial immiscibility for the n-butanol-water system. In detailed calculations using ASOG [27], Palmer indicates that the group contribution approach does very well in predicting this vapor composition and region of partial miscibility.

Table 1 Parameters for use in the analytical solution of groups [26]

Alcohol–hydrocarbon–water systems at 60°C

Group pair	Compounds	Assignments		
		CH_2	OH	FH
CH_2	Hexane	6.0		6.0
OH	Water		1.4	1.0
	Water		1.4	1.0
	Pentane	5.0		5.0
	Methanol	1.0	1.0	2.0
	Water		1.4	1.0
	Heptane	7.0		7.0
	Butanol	4.0	1.0	5.0
	Parameters			
CH_2		1.0	0.305	
OH		0.0147	1.0	

Methanol–glycerol systems

Group pair	Compounds	Assignments			
		CH_2	OH	GOH	FH
CH_2-	Glycerol			6.0	6.0
Glycerol				6.0	6.0
OH–	Methanol	1.0	1.0		2.0
Glycerol					
Calc.	Water		1.4		1.0
	Glycerol			6.0	6.0
	Parameters				
CH_2		1.0	0.305	0.353	
OH		0.0147	1.0	1.095	
Glycerol		0.0153	0.991	1.0	

Ketone–alcohol–water systems at 60°C

Group pair	Compounds	Assignments			
		CH_2	OH	C=O	FH
CH_2	Heptane	7.0			7.0
C=O	Acetone	2.0		1.0	3.0
OH–	Methanol	1.0	1.0		2.0
C–O	MEK	3.0		1.0	4.0
	Water	0	1.4		1.0
	Parameters				
CH_2		1.0	0.305	1.333	
OH		0.0147	1.0	0.668	
C–O		0.0503	0.840	1.0	

Table 1 Parameters for use in the analytical solution of groups [26] (*Continued*)

Aromatic–alcohol systems at 80°C

Group pair	Compounds	Assignments: CH_2	CA	OH	FH
CH_2–CA	Heptane	7.0			7.0
	Benzene		6.0		6.0
CA–OH	Ethanol	2.0		1.0	3.0
	Benzene		6.0		6.0
Calc.	Toluene	1.0	6.0		7.0
	Isoamyl alcohol	5.0		1.0	6.0
Parameters					
CH_2		1.0	0.734	0.32	
CA		1.24	1.0	0.534	
OH		0.0245	0.045	1.0	

Ester–alcohol systems at 50°C

Group pair	Compounds	Assignments: CH_2	COO	OH	FH
CH_2–COO	Propylformate	3.0	3.0		6.0
	Hexane	6.0			6.0
	Decane	10.0			10.0
	Ethylpropionate	5.0	3.0		8.0
COO–OH	Methylacetate	2.0	3.0		5.0
	Methanol	1.0		1.0	2.0
Parameters					
CH_2		1.0	0.072	0.305	
COO		1.096	1.0	1.1179	
OH		0.0147	0.165	1.0	

Ether–alcohol–water systems at 60°C

Group pair	Compounds	Assignments: CH_2	OH	–O–	FH
CH_2–O	Heptane	7.0			7.0
	Diethyl ether	4.0		1.0	5.0
CH–OH					
OH–O	Diethyl ether	4.0		1.0	5.0
	Ethanol	2.0	1.0		3.0
	Water		1.4		1.0
Calc.	Dioxane	4.0		2.0	6.0
	Methylal	3.0		2.0	5.0
	Methanol	1.0	1.0		2.0
Parameters					
CH_2		1.0	0.305	1.711	
OH		0.0147	1.0	0.355	
–O–		0.341	1.03	1.0	

Table 1 Parameters for use in the analytical solution of groups [26] (*Continued*)

Aromatic–ketone mixtures						
			Assignments			
Group pair	Compounds	Temp., °C	CH_2	CA	C=O	FH
CH_2	Heptane	|	7.0			7.0
C=O		60				
	Acetone	25	2.0		1.0	3.0
		60				
CH–	Heptane	25	7.0			
		60				
	Benzene	25		6.0		6.0
		60				
CA–	Benzene	25		6.0		6.0
		60				
C=O	Acetone	25	2.0		1.0	3.0
		60				
Calc.	Xylene	60	2.0	6.0		8.0
	Diisobutyl ketone	60	8.0		1.0	9.0
		Parameters				
CH_2		25	1.0	0.705	0.899	
		60	1.0	0.709	1.33	
		150	1.0	0.66	2.10	
CA		25	1.24	1.0	1.418	
		60	1.266	1.0	1.578	
		150	1.52	1.0	1.9	
C=O		25	0.0480	0.144	1.0	
		60	0.0503	0.162	1.0	
		150	0.075	0.21	1.0	

Aqueous nitrile systems at 75°C						
		Assignments				
Group pair	Compounds	CH_2	CH	CN	OH	FH
CH–CN	Acetonitrile	1.0		1.0		3.0
	Acrylonitrile	1.0	1.0	1.0		4.0
CN–OH	Acetonitrile					
	Water				1.4	1.0
		Parameters				
CH_2		1.0	0.825	0.4455	0.315	
CH		1.17	1.0	0.0338	0.680	
CN		0.047	0.385	1.0	0.291	
OH		0.021	0.025	0.549	1.0	

Table 1 Parameters for use in the analytical solution of groups [26] (*Continued*)

Chloride systems

Group pair	Compounds	Assignments				
		CH_2	OH	Cl	HBH	FH
CH_2–Cl	Cyclohexane	4.0	1.0			4.0
	CCl_4	1.0		4.0		5.0
OH–Cl	Methanol	1.0	1.0			2.0
	CCl_4	1.0		4.0		5.0
OH–HBH	Methanol	1.0	1.0			2.0
	Chloroform	1.0		3.0	1.0	4.0
	Ethanol	2.0	1.0			3.0
Parameters						
CH_2		1.0	0.29	0.837	1.0	
OH		0.007	1.0	0.0112	0.0722	
Cl		1.128	0.591	1.0	1.0	
HBH		1.0	5.077	1.0	1.0	

Fluorocompounds

Group pair	Compounds	Assignments			
		CH_2	CF_2	O	FH
CH_2–O	Cyclohexane	6.0			4.0
	Dioxane	4.0		2.0	4.65
CH_2–CF	Hexane	6.0			6.0
	Perfluorohexane		18.0		9.8
CF_2–O	Perfluoroheptane		12.0		11.4
	Dioxane				
HBH–O	HPDFH		21.0		11.4
	Dioxide	4.0		2.0	4.65

Parameters

	CH_2	CF_2	O	HBH
CH_2	1.0	0.424	0.00153	1.0
CF_2	1.541	1.0	0.604	1.0
O	0.495	0.334	1.0	0.0015
HBH	1.0	1.0	3.126	1.0

Table 1 Parameters for use in the analytical solution of groups [26] (*Continued*)

Anhydride–carboxylic acid mixtures

		Assignments				
Compound	Size groups	CH_2	OH	Benzene	Acid	Anhydride
Acetic acid	2	1			1	
Acetic anhydride	4	2				1
Benzene	6			6		
Benzoic acid	7			6	1	
n-Butanol	5	4	1			
o-Cresol	8	1	1	6		
Cyclohexane	6	6				
Ethanol	3	2	1			
Ethylbenzene	8	2		6		
Formic acid	2				1	
n-Octane	8	8				
Phenol	7		1	6		
Phthalic acid	8			6	2	
Phthalic anhydride	8			6		1
Propionic acid	3	2		1		
Toluene	7	1		6		
o-Toluic acid	8	1		6	1	
Water	1		1.4			

Parameters

	$-CH_2$	$-OH$	Benzene (=CH–)	Acid (–COOH)	Anhydride (–OCOO–)
CH_2	1.0	0.305	0.734	1.1333	0.7703
OH	0.0147	1.0	0.045	0.1960	
Benzene	1.24	0.534	1.0	1.1372	0.9841
Acid	0.0340	2.3518	0.0442	1.0	0.6205
Anhydride	0.0249		0.1876	0.5369	1.0

(b) UNIFAC Method

The second group contribution method to be considered is UNIFAC developed by Prausnitz and co-workers [29]. It has the advantage that mixture properties can be estimated from pure-component measurements without need for mixture data. In fashion similar to ASOG, the activity coefficient is considered to be composed of two parts: one due to differences in size and shape of the molecules in the mixture (the combinatorial portion) and one due to energy interactions (the residual portion).

$$\ln \gamma_i = \ln \gamma_i^C + \ln \gamma_i^R \qquad (10)$$

where R stands for residual and C for combinatorial.

The combinatorial portion of the activity coefficient is calculated using the relationship

$$\ln \gamma_i^C = \ln \frac{\Phi_i}{\tilde{x}_i} + \frac{z}{2} q_i \ln \frac{\Theta_i}{\Phi_i} + \ell_i - \frac{\Phi_i}{\tilde{x}_i} \sum_j \tilde{x}_j l_j \qquad (11)$$

where $l_i = z/2(r_i - q_i) - (r_i - 1)$, $r_i = \Sigma_K \nu_K^{(i)} R_K$ is a measure of molecular van der Waals volume, $z = 10$, $\nu_K^{(i)}$= number of K groups in molecule i, R_K, Q_K are values tabulated in Table 2a, Θ_i is the area fraction = $q_i x_i / \Sigma q_j \tilde{x}_j$, $q_i = \Sigma_K \nu_K^{(i)} Q_K$ is a measure of molecular surface area, $\Phi_i = V_i \tilde{x}_i / \Sigma r_j x_j$ is the segment fraction, similar to volume fraction.

The residual portion of the activity coefficient is calculated by

$$\ln \gamma_i^R = \sum_K \nu_K^{(1)} (\ln \Gamma_K - \ln \Gamma_K^{(i)}) \qquad (12)$$

where Γ_K is the group residual activity coefficient and $\Gamma_K^{(i)}$ the residual activity coefficient of group K in a solution containing only molecule i.

The group activity coefficients in Eq. (12) are calculated from

$$\ln \Gamma_K = Q_K \left[1 - \ln \left(\sum_m \Theta_m \Psi_{mK} \right) - \sum_m \frac{\Theta_m \Psi_{Km}}{\sum_n \Theta_n \Psi_{nm}} \right] \qquad (13)$$

where $\Theta_m = Q_m X_m / \Sigma Q_n X_n$ is the fraction of group m, X_m is the mole fraction of group m in the mixture, $\Psi_m = \exp(-a_{mn}/T)$ is the group interaction parameter, and a_{mn} is the group interaction parameter tabulated in Table 2b. Note $a_{mn} \neq a_{mn}$.

The chronological steps in solution of the UNIFAC equations are as follows:

1. Using Table 2a identify the kind and number of groups in each molecule in the solution.

2. Calculate the parameters for molecular volume and molecular surface area from

$$r_i = \sum_K \nu_K^{(i)} R_K$$

$$q_i = \sum_K \nu_K^{(i)} Q_K$$

3. Calculate the parameter ℓ_i from

$$\ell_i = \frac{z}{2}(r_i - q_i) - (r_i - 1.0)$$

4. Calculate the segment fractions and area fractions based on the specified composition

$$\Theta_i = \frac{q_i x_i}{\Sigma q_i x_i}$$

$$\Phi_i = \frac{r_i x_i}{\Sigma r_i x_i}$$

5. Calculate the combinatorial portion of the activity coefficient from Eq. (11).

6. For the pure components calculate the area fractions and group interaction parameters from Table 5.5.4(3). (Note that Θ_m is not the same as Θ_i that is used in calculating the combinatorial portion of the activity coefficient.)

$$\Theta_m = \frac{Q_m X_m}{\Sigma Q_m X_m}$$

$$\Psi_m = \exp \frac{-a_{mn}}{T}$$

7. Calculate the pure component residual activity coefficients from Eq. (13).

8. Repeat steps 6 and 7 for the specified solution concentration.

9. Calculate the residual portion of the activity coefficient using Eq. (12).

10. Calculate the activity coefficient for each component in the mixture using Eq. (10).

Example

Calculate the activity coefficients for the 20% *n*-butanol in water solution for the same conditions as used in the ASOG example.

From Table 5.5.4(2) the "groups" in the mixture are:

n-Butanol: 1 CH_3; 2 CH_2; 1 COH

Water: 1 H_2O

The volume and surface area parameters for these groups are:

Group	R_K	Q_K
CH_3	0.9011	0.848
CH_2	0.6744	0.540
COH	1.2044	1.124
H_2O	0.92	1.400

Denoting *n*-butanol as component 1 in the mixture and water as component 2, the parameters for molecular volume and molecular surface area may be calculated:

$$r_1 = 0.901\,1 + 2 \times 0.674\,4 + 1.204\,4 = 3.454\,3$$

$$r_2 = 0.920\,0$$

$$q_1 = 0.848 + 2 \times 0.540 + 1.124 = 3.055$$

$$q_2 = 1.40$$

The area fractions and volume fractions can now be calculated as can the parameter ℓ

$$\Theta_1 = \frac{3.052 \times 0.2}{3.052 \times 0.2 + 1.4 \times 0.8} = 0.352\,8$$

$$\Theta_2 = 0.647\,2$$

$$\Phi_1 = \frac{3.454\,3 \times 0.2}{3.454\,3 \times 0.2 + 0.92 \times 0.8} = 0.484\,2$$

$$\Phi_2 = 0.515\,8$$

$$l_1 = 5(3.454\,3 - 3.052) - (3.454\,3 - 1.0)$$

$$= -\,0.442\,8$$

$$l_2 = 5(0.92 - 1.4) - (0.92 - 1.0) = -\,2.32$$

The combinatorial portion of the activity coefficient can now be calculated from Eq. (11).

$$\ln \gamma_1^C = \ln\left(\frac{0.484\,2}{0.2}\right) + \left(5 \times 3.052 \ln \frac{0.352\,8}{0.484\,20}\right)$$

$$- 0.442\,8 - \frac{0.3528/0.2}{0.2(-\,0.442\,8) + 0.8(2.32)}$$

$$= -\,0.959\,7$$

$$\ln \gamma_2^C = \ln\left(\frac{0.515\,8}{0.8}\right) + \left[5 \times 1.4 \ln\left(\frac{0.647\,2}{0.515\,8}\right)\right]$$

$$- 2.32 - \frac{0.515\,8/0.8}{0.2(-\,0.442\,8) + 0.8(2.32)}$$

$$= 0.083\,5$$

The next step in the calculation is to determine the residual portion of the activity coefficient through use of Eqs. (12) and (13). The first step is to determine the residual activity coefficient for a solution containing only *n*-butanol and then only water. In this portion of the calculation subscript 1 will refer to CH_3, subscript 2 to CH_2, subscript 3 to COH, and subscript 4 to water. For pure *n*-butanol the area fractions for use in Eq. (13) are

$$\Theta_1 = \frac{1.0 \times 0.848}{0.848 + 2 \times 0.54 + 1 \times 1.124} = 0.277\,9$$

$$\Theta_2 = 0.353\,9$$

$$\Theta_3 = 0.368\,3$$

Group interaction parameters may be determined from Table 2b. Using these values and the temperature of 92.25 °C. reported for the azeotrope, the group interaction coefficients are

$$\Psi_{11} = \Psi_{22} = \Psi_{33} = \Psi_{12} = \Psi_{21} = 1.0$$

$$\Psi_{23} = \Psi_{13} = \exp\ -\frac{931.2}{365.4} = 0.078\,2$$

$$\Psi_{32} = \Psi_{31} = \exp\ -\frac{169.7}{365.4} = 0.628\,5$$

$$\Psi_{24} = \Psi_{14} = \exp\ -\frac{145.2}{365.4} = 0.018\,8$$

$$\Psi_{41} = \Psi_{42} = \exp\ -\frac{657.7}{365.4} = 0.165\,3$$

$$\Psi_{34} = \exp\ \frac{320.8}{365.4} = 2.405\,9$$

$$\Psi_{43} = \exp\ -\frac{287.5}{365.4} = 0.455\,3$$

Both the pure component activity coefficient and the activity coefficients in solution are calculated using Eq. (13). For the three groups in *n*-butanol (CH_3, CH_2, COH) the expanded form of Eq. (13) is

$$\ln \Gamma_1^{(1)} = Q_1 \left[1 - \ln(\Theta_1\Psi_{11} + \Theta_2\Psi_{21} + \Theta_3\Psi_{31}) - \frac{\Theta_1\Psi_{11}}{\Theta_1\Psi_{11} + \Theta_2\Psi_{21} + \Theta_3\Psi_{31}} - \frac{\Theta_2\Psi_{12}}{\Theta_1\Psi_{12} + \Theta_2\Psi_{22} + \Theta_3\Psi_{32}} - \frac{\Theta_3\Psi_{13}}{\Theta_1\Psi_{13} + \Theta_2\Psi_{23} + \Theta_3\Psi_{33}}\right]$$

$$\ln \Gamma_2^{(1)} = Q_2 \left[1 - \ln(\Theta_1\Psi_{12} + \Theta_2\Psi_{22} + \Theta_3\Psi_{32}) - \frac{\Theta_1\Psi_{21}}{\Theta_1\Psi_{11} + \Theta_2\Psi_{21} + \Theta_3\Psi_{31}} - \frac{\Theta_2\Psi_{22}}{\Theta_1\Psi_{12} + \Theta_2\Psi_{22} + \Theta_3\Psi_{32}} - \frac{\Theta_3\Psi_{23}}{\Theta_1\Psi_{13} + \Theta_2\Psi_{23} + \Theta_3\Psi_{33}}\right]$$

$$\ln \Gamma_3^{(1)} = Q_3 \left[1 - \ln(\Theta_1\Psi_{13} + \Theta_2\Psi_{23} + \Theta_3\Psi_{33}) - \frac{\Theta_1\Psi_{31}}{\Theta_1\Psi_{11} + \Theta_2\Psi_{21} + \Theta_3\Psi_{31}} - \frac{\Theta_2\Psi_{32}}{\Theta_1\Psi_{12} + \Theta_2\Psi_{22} + \Theta_3\Psi_{32}} - \frac{\Theta_3\Psi_{33}}{\Theta_1\Psi_{13} + \Theta_2\Psi_{23} + \Theta_3\Psi_{33}}\right]$$

Using these equations and the area fractions and interaction parameters calculated earlier, the pure component activity coefficients for *n*-butanol can be calculated as follows:

$$\ln \Gamma_1^{(1)} = 0.848 \Big[1 - \ln (0.277\,9 \times 1 + 0.353\,9 \times 1 + 0.368\,3 \times 0.628\,5) - \frac{0.277\,9 \times 1}{0.277\,9 \times 1 + 0.353\,9 \times 1 + 0.368\,3 \times 0.628\,5} - \frac{0.353\,9 \times 1}{0.277\,9 \times 1 + 0.353\,9 \times 1 + 0.368\,3 \times 0.628\,5} - \frac{0.368\,3 \times 0.078\,2}{0.277\,9 \times 0.078\,2 + 0.353\,9 \times 0.078\,2 + 0.368\,3 \times 1} \Big] = 0.293\,6$$

$$\ln \Gamma_2^{(1)} = 0.540[\quad] = 0.186\,9$$

$$\ln \Gamma_3^{(1)} = 1.124 \Big[1 - \ln (0.277\,9 \times 0.078\,2 + 0.353\,9 \times 0.078\,2 + 0.368\,3 \times 1) - \frac{0.277\,9 \times 0.628\,5}{0.277\,9 \times 1 + 0.353\,9 \times 1 + 0.368\,3 \times 0.628\,5} - \frac{0.352\,9 \times 0.628\,5}{0.277\,9 \times 1 + 0.353\,9 \times 1 + 0.368\,3 \times 0.628\,5} - \frac{0.368\,3 \times 1}{0.277\,9 \times 1 + 0.353\,9 \times 1 + 0.368\,3 \times 0.628\,5} \Big] = 0.597\,2$$

For pure water, $\ln \Gamma_4^{(1)} = 0.0$.

The residual portion of the activity coefficient must now be calculated for a 20% solution of *n*-butanol in water. For this solution the group mole fractions in the mixture are

$$X_1 = \frac{0.2 \times 1}{0.2 \times 4 + 0.8 \times 1} = 0.125$$

$$X_2 = 0.25 \qquad X_3 = 0.125 \qquad X_4 = 0.50$$

Based on these mole fractions the area fraction for each group can be calculated:

$$\Theta_1 = \frac{0.125 \times 0.848}{0.125 \times 0.848 + 0.25 \times 0.54 + 0.125 \times 1.124 + 0.5 \times 1.4} = 0.098$$

$$\Theta_2 = 0.124\,8$$

$$\Theta_3 = 0.129\,9$$

$$\Theta_4 = 0.647\,2$$

Earlier in the solution, Eq. (13) was expanded for the three groups occurring in *n*-butanol. This expansion shows a number of terms that recur in the calculations for each activity coefficient. For this reason these terms will be evaluated separately and utilized in the activity coefficient calculations which follow them.

$$\Theta_1 \Psi_{11} + \Theta_2 \Psi_{21} + \Theta_3 \Psi_{31} + \Theta_4 \Psi_{41} = 0.098 \times 1 + 0.124\,8 \times 1 + 0.129\,9 \times 0.628\,5 + 0.647\,2 \times 0.165\,3 = 0.411\,4$$

$$\Theta_1 \Psi_{12} + \Theta_2 \Psi_{22} + \Theta_3 \Psi_{32} + \Theta_4 \Psi_{42} = 0.098 \times 0.078\,2 + 0.124\,8 \times 1 + 0.129\,9 \times 0.628\,5 + 0.647\,2 \times 0.165\,3 = 0.411\,4$$

$$\Theta_1 \Psi_{13} + \Theta_2 \Psi_{23} + \Theta_3 \Psi_{33} + \Theta_4 \Psi_{43} = 0.098 \times 0.078\,2 + 0.124\,8 \times 0.078\,2 + 0.129\,9 \times 1 + 0.647\,2 \times 0.455\,3 = 0.442$$

$$\Theta_1 \Psi_{14} + \Theta_2 \Psi_{24} + \Theta_3 \Psi_{34} + \Theta_4 \Psi_{44} = 0.098 \times 0.018\,8 + 0.124\,8 \times 0.018\,8 + 0.129\,9 \times 2.405\,9 + 0.647\,2 \times 1 = 0.963\,9$$

$$\ln \Gamma_1 = 0.848 \left(1 - \ln 0.411\,4 - \frac{0.098}{0.411\,4} - \frac{0.124\,8}{0.411\,4} - \frac{0.129\,9 \times 0.078\,2}{0.442} \right) = 1.111\,7$$

$$\ln \Gamma_2 = 0.707\,9$$

$$\ln \Gamma_3 = 0.486\,9$$

$$\ln \Gamma_4 = 0.198\,9$$

The residual portion of the activity coefficient is calculated for each component by Eq. (12):

$$\ln \gamma_1^R = 1(1.111\,7 - 0.293\,6) + 2(0.707\,9 - 0.186\,9) + 1(0.486\,9 - 0.597\,3) = 1.749\,8$$

$$\ln \gamma_2^R = 0.198\,9$$

The activity coefficients for *n*-butanol and water in a 20% solution of *n*-butanol in water are calculated from Eq. (10):

$$\ln \gamma_1 = 1.749\,8 - 0.959\,7 = 0.790\,1 \qquad \gamma_1 = 2.203\,6$$

$$\ln \gamma_2 = 0.198\,1 + 0.835 = 0.281\,6 \qquad \gamma_2 = 1.325\,2$$

From the ASOG example we know that $P^0_{H_2O} = 572.43$ and $P^0_{nC_4OH} = 278.38$. If we assume that there is ideal behavior in the vapor phase, the composition of the vapor phase is

$$\pi \tilde{y}_i = \gamma_1 \tilde{x}_1 P_1^0 = 2.203\,6 \times 0.2 \times 278.38 = 122.7 \text{ mm Hg}$$

$$y_1 = 0.168$$

$$\pi \tilde{y}_2 = \gamma_2 \tilde{x}_2 P_2^0 = 1.325\,2 \times 0.8 \times 572.43$$
$$= 606.9 \text{ mm Hg}$$
$$y_2 = 0.832$$

The agreement here is not quite as good as with the ASOG procedure. However, the ASOG constants contain more "fitting" for specific behavior. For a generalized predictive procedure the UNIFAC procedure seems to give excellent results.

Although cumbersome for hand calculations, the group contribution procedures are readily programmable for digital computer solutions. If extensive use is to be made of any of them, computer solution is recommended.

Table 2a Parameters for use with UNIFAC

Group name	R_k	Q_k	Sample group assignment
CH_2			
CH_3	0.9011	0.848	Butane: 2 CH_3, 2 CH_2
CH_2	0.6744	0.540	
CH	0.4469	0.228	*i*-Butane: 3 CH_3, 1 CH
C	0.2195	0.000	2,2-Dimethylpropane 4 CH_3, 1 CH
C=C			
CH_2=CH	1.3454	1.176	1-Hexene: 1 CH_3, 3 CH_2, 1 CH_2=CH
CH=CH	1.1167	0.867	2-Hexene: 2 CH_3, 2 CH_2, 1 CH=CH
CH=C	0.8886	0.676	2-Methylbutene-2: 3 CH_3, 1 CH=C
CH_2=C	1.1173	0.988	2-Methylbutene-1: 2 CH_3, 1 CH_2, 1 CH_2=C
ACH			
ACH	0.5313	0.400	Benzene: 6 ACH
AC	0.3652	0.120	Styrene: 1 CH_2=CH, 5 ACH, 1 AC
$ACCH_2$			
$ACCH_3$	1.2663	0.968	Toluene: 5 ACH, 1 $ACCH_3$
$ACCH_2$	1.0396	0.660	Ethylbenzene: 1 CH_3, 5 ACH, 1 $ACCH_2$
ACCH	0.8121	0.348	Cumene: 2 CH_3, 5 ACH, 1 ACCH

Table 2a Parameters for use with UNIFAC (*Continued*)

Group name	R_k	Q_k	Sample group assignment
CCOH			
CH_2CH_2OH	1.8788	1.664	1-Propanol: 1 CH_3, 1 CH_2CH_2OH
$CHOHCH_3$	1.8780	1.660	2-Butanol: 1 CH_3, 1 CH_2, 1 $CHOHCH_3$
$CHOHCH_2$	1.6513	1.352	3-Octanol: 2 CH_3, 4 CH_2, 1 $CHOHCH_2$
CH_3CH_2OH	2.1055	1.972	Ethanol: 1 CH_3CH_2OH
$CHCH_2OH$	1.6513	1.352	Isobutanol: 2 CH_3, 1 $CHCH_2OH$
CH_3OH	1.4311	1.432	Methanol: 1 CH_3OH
H_2O	0.92	1.40	Water: 1 H_2O
ACOH	0.8952	0.680	Phenol: 5 ACH, 1 ACOH
CH_2CO			
CH_3CO	1.6724	1.488	Ketone group is 2nd carbon; 2-Butanone: 1 CH_3, 1 CH_2, 1 CH_3CO
CH_2CO	1.4457	1.180	Ketone group is any other carbon; 3-Pentanone: 2 CH_3, 1 CH_2, 1 CH_2CO
CHO	0.9980	0.948	Acetaldehyde: 1 CH_3, 1 CHO
COOC			
CH_3COO	1.9031	1.728	Butyl acetate: 1 CH_3, 3 CH_2, 1 CH_3COO
CH_2COO	1.6764	1.420	Butyl propanoate: 2 CH_3, 3 CH_2, 1 CH_2COO
CH_2O			
CH_3O	1.1450	1.088	Dimethyl ether: 1 CH_3, 1 CH_3O
CH_2O	0.9183	0.780	Diethyl ether: 2 CH_3, 1 CH_2, 1 CH_2O
CH–O	0.6908	0.468	Diisopropyl ether: 4 CH_3, 1 CH, 1 CH–O
FCH_2O	0.9183	(1.1)	Tetrahydrofuran: 3 CH_2, 1 FCH_2O
CNH_2			
CH_3NH_2	1.5959	1.544	Methylamine: 1 CH_3NH_2
CH_2NH_2	1.3692	1.236	*n*-Propylamine: 1 CH_3, 1 CH_2, 1 CH_2NH_2
$CHNH_2$	1.1417	0.924	Isopropylamine: 2 CH_3, 1 $CHNH_2$
CNH			
CH_3NH	1.4337	1.244	Dimethylamine: 1 CH_3, 1 CH_3NH
CH_2NH	1.2070	0.936	Diethylamine: 2 CH_3, 1 CH_2, 1 CH_2NH
CHNH	0.9795	0.624	Diisopropylamine: 4 CH_3, 1 CH, 1 CHNH
$ACNH_2$	1.0600	0.816	Aniline: 5 ACH_3, 1 $ACNH_2$
CCN			
CH_3CN	1.8701	1.724	Acetonitrile: 1 CH_3CN
CH_2CN	1.6434	1.416	Propionitrile: 1 CH_3, 1 CH_2CN
COOH			
COOH	1.3013	1.224	Acetic acid: 1 CH_3, 1 COOH
HCOOH	1.5280	1.532	Formic acid: 1 HCOOH
CCl			
CH_2Cl	1.4654	1.264	Butylchloride: 1 CH_3, 2 CH_2, 1 CH_2Cl
CHCl	1.2380	0.952	Isopropyl chloride: 2 CH_3, 1 CHCl
CCl	1.0106	0.724	*tert*-Butyl chloride: 3 CH_3, 1 CCl
CCl_2			
CH_2Cl_2	2.2564	1.988	Dichloromethane: 1 CH_2Cl_2
$CHCl_2$	2.0606	1.684	1,1-Dichloroethane: 1 CH_3, 1 $CHCl_2$
CCl_2	1.8016	1.448	2,2-Dichloropropane: 2 CH_3, 1 CCl_2
CCl_3			
$CHCl_3$	2.8700	2.410	Chloroform: 1 $CHCl_3$
CCl_3	2.6401	2.184	1,1,1-Trichloroethane: 1 CH_3, 1 CCl_3
CCl_4	3.3900	2.910	Carbon tetrachloride: 1 CCl_4
ACCl	1.1562	0.844	Chlorobenzene: 5 ACH, 1 ACCl
CNO_2			
CH_3NO_2	2.0086	1.868	Nitromethane: 1 CH_3NO_2
CH_2NO_2	1.7818	1.560	1-Nitropropane: 1 CH_3, 1 CH_2, 1 CH_2NO_2
$CHNO_2$	1.5544	1.248	2-Nitropropane: 2 CH_3, 1 $CHNO_2$
$ACNO_2$	1.4199	1.104	Nitrobenzene: 5 ACH, 1 $ACNO_2$
CS_2	2.057	(1.65)	Carbon disulfide: 1 CS_2

Table 2b Interaction parameters[a]

	CH_2	C=C	ACH	$ACCH_2$	CCOH	CH_3OH	H_2O	ACOH
CH_2	0	− 200.0	61.13	76.50	737.5	697.2	1318	(2789)
C=C	2520	0	340.7	4102	(535.2)	(1509)	599.6	n.a.
ACH	− 11.12	− 94.78	0	167.0	477.0	637.4	903.8	(1397)
$ACCH_2$	− 69.70	− 269.7	(− 146.8)	0	469.0	603.3	(5695)	(726.3)
CCOH	− 87.93	(121.5)	− 64.13	− 99.38	0	127.4	285.4	(257.3)
CH_3OH	16.51	(− 52.39)	− 50.00	− 44.50	− 80.78	0	− 181.0	n.a.
H_2O	580.6	511.7	362.3	(377.6)	− 148.5	289.6	0	442.0
ACOH	(311.0)	n.a.	(2043)	(6245)	(− 455.4)	n.a.	− 540.6	0
CH_2CO	26.76	− 82.92	140.1	365.8	129.2	108.7	605.6	n.a.
CHO	(505.7)	n.a.	n.a.	n.a.	n.a.	− 340.2	(− 155.7)	n.a.
COOC	114.8	n.a.	85.81	− 170.00	109.9	249.6	1135	853.6
CH_2O	83.36	76.44	52.13	65.69	42.00	(339.7)	634.2	n.a.
CNH_2	− 30.48	(79.40)	− 44.85	n.a.	(− 217.2)	(− 481.7)	− 507.1	n.a.
CNH	65.33	− 41.32	− 22.31	(223.0)	− 243.3	(− 500.4)	− 547.7	n.a.
$ACNH_2$	5339	n.a.	650.4	3399	(− 245.0)	n.a.	− 339.5	n.a.
CCN	35.76	26.09	(− 22.97)	− 138.4	n.a.	(168.8)	242.8	n.a.
COOH	315.3	(349.2)	62.32	268.2	− 17.59	1020	− 292.0	n.a.
CCl	(91.46)	(− 24.36)	(4.680)	(122.9)	368.6	529.0	698.2	n.a.
CCl_2	(34.01)	(− 52.71)	n.a.	n.a.	601.6	(669.9)	708.7	n.a.
CCl_3	36.70	(− 185.1)	288.5	(33.61)	491.1	649.1	826.8	n.a.
CCl_4	− 78.45	(− 293.7)	− 4.700	134.7	570.7	860.1	1201	(1616)
ACCl	− 141.3	n.a.	(− 237.7)	n.a.	(134.1)	n.a.	920.4	n.a.
CNO_2	− 32.69	(− 49.92)	10.38	− 97.05	n.a.	(252.6)	614.2	n.a.
$ACNO_2$	(5541)	n.a.	(1825)	n.a.	n.a.	n.a.	360.7	n.a.
CS_2	(11.46)	n.a.	− 18.99	n.a.	442.8	914.2	1081	n.a.

	CH_2CO	CHO	COOC	CH_2O	CNH_2	CNH	$ACNH_2$	CCN
CH_2	476.4	(677.0)	232.1	251.5	391.5	255.7	1245	612.0
C=C	524.5	n.a.	n.a.	289.3	(396.0)	273.6	n.a.	370.9
ACH	25.77	n.a.	5.994	32.14	161.7	122.8	668.2	(212.5)
$ACCH_2$	− 52.10	n.a.	5688	213.1	n.a.	(− 49.29)	612.5	6096
CCOH	48.16	n.a.	76.20	70.00	(110.8)	188.3	(412.0)	n.a.
CH_3OH	23.39	306.4	− 10.72	(− 180.6)	(359.3)	(266.0)	n.a.	(45.54)
H_2O	− 280.8	(649.1)	− 455.4	− 400.6	357.5	287.0	213.0	112.6
ACOH	n.a.	n.a.	− 713.2	n.a.	n.a.	n.a.	n.a.	n.a.
CH_2CO	0	− 37.36	− 213.7	(5.202)	n.a.	n.a.	n.a.	428.5
CHO	128.0	0	n.a.	n.a.	n.a.	n.a.	n.a.	n.a.
COOC	372.2	n.a.	0	− 235.7	n.a.	(− 73.50)	n.a.	533.6
CH_2O	(52.38)	n.a.	461.3	0	n.a.	(141.7)	n.a.	n.a.
CNH_2	n.a.	n.a.	n.a.	n.a.	0	(63.72)	n.a.	n.a.
CNH	n.a.	n.a.	(136.0)	(− 49.30)	(108.8)	0	n.a.	n.a.
$ACNH_2$	n.a.	n.a.	n.a.	n.a.	n.a.	n.a.	0	n.a.
CCN	− 275.1	n.a.	− 297.3	n.a.	n.a.	n.a.	n.a.	0
COOH	− 297.8	n.a.	− 256.3	− 338.5	n.a.	n.a.	n.a.	n.a.
CCl	(286.3)	n.a.	n.a.	225.4	n.a.	n.a.	n.a.	n.a.
CCl_2	(423.2)	n.a.	(− 132.9)	(− 197.7)	n.a.	n.a.	n.a.	n.a.
CCl_3	552.1	n.a.	176.5	− 20.93	n.a.	n.a.	n.a.	(74.04)
CCl_4	372.0	n.a.	129.5	n.a.	n.a.	91.13	(1302)	(492.0)
ACCl	n.a.	n.a.	− 299.2	n.a.	203.5	− 108.4	n.a.	n.a.
CNO_2	(− 142.6)	n.a.	n.a.	(− 94.49)	n.a.	n.a.	n.a.	n.a.
$ACNO_2$	n.a.	n.a.	n.a.	n.a.	n.a.	n.a.	(5250)	n.a.
CS_2	298.7	n.a.	233.7	79.79	n.a.	n.a.	n.a.	n.a.

Table 2b Interaction parameters (*Continued*)

	COOH	CCl	CCl_2	CCl_3	CCl_4	ACCl	CNO_2	$ACNO_2$	CS_2
CH_2	663.5	(35.93)	(53.76)	24.9	104.3	321.5	661.5	(543.0)	(114.1)
C=C	(730.4)	(99.61)	(337.1)	(4583)	(5831)	n.a.	(542.1)	n.a.	n.a.
ACH	537.4	(− 18.81)	n.a.	− 231.9	3.000	(538.2)	168.1	(194.9)	97.53
$ACCH_2$	603.8	(− 114.1)	n.a.	(− 12.14)	− 141.3	n.a.	3629	n.a.	n.a.
CCOH	77.61	− 38.23	− 185.9	− 170.9	− 98.66	(290.0)	n.a.	n.a.	73.52
CH_3OH	− 289.5	− 38.32	(− 102.5)	− 139.4	− 67.80	n.a.	(75.14)	n.a.	− 31.09
H_2O	225.4	325.4	370.4	353.7	497.5	678.2	− 19.44	399.5	887.1
ACOH	n.a.	n.a.	n.a.	n.a.	(4894)	n.a.	n.a.	n.a.	n.a.
CH_2CO	669.4	(− 191.7)	(− 284.0)	− 354.6	− 39.20	n.a.	(137.5)	n.a.	162.3
CHO	n.a.	n.a.	n.a.	n.a.	n.a.	n.a.	n.a.	n.a.	n.a.
COOC	660.2	n.a.	(108.9)	− 209.7	54.47	808.7	n.a.	n.a.	162.7
CH_2O	664.6	301.1	(137.8)	− 154.3	n.a.	n.a.	(95.18)	n.a.	151.1
CNH_2	n.a.	n.a.	n.a.	n.a.	n.a.	68.81	n.a.	n.a.	n.a.
CNH	n.a.	n.a.	n.a.	n.a.	71.23	(4350)	n.a.	n.a.	n.a.
$ACNH_2$	n.a.	n.a.	n.a.	n.a.	(8455)	n.a.	n.a.	(− 62.73)	n.a.
CCN	n.a.	n.a.	n.a.	− 15.62	(− 54.86)	n.a.	n.a.	n.a.	n.a.
COOH	0	44.42	− 183.4	n.a.	212.7	n.a.	n.a.	n.a.	n.a.
CCl	326.4	0	108.3	(249.2)	62.42	n.a.	n.a.	n.a.	n.a.
CCl_2	1821	− 84.53	0	(0)	(56.33)	n.a.	n.a.	n.a.	n.a.
CCL_3	n.a.	(− 157.1)	(0)	0	− 30.10	n.a.	n.a.	n.a.	256.5
CCl_4	689.0	11.80	(17.97)	51.90	0	(475.8)	(490.9)	(534.7)	132.2
ACCl	n.a.	n.a.	n.a.	n.a.	(− 255.4)	0	(− 154.5)	n.a.	n.a.
CNO_2	n.a.	n.a.	n.a.	n.a.	(− 34.68)	·(794.4)	0	n.a.	n.a.
$ACNO_2$	n.a.	n.a.	n.a.	n.a.	(514.6)	n.a.	n.a.	0	n.a.
CS_2	n.a.	n.a.	n.a.	− 125.8	− 60.71	n.a.	n.a.	n.a.	0

[a] n.a., not available. Parameters in parentheses are based on limited data.

4.3. Thermophysical Properties

Estimating the properties of mixtures requires, in general, knowledge of the properties of the pure components of which the mixture is comprised. The effort is to combine these pure component properties in such a way as to yield the properties of the mixture. In many instances, mixture properties are "well behaved," and reasonably accurate prediction of properties is easily accomplished. There are cases, however, in which mixture properties behave badly and prediction is difficult. There are mixtures of relatively common materials for which properties go through a maximum or minimum that cannot be predicted a priori, and that are impossible to predict from pure component properties. As a simple example [30] consider that the density of a mixture of acetic acid and water goes through a maximum with composition. The author's experience indicates that this behavior will be at all temperatures to the mixture critical, although the composition at which the maximum/minimum exists may shift with changes in temperature.

The methods discussed below will deal with the problem of predicting mixture properties from those of the pure components. However, when methods are available, attention will also be given to techniques for predicting mixture properties directly.

A. Liquid density

Mixture density predictions are usually based on weighted averages from composition. The correct procedure requires only that attention be given to keeping dimensional consistency in the calculations. If pure component densities are available on a mass/volume basis, the correct procedure is

$$\rho_{mix} = \Sigma v_i \rho_i \tag{1}$$

If, on the other hand, pure component values are available for molar volumes, then

$$\rho_{mix} = \Sigma \tilde{x}_i \rho_i \tag{2}$$

The dimensional consistency of each is clear. For Eq. (1):

$$\left(\frac{\text{kg}}{\text{m}^3}\right)_{\text{mix}} = \sum \frac{v_i}{V_{\text{mix}}} \frac{kg_i}{v_i}$$

For Eq. (2):

$$\left(\frac{\text{N}}{\text{m}^3}\right)_{\text{mix}} = \sum \frac{V_i}{V_{\text{mix}}} \frac{N_i}{\text{m}^3}$$

where v = volume fraction of one component in mixture
V = volume of one component
i = any component
mix = mixture

The Rackett equation [31] can be used for mixture density. It does require a limited amount of mixture density data for application. In this use it is written as

$$\rho = AB^{-(1-T/C)^{2/7}} \tag{3}$$

where A, B, and C are constants. If sufficient data are available to obtain reliable values for the constants by nonlinear regression, Eq. (3) can be used to extend density data over a wide temperature range. The constants are composition specific and can be used for only one mixture composition.

There are no a priori ways to estimate the density of hydrocarbon mixtures of undefined composition. Generally, the density at ambient conditions and the boiling range of the fraction are known. Figure 1 [32] can be used to determine the density of the undefined mixture at other temperatures.

Unless data are collected in the critical region, density is influenced very little by pressure change. For this reason process engineering requirements will normally be satisfied by taking into account temperature changes while neglecting pressure changes.

B. Viscosity

There are no reliable techniques for estimating mixture viscosity in the absence of experimental data. If the system is well behaved and the mixture viscosity varies monotonically between the viscosities of the pure components, a mole fraction or mass fraction average will generally come within 15% of the mixture viscosity. The great danger, however, is that the mixture will exhibit either a maximum or a minimum. If this happens, the values estimated by averaging may be in error by a substantial margin. The simple acetone-water binary mixture displays a viscosity maximum with composition in which the mixture viscosity is

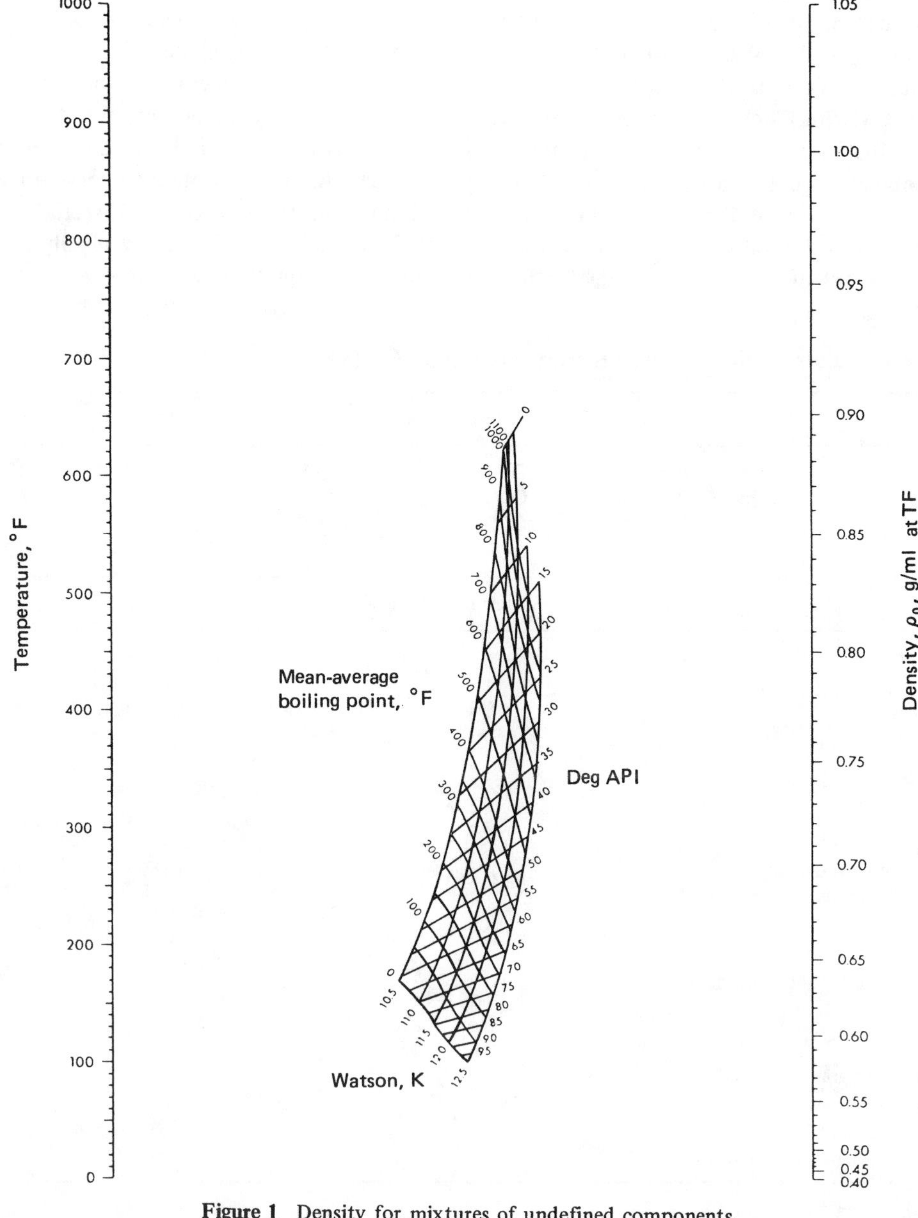

Figure 1 Density for mixtures of undefined components.

more than 100% higher than that for either pure component. Simple mass fraction averaging obviously cannot be used on such a system. Unfortunately, there is no way to predict such behavior except for experimental measurements.

Reid and Sherwood [33] suggest for binaries the use of a modification of the procedure suggested for pure components by Souders [31]:

$$\log (\log 10\, \eta_{Lm}) = \rho_{Lm}\left(\frac{\tilde{x}_1 T_1 \tilde{x}_2 T_2}{\tilde{x}_1 M_1 + \tilde{x}_2 M}\right) - 2.9 \qquad (4)$$

where η_{Lm} = mixture viscosity, cP
$\tilde{x}$ = mole fraction
I = viscosity constant calculated from Table 1 [34]
ρ_{Lm} = mixture density in g mole/cm^3
M = molecular weight

Equation (4) can be used to predict the viscosity of binary mixtures in the absence of either pure component or mixture data. If there is no maximum or minimum in the viscosity-composition curve, errors should be in the 20-25% range for mixtures of hydrocarbons and nonpolar, nonassociating materials; mixtures of polar and/or associating components will display larger errors, and may be completely unreliable.

If a limited amount of experimental data for the mixture is available, it may be extrapolated to higher or lower temperature by plotting log viscosity as a function of reciprocal temperature. This technique must be used with caution because the log $\eta - 1/T$ relationship is truly linear only for a very few pure materials, and even fewer mixtures.

For mixtures of undefined composition, minimal information is necessary before the viscosity can be estimated. If the density and true boiling point for the fraction are known, the procedure recommended by the American Petroleum Institute [35] can be used. In this procedure, Fig. 2 is used to estimate the kinematic viscosity at temperatures of 100° and 200°F. These kinematic viscosities are then plotted on graph paper (very nearly log ν versus $1/T$) to obtain viscosities at other temperatures.

Amin and Maddox [36-39] made an extensive study of viscosity-temperature behavior for crude oil fractions. They concluded that oil mixtures followed the log $\nu - 1/T$ linear relationship, but that other materials, even pure components, did not. By regression analysis of data from the literature and other sources, Amin and Maddox developed an equation relating kinematic viscosity for crude oil fractions to easily measured and generally readily available information. Their equation is

Table 1 Values for determining viscosity constant in Eq. (4)

Group values					
CH_2	55.6	OH	57.1	N	37
H	2.7	COO	90	Cl	60
C	50.2	COOH	104.4	Br	79
O	29.7	NO_2	80	I	110

Structural values[a]					
Double bond	− 15.5	$(R)_2$CHCH$(R)_2$ (R\\CHCH/R with R/ and \\R)	+ 8	−CH=CHCH$_2$X	+ 4
5-C ring	− 24				
6-C ring	− 21				
Side group on 6-C ring: Molecular weight less than 17	− 9	R–C(R)(R)–R	+ 13	$(R)_2$CHX	+ 6
Molecular weight more than 16	− 17	H–C(=O)–R	+ 10		
Ortho and para	+ 3				
Meta	+ 1	CH_3–C(=O)–R	+ 5		

[a]X is a negative group.

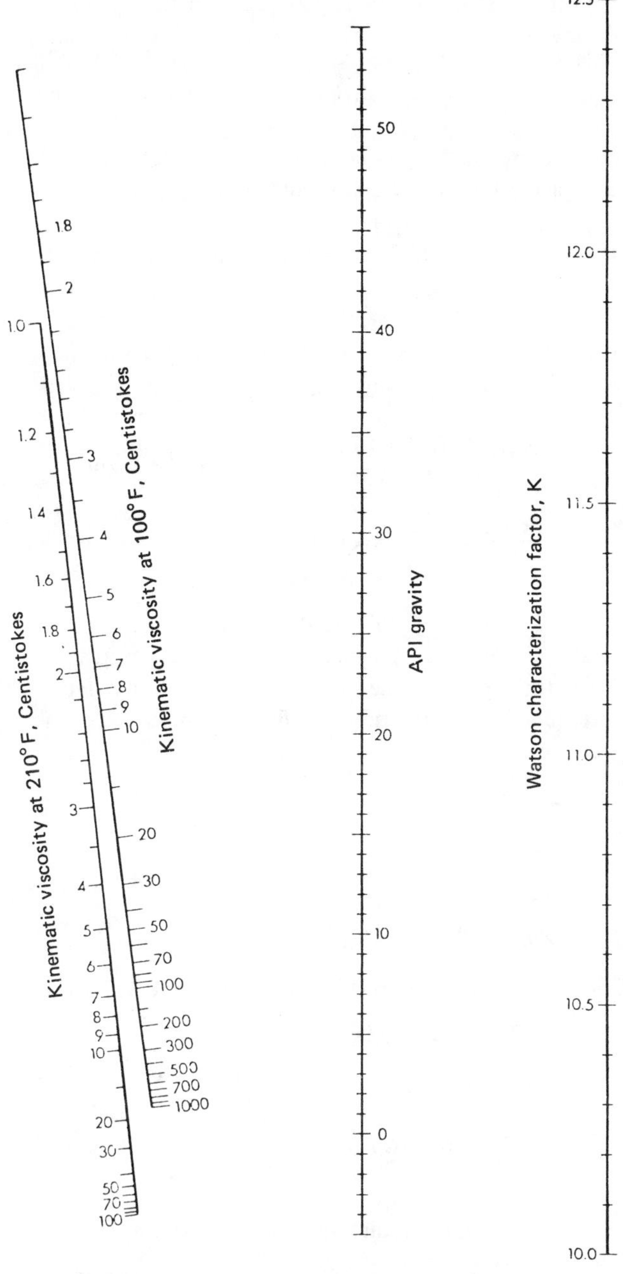

Figure 2 Viscosity for mixtures of undefined components.

$$\nu = A_{\exp} \frac{B}{T} \tag{5}$$

where ν = kinematic viscosity in centistokes
A = constant = $(223\ T_b^{-0.568} - 4.038)\ K/B$
B = constant $\ln B = 4.924 + 0.004\ 54 T_b$
T_b = 50% boiling point in K
T = temperature, K

Example

Estimate the viscosity of a 200°-500°F boiling range cut of a crude oil at a temperature of 93.4°C. The cut is from a typical mid-continent U.S. crude oil, has a 50% boiling point of 415.22 K, and a UOPK of 11.95 [38].

Equation (5) will be used to estimate the viscosity of the crude. The constants A and B in the equation are evaluated as:

$$B = \exp(4.924 + 0.004\ 54 \times 415.22) = 906.05$$

$$A = (223 \times 415.22^{-0.568} - 4.038)\left(\frac{11.95}{906.05}\right) = 0.042\ 5$$

Using these constants in Eq. (5) the kinematic viscosity of the crude oil is estimated as

$$\nu = 0.042\ 5 \exp \frac{906.05}{93.4 + 273.15} = 0.503\ 3 \text{ cS}$$

The value of kinematic viscosity based upon experimental measurements of the absolute viscosity and density is 0.481 3 cS [38].

If the viscosity of the fraction were desired at 10 000 psia, Fig. 3 would be employed. The experimentally determined density for the fraction at 93.4°C is 0.707 g/cm^3. Using this, the absolute viscosity is

$$\eta = \nu\rho = 0.503\ 2 \times 0.707 = 0.356 \text{ cP}$$

On Fig. 3, go vertically from 10 000 psia until a viscosity of 0.356 cP is reached and then proceed horizontally until the reference line is reached. From the reference line proceed parallel to the curved lines until a UOPK factor of 11.95 is reached. From a K of 11.95 proceed horizontally to the left-hand axis and

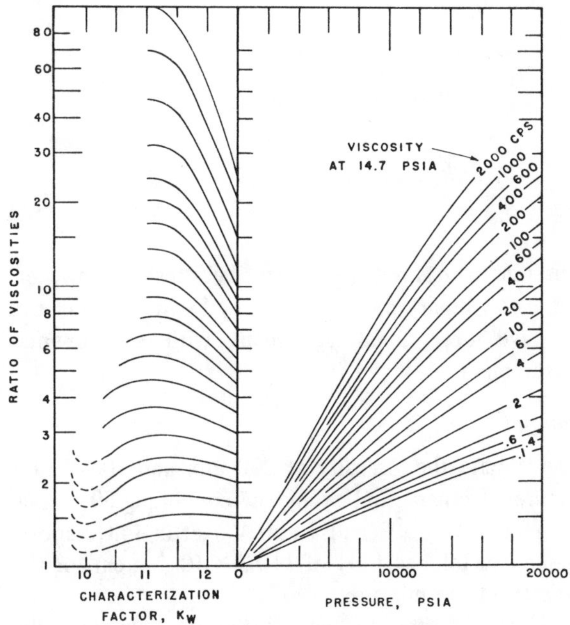

Figure 3 Effect of pressure on viscosity of mixtures of undefined liquid components. From Lockhart and Lenoir [40].

read off the viscosity ratio of 1.75. The absolute viscosity of the fraction at 93.4°C and 10 000 psia is

$$\eta = 0.356 \times 1.75 = 0.623 \text{ cP}$$

At moderate conditions the influence of temperature on gas viscosity is more significant than that of pressure. However, at extreme pressure there can be a significant influence of pressure on gas mixture viscosity. The procedure recommended by Dean and Stiel [14] appears to be the best way for predicting viscosity of gas mixtures under high-pressure conditions. Their recommended equation is

$$(\eta_m - \eta_m^0)\xi_m = 1.08 \left[\exp\left(1.439\rho_{rm}\right) - \exp\left(-1.111\rho_{rm}^{1.858}\right)\right] \quad (6)$$

where η_m = high-pressure mixture viscosity in μP
η_m^0 = low-pressure mixture viscosity in μP
ρ_{rm} = pseudoreduced mixture density
ρ_m = mixture density in g mole/cm^3
ρ_{cm} = pseudocritical mixture density in g mole/cm^3
$\xi_m = T_{cm}^{1/6}/M_m^{1/2}P_{cm}^{2/3}$

In their evaluation of Eq. (6) Dean and Stiel evaluated the pseudocritical values for the mixture according to the modified Prausnitz and Gunn rules [42].

$$T_{cm} = \sum_i y_i T_{ci} \quad (7)$$

$$Z_{cm} = \sum_i y_i Z_{ci} \quad (8)$$

$$V_{cm} = \Sigma y_i V_{ci} \quad (9)$$

$$P_{cm} = \frac{Z_{cm} R T_{cm}}{V_{cm}} \quad (10)$$

$$\rho_{cm} = \frac{\sum_i \tilde{x}_i M_i}{\Sigma \tilde{x}_i V_{ci}} \quad (11)$$

Equation (6) has not been extensively tested. However, on the few binary mixtures used by Dean and Stiel Eq. (6) gave exceptional agreement with experimental measurement.

Example

Estimate the viscosity at 200 atm and 100°C of a mixture of ethane and propylene containing 50.1 mole % ethane. At 100°C and 1 atm Vargaftik [43] reports a value by Golubev [44] of 110.3 × 10^{-6} g/cm for the viscosity of the mixture.

The first step in the solution is to obtain the critical parameters for the two components in the gas mixture:

	T_c, K	V_c, m^3/mole	Z_c	M
Ethane	305.4	0.148	0.285	30.07
Propylene	305.0	0.181	0.273	42.081

From these values properties of the mixture are calculated according to the stated mixing rules:

$$M_m = 0.501 \times 30.07 + 0.499 \times 42.081 = 36.06$$

$$T_{cm} = 0.501 \times 305.4 + 0.499 \times 365 = 335.1$$

$$V_{cm} = 0.501 \times 0.148 + 0.499 \times 0.181 = 0.165$$

$$Z_{cm} = 0.501 \times 0.285 + 0.499 \times 0.273 = 0.279$$

$$P_{cm} = \frac{0.279 \times 0.083\,14 \times 335.1}{0.016\,5} = 218.55 \text{ kg/m}^3$$

$$\rho_{cm} = \frac{36.06}{0.165} = 218.55 \text{ kg/m}^3$$

In order to use Eq. (6) the density of the gas mixture at 100°C and 200 atm must be estimated. This can be done through the use of the compressibility factor determined by the use of reduced temperature and pressure:

$$T_r = \frac{373.15}{335.1} = 1.11 \qquad P_r = \frac{200}{46.5} = 4.30$$

$$Z = 0.635$$

$$PV = ZnRT$$

$$200 \times 1.013\,25 \times V = 0.083\,14 \times 0.635 \times 373.15$$

$$V = 0.097\,2 \text{ m}^3\text{/kg mole}$$

$$\rho_m = 10.29 \text{ kg mole/m}^3 = 370.94 \text{ kg/m}^3$$

$$\rho_{Rm} = \frac{370.94}{218.55} = 1.697\,3$$

From Eq. (6) the mixture viscosity at 200 atm is

$$(\eta_m - 110.3)0.033\,4 = 1.08[\exp 1.439 \times 1.697\,3)$$

$$- \exp(-1.111 \times 1.697\,3^{1.858})]$$

$$= 1.08(11.501 - 0.051)$$

$$\eta_m - 110.3 = 370.24$$

$$\eta_m = 480.5$$

The experimental value reported by Golubev is 500 × 10^{-6} g/cm.

C. Thermal conductivity

Early efforts at estimating mixture thermal conductivities used weighted averages based on mass or molecular fractions. As these were shown to be inadequate, more sophisticated (usually empirical) models were suggested. Jamieson and Irving [45] suggested a procedure based on weight fractions:

$$\lambda_{mix} = x_1\lambda_1 + x_2\lambda_2 - (\lambda_2 - \lambda_1)(1 - \sqrt{x_2})x_2 \quad (12)$$

where 2 represents the component with larger thermal conductivity. Equation (12) gives better accuracy with nonaqueous mixtures. Errors will normally be less than 10% unless the mixture displays a minimum or maximum with composition.

Li [46] assumed: (1) Energy transport in the liquid state is by collision among molecules. Collision is due primarily to oscillation of neighboring molecules and the frequency of collision is approximately proportional to the number and size of neighboring molecules. (2) The interaction thermal conductivity is the harmonic mean of pure component values. The thermal conductivity then is

$$\lambda_m = \sum_i \sum_j \phi_i \phi_j \lambda_{ij} \quad (13)$$

where

$$\lambda_{ij} = 2(\lambda_i^{-1} + \lambda_j^{-1})^{-1}$$

$$\phi_i = \frac{\tilde{x}_i V_i}{\Sigma \tilde{x}_j V_j}$$

$$\Sigma \phi_i = 1.0$$

Li indicates that, for nonaqueous mixtures, critical volume may be substituted for molar volume without seriously influencing the accuracy of Eq. (13).

Example

Estimate the thermal conductivity of a mixture of methanol and water which contains 20 mole % water at 60°C. The molecular weight of methanol is 32.042 and the density at 60°C is 0.755 5 g/cm [47]. The molecular weight of water is 18.016 and the density at 60°C is 0.983 g/cm^3 [48]. Equation (13) in expanded form for a binary mixture is

$$\lambda_m = \Sigma\phi_i - \phi_j\lambda_{ij} = \phi_1\phi_1\lambda_{11} + \phi_1\phi_2\lambda_{12} + \phi_2\phi_2\lambda_{12}$$

Calling methanol component 1 and water component 2 the molecular volumes are

$$V_1 = \frac{32.042}{0.755\ 5} = 42.44 \text{ cm}^3/\text{g mole}$$

$$V_2 = \frac{18.016}{0.983\ 2} = 18.32 \text{ cm}^3/\text{g mole}$$

The superficial volume fractions for each component are

$$\phi_2 = \frac{0.2 \times 18.32}{0.2 \times 18.32 + 0.8 \times 42.44} = 0.097\ 4$$

$$\phi_1 = 0.902\ 6$$

The thermal conductivity interactions are

$$\lambda_{11} = \frac{2(1)}{0.188^{-1} + 0.188^{-1}} = 0.188$$

$$\lambda_{12} = \frac{2(1)}{1(0.188^{-1} + 0.652^{-1})} = 0.291\ 9$$

$$\lambda_{13} = 0.652$$

The mixture thermal conductivity then becomes

$$\lambda_m = 0.902\ 6 \times 0.902\ 6 \times 0.188 + 0.902\ 6 \times 0.097\ 4 \times 0.291\ 9 + 0.097\ 4 \times 0.902\ 6 \times 0.291\ 9 + 0.097\ 4 \times 0.097\ 4 \times 0.652 = 0.210\ 7$$

According to Li [46] the experimental thermal conductivity determined by Rastorguev and Ganier [49] was 0.216 W/mK.

If the procedure of Jamieson and Irving in Eq. (12) is to be used the first step is to calculate the weight fractions: for methanol

$$X_1 = \frac{0.8 \times 32.042}{0.2 \times 18.016 + 0.8 \times 32.042} = 0.876\ 8$$

For water

$$X_2 = \frac{0.2 \times 18.016}{0.2 \times 18.016 + 0.8 \times 32.042} = 0.123\ 2$$

The thermal conductivity for the mixture then becomes

$$\lambda_m = 0.123\ 2 \times 0.652 + 0.876\ 8 \times 0.188 - (0.652 - 0.188)(1 - 0.123\ 2)0.123\ 2$$

$$\lambda_m = 0.020\ 8$$

Li indicates that, for nonaqueous mixtures, critical volume may be substituted for molar volume without seriously influencing the accuracy of Eq. (13).

For mixtures of undefined hydrocarbon components the thermal conductivity may be estimated by

$$\lambda = 1.744 \times 10^{-2} - 1.493 \times 10^{-5}\ T \quad (14)$$

where λ is the thermal conductivity in(W/m^2)/(K/m) and T is the temperature in K.

Equation (14) provides only one thermal conductivity value regardless of boiling range, density, or Watson characterization factor. This seems an oversimplification and has been shown experimentally to be incorrect [50]. However, it will provide estimates of thermal conductivity for crude fractions that are generally within ± 25% of the true thermal conductivity of the mixture. It is limited to boiling range cuts distilled from raw crude. It should not be used for internal refinery streams that contain substantial quantities of unsaturated hydrocarbons.

The effect of pressure on liquid thermal conductivity is small over normal pressure ranges.

D. Heat capacity

There are no techniques for estimating the heat capacity of mixtures without knowledge of pure component values. When pure component data are available, the mixture value is taken as the unweighted average of the pure component values, keeping in mind the necessity for dimensional consistency. This approach ignores heat of mixing and similar effects. For this reason, large errors could be generated when the heat of mixing is nonzero. For hydrocarbons and nearby homologs, the assumption is good and heat capacities should be within 10% although few mixture heat capacity data are available for comparison.

For undefined mixtures, liquid heat capacity may be estimated by use of Fig. 4 which was adapted from a chart by Maxwell [51]. Density and true boiling point range for the mixture are required for use of Fig. 4. At pseudo-reduced temperatures below 0.85, the isobaric heat capacity of undefined mixtures can be estimated as [52]:

$$C_p = A_1 + A_2 T + A_3 T^2$$

where C_p is the isobaric heat capacity for the liquid petroleum fraction in Btu/lb °F.

$$A_3 = -\,1.171\,26 + (0.023\,722 + 0.024\,907\ \mathrm{SG})K + \frac{(1.149\,82 - 0.046\,535\,K)}{\mathrm{SG}}$$

$$A_2 = 10^{-4}(1.0 + 0.824\,63dK)\left(1.121\,72 - \frac{0.276\,34}{\mathrm{SG}}\right)$$

$$A_2 = -\,10^{-8}(1.0 + 0.824\,63K)\left(2.902\,7 - \frac{0.709\,58}{\mathrm{SG}}\right)$$

T_r = reduced temperature, T/T_{pc}
T = temperature in °R
T_{pc} = pseudocritical temperature, °R
K = Watson characterization factor (see p. 5.2.1-7)
SG = specific gravity 60F/60F

E. Specific heat capacity of gas mixtures

For mixtures of ideal gases the heat capacity can be estimated as either the mole fraction or weight fraction average depending upon which heat capacity is available. The methods suggested in Sec. 5.4 can be used to obtain the pure component values.

For mixtures of real gases the departure from ideal gas behavior must be accounted for. Figure 5 shows the molal heat capacity departure of real gases from ideal gas behavior as a function of reduced temperature and pressure [53].

Example

Estimate the heat capacity of a gas mixture that is 40 mole % methane and 60 mole % ethane. The mixture is at 40°C and at a pressure of 7 MPa. From the GPA data book [54] the molal ideal gas state heat capacity for methane is 36.21 and for ethane 54.21 kJ/kg mole K.

The solution first requires estimating the mole fraction average heat capacity, critical temperature, and critical pressure. The calculation is shown in the following table.

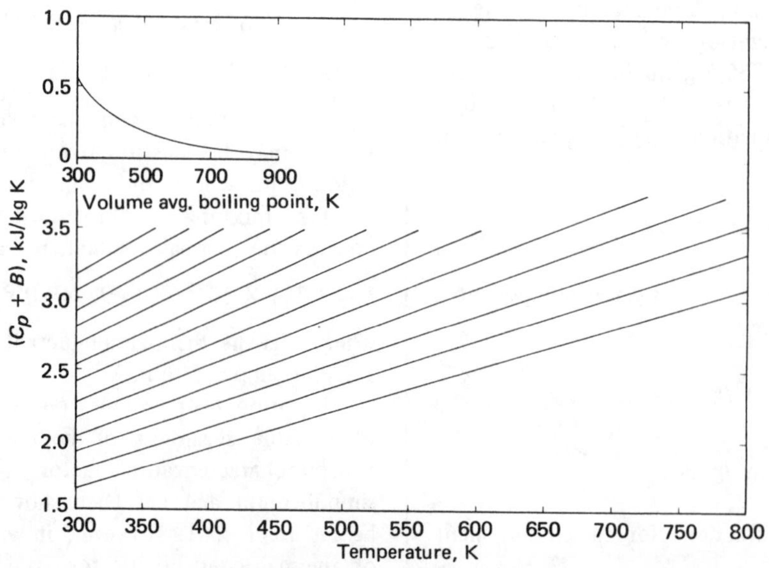

Figure 4 Specific heat capacity of hydrocarbon liquid mixtures $C_p = (C_p + B) - B$.

Comp	Mole fraction	C_p, kj/kg mole K	C_{pm}	T_c, K	T_{cm}, K	P_c, MPa	P_{cm}, MPa
C_1	0.4	36.21	14.48	190.6	76.2	4.60	1.84
C_2	0.6	54.21	32.53	305.9	183.5	4.88	2.93
			47.01		259.7		4.77

Critical temperature and pressure values were taken from the American Petroleum Institute [55].

The reduced temperature is

$$T_r = \frac{313.2}{259.7} = 1.21$$

The reduced pressure is

$$P_r = \frac{7}{4.77} = 1.47$$

Using these values at reduced temperature and pressure the heat capacity departure as read from Fig. 5 is

$$C_p - C_p^* = 9.7 \text{ Btu/lb mole °R} = 40.6 \text{ kJ/kg mole K}$$

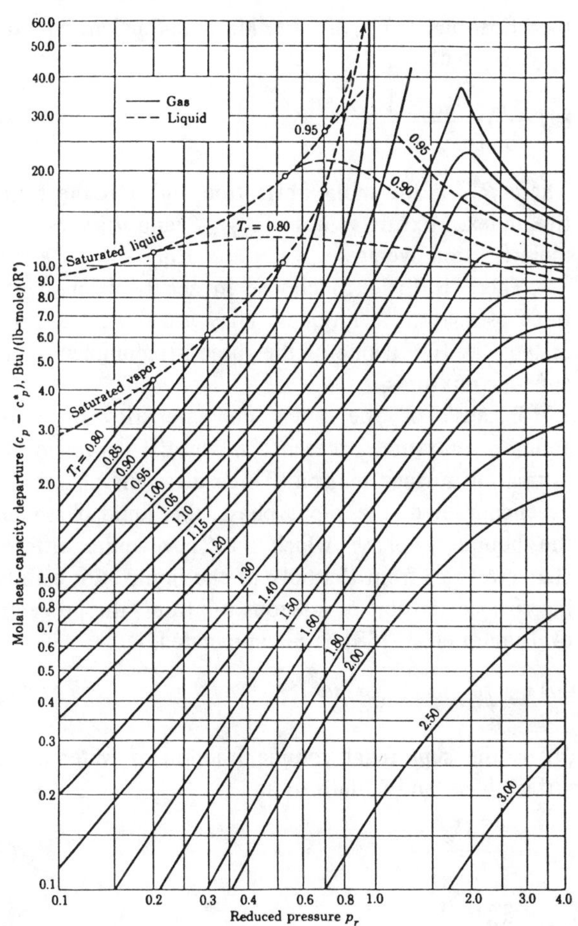

Figure 5 Real gas molal heat capacity departure. From Leyderson et al. [55].

The real gas mixture heat capacity is

$$C_{p_m} - 47.01 = 40.6 = 87.61 \text{ kJ/kg mole K}$$

F. Mixtures of undefined components

Figure 6 is taken after the work of Fallon and Watson [56] as presented in Standards of Tubular Exchanger Manufacturers Association [57]. It provides the specific heat capacity of petroleum fractions as a function of the Watson characterization factor and temperature. The specific heat capacity values from Fig. 6 are for low-pressure use and compare with the ideal gas state heat capacity for pure components. Figure 5 can be used to adjust these values for real gas pressure effects if pseudocritical values are estimated as outlined in Sec. 5.2.1(B).

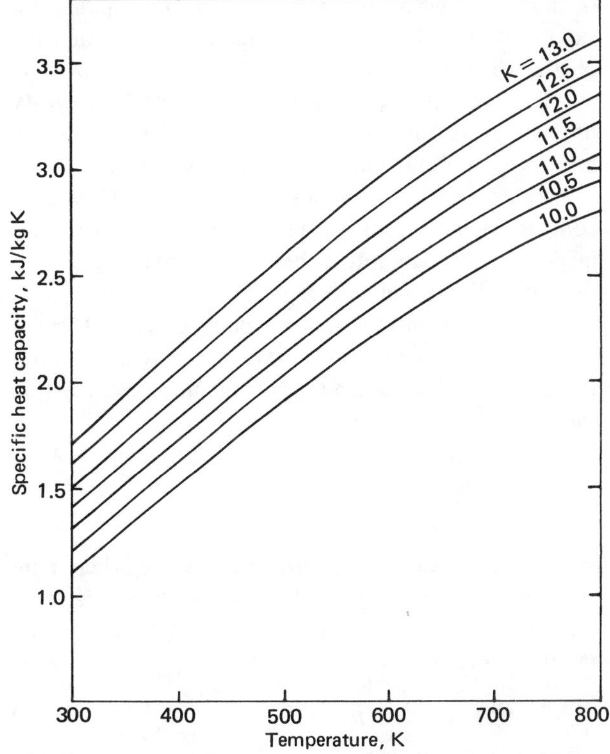

Figure 6 Specific heat capacities of vapor mixtures of undefined hydrocarbon components.

4.4. Interfacial Tension

In the literature, there is great confusion in using terms to describe the "pulling" on molecules at the interface between two phases. In this discussion "surface tension" will be applied to those instances when the gas phase is air; interfacial tension will apply to instances when the gas phase is the equilibrium vapor of the liquid. Liquid-liquid and liquid-solid phases will not be considered.

For most process design applications, the quantity of interest is the interfacial tension. Seldom do we have process fluids under conditions such that they are in contact with air at atmospheric pressure. Most pure component measurements, until recently at least, were made on the laboratory bench and the engineer is limited to the use of surface tension.

For binary mixtures, when pure component values of surface tension are known, the mixture surface tension may be estimated by simple mole fraction average:

$$\sigma_m = \sum_i x_i \sigma_i \tag{1}$$

When the mixture is at high pressure, or when pure component values are not available, the Macleod-Sugden [56] relationship can be used to estimate mixture surface tension:

$$\sigma_m^{1/4} = {}_i(P_i)(\rho_{Lm}\tilde{x}_i - \rho_{Vm}\tilde{y}_i) \tag{2}$$

where P is the Parachor number and m is the mixture property. If density values for the mixture are not available, but it is made up of defined components, a modification [57] of the Macleod-Sugden equation may be used:

$$\sigma_m \; \frac{P_m}{R_{Dm}} \frac{n_m^2 - 1}{n_m^2 + 2} \tag{3}$$

where R_D is the molar refraction and n is the refractive index. Mixture values of R_D, P, and n are taken as mole fraction averages.

There is little to choose in accuracy among the three estimating techniques. They are subject to errors as large as 10%. None of the three is recommended for aqueous mixtures.

Because of the polar characteristics of water, aqueous mixtures present a special problem for estimating surface tension. Three different phases may be encountered—the vapor phase, the liquid phase, and the boundary phase—which may have quite different characteristics from the bulk of the liquid. For estimating surface tension of aqueous mixtures, the procedure of Tamura et al. [58] is recommended.

$$\sigma_m^{1/4} = \psi_w^\sigma \sigma_w^{1/4} + \psi_o^\sigma \sigma_o^{1/4} \tag{4}$$

ψ_w^σ is the superficial volume fraction of water in the surface layer and is defined as

$$\psi_w^\sigma = \frac{\tilde{x}_w^\sigma V_w}{V^\sigma} \tag{5}$$

$$\psi_o^\sigma = \frac{\tilde{x}_o^\sigma V_o}{V^\sigma} \tag{6}$$

For calculating the superficial surface volume fractions,

equilibrium between the surface and bulk phases is assumed.

$$C = B + D \tag{7}$$

$$B = \log \frac{\psi_w^q}{\psi_o} \tag{8}$$

$$C = \log \frac{(\psi_w^\sigma)^q}{\psi_o^\sigma} \tag{9}$$

$$D = 0.441 \frac{q}{T}\left(\frac{\sigma_o V_o^{2/3}}{q}\right) - \sigma_w V_o^{2/3} \tag{10}$$

$$\psi_w = \frac{\tilde{x}_w V_w}{\tilde{x}_w V_w + \tilde{x}_o V_o} \tag{11}$$

$$\psi_o = \frac{x_o V_o}{\tilde{x}_w V_w + \tilde{x}_o V_o} \tag{12}$$

where $\tilde{x}$ = bulk mole fraction of component
V = molar volume of component in cm^3/g mole
σ = surface tension
T = temperature, K
q = constant depending on type and size of organic component. Values in Table 1.
ψ = superficial volume fraction
o = a subscript denoting organic component
w = a subscript denoting water
σ = a superscript denoting the surface zone between the liquid and vapor phase

Reid et al. [58] report errors less than 10% when q is less than 5 with errors as large as 20% for values of q larger than 5. Equation (4) can be applied to organic mixtures by letting q equal the ratio of molar volumes of solute to solvent.

Example

Calculate the interfacial tension of a 30 mole % acetone-water mixture at 150°F. According to Miller [59] the surface tension of water is 66.84 dynes/cm and the surface tension of acetone is 17.62 dynes/cm at 150°F (65.6°C). The molecular weight of acetone and water are 58.1 and 18.016, while the densities are 0.737 9 and 0.980 3 [60].

The molar volumes are calculated from the molecular weight and density:

$$V_o = \frac{58.1}{0.737\,9} = 78.74 \text{ cm}^3/\text{g mole}$$

$$V_w = \frac{180\,16}{0.980\,3} = 18.38 \text{ cm}^3/\text{g mole}$$

The superficial volume fractions for water and acetone are calculated from Eqs. (11) and (12):

$$\psi_w = \frac{0.7 \times 18.38}{0.7 \times 18.38 + 0.3 \times 78.74} = 0.352\,6$$

$$\psi_o = \frac{0.3 \times 78.74}{0.7 \times 18.38 + 0.3 \times 78.74} = 0.647\,4$$

Using a value of $q = 2$ for acetone from Table 1 the constant B is evaluated from Eq. (8):

$$B = \log \frac{(0.352\,4)^2}{0.647\,6} = \log 0.191\,8 = -\,0.717\,2$$

and the constant D from Eq. (1):

$$D = 0.441 \frac{2}{338.7}\left[\frac{17.62(78.74)^{2/3}}{2} - 66.84(78.74)^{2/3}\right]$$

$$= 0.002\,604(161.85 - 1\,227.9) = -\,2.776$$

The constant C is evaluated from Eq. (7):

$$C = B + D = -\,0.717\,2 - 2.776 = -\,3.493\,2$$

However, constant C is also expressed in terms of the interphase composition in Eq. (9). Since by definition the superficial volume fractions must sum to unity

$$\psi_o^\sigma + \psi_w^\sigma = 1.0$$

The superficial volume fractions at the interface can be calculated by combination of Eqs. (7) and (9).

$$\frac{\log\,(\psi_w^\sigma)^q}{\psi_o} = -\,3.493\,2$$

$$\frac{\log\,(\psi_w^\sigma)^2}{1 - \psi_w^\sigma} = -\,3.493\,2$$

Solving for the superficial volume fractions at the interface:

$$\psi_w^\sigma = 0.017\,8$$

$$\psi_o^\sigma = 0.982\,2$$

Table 1 Values of constant q for use with procedure of Tamura et al. [58]

Materials	q	Example
Fatty acids, alcohols	Number of carbon atoms	Acetic acids, $q = 2$
Ketones	One less than the number of carbon atoms	Acetone, $q = 2$
Halogen derivatives of fatty acids	Number of carbons times ratio of molal volume of halogen derivative to parent fatty acid	Chloroacetic acid, $q = 2\frac{V_b\text{ (chloroacetic acid)}}{V_b\text{ (acetic acid)}}$

The mixture interfacial tension at 150°F can now be calculated from Eq. (4);

$$\sigma_m^{1/4} = 0.017\,8(66.84)^{1/4} + 0.982\,2(17.62)^{1/4}$$

$$\sigma_m = 18.12 \quad \text{(experimental value is 23.1)}$$

Miller [59] gives an experimentally determined value of 23.1 dynes/cm which differs by more than 20% from the calculated value.

Surface tensions for mixtures of undefined components can be estimated through the Parachor number. Equation (2) is used with the Parachor number being determined by [61]

$$P = 40 + 2.38 \text{ (mol. wt.)} \tag{13}$$

4.5. Diffusion Coefficients

A. Introduction

Diffusion refers here to transport of a material within one phase due to molecular characteristics alone: there is no external force or turbulence causing the transport. Diffusion can be caused by gradients within the phase created by a number of variables. The only case considered here is diffusion caused by a concentration gradient. In a binary mixture, if component A diffuses from position 1 to position 2 at a given rate, then to maintain equal molal flow, component B must diffuse at the same rate in the opposite direction. If F^m is the molal flux, the diffusion coefficient is defined as

$$F_A^m = -CD_{AB}\nabla x_A \tag{1}$$

$$F_B^m = -CD_{BA}\nabla x_B \tag{2}$$

where C = total molar concentration
F^m = the molar flux due to diffusion
D = the diffusion coefficient

The activity corrected diffusion coefficient is

$$\Delta_{AB} = \frac{D_{AB}}{(\partial \ln a_A)/(\partial \ln \tilde{x}_A)} \tag{3}$$

For gases, $D_{AB} = \Delta_{AB}$ because concentration effects are much less for gases than for liquids.

B. Diffusion in liquids

For diffusion of components through dilute solutions, the equation proposed by Wilkie and Chang [62] is recommended:

$$D_{AB} = \frac{1.17 \times 10^{-16}\,(\phi \tilde{M}_B)^{1/2} T}{\eta_B \tilde{v}_A^{0.6}} \tag{4}$$

where D_{AB} = diffusion coefficient of solute A in low concentrations in solvent B (m^2/s)
$\tilde{M}_B$ = molecular weight of solute B
T = temperature (K)
η_B = viscosity of the solvent B at temperature T (Ns/m^2)
ϕ = the associated factor for solvent B (= 2.6 for water; = 1.0 for nonassociated solvents)
$\tilde{v}_A$ = molar volume of solute at its normal boiling point ($m^3/kmol$)

For nonassociated solvents, Eq. (4) can be used with a value of 1.0 for ϕ.

Tyn [64] developed a nomograph for estimating diffusion coefficients for various solute-solvent combinations at different temperatures. His approach was to assume a linear relationship between the temperature at which a liquid mixture will have a given diffusion coefficient and the temperature at which water exhibits the same coefficient. His nomograph is shown in Fig. 1. X and Y coordinates for different components are shown in Table 1. The use of the nomograph is best illustrated with an example (see below).

Example

Calculate the liquid diffusion coefficient for methanol solute in water solvent at 15°C. Here, $\tilde{M}_B = 18.016, \eta_B = 1.146 \times 10^{-3}, \phi = 2.6, T = 288.15\text{K}$

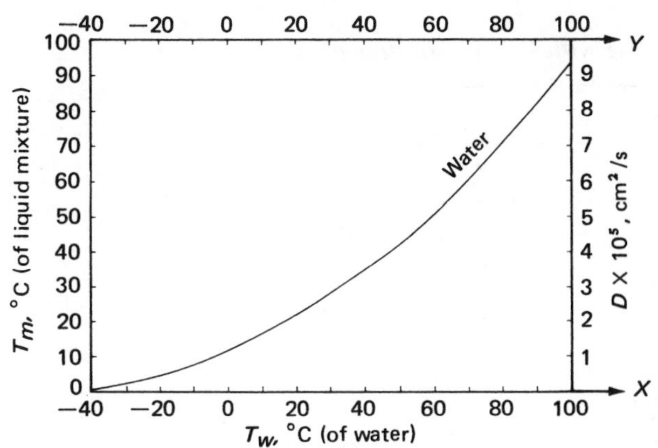

Figure 1 Nomograph for estimating liquid diffusion coefficients.

and (from the table for methanol in Sec. 5.5.1) $\tilde{v}_A = 32.00/751.0 = 0.042\,6$ m³/k mole.

Using Eq. (4), the diffusion coefficient is:

$$D_{AB} = \frac{(1.17 \times 10^{-16})(2.6 \times 18.016)^{1/2}\ (2.88.15)}{(1.146 \times 10^{-3})(0.0426)^{0.6}}$$

$$= 1.34 \times 10^{-9}\ \text{m}^2/\text{s}$$

According to Bretsznajder [65] the experimental value is 1.28×10^{-9} m²/s.

If Fig. 1 is to be used, the values of X and Y must be determined from Table 1. For the methanol-water system the X coordinate is − 12.0 and the Y coordinate is 80. These two points are connected with a straight line. From the intersection of that straight line

Table 1 Mixture coordinates for use with Fig. 1

Binary liquid mixture				
No.	Solute	Solvent	X	Y
1	Methane	Water	− 11.5	76.5
2	Ethane	Water	− 16.0	63.8
3	Propane	Water	− 21.0	62.5
4	Butane	Water	− 22.0	56.8
5	Pentane	Water	− 23.0	51.0
6	Cyclohexane	Water	− 21.5	41.0
7	Cyclopentane	Water	− 19.0	46.0
8	Methylcyclopentane	Water	− 22.0	41.0
9	Acetylene	Water	0.0	70.0
10	Benzene	Water	− 15.0	55.0
11	Toluene	Water	− 21.5	45.5
12	Ethyl benzene	Water	− 21.5	41.0
13	Methanol	Water	− 12.0	80.0
14	Ethanol	Water	− 14.0	50.5
15	*n*-Propanol	Water	− 19.5	56.0
16	*i*-Propanol	Water	− 20.0	53.0
17	*n*-Butanol	Water	− 21.0	50.0
18	*t*-Butanol	Water	− 27.0	60.0
19	Ethyleneglycol	Water	− 16.0	51.0
20	Acetone	Water	− 11.0	44.5
21	Manitol	Water	− 27.0	24.0
22	Alanine	Water	− 22.0	40.5
23	Acetamide	Water	− 18.0	64.0
24	Butylamide	Water	− 22.5	59.5
25	Foramide	Water	− 11.0	73.5
26	Glycoamide	Water	− 17.0	49.0
27	Propioamide	Water	− 22.0	63.5
28	Urea	Water	− 13.5	58.0
29	Triethyleneglycol	Water	− 28.0	41.0
30	Iodine	Ethanol	− 10.5	47.5
31	Water	Ethanol	− 19.0	53.5
32	Acetic acid	Carbon tetrachloride	− 2.5	35.5
33	Formic acid	Carbon tetrachloride	+ 4.5	44.5
34	Benzene	Carbon tetrachloride	− 6.5	41.0
35	Cyclohexane	Carbon tetrachloride	− 8.0	36.5
36	Heptane	Benzene	+ 3.0	63.0
37	Methanol	Benzene	+ 25.5	100.0
38	Ethanol	Benzene	+ 12.0	92.0
39	Benzoic acid	Benzene	− 7.5	43.0

Table 1 Mixture coordinates for use with Fig. 1 (*Continued*)

Binary liquid mixture				
No.	Solute	Solvent	X	Y
40	Acetic acid	Toluene	+ 7.0	64.5
41	Formic acid	Toluene	+ 19.5	52.5
42	Cyclohexane	Toluene	+ 1.5	81.0
43	Methylcyclohexane	Toluene	+ 5.0	69.0
44	Aniline	Toluene	+ 1.0	67.0
45	Acetic acid	Acetone	+ 23.0	84.0
46	Water	Acetone	+ 49.5	97.0
47	Chloroform	Acetone	+ 26.5	92.0
48	Iodine	*n*-Octane	+ 16.0	69.0
49	Toluene	Aniline	− 36.0	28.0
50	Benzene	Heptane	+ 21.5	93.0
51	Toluene	Heptane	+ 28.5	87.0
52	Acetone	Chloroform	+ 8.0	66.5
53	Toluene	Methylcyclohexane	− 3.0	50.0
54	Carbon tetrachloride	Cyclohexane	− 7.5	50.0
55	Iodine	Cyclohexane	0.0	53.0
56	Benzene	Cyclohexane	− 1.5	63.0
57	Toluene	Cyclohexane	− 2.5	42.0
58	Chlorobenzene	Bromobenzene	− 7.5	33.5
59	Toluene	Chlorobenzene	− 1.0	49.0

with the 15°C horizontal line, extend vertically to the curve labeled "water" and then horizontally to the right-hand axis to read a diffusion coefficient of:

$$D_{AB} = 1.25 \times 10^{-5}\ \text{cm}^2/\text{s}$$

C. Diffusion in gases

For gaseous diffusion coefficients in binary mixtures, the procedure of Fuller et al. [67] is recommended. They utilized a group contribution approach to define special "diffusion volumes" which are used to predict diffusion coefficients from:

$$D_{AB} = \frac{1.013 \times 10^{-7} T^{1.75} (1/\tilde{M}_A + 1/\tilde{M}_B)^{1/2}}{P\left[\left(\sum_A v_i\right)^{1/3} + \left(\sum_B v_i\right)^{1/3}\right]^2} \qquad (8)$$

where D_{AB} = binary diffusion coefficient, m²/s
v_i = special diffusion parameters to be summed over atoms, groups, and structural features of the diffusing species
$\tilde{M}$ = molecular weight, kg/kmole
P = pressure, bar

Example

Determine the gas phase diffusivity of methane through water vapor at a temperature of 352.3 K and one atmosphere (1.013 bar) pressure. The molecular weight of water is 18.016 and the molecular weight of methane is 16.043. Using values from Table 2 the special molecular diffusion volumes for methane and water are calculated as:

$$v_{H_2O} = 12.7$$

$$v_{CH_4} = v_C + 4\, v_H = 16.5 + 4(1.98) = 24.42$$

Using Eq. (8) the diffusivity of methane gas through water vapor is calculated as

Table 2 Diffusion volumes for use with Eq. (8)

Atomic and structural diffusion volume increments			
C	16.5	(Cl)	19.5
H	1.98	(S)	17.0
O	5.48	Aromatic or heterocyclic rings	− 20.2
(N)	5.69		
Diffusion volumes of simple molecules			
H_2	7.07	CO_2	26.9
D_2	6.70	N_2O	35.9
He	2.88	NH_3	14.9
N_2	17.9	H_2O	12.7
O_2	16.6	(CCl_2F_2)	114.8
Air	20.1	(SF_6)	69.7
Ne	5.59		
Ar	16.1	(Cl_2)	37.7
Kr	22.8	(Br_2)	67.2
(Xe)	37.9	(SO_2)	41.1
CO	18.9		

$$D_{AB} = \frac{1.013 \times 10^{-7}\ (352.3)^{1.75}(1/18.016 + 1/16.043)^{1/2}}{1.013(12.7^{1/3} + 24.42^{1/3})^2}$$

$$= 0.358\,9 \times 10^{-4}\ \text{m}^2/\text{s}$$

Fuller et al. report an experimental measurement by Schwertz and Brown of 0.356×10^{-4} m^2/s.

REFERENCES

1. Gibbs, J. W., On the Equilibrium of Heterogenous Substances, *Trans. Connecticut Acad. Sci.*, III, pp. 108–248, Oct. 1875–May 1976, and pp. 343–524, May 1877–July 1878.
2. Maddox, R. N., and Erbar, J. H., *Advanced Gas Conditioning and Processing,* p. 44, Campbell Petroleum Series, Norman, Okla., 1979.
3. Gilliland, E. R., and Reed, C. E., Degrees of Freedom in Multicomponent Absorption and Rectification Columns, *Ind. Eng. Chem.*, vol. 34, pp. 551–557, 1942.
4. Kwauk, M., A System for Counting Variables in Separation Processes, *AIChE J.*, vol. 2, pp. 240–248, 1956.
5. Maadah, A. G., and Maddox, R. N., Thermodynamics, Equilibrium and Efficiency of the Claus Process, *Proc. 5th Canad. Symp. Catalysis, Calgary, Alberta,* 1977.
6. Chao, K. C., and Seader, G. D., A General Correlation of Vapor-Liquid Equilibrium in Hydrocarbon Mixtures, *AIChE J.*, vol. 7, p. 598, 1961.
7. Curl, R. F., Jr., and Pitzer, K. S., Volumetric and Thermodynamic Properties of Fluids Enthalpy, Entropy and Free Energy, *Ind. Eng. Chem.*, vol. 50, p. 265, 1958.
8. Katz, D. L., Cornell, D., Kobayashi, R., Poettman, F. H., Vary, J. A., Elenbaas, J. R., and Weinaug, C. F., *Handbook of Natural Gas Engineering,* p. 82, McGraw-Hill, New York, 1959.
9. Pitzer, K. S., Lippman, D. Z., Curl, R. F., Jr., Huggins, C. M., and Paterson, D. E., Vapor Pressure and Entropy of Vaporization, *J. Am. Chem. Soc.*, vol. 77, pp. 3427–3440, 1955.
10. Edmister, W. C., *Applied Hydrocarbon Thermodynamics,* vol. I, p. 27, Gulf, Houston, 1961.
11. Maddox, R. N., and Erbar, J. H., Low Pressure Retrograde Condensation, *Oil Gas J.*, p. 65, July 11, 1977.
12. Wilson, A., Erbar, J. H., and Maddox, R. N., C_6+ Fractions Affect Phase Behavior, *Oil Gas J.*, p. 76, August 21, 1978.
13. Maddox, R. N., and Erbar, J. H., *Advanced Gas Conditioning and Processing,* p. 146, Campbell Petroleum Series, Norman, Okla., 1979.
14. Kesler, M. G., and Lee, B. I., Improve Prediction of Enthalpy of Fractions, *Hydrocarbon Proc.*, vol. 55, no. 3, March 1976.
15. Johnson, R. C., and Maddox, R. N., Densities for North Sea, Arabian Light and Mid-Continent U.S. Crude Fractions, Rept. LDR-8, Fluid Properties Research, Inc., 1976.
16. *Technical Data Book–Refining,* 2d ed., pp. 4–43, American Petroleum Institute, Washington, D.C., 1970.
17. *Engineering Data Book,* pp. 18-1 to 18-112, Gas Processors Association, Tulsa, Okla., 1972.
18. Redlich, O., and Kwong, J. N. S., The Thermodynamics of Solutions V: An Equation of State, *Chem. Rev.*, vol. 44, pp. 233–244, 1949.
19. Soave, G., Equilibrium Constants from a Modified Redlich-Kwong Equation of State, *Chem. Eng. Sci.*, vol. 27, pp. 1197–1203, 1972.
20. Erbar, J. H., *Documentation for GPA K&H Mod II,* Gas Processors Association, Tulsa, Okla., 1973.
21. Starling, K. E., and Han, M. S., Thermodata Refined for LPG Industrial Applications, *Hydrocarbon Proc.*, vol. 51, pp. 107–115, 1972.
22. Peng, D. Y., and Robinson, D. B., A New Two-Constant Equation of State, *Ind. Eng. Chem. Fundam.*, vol. 15, pp. 59–64, 1976.
23. Erbar, J. H., and Maddox, R. N., private communication, 1977.
24. Pierotti, G. J., Deal, C. H., and Derr, E. L., Activity Coefficients and Molecular Structure, *Ind. Eng. Chem.*, vol. 51, p. 95, 1959.
25. Deal, C. H., Derr, E. L., Group Contributions in Mixtures, *Ind. Eng. Chem.*, vol. 60, no. 4, p. 28, 1968.
26. Derr, E. L., and Deal, C. H., Analytical Solution of Groups: Correlation of Activity Coefficients through Structural Group, *Int. Symp. Distillation,* 1969, Brighton, England, Proc., *Inst. Chem. Eng.*, London.
27. Palmer, D. A., Predicting Equilibrium Relationships for Maverick Mixtures, *Chem. Eng.*, p. 80, June 9, 1975.
28. Fredenslund, A., Gmehling, J., Michelson, J. L., Rasmussen, P., and Prausnitz, J., Computerized Design of Multicomponent Distillation Columns Using the UNIFAC Group Contribution Method for Calculation of Activity Coefficients, *I&EC Process Design Develop.*, vol. 16, p. 450, 1977.
29. Reid, R. C., Prausnitz, J. M., and Sherwood, T. K., *Properties of Gases and Liquids,* 3d ed., p. 347, McGraw-Hill, New York, 1977.
30. Perry, J. H., *Chemical Engineering Handbook,* 3d ed., pp. 185–186, McGraw-Hill, New York, 1950.
31. Rackett, H., Equation of State for Saturated Liquids, *J. Chem. Eng. Data,* vol. 15, p. 514, 1970.
32. *Technical Data Book–Refining,* 2d ed., pp. 6–61, American Petroleum Institute, Washington, D.C., 1970.
33. Reid, R. C., and Sherwood, T. K., *The Properties of Gases and Liquids,* 2d ed., p. 448, McGraw-Hill, New York, 1966.
34. Souders, M., Viscosity and Chemical Constitution, *J. Am. Chem. Soc.*, vol. 60, p. 154, 1938.

35. *Technical Data Book–Refining,* 2d ed., pp. 11-31, American Petroleum Institute, Washington, D.C., 1970.
36. Amin, M., Temperature Dependence of the Viscosity of Some Crude Oils and Pure Liquids, Ph.D. thesis, Oklahoma State University, Stillwater, Okla.
37. Amin, M. B., and Maddox, R. N., Predicting Viscosity of Crude Oil Fractions, *72nd Annu. Meet., Amer. Inst. Chem. Eng.,* San Francisco, Calif., November 25-29, 1979.
38. Amin, M. B., and Maddox, R. N., Temperature Dependence of the Viscosity of Some Crude Oils and Pure Liquids, Liquid Viscosity Rept. LVR-15, Fluid Properties Research, Inc., Stillwater, Okla., 1979.
39. Maddox, R. N., Predicting Viscosity of Crude Oil Fractions, Liquid Viscosity Rept. LVR-13, Fluid Properties Research, Inc., Stillwater, Okla., 1979.
40. Lockhart, F. J., and Lenoir, J. M., Liquid Viscosities at High Pressure, *Pet. Ref.,* vol. 40, no. 3, p. 209, March 1961.
41. Dean, D. E., and Stiel, L. I., The Viscosity of Nonpolar Gas Mixtures at Moderate and High Pressures, *AIChE J.,* vol. 11, pp. 526-532, 1965.
42. Reid, R. C., Prausnitz, J. M., and Sherwood, T. K., *The Properties of Gases and Liquids,* 3d ed., p. 74, McGraw-Hill, New York, 1977.
43. Vargaftik, N. B., *Tables on the Thermophysical Properties of Liquids and Gases,* 2d ed., p. 676, Hemisphere, New York, 1975.
44. Golubev, I. F., *Viscosity of Gases and Gas Mixtures,* Fizmat Press, 1959.
45. Jamieson, D. T., and Irving, J. B., Thermal Conductivity of Binary Liquid Mixtures, Rept. no. 567, National Engineering Laboratory, East Kilbride, Glasgow, 1974.
46. Li, C. C., Thermal Conductivity of Liquid Mixtures, *AIChE J.,* vol. 22, no. 5, p. 927, September 1976.
47. Johnson, R. C., Ratcliffe, A. E., and Maddox, R. N., Experimental Liquid Densities for Pure Paraffin Alcohols, Rept. LDR-10, Fluid Properties Research, Inc., Stillwater, Okla., 1977.
48. Maddox, R. N., Design Data Standard–Liquid Viscosity of Water, Fluid Properties Research, Inc., Rept. DDS-LV-12, Stillwater, Okla., 1979.
49. Rastorguev, Y. L., and Ganier, Y. A., Thermal Conductivity of Non-Electrolyte Solutions, *Russ. J. Phys. Chem.,* vol. 41, no. 6, p. 717, 1967.
50. Lee, M. C., and Maddox, R. N., Comparison of Experimental and Calculated Thermal Conductivity for Fractions of Three Crude Oils, Thermal Conductivity Rept. LCR-5, Fluid Properties Research, Inc., Stillwater, Okla., 1979.
51. Maxwell, J. B., *Data Book on Hydrocarbons,* p. 93, van Nostrand, Princeton, 1950.
52. *Technical Data Book–Refining,* 2d ed., p. 7-167, American Petroleum Institute, Washington, D.C., 1970.
53. Lyderson, A. L., Greenkorn, R. A., and Hougen, O. A., *Generalized Thermodynamic Properties of Pure Fluids,* Engineering Experiment Station Report No. 4, University of Wisconsin, October, 1955.
54. *Engineering Data Book,* p. 4-1, Gas Processors Association, Tulsa, Okla., 1972.
55. *Technical Data Book–Refining,* 2d ed. p. 1-22, American Petroleum Institute, Washington, D.C., 1970.
56. Macleod, D. B., On a Relation Between Surface Tension and Density, *Trans. Faraday Soc.,* vol. 19, pp. 38-41, 1923.
57. Meissner, H. P., and Michaels, A. S., Surface Tensions of Pure Liquids and Liquid Mixtures, *Ind. Eng. Chem.,* vol. 41, pp. 2782-2787, 1949.
58. Reid, R. C., Prausnitz, J. M., and Sherwood, T. K., *The Properties of Gases and Liquids,* 3d ed., p. 623, McGraw-Hill, New York, 1977.
59. Miller, M. B., Experimental Determination of the Interfacial Tension of Binary Systems, M.Sc. thesis, Oklahoma State University, Stillwater, Okla., 1972.
60. Vargaftik, N. B., *Tables on the Thermophysical Properties of Liquids and Gases,* 2d ed., pp. 43 and 427, Hemisphere, New York, 1975.
61. Baker, O., and Swerdloff, W., Calculations of Surface Tension–3: Calculation of Surface Tension Parachor Values, *Oil Gas J.,* vol. 53, p. 87, 1955.
62. Wilke, C. R., and Chang, P., Correlation of Diffusion Coefficients in Dilute Solutions, *AIChE J.,* vol. 1, pp. 264-270, 1955.
63. Kuong, J. F., Nomograph gives Diffusion Rate in Dilute Solutions, *Chem. Eng.,* vol. 68, no. 12, pp. 258-260, 1961.
64. Tyn, M. T., Estimating Diffusion Coefficients at Temperature, *Chem. Eng.,* p. 106, June 9, 1975.
65. Bretsznajder, S., *Prediction of Transport and Other Phsyical Properties of Fluids,* p. 359, Pergamon Press, New York, 1971.
66. Wasan, D. T., Subramaniam, T. K., and Randhave, S. S., Estimation of Eddy Diffusion Coefficients, *Chem. Eng.,* p. 165, Feb. 14, 1966.
67. Fuller, E. N., Schettler, P. D., and Giddings, J. C., A New Method for Prediction of Binary Gas-Phase Diffusion Coefficients, *Ind. Eng. Chem.,* vol. 58, no. 5, p. 19, May 1966.

CHAPTER 5
Properties of Solids

In contrast to Chapter 1 on the properties of fluids, this section deals mainly with the materials from which heat exchangers could be fabricated and only selectively with solid materials that could also be handled in heat exchangers, e.g., pebble heaters. This narrows considerably the range of materials to be treated.

For the more common materials such as metals and alloys, graphite and silicon carbide, refractories and glasses as well as organic polymers, most of the required values are tabulated in the suppliers' catalogs and in the literature. Where such knowledge is lacking (e.g., where the tables do not cover the materials or the temperature range of interest), the problems of predicting physical properties begin.

The solid state is a state of much higher regularity than that of gases and liquids. Therefore even the prediction of simple properties such as density or thermal conductivity requires the knowledge of other properties–the availability of which is far less probable (e.g., lattice constants of the crystal) than that of the property in question itself.

Methods of predicting physical properties on such a basis are very interesting from a scientific point of view, but they are obviously of no use for the engineer who is in the course of designing heat exchanger equipment. The objective of this section is therefore to provide general rules of material behavior rather than predicting exact values.

5.1. Density of Solids

For a material with unknown density (i.e., mass per unit volume, ρ, in kg/m^3) rarely more than the simplest facts will be available. Thus these have to be taken as the basis for deriving an estimated density.

A. Metal alloys

The densities of the pure metals are related to their places in the Periodic table but are not strictly parallel to their atomic masses. The density of alloys can be estimated as the composed density according to the mass fractions of the single elements in the alloy:

$$\rho_A = \frac{1}{x_1/\rho_1 + x_2/\rho_2 + x_3/\rho_3} \qquad (1)$$

where ρ_i is the density of the single component and ρ_A the density of the alloy.

However, it has to be noted that the same alloy often can exist in different crystal structures such as face-centered and body-centered cubic patterns. For the same or nearly same composition, this results not only in slightly different densities but also in largely different coefficients of thermal expansion.

The coefficient of linear thermal expansion (the fraction of elongation caused by the unit of temperature change, α, in m/m K) ranges from 0.02×10^{-4} (Fe + 36% Ni) to 0.26×10^{-4} m/m K (Mg) with most values around 0.1×10^{-4} m/m K. α is generally larger close to the melting point. In alloys, irregularities of the coefficient of thermal expansion are to be expected.

B. Graphite and carbon

The densities of graphite and carbon range from 1 500 to 2 000 kg/m^3 depending upon the degree of regularity and of porosity resulting from the methods of manufacturing. Graphite and carbon can be made impervious by filling the pores with resins. Such an impregnation will increase the density up to about 2 000 kg/m^3. For heat exchangers serving the temperature range up to 150°C, these are the commonly used materials.

Within certain limitations, the carbonizing or graphitizing process can be repeated after impregnating in order to produce an essentially impervious or even glassy material with a density about equal to that of the normal impregnated material. The coefficient of linear thermal expansion is about

$$\alpha = 10^{-4} a + 0.4 \times 10^{-8} T \quad \text{m/m K} \qquad (2)$$

with T in °C and where $a = 0.01$ for carbon, $a = 0.02$ for impervious carbon, and $a = 0.04$ for pure and impervious graphite.

C. Refractories (oxides and silicon carbide)

Refractories consist mostly of oxides of Al, Si, Mg, Fe, Ti, Cr, Zr, but also of SiC and ZrC. The densities of the pore-free materials vary between about 2 600 kg/m^3 (SiO_2) and 3 700 kg/m^3 (Al_2O_3) to over 4 200 kg/m^3 (40% Cr_2O_3) and 4 700 kg/m^3 (90% ZrO_2). The porosity is normally about 25% and in insulating materials up to 80%.

The mean coefficient of linear thermal expansion is normally in the range 0.05×10^{-4}–0.08×10^{-4} m/m K, but is sometimes about 20% greater in an intermediate temperature range between 300 and 700°C. Refractories containing MgO have higher values (0.15×10^{-4}–0.20×10^{-4} m/m K) while those containing SiC have lower values (0.04×10^{-4} m/m K). For illustration, see Fig. 1.

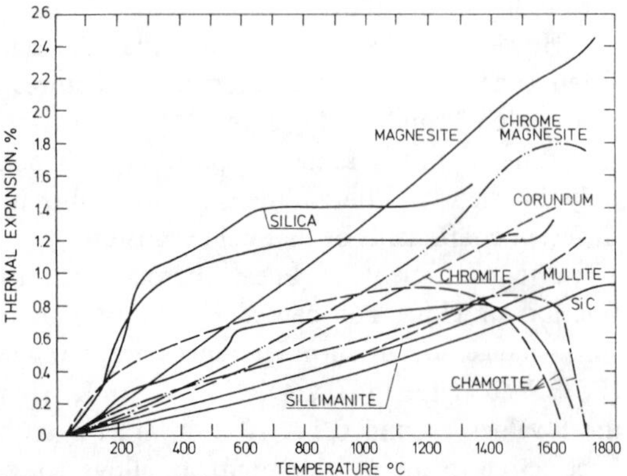

Figure 1 Linear thermal expansion of different refractory materials as a function of temperature [1].

It must be mentioned that leaps (or steps) in the thermal expansion of some materials are observed because of the reversible or permanent changes in the lattice structure of the crystals that occur with high-temperature and low-temperature modifications. SiO_2 has many modifications with densities between 2 200 and 2 650 kg/m^3 (see Fig. 2). Clay brick shows a shrinkage at higher temperatures due to sintering and glassifying of the material. This can sometimes be compensated by a sufficient content of SiO_2 which at the same time will expand.

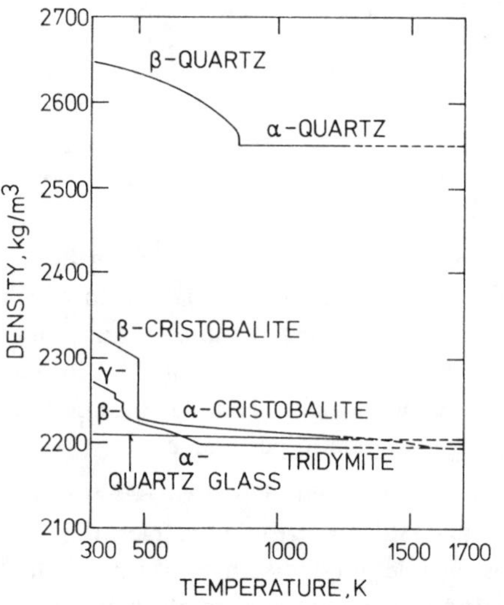

Figure 2 Change of density for different modifications of SiO_2 during heating, cooling, and transformation processes. Modified from [1].

D. Organics

Organic solids that could be used in the construction of heat exchangers, such as PTFE, have densities around 1 100 kg/m^3. The coefficient of linear thermal expansion is usually around 1×10^{-4} m/m K.

5.2. Specific Heat of Solids

Every solid–metallic, nonmetallic, crystalline, or even amorphous–can be considered as a more or less regular three-dimensional lattice of atoms. Each of them is fixed in its position by elastic forces that are functions of the atomic distances and the properties of the neighboring atoms. By far the most important contribution to the internal energy of a solid is the energy stored in the thermal vibration of its atoms in the lattice. These vibrations of an average atom are three dimensional and may be resolved into three independent vibrations parallel to the three coordinate axes.

Similar to the kinetic theory of gases, the kinetic theory of solids assigns to each component the energy

$$\frac{\tilde{R}T}{\tilde{L}} = kT \quad \text{joules per average atom}$$

where $\tilde{R}/\tilde{L} = k$ is the Boltzmann atomic constant with the value 1.380×10^{-23} J/K per single atom. (In gases the energy assigned to each degree of freedom is $kT/2$ due to the absence of elastic forces.)

In consequence, a solid with the mass M, containing $M\tilde{L}/\tilde{M}_{at}$ atoms, should have the internal energy content

$$uM = 3\frac{M\tilde{L}}{\tilde{M}_{at}}kT = 3\frac{M\tilde{R}T}{\tilde{M}_{at}} \quad \text{J} \tag{1}$$

The internal energy content per unit of atomic mass, $\tilde{u}_{at}$, is then

$$\tilde{u}_{at} = u\tilde{M}_{at} = 3\tilde{R}T \tag{2}$$

At constant volume, all added heat will increase the internal energy, and thus the specific heat at constant volume per unit of atomic mass, $\tilde{c}_{v\,at}$, should be

$$\tilde{c}_{v\,at} = \left(\frac{\partial \tilde{u}_{at}}{\partial T}\right)_v = 3\tilde{R} = 24.95 \text{ J/atom K} \tag{3}$$

For calculating c_p from c_v the work of thermal expansion of the solid has to be taken into account:

$$c_p = c_v + \frac{\beta^2 vT}{\kappa} \quad \text{J/kg K} \tag{4}$$

This expression is in analogy with the similar expression for ideal gases:

$$c_p = c_v + P\left(\frac{\partial v}{\partial T}\right)_p \tag{5}$$

where $\partial v/\partial T = \beta v$

P represents $(\beta/\kappa)T$

β is the volumetric thermal expansivity = $1/v(\partial v/\partial T)_p$ in K^{-1}

κ is the compressibility = $1/v(\partial v/\partial p)_s$ in m^2/N

v is the specific volume in m^3/kg

Since it is very unlikely that β and κ are known for a solid with unknown specific heat, this formula can serve to provide a better understanding only. Depending mostly upon the compressibility, the specific heat capacity at constant pressure for most solids is in the range of about 0–15% higher than that at constant volume. This is in good agreement with the Dulong-Petit law, first published in 1819. Dulong and Petit had discovered that the atomic heat of many solids were about the same, with an average value of about $\tilde{c}_{p\,at}$ = 26.7 J/atom K.

The Dulong-Petit law, simple as it is, applies only in the range above room temperature with exceptions even there. At lower temperature the specific heat of solids decreases rapidly. This is illustrated in Fig. 1 which shows atomic heats at constant volume for some materials in a temperature range from 0 to 400 K.

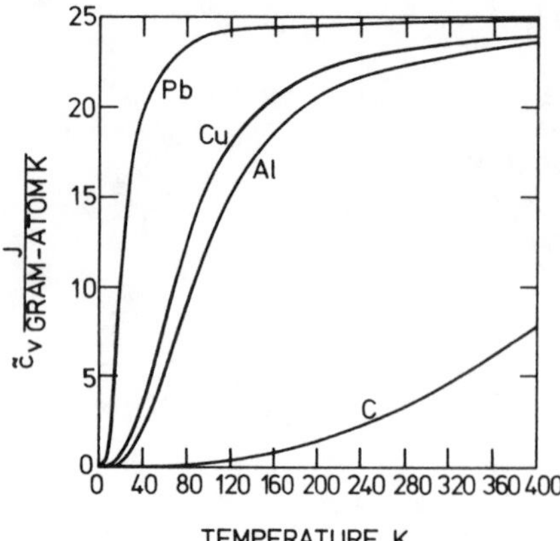

Figure 1 Atomic heats at constant volume as a function of temperature. From Worthing [2].

Improvements in the theory were made by Einstein who introduced Planck's constant $h = 6.625 \times 10^{-34}$ J s and a characteristic frequency ν in s^{-1}, replacing kT in Eq. (1) by the expression $h\nu/(e^{h\nu/kT} - 1)$, which becomes equal to kT in the higher-temperature range but approaches zero together with temperature.

At low temperatures, Einstein's method gave too low specific heats, but its relative success proved that the quantum theory also applies to lattice waves. In consequence, there is a quantum of energy for a lattice wave, now named a *phonon* in analogy to the photon which is the quantum of energy of an electromagnetic wave.

Debye improved Einstein's theory by a more elaborate method. He also employed the quantum value of vibrational energy–the phonon–but as degrees of freedom he defined the possible number of standing-wave trains per unit volume and frequency. The theoretical deduction is beyond the scope of this book (see Kittel [3]), however, Debye's theory results in a diagram $\tilde{c}_{v\,at}$ versus a dimensionless temperature, T/Θ, with Θ representing the Debye temperature, which is in very good agreement with experimental values of $\tilde{c}_{v\,at}$ for various materials. (See Fig. 2.)

There are methods of calculating Θ from other material properties, but the basic values for these will rarely be available. For the present purpose it is therefore recommended that Θ be obtained by the rule

$$\Theta = 3T_{0.7}$$

where $T_{0.7}$ is the temperature where $\tilde{c}_{v\,at}$ is 70% of its final value or $\tilde{c}_{v\,at} = 17.5$ J/atom K.

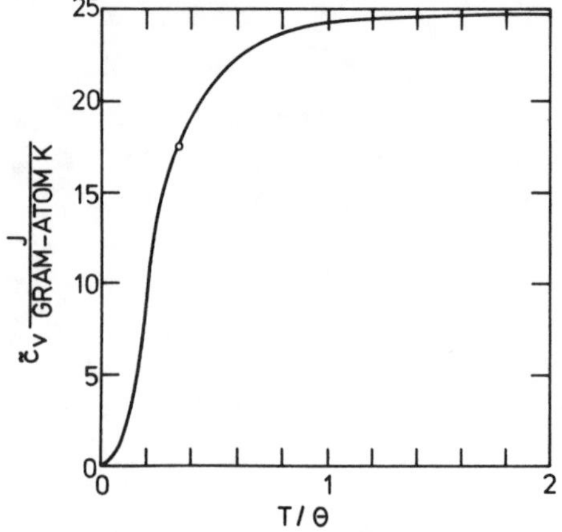

Figure 2 Atomic heat at constant volume, according to Debye's theory, as a function of T/Θ where Θ is the Debye temperature. Simplified from Schroedinger [4].

For low temperatures, approximately up to $T = 0.1\ \Theta$ (sometimes only up to $T = 0.02\ \Theta$), c_v rises as T^3 ("T^3 law"). The Debye temperatures for some materials are given in Table 1. The rule seems to be valid also for alloys and chemical compounds. Deviations of normally ±10% for c_v and of ±15% for c_p are to be expected, but there are some materials for which the deviations are much greater, e.g., Li.

Debye's theory considers only the most important contribution to the specific heat of solids, that of the lattice vibrations. However, there are many other phenomena that may cause a solid to absorb additional heat, e.g., the changes in modification of a crystal or other changes in the atomic structure (order-disorder phenomena).

Table 1 Debye temperatures Θ for some materials[a]

Metal	Θ	Metal	Θ
Pb	88	Al	398
Hg	97	Fe	453
Cd	168	Other	
Na	172	KBr	177
Ag	215	KCl	230
Ca	226	NaCl	281
Zn	235	C	1 860
Cu	315		

[a]From Schroedinger [4].

5.3. Thermal Conductivity of Solids

Thermal conductivity is defined as the ratio of the flux of thermal energy in a given direction to the temperature gradient in the same direction:

$$\lambda = \frac{\dot{q}}{\partial T/\partial l} \quad \text{W/m K} \tag{1}$$

where $\dot{q}$ is the flux of thermal energy in W/m^2. This is, of course, also true for a fluid as long as no convective heat transport has to be considered. In a gas, the kinetic theory of gases provides an easy explanation for the thermal conductivity λ, relating it to the heat capacity per unit volume of the gas c_p, the average particle velocity v, and the mean free path l:

$$\lambda = \tfrac{1}{3} c_p v l \tag{2}$$

In solids, as in gases and liquids, the thermal energy is mainly stored in the motional energy of the atoms, but in solids they are fixed to their respective positions within the lattice by elastic forces. Explaining thermal conductivity is therefore explaining the different kinds of interaction by which the lattice vibrations are transmitted throughout a crystalline or amorphous solid when the equilibrium is distorted by a temperature gradient.

At moderate temperatures, i.e., up to about 1 000 K, there are mainly two possible vehicles for the transport of heat through a solid body:

1. The elastic forces in the lattice which cause the atoms to arrange their vibrations in trains of standing waves. The quantum of energy of such a standing wave is called a phonon. Since an otherwise undisturbed lattice wave travels through a regular lattice without change of momentum, the phonons can be treated much like particles.

2. Free electrons which as "electron gas" circulate between the atoms, picking up energy in collisions with vibrating atoms and delivering energy to other vibrating atoms. Free electrons are present in electric conductors (metals, graphite, silicon carbide) and, to some degree, in semiconductors.

In the temperature range above about 1 000 K, enough photons are stimulated to take part in the transport of heat, especially in materials transparent to infrared radiation. The photons are also to be considered as particles, and the contribution of each of the three types of carriers to the thermal conductivity would then be

$$\lambda_i = \tfrac{1}{3} c_i v_i l_i \tag{3}$$

Since there are no elementary means of calculating the mean free path l for the different types of carriers, the above relationship can be used in a very general way only.

For further discussion, the solid materials will be divided into electrical conductors, electrical nonconductors or insulators, and semiconductors.

A. Thermal conductivity of electrical insulators

In electrical insulators (e.g., refractory materials such as SiO_2, Al_2O_3, etc.) up to about 1 000 K, phonons are the only carriers of thermal energy. At low temperatures (say below 30 K) they are nearly independent of each other. This means that their average free path is only limited by irregularities of the crystalline structure. In a high-purity single crystal the mean free path in this temperature range can reach a few millimeters, but even the presence of different isotopes can scatter the phonons or limit their free path.

As long as the mean free path of the phonons is a function of the configuration of the crystal, l can be considered a constant. The velocity of sound, v, is essentially independent of temperature, and thus the conductivity

$$\lambda = \tfrac{1}{3} cvl$$

will vary with the specific heat c, which in this range varies as T^3 according to Debye's theory. More-regular crystals will show higher values of λ than less-regular ones, and in glasses λ will be much lower than in any crystal.

With increasing temperature the mean free path of the phonons will be limited more and more by collisions with other phonons. These interactions between phonons and phonons, according to Peierls [5] can be classified as "normal" processes (which do not give rise to thermal resistance) and Umklapp (reversing) processes in which momentum is lost. The probability of the occurrence of Umklapp processes rises as T, thus decreasing the mean free path and the thermal conductivity as T^{-1}.

With increasing importance of phonon-phonon interactions at higher temperatures, the effect of crystal regularity on the thermal conductivity decreases. At about 1 000 K the thermal conductivity of a very regular crystal and of a glass are in the same order of magnitude. On the other hand, the very disordered crystal structure of a glass will result in comparatively very short mean free paths even at low temperatures, and in a wide range of temperature λ will therefore vary as c, i.e., increase approximately according to Debye's theory (see Sec. 5.2).

At temperatures of about 1 000 K the thermal conductivity of ceramic materials in their nonporous body is in the order of magnitude of 2 W/m K, glasses generally somewhat lower. Apparently the lighter elements in the compounds are more favorable to a high thermal conductivity than the heavier ones.

At temperatures above 1 700 K the transport of heat due to internal radiation becomes important, especially in refractories containing a large amount of glassy matter. The thermal conductivities of some nonporous refractory materials are shown in Fig. 1.

Very often the thermal conductivity of refractory material is reduced by a certain degree of porosity—the effect of which is mainly determined by the distribution and shape of the pores in the solid body and by the thermal conductivity of the gas in the pores. As a rough rule, a 1% increase in porosity will increase the thermal resistance (λ^{-1}) by about 5%. This rule is valid up to a porosity of about 60%, and for a temperature up to

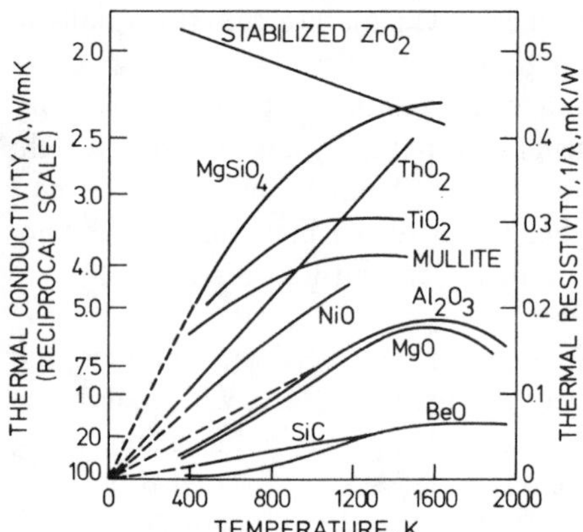

Figure 1 Thermal conductivities of some (nonporous) oxidic materials as functions of temperature. From Kingery [6].

about 1 200 K. At higher temperatures the radiative heat transfer through the pores becomes important, contributing an apparent thermal conductivity that varies as T^3. At temperatures above 1 700 K an increase in porosity will therefore cause an increase of thermal conductivity.

Hydrogen and helium have much greater thermal conductivities than all other gases. Their presence in the pores will therefore markedly increase the thermal conductivity of a porous refractory material. The effect of gas convection in the pores increases with the size of the single pore and with the pressure of the gas. It should also be noted that any condensation and vaporization within the pores of an otherwise insulating material will considerably contribute to the transmission of heat.

B. Thermal conductivity of electrical conductors

In metal lattices, the valence electrons can circulate more or less freely among the atoms, thus being able to transmit electric charge, i.e., to produce electrical conductivity. Their mean free path is limited at very low temperatures mainly by impurities and irregularities of the lattice. At higher temperatures the interactions between electrons and vibrations of the atoms—phonons—also become more and more important. In very pure crystals, this effect begins at very low temperatures. Metal alloys, due to their less-regular lattice, show a much lower electrical conductivity at low temperatures than pure metals do. So the decrease due to electron-phonon interactions becomes obvious in a higher temperature range only. The change of electrical conductivity between room temperature and the temperature of liquid helium (4 K) can vary between about a factor of 10 000 for very pure metals and 2 in specific alloys.

By their ability to interact with lattice vibrations, the free electrons will pick up energy from vibrating atoms and transmit it to other atoms, thus increasing their vibrational energy. On the average, this results in a transmission of heat in the direction of the negative temperature gradient. Due to the relatively high velocity of the electrons and their relatively large mean free path, their contribution to the thermal conductivity is normally much higher (5-50 times) than that of the phonons, in spite of their very small contribution to the specific heat.

The exchange of energy between phonons and electrons is subject to quantum effects which allow the number of electrons participating in the exchange of energy with phonons increase as T. Even at room temperature only about 1% of the free electrons of a metal are engaged in the transport of heat. Therefore, at very low temperatures (up to 15 or 30 K) where the electron free path is only limited by lattice imperfections, the *thermal conductivity* of metals varies as temperature, whereas in the same temperature range the *electrical conductivity* varies as T^{-5}. The state of electrical superconductivity is not accompanied, for a similar reason, by an infinite thermal conductivity because the fraction of electrons participating in the superconduction no longer contributes to the thermal conductivity.

At higher temperatures the number of electrons contributing to the transport of thermal energy continues to rise as T, but at the same time their mean free path decreases by phonon-electron interactions. The first phenomenon will dominate over the whole temperature range in metals that start out with a high density of lattice imperfections, thus resulting in a permanent rise of thermal conductivity with increasing temperature. On the contrary, in pure metals the thermal conductivity reaches a maximum at the temperature of commencement of phonon-electron interactions followed by a decrease of thermal conductivity over the rest of the temperature range (Table 1). For temperatures over about 150 K the thermal conductivity λ and the electrical conductivity σ are related to each other by the Wiedemann-Franz-Lorenz law

$$\frac{\lambda}{T} = L\sigma \tag{4}$$

where L is the Lorenz number = 2.45×10^{-8} W Ω/K² from theoretical considerations. Practical values of L range between 2.3×10^{-8} and 3.2×10^{-8} W Ω/K².

The fact that λ/T and not λ must be compared with σ is explained by the temperature dependency of the fraction of electrons participating in the energy exchange with phonons. Some values of L are tabulated in Table 1.

Table 1 Experimental values[a] of the Lorenz number $L = \lambda/(\sigma T)$

Metal	Temperature (K)	L × 10⁸
Al	21.2	1.77
	373.2	2.23
Mo	90.2	1.76
	273.2	2.61
W	90.2	2.00
	273.2	3.06
Fe	90.2	1.60
	273.2	2.47
Pt	20.7	1.09
	273.2	2.51
Cu	21.2	0.77
	83.2	1.57
	273.2	
Ag	90.2	1.62
	273.2	2.31
Au	21.2	1.85
	83.2	2.05
	273.2	2.35
Pb	21.8	1.48
	291.2	2.45

[a]From D'Ans and Lax [7].

C. Thermal conductivity of electrical semiconductors

Electrical semiconductors take an intermediate position between metals and insulators. Thus their thermal conductivity can be estimated as the sum of their lattice conductivity λ_l and their electron conductivity λ_e. The latter can be calculated by the Wiedemann-Franz-Lorenz law:

$$\lambda = \lambda_l + \lambda_e = \lambda_l + L\sigma T \tag{5}$$

with L = 2.45×10^{-8} W Ω/K².

It should be added that there is an additional effect in semiconductors which can also contribute to the thermal conductivity: bipolar diffusion. At sufficiently high temperatures (above room temperature, but markedly only over 800 K) an increasing number of electron-hole pairs can be thermally generated. These pairs will recombine on the average at places of lower temperature, thus delivering their energy of generation.

D. Data Tables

In the tables that follow an attempt is made to give, as comprehensively as possible, a tabulation of the temperature dependence of thermal conductivity of elements, alloys, and other solids. In addition, a further tabulation is given of typical room temperature values of this property for other solids. In using the tables the following remarks should be kept in mind:

1. In structured materials such as crystals, wood, mica, etc., the property may be an acute function of direction of heat transfer.
2. Impurities, thermal history of the specimen, environmental effects (such as applied magnetic field) can sometimes markedly affect the property.
3. For low density insulation, thermal radiation should also be considered.
4. Exposure of the material to the actual environment can markedly affect the property due to such factors as contact resistance, degradation, oxidation, moisture seepage, etc.
5. At low temperatures, the thermal conductivity can vary rapidly (see Fig. 1) and also be a sensitive function of specimen purity. The tabulated values for the pure elements are for materials of highest purity that are generally available. The actual property for real substances may be much less.
6. Where there may be more than one state of the material, the tabulation given is for the polycrystalline state. Limitations of space have prevented a more detailed tabulation. Readers interested in exploring the subject further may find the publications given in [8–13] helpful.

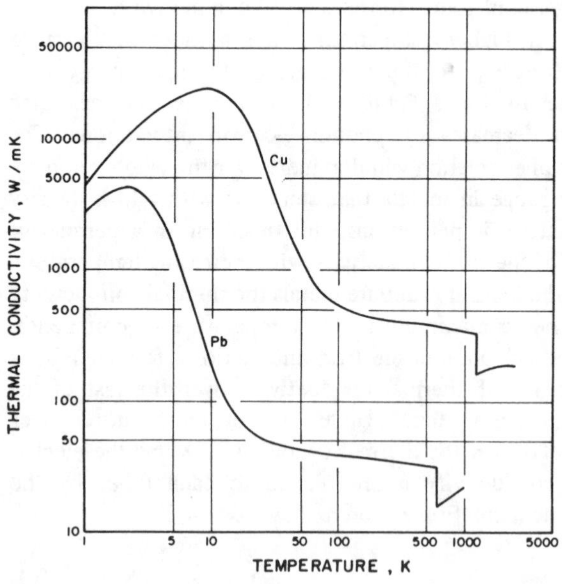

Figure 1 Relationship between thermal conductivity and temperature for copper and lead.

Table 1 Thermal conductivity of solids as a function of temperature, W/m K

Material	Thermal conductivity at different temperatures, K																	
	10	20	40	60	80	100	150	200	250	300	400	500	600	800	1 000	1 200	1 400	1 500
Alumina	3 000	12 500	13 500	3 100	1 100	520	155	80	60	42	27	20	16	10	7.6	6.3	5.7	5.4
Aluminum	38 000	13 000	2 400	850	430	300	250	237	235	237	240	236	231	218	–	–	–	–
Antimony	480	230	110	75	58	46	36	30	26.7	24.3	21.3	19.5	18.3	16.8	–	–	–	–
Argon	5.6	1.5	0.7	0.3	–	–	–	–	–	–	–	–	–	–	–	–	–	–
Arsenic								69	58	50	41	35	–	–	–	–	–	–
Barium							20.5	19.4	18.6	18.2					–	–	–	–
$BaTiO_3$	10	18	21	19	16	13	8	5.9	4.8	4.1	3.5	3.0						
Beryllium	1 800	3 480	4 620	2 980	1 620	890	450	300	235	200	160	139	126	106	91	79	69	64
Beryllia											198	153	114	60	33	22	17	17
Bismuth	225	100	46	30	24	19	12	9.7	8.6	7.9	7.0	6.6	–	–	–	–	–	–
Boron	175	250	430	360	260	190	94	55										
Brass	5	12	24	33	40	46	62	74	92	114	134	143	146	150				
Cadmium	1 200	270	140	115	106	103	101	99	98	97	95	92	–	–	–	–	–	–
Calcium								220	210	201	189	182	178	153	116	–	–	–
CaF_2	2 800	2 500	800	125	60	40	23	18	15	12								
CaO											129	99	85	79	74	72	70	69
Carbon	0.07	0.16	0.31	0.44	0.56	0.67	0.94	1.18	1.40	1.60	1.87	2.06	2.19	2.37	2.5	2.8	3.2	3.5
Cerium	1.1	1.9	3.2	4.3	5.2	6.0	7.7	9.0	10.3	11.4	13.3	15.0	16.5	19.3	21.8	–	–	–
Cesium	69	55	47	43	41	40	38	37	36	36	–	–	–	–	–	–	–	–
Chromium	385	595	425	250	185	160	130	110	100	94	91	86	81	71	65	62	59	57
Cobalt	265	440	375	250	195	165	140	120	110	100	85	75	67	58	52	49	42	42
Constantan						19	20	21	22	23	27							
Copper	24 500	10 800	2 170	830	560	480	429	413	406	401	393	386	379	366	352	339	–	–
Duralumin	16	30	56	74	88	101	123	138	151	174	187	188						
Dysprosium	10	14	14	13	12	10	9	10	10	11	11	12	12	14	15	17	18	19
Erbium	7.1	7.8	9.3	9.8	11	12	14	15	15	14	14	14	14	15	16	17	19	19
Eureka								20	21	22	24	25						
Gadolinium	31	32	23	18	15	15	13	12	11	10								
German silver	4.8	7.5	11.5	13.8	15.3	17	19	21	23	24	25	27						
Germanium	1 770	1 490	800	490	325	230	130	97	75	60	43	34	27	20	17	17	–	–
Gold	3 250	1 580	515	374	332	327	325	323	321	317	311	304	298	284	270	255	–	–
Hafnium	10	18	24	26	26	26	25	24	24	23	23	22	21	21	21	21	21	21
Holmium	12	14	16	16	15	14	14	15	16	16	17	18						
Indium	590	190	170	100	99	98	93	90	86	82	75	–	–	–	–	–	–	–
Iridium	1 270	1 900	750	330	210	172	159	153	150	147	144	141	138	132	126	120	114	111
Iron	1 480	1 540	625	285	175	134	104	94	87	80	70	61	55	43	32	28	31	32
Lanthanum	18	17	10	9	9	10	11	12	13	14	15	16	18	21	23	–	–	–
Lead	180	60	45	43	41	39.7	37.9	36.7	36.0	35.3	34.0	32.8	31.4	–	–	–	–	–
Lithium	610	720	345	175	120	105	95	90	87	85	80	–	–	–	–	–	–	–
LiF	11 000	13 500	1 300	280	146	95	42	27										
Limestone						3.2	2.7	2.3	2.0	1.8	1.6	1.5						
Lutetium	36	41	32	30	29	28	26	25	24	23	22	22						
Magnesium	5 600	2 700	720	325	200	169	161	159	157	156	153	151	149	146				
Manganese	1.6	2.4	3.6	4.5	5.3	5.8	6.6	7.2	7.5	7.8	7.9	8.0						
Manganin								14	18	21	23	27						
Mercury	46	40	36	34	33	32	30	29	–	–	–	–	–	–	–	–	–	–
Magnesium oxide											32.4	25.5	19.9	12.8	9.2	7.2	6.2	5.9
Molybdenum	150	285	355	260	210	180	149	143	140	138	134	130	126	118	112	105	100	98

(Continued)

Monel	3.5	7.1	11.0	13.3	14.9	16	18	20	22	23	25	27						
Naphthalene					1.20	0.96	0.62	0.50	0.42	0.35	0.29	0.27						
Nickel	1 810	1 650	580	310	210	165	120	105	98	91	80	72	66	68	72	76	80	83
Nimonic 75										13	15	17	18	21	25			
Nimonic 80											12	14	15	18	23			
Nimonic 90										12	13	15	16	19	23			
Nimonic 95											13	15	16	18	22			
Nimonic 100											12	13	14	16	20			
Niobium	295	250	95	66	58	55	53	53	53	54	55	57	58	61	64	68	71	72
Nylon	0.033	0.099	0.170	0.206	0.230	0.242												
Osmium	1 020	1 600	640	220	140	115	96	91	89	88	87	87						
Palladium	1 150	600	175	98	81	77	73	72	72	72	74	76	80	87	94	102	107	110
Phosphorus	7	27	44	40	35	31	23	18	14	12	–	–	–	–	–	–	–	–
Platinum	1 230	500	140	95	82	78	74.0	72.6	71.8	71.6	71.8	72.3	73.2	75.6	79	83	87	90
Plutonium					3.1	3.3	4.0	4.8	5.7	6.7	9.6							
Porcelain												1.45	1.7	2.0	2.2	2.4		
Potassium	460	165	115	110	108	107	105	104	104	102								
KCl	440	400	105	52	32	25	13	11	8	7	6	5						
Praseodymium					7	8	9	10	12	13	14	15	16	18	22	–	–	–
Pyrex				0.34	0.46	0.57	0.76	0.89	0.95									
Quartz	8 000	3 600	850	410	240	200												
Rhenium	3 550	1 150	155	77	63	59	54	51	49	48	46	45	44	44	45	46	47	48
Rhodium	2 780	3 650	1 020	380	240	190	158	154	152	150	146	141	136	127	121	116	112	110
Rubber						0.14	0.15	0.15	0.16	0.16								
Rubidium	150	69	64	62	61	60	59	59	59	58								
Ruby	650	1 335	1 700	1 000	500	290												
Ruthenium	1 500	2 260	950	310	185	155	128	118	117	117	114	111	108	102	98	94	91	90
Salt	770	330	82	44	30	24	15	11	8.9	7.1	4.4	3.4						
Sapphire	2 600	14 600	16 500	4 400	1 175	480	150	82	58	46	32	24						
Scandium	7	12	14	14	14	14	15	15	16	16	16	17						
Silicon	2 330	4 980	3 530	2 110	1 340	884	410	260	190	150	99	76	62	42	31	26	24	23
Silumin												177	181					
Silver	16 800	5 100	1 050	550	470	450	432	420	429	429	425	419	412	396	379	361	–	–
Silver solder	1 650	720	179	85	54	39	23	16	13	10	7.6	6.0						
Sodium	2 200	610	190	145	135	136	140	142	143	141	–	–	–	–	–	–	–	–
NaF	4 500	3 100	560	240	120	85	55	45	38	36								
Sulfur	8.2	2.4	1.1	0.8	0.72	0.56	0.43	0.36	0.31	0.27	–	–	–	–	–	–	–	–
Tantalum	107	142	87	65	60	59	58	58	57	58	58	59	59	59	60	61	62	62
Technetium										51	50	50	50	52	55			
Teflon	0.094	0.142	0.194	0.220	0.23	0.23	0.24	0.25	0.25	0.26								
Terbium	19	23	19	17	15	14	12	10	10	11								
Thallium	190	80	65	61	58	56	52	49	48	46	44	42	–	–	–	–	–	–
Thoria									16	13	10	8	6	5	4	3	3	2
Thorium	470	170	85	69	63	60	56	55	54	54	55	55	56	57	58	59	60	61
Thulium	23	18	11	11	13	14	15	16	17	17								
Tin	1 900	320	135	104	92	85	78	73	70	67	62	60						
Titanium	14	28	39	36	33	31	27	25	23	21	20	20	19	19	21	22	24	25
Titanium oxide	1 000	850	130	60	31	20	11	9	8.5	8	6	5	4	3	3	3	3	3
Tungsten	9 700	4 050	690	310	230	208	192	185	180	174	159	146	137	125	118	112	108	106
Uranium	10	16	18	20	21	22	24	25	26	28	30	33	34	39	44	49	56	–
Vanadium	14	26	39	41	39	36	32	31	31	31	31	32	33	36	38	41	43	45
Yttrium	7	13	15	16	16	16	16	17	17	17	18	19	21	23	25			
Zinc	4 750	1 000	280	160	130	117	117	118	118	116	111	107	103	–	–	–	–	–
Zirconium	99	108	59	44	37	33	28	25	24	23	22	21	21	21	23	26	28	29
Zirconium oxide												2.1	2.1	2.2	2.2	2.3	2.3	2.4

Table 2 Thermal conductivity of materials at room temperature, W/m K

Material		Value	Material		Value
Stainless steel	201	16.3	Hastelloy	B	10.4
	202	16.3		C	8.8
	301	16.2		D	20.9
	302	16.2		G	10.8
	304	16.3		X	9.0
	304L	16.3	Incoloy	800	12.2
	309	13.8		901	13.4
	310	13.8	Nimonic	80A	12.2
	316	16.3	Inconel	600	15.0
	318L	16.3		610	15.0
	321	15.8		625	10.8
	330	13.0		700	12.4
	347	15.8		705	15.0
	410	24.9		722	14.7
	414	24.9		X750	14.7
	420	24.9	Monel	400	25.1
	430	25.9		401	19.2
	431	20.2		R405	25.9
	446	20.9		404	25.0
Cast stainless	CA15	24.5		411	26.8
	CA40	24.5		505	19.6
	CB30	22.1		506	20.9
	CC50	21.8	Aluminum alloy	1100	340
	CF8	15.9		3003	192
	CF8M	16.2		5052	137
	CF20	15.9		5086	135
	CK20	14.2		6063	186
	CN7M	20.9		7075	120
	HU	15.4			
	HW	13.3			
	HA	29.4			
	HC	30.9			
	HD	30.9			

Further data on commercial fabricating materials are given in Section 5.5.

5.4. Elastic Properties

A. Introduction

Solids deform when subjected to stresses. The strain is defined as the degree of distortion per unit length, e.g., the change in length per unit length of a wire when subjected to a tensile stress. The stress is defined as the applied force per unit area. A material is said to show perfectly *elastic* behavior when the stress–strain relationship is perfectly reversible. Furthermore if the strain is proportional to the applied stress then the behavior is known as linear elastic. Most engineering materials show approximately linear elastic behavior up to the onset of plastic deformation. Such behavior was originally described by Hooke and is therefore sometimes known as Hookean.

An important part of mechanical engineering design involves the computation of the distribution of strains in a structure when subject to the stresses imposed in service. Stresses may be imposed by fluid pressure or by fluid motion and also by nonuniform thermal expansion during changes in temperature. Elastic properties are traditionally regarded as being non-structure sensitive but there is one important aspect in which this is not true. The individual grains or crystals of metals are elastically anisotropic. Thus the elastic constants are a function of the orientation of the grain with respect to the orientation of the imposed stresses. The process of manufacture of components tends to introduce a certain degree of preferred orientation of the individual grains composing the structure and thus to introduce elastic anisotropy. It is probable that the existence of various degrees of preferred orientation in test specimens has led to the rather wide scatter in data for the elastic properties of metals and alloys. Because this scatter can introduce errors of as much as 20% in some cases in computing strains, the subject is dealt with in depth in this section. Table 3 should be regarded only as an example of the type of information in the literature. There is no reason to suppose, for example, that steels with 5–9% chromium should differ significantly in Young's modulus from those containing slightly smaller or larger amounts of chromium as shown in that table.

Design codes require that stresses are less than the yield stress in a range in which structural materials are assumed to show linear elastic behavior. The behavior of real materials, however, is only approximately elastic so that on loading and unloading below the yield stress a narrow hysteresis loop is generated. For this reason materials have a nonzero damping capacity. The extent of the departure from elastic behavior becomes greater as the stress increases. Increases in the duration of loading and rise in temperature usually cause increased departure from elastic behavior. For many design purposes, perfect linear elastic behavior is assumed. The finer points of stress–strain behavior are included in this section since they could become very important. For example, the damping capacity of a heat exchanger tube might increase by an order of magnitude when the tube is pressurized. Similarly the elastic constants and damping capacity show significant changes when the temperature is increased in service, causing discrepancies between the behavior during testing cold and unpressurized and in service.

The present section therefore not only describes and defines the various ideal linear elastic moduli that are used in elementary engineering design but also indicates the major sources of error that can arise when the elastic behavior of real materials is not understood.

B. Static and dynamic properties

Since the velocity of sound is not infinite in metals, the sudden application of a stress causes initially only a local response. The stress must be maintained for a time that is long compared with the time for the stress wave to travel throughout the specimen and for the vibrations to die out before the external strain becomes steady in a perfectly elastic solid. Where a problem specifically involves the rapid application of stress for short times or the application of rapidly alternating stresses to a perfectly elastic solid, the problem can be treated by considering the specimen to be divided into small elements of volume, each sufficiently small for the stress and the strain throughout each element of volume to be regarded as uniform.

It is commonly supposed that the elastic constants of real materials are physical properties which are not structure sensitive and therefore depend only on a linear average of the interatomic force–interatomic distance relationship (interatomic potentials). Inspection of tabulated data in standard reference books indicates that this assumption is only approximate. For example, there is a consistent difference in the elastic properties of materials in the annealed and in the cold worked conditions. A significant part of the contribution to the total strain suffered when a stress is applied arises from time-dependent, thermally activated relaxation processes involving movements of some atoms by one or more interatomic distances. For a stress applied in the nominally elastic range, most of the atoms would move with respect to their neighbors by a much smaller distance—typically less than 0.1% of the interatomic distance.

The relative contribution to the total strain of the atoms undergoing large movement is greater at high stresses and at elevated temperatures. The structure-sensitive and time-dependent nature of the elastic properties under such conditions becomes even more obvious. Values for Young's modulus of steel at 600°C may differ by a factor ~2 between quasistatic and acoustical conditions, or between creep-resistant steel and soft steel [14].

All the elastic properties of a given material vary with temperature. When a material is heated it expands and the overall interatomic distance increases, with a corresponding reduction in the interatomic forces. Thus, part of the reduction in elastic moduli is not time dependent. In addition, however, the increased availability of thermal energy at elevated temperatures increases the number of relaxation processes that can occur, and so there is a further increased time-dependent strain contribution. Thus the dynamic properties change less rapidly with temperature than the quasistatic properties. For example, the quasistatic value of Young's modulus for steel drops by 50% between room temperature and 600°C, while the drop is only about 25% when measured by an acoustical technique [14].

C. The stress–strain curve

Two important types of load extension curve are illustrated in Figs. 1 and 2. Such curves are generated typically by applying a gradually increasing tensile load to a standard specimen (Fig. 3). For each applied load the extension (elongation) of the specimen is measured between gauge marks on the plain shank of the specimen. The load can be converted to stress by dividing by the current value of the cross section for each load as the load increases. The extension can be converted to a single value of strain only while the strain remains uniform along and across the specimen.

A specimen is said to show elastic behavior when for each stress state there is a unique set of strains. Thus for example, the length of a tensile specimen showing perfect elastic behavior is the same for a given stress whether approached from a higher or lower stress. In real materials the behavior can only approximate true elastic behavior since the damping capacity inevitably has a nonzero value, and therefore there must be some elastic hysteresis. Above the elastic limit the unloading curve can be distinguished experimentally from the loading curve.

In the following definitions *stress* is equivalent to intensity of stress and is the load (force) applied to unit

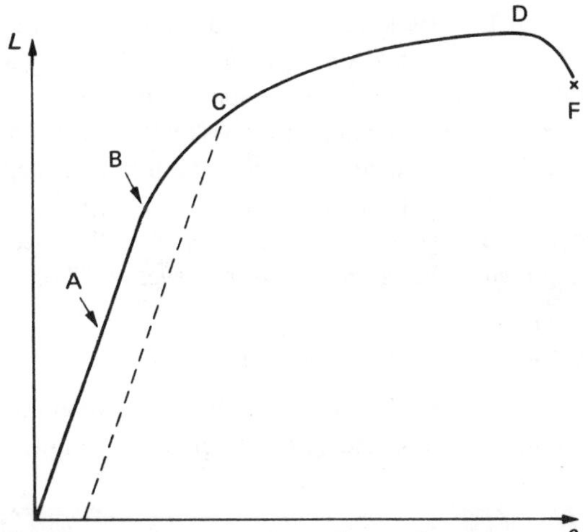

Figure 1 Applied tensile load L versus resulting extension e. A–linear elastic region; B–limit of proportionality; C–0.1% proof stress = load/cross sectional area of specimen. The broken line is drawn parallel to A to intersect the extension axis at 0.1% extension (strain). D–maximum load; F–point of fracture.

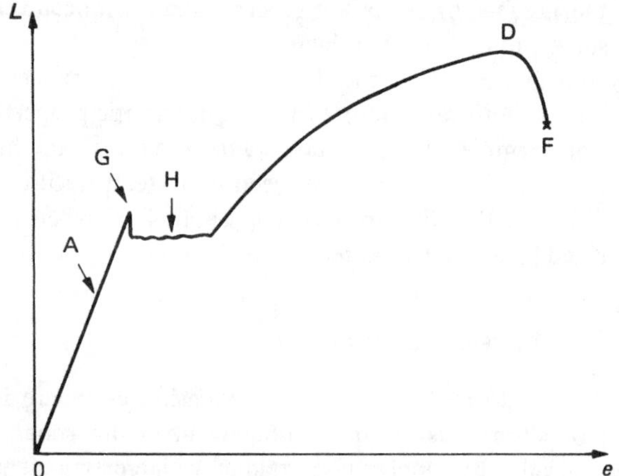

Figure 2 Load-extension curve for a material showing an upper yield point G and lower yield extension H. Other labels are as for the previous figure.

area. The dimensions of stress are the same as those of pressure and hence in SI units are measured in pascals. *Strain* is equivalent to intensity of strain and is the displacement per unit length caused by applying a stress. It is a dimensionless quantity.

On the load extension curves, Figs. 1 and 2, A is the *linear elastic* region. In this region strain is proportional to stress; i.e., the material obeys Hooke's law. Point B, Fig. 1–the lowest stress at which departure from a straight line can be detected–is the *limit of proportionality*. Beyond B the material may still show approximate elastic behavior. In many materials departure from linearity is gradual and therefore it is not easy to detect the point B experimentally. For such materials a point C, called the *proof stress* is defined, where the strain exceeds that for linear elastic behavior by an arbitrary amount such as 0.1% (0.1% proof stress).

Other types of material, particularly the ferritic steels, show a sudden departure from linear elastic behavior when the stress is increased to the point G, Fig. 2, called the *upper yield stress*, where there is a drop in load followed by a region of constant load, the *lower yield stress*, a regime of inhomogeneous plastic strain. The upper yield stress is increased as the duration of the applied load decreases [15]. The maximum load (point D) divided by the original cross-sectional area of the specimens is known as the *ultimate tensile strength* (UTS).

D. Elastic properties of isotropic materials

Young's modulus E, also known as the tensile modulus, the modulus of extensibility, or the elastic modulus, is the tensile stress divided by the strain in the direction of the applied stress, measured in the stress range in which behavior is linear elastic. It is the constant of proportionality in Hooke's law and is also the slope of the linear part of the true stress–strain curve. The dimensions are the same as those for stress (pressure).

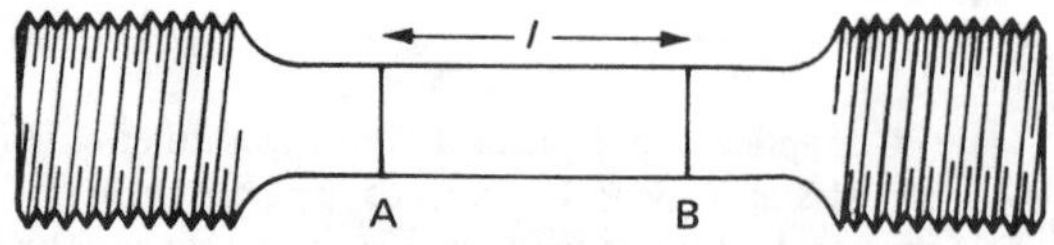

Figure 3 Standard, plain, round tensile test specimen. l is the gauge length; A and B are gauge marks.

The shear modulus G, also known as the torsion modulus or the modulus of rigidity is given by the shear stress divided by the shear strain. A shear stress applied to an isotropic specimen or a cubic metal causes change in shape without change in volume. The shear stress is taken as the magnitude of the applied force divided by the area to which the shearing force is applied. The strain is measured as the tangent of the angle θ in Fig. 4.

The bulk modulus K, or modulus of compressibility, is defined as the applied hydrostatic pressure increase divided by the resulting proportional decrease in volume, $-dV/V$. It differs entirely from the shear modulus in that for isotropic material and for cubic metals it is a measure of resistance to change in volume with no change in shape.

Poisson's ratio ν is a dimensionless quantity given by the change in lateral strain divided by the change in longitudinal strain when a uniaxial stress is applied along the length of a specimen. Values of ν range in principle from zero, that is, no lateral contraction when a tensile stress is applied, up to a value of $\frac{1}{2}$, that is, no change in volume when a tensile stress is applied. No metals possess the extreme values. A value of zero would imply a highly directional chemical bonding. Beryllium does approach this value in that $\nu = 0.06$. A value of $\frac{1}{2}$ would imply a material with a rigidity modulus equal to zero. The very ductile metals such as gold, silver, and lead have values of about 0.4. Liquids would have the values of $\frac{1}{2}$.

For fully isotropic materials there are only two independent parameters in this group. Given values of any two of them, it is possible to calculate the remainder using the relationships given below.

For the shear modulus,

$$G = \frac{3K(1-2\nu)}{2(1+\nu)} = \frac{E}{2(1+\nu)} = \frac{3EK}{9K-E} \tag{1}$$

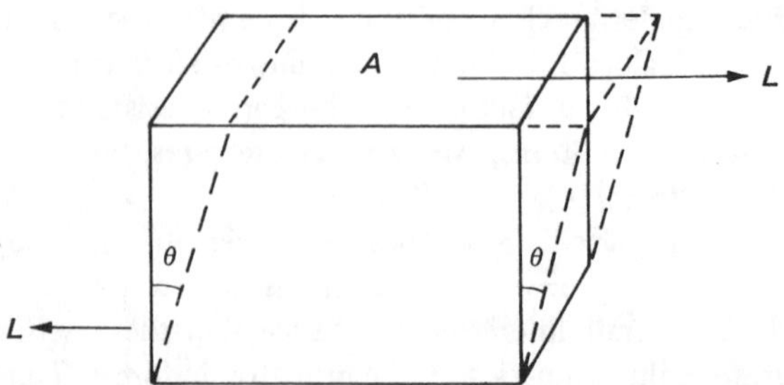

Figure 4 Broken line indicates the new position after the shearing load L is applied to the area A. The shear modulus $G = L/A \tan\theta$.

For Young's modulus,

$$E = 3K(1 - 2\nu) = 2G(1 + \nu) = \frac{9KG}{G + 3K} \quad (2)$$

For the bulk modulus,

$$K = \frac{2G(1 + \nu)}{3(1 - 2\nu)} = \frac{E}{3(1 - 2\nu)} = \frac{EG}{9G - 3E} \quad (3)$$

For Poisson's ratio,

$$\nu = \frac{E - 2G}{2G} = \frac{3K - E}{6K} = \frac{3K - 2G}{6K + 2G} \quad (4)$$

In materials not showing a marked yield point, just beyond the limit of proportionality the slope of the curve in Fig. 1 is less than the slope in the linear region. A tangent to a true stress-strain curve would indicate the response when small stresses are superimposed on a large static stress. The local value of the slope is known as the *tangent modulus.* It might be of some significance in heat exchanger design.

The *secant modulus* is the slope of a line joining a point on the stress-strain curve to the origin. For very soft materials some of the low static values of Young's modulus that have been published are probably more correctly described as secant moduli.

E. Elastic anisotropy

As mentioned in Sec. A, the properties of a commercial material are determined by the combined effects of the properties of the individual grains–individual single crystals–of which the material is composed. The individual grains are generally anisotropic in their elastic properties [16], i.e., the strain generated depends on the orientation of the applied stress with respect to the crystal orientation of the grain. The degree of anisotropy varies from one group of materials to another. Also fabrication processes including deformation and annealing stages introduce varying degrees of preferred orientation of the grains and hence the final products themselves may show varying degrees of elastic anisotropy. The nature of elastic anisotropy in metals is discussed here.

For a general single crystal of low symmetry the complete stress tensor relates stress to strain as a function of crystal directions.

A special notation is used for single crystal elastic constants. The three orthogonal crystal axes are represented by numbers:

$$x = 1 \qquad y = 2 \qquad z = 3$$

The shorthand notation for the coefficients relating stress to strain is

$$11 = 1 \qquad 22 = 2 \qquad 33 = 3$$

$$23 = 4 \qquad 31 = 5 \qquad 12 = 6$$

Thus $C_{11} = C_{1111}$, $C_{44} = C_{2323}$, $C_{66} = C_{1212}$, etc. Hooke's law for the xy shear is

$$\sigma_{12} = C_{1212}\,\epsilon_{12}$$

i.e.,

$$\sigma_{xy} = C_{66}\,\epsilon_{xy}$$

The general form of the stress tensor in the shorthand notation for a crystal of the lowest symmetry is

$$\sigma_i = C_{ij}\epsilon_j$$

where i takes the values 1-6 down the rows and j takes the values 1-6 across the columns.

For single crystals of most metals the symmetry is such that some of the coefficients are zero while others have common values. For example, for single crystals of a close-packed hexagonal metal such as titanium the coefficients reduce to

$$\begin{bmatrix} C_{11} & C_{12} & C_{13} & 0 & 0 & 0 \\ C_{12} & C_{11} & C_{13} & 0 & 0 & 0 \\ C_{13} & C_{13} & C_{33} & 0 & 0 & 0 \\ 0 & 0 & 0 & C_{44} & 0 & 0 \\ 0 & 0 & 0 & 0 & C_{44} & 0 \\ 0 & 0 & 0 & 0 & 0 & C_{66} \end{bmatrix}$$

i.e., $C_{44} = C_{55}$, $C_{22} = C_{11}$, etc., and where $C_{66} = \frac{1}{2}(C_{11} - C_{12})$. The Hooke's law equations are for hexagonal metals:

$$\sigma_{xx} = C_{11}\epsilon_{xx} + C_{12}\epsilon_{yy} + C_{13}\epsilon_{zz}$$

$$\sigma_{yy} = C_{12}\epsilon_{xx} + C_{11}\epsilon_{yy} + C_{13}\epsilon_{zz}$$

$$\sigma_{zz} = C_{13}\epsilon_{xx} + C_{13}\epsilon_{yy} + C_{33}\epsilon_{zz}$$

$$\sigma_{yz} = C_{44}\epsilon_{yz} \qquad \sigma_{zx} = C_{44}\epsilon_{zx}$$

$$\sigma_{xy} = C_{66}\epsilon_{xy} \qquad C_{66} = \tfrac{1}{2}(C_{11} - C_{12})$$

The elastic constants are unchanged if the axes are rotated about the hexad, z axis, which has sixfold symmetry. Young's modulus takes extreme (different) values for tension in the direction of the hexad axis and in directions at right angles to it.

Three different shear moduli are possible:

1. C_{44}, the mode for shear on the basal plane;
2. $C_{66} = \frac{1}{2}(C_{11} - C_{12})$, the mode for prismatic shear;
3. $C^* = C_{11} + C_{12} + 2C_{33} - 4C_{13}$, a complex shear corresponding to a change in axial ratio at constant volume.

Values for the c coefficients of hexagonal metal single crystals are quoted in Table 1.

The face-centered-cubic and body-centered-cubic metal single crystals have a higher degree of symmetry.

The c coefficients take only three different values, because the x, y, and z axes are equivalent. Thus,

$$C_{11} = C_{33} \qquad C_{12} = C_{13} \qquad C_{44} = C_{66}$$

Anisotropy of the shear modes are still possible; thus $\frac{1}{2}(C_{11} - C_{12}) \neq C_{44}$. The ratio $\frac{1}{2}(C_{11} - C_{12})/C_{44}$ is known as the anisotropy ratio, and for isotropic material equals unity. Some values of the c coefficients for cubic metals are shown in Table 2.

If materials for structural applications consist of a fine-grain structure in which individual crystals have random crystal orientations a pseudomacroscopic elastic isotropy is achieved. Working processes, such as those involved in the drawing of tube or wire, introduce some degree of preferred orientation and hence some elastic anisotropy is introduced.

Since the individual grains of a polycrystalline aggregate of most materials are elastically anisotropic, application of an external stress gives rise to an internal stress pattern. Thus the stress and strain states of the individual grains are not the same as the macroscopic average state of the specimen. In addition, most structural materials contain distributed second-phase particles having different elastic moduli from those of the matrix phase, which give rise to further internal stresses.

F. Anelasticity–damping capacity

When a metal specimen is subjected to stress cycles, energy is absorbed. Similarly, a single impact sets up vibrations that gradually die away because energy is absorbed during each cycle.

Some of the energy-absorbing processes arise from the movement of solute atoms and crystal defects. Thus damping is both a structure-sensitive and temperature-sensitive property. The processes giving rise to damping have been described in detail by Zener [17]. While many of the processes involve a monotonic increase in damping with increase in temperature and absolute stress level [17, 18] others show sharp peaks as a function of temperature at constant frequency or as a function of frequency at constant temperature [17, 19]. One of the major contributions to damping in metals involves relaxation processes at grain boundaries. For this reason a single crystal even of a very soft metal may have very low damping capacity. Contributions to damping can arise from the flow of heat evolved when a metal undergoes elastic strain, and also from thermally activated processes including hopping of solute atoms from one lattice position to another, the bowing of dislocations, and atomic movements at the grain boundary.

Damping capacity can be expressed as dE/E, the fraction of the maximum strain energy dissipated per cycle; or λ, the logarithmic decrement; or as $\tan\delta$ where δ is the angle of lag between strain and stress; or as Q^{-1}, defined for oscillatory electrical circuits; or as $\Delta f/f$, the ratio of the width of the resonance at half amplitude to the resonant frequency. When the damping capacity is small these quantities are all simply related as follows:

$$\frac{\Delta E}{E} = 2\lambda = 2\pi \tan\delta = 2\pi Q^{-1} = \frac{2\pi\,\Delta f}{\sqrt{3}f}$$

For structural materials $\Delta E/E$ may range from about 10^{-3} to 5×10^{-2} but for single crystals of pure metal it may be near 10^{-5}. It is not possible to make generalized statements since the damping capacity is a function of vibration frequency, temperature, and the absolute stress level imposed. Typically, in structural materials there may be an order of magnitude increase in damping capacity of a given material between room temperature and about 600°C. At room temperature there may be an increase in damping capacity by an order of magnitude when the static stress level is raised from 0 to 240 MPa [18].

Since damping is a structure-sensitive property, the damping capacity of a material will be a distributed quantity. However, there are not sufficient data available, particularly at working temperatures and stresses, to derive the appropriate statistics.

Following the fabrication of a heat exchanger, there may be changes in the damping capacity of the structure due to processes such as shakedown when the internal stresses are redistributed and from aging processes in the metal. However, the absorption of energy in materials of construction of a heat exchanger is unlikely to make a major contribution to the total damping from all other sources.

Some of the energy-absorping processes during vibration may give rise to cumulative changes in structure. At the tips of flaws and other stress-concentrating features the structural changes may constitute the first stages in fatigue crack growth.

G. Data Tables

Comparison of the values for an elastic property of a particular metal or alloy quoted in the various data books and published in original papers indicates a remarkably wide spread. For any given material, values differing by 20% are quite common. Therefore any calculation using elastic properties as input data should explicitly declare the values that have been taken.

The sources of error in measurement of elastic properties arise partly from difficulties in measurement of elastic strain. The quantity to be measured is at most a few parts per thousand change in length and to be measured with two to three figure accuracy. Other sources of error are more fundamental. For example, in the preparation of specimens for measurement of elastic properties, a certain degree of preferred orientation of the individual grains is introduced, thus affecting the properties in the ways described in Sec. A & E. It should also be realized that data for isotropic material may not best represent the properties of components that might have developed preferred orientation of the grains during fabrication. Finally, both at room temperature and to a greater extent at elevated temperatures and at high stress levels stress-strain response of a specimen is time dependent and structure sensitive.

Where high accuracy is required as for example, in attempting to understand the performance of a particular fabricated heat exchanger, the best procedure would be to measure the elastic properties on samples of the actual materials at the appropriate temperature and strain rate.

Tables 1 and 2 show single crystal elastic coefficients for hexagonal and cubic metals, respectively [20, 21] They are included here to indicate the extent to which elastic anisotropy can be expected in wrought polycrystalline materials with preferred orientation of the grains.

The only values for Young's modulus tabulated here are derived from [22], which was produced specifically for tubular heat exchanger designers in the U.S.

Some of the reference books containing tabulated elastic constants data for structural materials are listed below with an indication of the types of data available, [23 to 29].

The data in Tables 1-3 are in the SI units, pascals (Pa), where 1 Pa = 1 N/m^2; 1 pound per square inch = 6.894757×10^3 Pa; and 1 dyne/cm^2 = 0.1 Pa.

Table 1 Single crystal elastic constants of some hexagonal metals in gigapascals[a]

			Shear constants				
Metal	C_{11}	C_{33}	C_{12}	C_{13}	C_{44}	C_{66}	$C^b/6$
Beryllium	299	342	28	11	166	136	161
Zinc	161	66	34	50	40	63	21
Magnesium	57	62	23	19	17	17	21
Zirconium	144	165	73	65	33	36	47
Titanium	163	180	93	62	47	35	62
Cadmium	115	51	40	40	20	38	16

[a]From Tang et al. [1]. (Giga = 10^9.)
[b]$C = C_{11} + C_{12} + 2C_{33} - 4C_{13}$

Table 2 "c" Coefficients for some cubic metals in gigapascals[a]

Metal	C_{11}	C_{12}	C_{44}	Anisotropy ratio
Silver	120	90	44	0.34
Aluminum	112	66	28	0.82
Copper	168	121	75	0.31
Brass, 28% Zn	147	111	72	0.25
α-Iron (ferrite)	237	141	116	0.41
Molybdenum	455	176	110	1.27
Nickel	251	157	121	0.39

[a]From Smithells [21]

Table 3 Quasistatic Young's modulus values in gigapascals as a function of temperature[a]

Material	Value at different temperatures, K												
	300	325	350	425	475	525	575	650	700	750	800	850	925
Carbon steel	200		197	194	190	185	180	169	159	145	128	–	–
Aust. stainless steel	193	–	187	183	179	174	170	164	159	154	150	–	–
Carbon-Mo and low chrome steels (through 3% chrome)	206	–	203	200	197	193	190	183	177	169	161	145	108
Intermediate Cr-Mo steels (5–9% Cr)	189	–	187	185	182	180	176	172	167	162	158	153	143
Straight chrome steels (12, 17, 27% Cr)	201	–	197	195	191	187	180	171	159	145	131	113	84
Gray cast iron	92	–	91	89	87	84	81	76	70				
Nickel-copper alloy 400	179	–	179	178	177	175	171	159	145	128	113	102	90
Cupro nickels (70-30, 80-20, 90-10)	130	–	127	124	121	119	116	112	105				
Aluminum	73	–	73	72	70	66	59						
Ni-Cr-Fe alloy 600	219	–	213	210	207	204	201	197	192	–	174	–	138
Ni-Fe-Cr alloy 800	196	–	191	188	185	181	178	174	170				
Ni-Fe-Cr-Mo-Cu alloy 825	195	–	–	–	185	–	–	174					
Ni-Mo alloy B	203	–	–	–	–	–	190						
Ni-Mo-Cr alloy C-276	–	205	–	–	188	–	–	–	175				
Nickel	207	–	204	201	197	193	189	185	181	177	173	169	163
Copper	110	109	108	106	104	101	98	94					
Commercial brass	97	96	94	93	90	88	85	81					
Leaded tin bronze	90	89	87	85	83	81	78	75					
Phosphor bronze	103	102	99	96	93	89	82	72	60				
Muntz	105	103	96	89	81	75							
Admiralty	110	–	107	101	93	86	76	66	52				
Titanium grade 2	–	117	100	–	90	–	79						
Zirconium	–	99	–	92	–	80	–	68					
Cr-Mo alloy XM-27	215	–	198	188	187								
Cr-Ni-Fe-Mo-Cu-Cb alloy 20 Cb	193												

[a]From [22]

Further data on commercial fabrication materials are given in Section 5.5.

5.5. Thermal and Mechanical Properties of Heat Exchanger Construction Material

This section provides data on the three essential properties of materials required for the design of heat exchangers, namely thermal conductivity, the mean coefficient of expansion (from ambient temperature to required temperature), and the modulus of elasticity. A wide variety of materials is used in practice, and the materials listed in the Standards of the American Society of Mechanical Engineers (ASME) have been taken as a basis for this report, with a few additions. The Unified Numbering System (UNS) and the Werkstoff-Nummer (WN) or DIN-Norm (DIN) have been given where applicable.

Different data sources give large variations (frequently up to 25%) in the values of these properties at high and/or low temperatures. In particular, significant disagreement between ASME and TEMA (Tubular Exchanger Manufacturers' Association) tables is frequent. Where alternative sources have been referenced, preferred values have been adopted.

Data for many of the materials are available only at ambient temperatures, and values at other temperatures have been estimated by assuming the temperature variation to be similar for similar materials. Curves of property versus temperature sometimes exhibit maxima and minima, for example, in the thermal conductivity of some chromium steels or of aluminum alloys and in the coefficient of expansion of steels at high temperatures. In such cases it was not possible to extrapolate values from data over a limited temperature range because the point of inflection was unknown.

Values are given for carbon and low alloy steels (Table 1), high chrome steels (Table 2), nickel and nickel alloys (Table 3), copper and copper alloys (Table 4), aluminum alloys (Table 5), and titanium, zirconium, and cast iron (Table 6). Quantities and units used are: ρ, density (g/cm^3); TC, thermal conductivity (W/m K); CE, coefficient of expansion (1/K), and ME, modulus of elasticity (GPa) (GPa = 10^9 Pa; 1 Pa = 1 N/m^2; 1 psi = 6.894 757 × 10^3 Pa).

Tabular information is given for the following materials:

	UNS	WN*
Carbon steels		
Carbon/silicon		
Carbon/manganese		
Carbon/manganese/silicon		
C–1/2Mo; 1/2Cr–1/2Mo; 1/2Cr–1/5Mo–V; 1-3/4Cr–1/2Mo–Cu		1.5423
1/2Cr–1/4Mo–Si; 1Cr–1/2Mo–V; 1-1/4Cr–1/2Mo; 1-1/4Cr–1/2Mo–Si		1.7335
1Cr–1/5Mo; 1Cr–1/2Mo		1.7335
2Cr–1/2Mo		—
1Cr–1/5Mo–Si		1.7335
2-1/4Cr–1Mo		1.7380
3Cr–1Mo		—
5Cr–1/2Mo		1.7362
5Cr–1/2Mo–Si		1.7362
5Cr–1/2Mo–Ti		—
7Cr–1/2Mo		1.7368
9Cr–1Mo		1.7386
1/2Ni–1/2Cr–1/4Mo–V		—

	UNS	WN*
3/4Cr-1/2Ni-Cu		—
3/4Cr-3/4Ni-Cu-Al		—
3/4Ni-1/2Cu-Mo		—
1Ni-1/2Cr-1/2Mo		—
1-3/4Ni-3/4Cr-1/4Mo		—
2-1/2Ni		1.5635
3/4Ni-1/2Mo-1/3Cr-V		—
1/2Ni-1/2Mo-V; 3/4Ni-1/2Cr-1/2Mo-V; 3/4Ni-1/2Mo-Cr-V; 3/4Ni-1Mo-3/4Cr		—
1/2Cr-1/2Ni-1/5Mo		—
2Ni-3/4Cr-1/3Mo		—
2Ni-3/4Cr-1/4Mo		—
3-1/2Ni		1.5637
3-1/2Ni-1-3/4Cr-1/2Mo-V		—
5Ni-1/4Mo		—
8Ni		—
9Ni		1.5662
Mn-1/2Mo		—
Mn-1/2Mo-1/4Ni		—
Mn-1/2Mo-1/2Ni		—
Mn-1/2Mo-3/4Ni		1.6310
Mn-1/4Mo		—
Mn-V		—
High chrome steels		
12Cr-1Al	S40500	1.4002
13Cr	S41000	1.4006
15Cr	S42900	—
17Cr	S43000	1.4016
18Cr/8Ni	S30400	1.4301
23Cr/12Ni	S30900	1.4828
25Cr/20Ni	S31000	1.4841
16Cr/12Ni/2Mo	S31600	1.4401
18Cr/13Ni/3Mo	S31700	—
18Cr/10Ni/Ti	S32100	1.4541
18Cr/10Ni/Cb	S34700	1.4550
18Cr/18Ni/2Si	S38100	—
Nickel and nickel alloys		
Nickel 200	N02200	2.4066
Nickel 201	N02201	2.4068
Nickle 400	N04400	2.4360
Nickel 405	N04405	—
Nickel 600	N06600	2.4816
Nickel 800	N08800	1.4876
Nickel 825	N08825	2.4858
Alloy X	N06002	2.4606
Alloy C4	N06455	2.4610
Alloy 625	N06625	2.4856
Alloy G	N06007	—
Alloy G3	N06985	2.4603
Alloy B	N1001	2.4810
Alloy B2	N10665	2.4617
Alloy 276	N10276	2.4819
Alloy C22	—	—
Copper and copper alloys		
Copper	C10200	2.0040
Muntz metal	C36500	2.0372
Admiralty brass	C44300	2.0470
Naval brass	C46400	2.0530
Phosphor bronze A	C51000	2.1016
Aluminum bronze D	C61400	2.0932
Aluminum brass	C68700	2.0460
Copper nickel 90/10	C70600	2.0872
Copper nickel 80/20	C71000	2.0878
Copper nickel 70/30	C71500	2.0882
Aluminum alloys (annealed condition)		
Alloy 1100	A91100	3.0205
Alloy 3003	A93003	3.0517
Alloy 5052	A95052	3.3523
Alloy 5083	A95083	3.3547
Alloy 6061	A96061	3.3211
Alloy 6063	A96063	3.3206
Miscellaneous materials		
Unalloyed titanium (grade 2)	R50400	3.7055
Zirconium	—	—
Grey cast iron	—	—

**Note*: Where no Werkstoff-Nummer is given, there is no equivalent to the ASME materials quoted.

Table 1 Carbon and low alloy steels

	Temperature										
	−100 173	0 273	100 373	200 473	300 573	400 673	500 773	600 873	700 973	800 1 073	900°C 1 173 K
Carbon steel, $\rho = 7.9$											
TC	64.9	61.8	57.8	53.5	49.0	44.5	40.2	35.7	31.2	27.3	(26.0)
CE	(10.8)	(11.4)	12.1	12.7	13.3	13.9	14.4	14.8	15.0	14.8	12.6
ME	(208)	(204)	199	192	185	172	(141)				
Carbon/silicon, $\rho = 7.9$											
TC	(53.0)	(52.0)	50.2	47.8	44.6	41.5	38.1	34.8	30.6	26.6	(26.0)
CE	(8.6)	(9.8)	11.1	12.1	13.0	13.7	14.2	14.6	(14.8)	(14.6)	(12.6)
ME	(194)	(192)	(191)	(187)	(179)	(168)	(119)				
Carbon/manganese, $\rho = 7.9$											
TC	(45.9)	(47.2)	47.8	46.4	43.8	40.8	37.6	33.7	29.8	26.3	(26.0)
CE	(10.8)	(11.4)	12.1	12.7	13.3	13.9	14.4	14.8	15.0	14.8	12.6
ME	(216)	207	198	188	179	170	161	152			
Carbon/manganese/silicon, $\rho = 7.9$											
TC	(36.7)	(40.0)	42.2	41.9	40.3	38.2	35.8	33.1	28.9	26.1	(26.0)
CE	(8.0)	(9.4)	10.7	11.9	12.8	13.6	14.2	(14.6)	(14.8)	(14.6)	(12.6)
ME	(219)	(214)	210	202	194	185	176	165	(152)		
C-1/2Mo; 1/2Cr–1/2Mo; 1/2Cr–1/5Mo–V; 1-3/4Cr–1/2Mo–Cu, $\rho = 7.9$											
TC	(40.2)	(42.4)	43.6	43.1	41.4	38.9	36.3	33.4	29.2	26.0	(26.0)
CE	(8.6)	(9.8)	11.1	12.1	13.0	13.7	14.2	14.6	(14.8)	(14.6)	(12.6)
ME	(210)	207	203	197	190	180	166	138			
1/2Cr–1/4Mo–Si; 1Cr–1/2Mo–V; 1-1/4Cr–1/2Mo; 1-1/4Cr–1/2Mo–Si, $\rho = 7.9$											
TC	(34.4)	(36.3)	37.9	38.1	37.1	35.7	33.7	31.7	28.6	25.8	(25.4)
CE	(8.0)	(9.4)	10.7	11.9	12.8	13.6	14.2	(14.6)	(14.8)	(14.6)	(12.6)
ME	(210)	207	203	197	190	180	166	138			
1Cr–1/5Mo; 1Cr–1/2Mo, $\rho = 7.85$											
TC	(40.3)	(41.5)	42.2	41.5	39.6	37.6	35.1	32.7	29.4	26.1	(25.6)
CE	(8.6)	(9.5)	11.1	12.1	13.0	13.7	14.2	14.6	(14.8)	(14.6)	(12.6)
ME	(200)	(198)	194	190	184	177	161	(90)			
2Cr–1/2Mo, $\rho = 7.9$											
TC	(40.3)	(41.5)	42.2	41.5	39.6	37.6	35.1	32.7	29.4	26.1	(25.6)
CE	(8.0)	(9.4)	10.7	11.9	12.8	13.6	14.2	(14.6)	(14.8)	(14.6)	(12.6)
ME	(200)	(198)	(194)	(190)	(184)	(177)	(161)	(90)			
1CR–1/5Mo–Si, $\rho = 7.8$											
TC	(42.9)	(44.3)	44.7	43.1	40.8	38.4	36.0	33.4	29.6	26.3	(26.0)
CE	(8.6)	(9.8)	11.1	12.1	13.0	13.7	14.2	14.6	(14.8)	(14.6)	(12.6)
ME	(200)	(198)	(194)	(190)	(184)	(177)	(161)	(90)			
2-1/4Cr–1Mo, $\rho = 7.8$											
TC	(35.0)	(36.0)	37.0	37.0	36.7	35.5	33.7	32.0	29.8	26.7	(26.0)
CE	(10.9)	(11.5)	12.1	12.6	13.1	13.7	14.2	(14.6)	(14.8)	(14.6)	(12.6)
ME	(217)	(213)	208	201	194	185	174	(156)			

Table 1 Carbon and low alloy steels (*Continued*)

	Temperature										
	−100 173	0 273	100 373	200 473	300 573	400 673	500 773	600 873	700 973	800 1 073	900°C 1 173 K
3Cr–1Mo, $\rho = 7.8$											
TC	(27.3)	(31.3)	35.0	36.2	36.2	35.0	33.4	31.7	29.6	27.5	(28.0)
CE	(8.0)	(9.4)	10.7	11.9	12.8	13.6	14.2	(14.6)	(14.8)	(14.6)	(12.6)
ME	(217)	(213)	(208)	(201)	(194)	(185)	(174)	(156)			
5Cr–1/2Mo, $\rho = 7.8$											
TC	(23.9)	(28.2)	31.7	32.9	32.9	32.5	31.7	30.3	28.7	27.0	(27.9)
CE	(10.5)	(11.4)	12.1	12.5	12.8	13.1	13.3	13.5	13.7		
ME	(220)	(214)	208	201	194	184	170	153	(131)		
5Cr–1/2Mo–Si, $\rho = 7.7$											
TC	(13.8)	(19.9)	23.0	25.4	27.1	28.2	28.7	28.6	27.9	26.3	(26.9)
CE	(10.5)	(11.4)	12.1	12.5	12.8	13.1	13.3	13.5	13.7		
ME	(220)	(214)	208	201	194	184	170	153	(131)		
5Cr–1/2Mo–Ti, $\rho = 7.7$											
TC	(23.8)	(26.7)	29.6	31.5	32.0	31.8	31.2	30.0	28.5	27.0	(27.7)
CE	(10.5)	(11.4)	12.1	12.5	12.8	13.1	13.3	13.5	13.7		
ME	(220)	(214)	208	201	194	184	170	153	(131)		
7Cr–1/2Mo, $\rho = 7.7$											
TC	(20.8)	(23.7)	26.5	28.4	29.5	29.9	29.8	29.2	28.0	26.7	(27.3)
CE	(9.7)	(10.3)	10.8	11.3	11.7	12.0	(12.3)	(12.5)	(12.7)		
ME	(220)	(214)	208	201	194	184	170	153	(131)		
9Cr–1Mo, $\rho = 7.78$											
TC	(18.6)	(21.6)	24.3	26.3	27.5	27.8	27.9	27.5	27.0	25.9	(26.1)
CE	(9.7)	(10.3)	10.8	11.3	11.7	12.0	(12.3)	(12.5)	(12.7)		
ME	(220)	(214)	208	201	194	184	170	153	(131)		
1/2Ni–1/2Cr–1/4Mo–V, $\rho = 7.8$											
TC	(38.7)	(41.5)	42.7	41.9	40.0	37.9	35.7	33.1	29.6	26.1	(25.8)
CE	(10.8)	(11.4)	12.1	12.7	13.3	13.9	14.4	14.8	15.0	14.8	12.6
M	(200)	(193)	187	181	174	168	(159)				
3/4Cr–1/2Ni–Cu, $\rho = 7.8$											
TC	(48.5)	(49.2)	47.8	45.7	43.3	40.5	37.4	34.3	30.5	26.5	(26.0)
CE	(10.8)	(11.4)	12.1	12.7	13.3	13.9	14.4	14.8	15.0	14.8	12.6
M	(200)	(193)	187	181	174	168	(159)				
3/4Cr–3/4Ni–Cu–Al, $\rho = 7.8$											
TC	(39.3)	(40.3)	40.8	40.2	38.7	37.0	35.0	32.7	29.4	26.1	(25.6)
CE	(10.8)	(11.4)	12.1	12.7	13.3	13.9	14.4	14.8	15.0	14.8	12.6
ME	(200)	(193)	187	181	174	168	(159)				
3/4Ni–1/2Cu–Mo, $\rho = 7.8$											
TC	(47.8)	(47.8)	47.2	45.3	42.8	40.7	36.9	33.7	29.8	26.5	(26.0)
CE	(10.8)	(11.4)	12.1	12.7	13.3	13.9	14.4	14.8	15.0	14.8	12.6
ME	(200)	(193)	187	181	174	168	(159)				

	Temperature										
	−100 173	0 273	100 373	200 473	300 573	400 673	500 773	600 873	700 973	800 1 073	900°C 1 173 K
1Ni–1/2Cr–1/2Mo, $\rho = 7.8$											
TC	(38.6)	(40.5)	41.7	41.3	39.5	37.4	35.1	32.4	29.2	26.0	(25.6)
CE	(10.8)	(11.4)	12.1	12.7	13.3	13.9	14.4	14.8	15.0	14.8	12.6
ME	(200)	(193)	187	181	174	168	(159)				
1-3/4Ni–3/4Cr–1/4Mo, $\rho = 7.8$											
TC	(32.5)	(35.0)	36.9	37.2	36.7	35.6	34.1	32.4	29.1	26.1	(26.1)
CE	(10.1)	(11.0)	11.8	12.6	13.1	13.6	(13.9)	(14.2)	(14.4)		
ME	(200)	(193)	187	181	174	168	(159)				
2-1/2Ni, $\rho = 7.8$											
TC	(42.9)	(44.5)	45.5	44.7	42.2	39.6	37.0	34.1	30.1	26.7	(26.7)
CE	(10.1)	(11.0)	11.8	12.6	13.1	13.6	(13.9)	(14.2)	(14.4)		
ME	(200)	(193)	187	181	174	168	(159)				
3/4Ni–1/2Mo–1/3Cr–V, $\rho = 7.8$											
TC	(39.8)	(40.7)	41.4	41.0	39.3	37.2	35.0	32.6	29.2	26.1	(25.6)
CE	(10.8)	(11.4)	12.1	12.7	13.3	13.9	14.4	14.8	15.0	14.8	(12.6)
ME	(200)	(193)	187	181	174	168	(159)				
1/2Ni–1/2Mo–V; 3/4Ni–1/2Cr–1/2Mo–V; 3/4Ni–1/2Mo–Cr–V; 3/4Ni–1Mo–3/4Cr, $\rho = 7.8$											
TC	(35.1)	(37.2)	38.6	38.6	37.4	35.8	34.1	32.0	29.1	26.0	(26.0)
CE	(10.8)	(11.4)	12.1	12.7	13.3	13.9	14.4	14.8	15.0	14.8	12.6
ME	(200)	(193)	187	181	174	168	(159)				
1/2Cr–1/2Ni–1/5Mo, $\rho = 7.8$											
TC	(39.3)	(40.3)	40.8	40.2	38.7	37.0	35.0	32.7	29.4	26.1	(25.6)
CE	(8.6)	(9.8)	11.1	12.1	13.0	13.7	14.2	14.6	(14.8)	(14.6)	(12.6)
ME	(200)	(193)	187	181	174	168	(159)				
2Ni–3/4Cr–1/3Mo; 2Ni–3/4Cr–1/4Mo, $\rho = 7.8$											
TC	(27.3)	(32.2)	36.0	37.0	36.8	35.8	34.3	32.4	29.8	25.8	(26.5)
CE	(10.1)	(11.0)	11.8	12.6	13.1	13.6	(13.9)	(14.2)	(14.4)		
ME	(200)	(193)	187	181	174	168	(159)				
3-1/2Ni, $\rho = 7.85$											
TC	(33.7)	(37.6)	41.3	41.5	39.9	38.1	35.8	33.2	29.4	26.5	(26.1)
CE	(10.1)	(11.0)	11.8	12.6	13.1	13.6	(13.9)	(14.2)	(14.4)		
ME	(200)	(193)	187	181	174	168	(159)				
3-1/2Ni–1-3/4Cr–1/2Mo–V, $\rho = 7.85$											
TC	(33.7)	(36.5)	39.8	40.4	38.9	37.2	35.1	32.5	29.2	26.1	(26.1)
CE	(10.1)	(11.0)	11.8	12.6	13.1	13.6	(13.9)	(14.2)	(14.4)		
ME	(200)	(193)	(187)	(181)	(174)	(168)	(159)				
5Ni–1/4Mo, $\rho = 7.9$											
TC	(29.4)	(32.9)	36.0	37.0	36.8	35.7	34.1	31.8	28.7	26.3	(26.3)
CE	(10.3)	(11.0)	11.7	12.3	12.8	13.2	(13.6)	(13.8)	(14.2)		
ME	(200)	(193)	(187)	(181)	(174)	(168)	(159)				

Table 1 Carbon and low alloy steels (*Continued*)

	Temperature										
	−100 173	0 273	100 373	200 473	300 573	400 673	500 773	600 873	700 973	800 1 073	900°C 1 173 K
8Ni, $\rho = 7.9$											
TC	(27.9)	(29.9)	32.0	33.6	34.3	34.1	32.9	30.5	27.6	26.3	(27.5)
CE	(8.7)	(9.7)	10.7	11.4	12.0	12.3	(12.7)	(13.0)	(13.2)		
ME	(200)	(193)	(187)	(181)	(174)	(168)	(159)				
9Ni, $\rho = 7.86$											
TC	(26.0)	(29.4)	32.7	34.1	34.4	33.9	32.9	30.8	27.3	26.1	(26.8)
CE	(8.7)	(9.7)	10.7	11.4	12.0	12.3	(12.7)	(13.0)	(13.2)		
ME	(200)	(193)	(187)	(187)	(174)	(168)	(159)				
Mn–1/2Mo, $\rho = 7.9$											
TC	(36.2)	(39.6)	42.2	42.7	41.2	38.9	36.4	33.7	30.6	26.7	(28.0)
CE	(11.9)	(12.5)	13.1	13.6	14.0	14.4	(14.7)	(14.9)	(15.1)		
ME	(210)	(202)	195	190	183	170	(150)				
Mn–1/2Mo–1/4Ni. $\rho = 7.9$											
TC	(35.1)	(38.6)	41.4	41.9	40.5	38.2	36.0	33.4	30.6	26.5	(27.2)
CE	(11.9)	(12.5)	13.1	13.6	14.0	14.4	(14.7)	(14.9)	(15.1)		
ME	(210)	(202)	195	190	183	170	(150)				
Mn–1/2Mo–1/2Ni, $\rho = 7.9$											
TC	(34.6)	(37.9)	40.6	41.4	40.0	38.1	36.0	33.4	30.3	26.3	(27.5)
CE	(11.9)	(12.5)	13.1	13.6	14.0	14.4	(14.7)	(14.9)	(15.1)		
ME	(210)	(202)	195	190	183	170	(150)				
Mn–1/2Mo–3/4Ni, $\rho = 7.9$											
TC	(33.2)	(36.9)	39.8	40.4	39.6	37.9	35.7	33.2	30.3	26.3	(27.7)
CE	(11.9)	(12.5)	13.1	13.6	14.0	14.4	(14.7)	(14.9)	(15.1)		
ME	(210)	(202)	(195)	(190)	(183)	(170)	(150)				
Mn–1/4Mo, $\rho = 7.9$											
TC	(29.2)	(33.0)	36.2	37.0	36.7	35.5	33.8	31.7	29.1	25.3	(26.7)
CE	(11.9)	(12.5)	13.1	13.6	14.0	14.4	(14.7)	(14.9)	(15.1)		
ME	(210)	(202)	195	190	183	170	(150)				
Mn–V, $\rho = 7.9$											
TC	(38.9)	(41.5)	43.8	44.3	42.9	40.8	38.2	35.1	31.2	26.8	(27.3)
CE	(10.9)	(11.7)	12.5	13.1	13.5	13.8	(14.1)	(14.3)	(14.5)		
ME	(210)	(202)	195	190	183	170	(150)				

Table 2 High chrome steels

	Temperature										
	−100 173	0 273	100 373	200 473	300 573	400 673	500 773	600 873	700 973	800 1 073	900°C 1 173 K
12 Cr-1 AC (S40500), $\rho = 7.8$											
TC	(19)	(24)	24.8	25.0	25.3	25.4	25.4	25.5	25.6	25.7	
CE	(9.5)	(10.0)	10.3	10.7	11.1	11.4	(11.8)	(12.1)	(12.4)	(12.6)	(10.8)
ME	(219)	(218)	214	209	201	193	182	168	145		
13 Cr (S41000), $\rho = 7.73$											
TC	19	24.6	26.8	27.3	27.5	27.5	27.4	27.0	26.1	26.0	
CE	9.5	10.0	10.3	10.7	11.1	11.4	11.8	12.1	12.4	12.6	10.8
ME	(207)	(203)	198	191	181	165	141	106	(55)		
15 Cr (S42900), $\rho = 7.8$											
TC	(19)	(24)	24.9	25.3	25.5	25.6	25.6	25.6	25.6	25.6	
CE	(9.1)	(9.5)	9.9	10.3	10.7	11.0	(11.2)	(11.4)	(11.5)	(11.5)	
ME	(207)	(203)	198	191	181	165	141	106	(55)		
17 Cr (S43000), $\rho = 7.8$											
TC	(17)	(21.6)	22.3	22.7	22.9	23.3	23.5	23.9	24.3	25.0	
CE	(9.1)	(9.5)	9.9	10.3	10.7	11.0	(11.2)	(11.4)	(11.5)	(11.5)	
ME	(207)	(203)	198	191	181	165	141	106	(55)		
18 Cr/8 Ni (S30400), $\rho = 7.9$											
TC	(14.7)	15.6	16.5	17.3	17.7	19.7	21.3	22.7	24.1	25.4	26.8
CE	(14.2)	(15.1)	15.8	16.5	17.1	17.6	18.0	18.4	18.8	19.1	19.2
ME	(203)	(197)	190	183	177	169	161	154	146	138	(131)
23 Cr/12 Ni (S30900), $\rho = 7.9$											
TC	(12.5)	(13.8)	15.2	16.6	18.0	19.6	20.9	22.3	23.7	25.1	(26.6)
CE	(15.3)	(15.8)	16.2	16.4	16.6	16.7	(16.7)	(16.8)	(16.9)	(17.0)	(17.1)
ME	(203)	(197)	(190)	(183)	(177)	(169)	(161)	(154)	(146)	(138)	(131)
25 Cr/20 Ni (S31000), $\rho = 7.9$											
TC	(11.0)	(12.5)	14.0	15.6	17.1	18.7	20.2	21.7	23.2	24.7	(26.5)
CE	(15.3)	(15.8)	16.2	16.4	16.6	16.7	(16.7)	(16.8)	(16.9)	(17.0)	(17.1)
ME	(203)	(197)	190	183	177	169	161	154	146	138	(131)
16 Cr/12 Ni/2 Mo (S31600), $\rho = 7.96$											
TC	(11.4)	12.7	14.0	15.3	16.6	17.8	19.2	20.5	21.8	23.0	24.4
CE	(15.2)	(15.6)	16.0	16.4	16.9	17.5	18.2	18.8	19.3	19.5	19.6
ME	(203)	(197)	190	183	177	169	161	154	146	138	(131)
18 Cr/13 Ni/3 Mo (S31700), $\rho = 8.0$											
TC	(11.2)	(13.0)	14.7	16.3	17.8	19.3	20.8	22.3	23.7	25.1	(26.5)
CE	(14.2)	(15.0)	15.8	16.6	17.2	17.7	(18.1)	(18.5)	(18.9)	(19.3)	(19.7)
ME	(196)	(190)	(185)	(179)	(174)	(168)	(163)	158	152	147	(142)
18 Cr/10 Ni/Ti (S32100), $\rho = 7.90$											
TC	(11.2)	(13.0)	14.7	16.3	17.8	19.3	20.8	22.3	23.7	25.1	(26.5)
CE	(15.6)	(16.0)	16.5	16.8	17.0	17.2	(17.4)	(17.6)	(17.8)	(18.0)	(18.2)
ME	(203)	(197)	190	183	177	169	161	154	146	138	(131)

Table 2 High chrome steels (*Continued*)

	Temperature										
	−100 173	0 273	100 373	200 473	300 573	400 673	500 773	600 873	700 973	800 1 073	900°C 1 173 K
18 Cr/10 Ni/Cb (S34700), $\rho = 7.93$											
TC	(12.5)	13.8	15.2	16.6	17.9	19.2	20.6	22.0	23.4	24.7	26.0
CE	(15.6)	(16.0)	16.5	16.8	17.0	17.2	(17.4)	(17.6)	(17.8)	(18.0)	(18.2)
ME	(203)	(197)	190	183	177	169	161	154	146	138	(131)
18 Cr/18 Ni/2 Si (S38100), $\rho = 8.02$											
TC	(9.0)	(10.7)	12.4	14.1	15.7	17.3	19.0	20.7	22.3	24.0	25.6
CE	(15.3)	(15.8)	(16.2)	(16.4)	(16.6)	(16.7)	(16.7)	(16.8)	(16.9)	(17.0)	
ME	(204)	(197)	190	183	175	168	161	(154)	(146)	(138)	(131)

Table 3 Nickel and nickel alloys

	Temperature											
	−200 73	−100 173	0 273	100 373	200 473	300 573	400 673	500 773	600 873	700 973	800 1 073	900°C 1 173 K
Nickel 200 (N02200), $\rho = 8.89$ g/cm³												
TC		77.4	72.2	70.0	61.4	56.6	55.7	57.6	59.7	61.8	63.9	65.9
CE	9.3	10.9	12.2	13.3	13.8	14.3	14.8	15.2	15.5	15.8	16.1	(16.4)
ME		(217)	(214)	210	207	203	198	190	182	174	163	141
Nickel 201 (N02201), $\rho = 8.89$												
TC		88.6	81.0	73.6	66.6	59.9	56.8	58.2	60.5	62.7	65.1	67.3
CE	9.3	10.9	12.2	13.3	13.8	14.3	14.8	15.2	15.5	15.8	16.1	(16.4)
ME		(210)	207	(203)	(200)	(196)	(191)	(183)	(175)	(167)	(156)	(134)
Nickel 400 (N04400), $\rho = 8.83$												
TC	15.5	19.7	21.4	24.3	27.6	30.6	33.5	36.8	39.8	42.7	(45.6)	(49.0)
CE	10.9	12.0	13.1	14.1	15.5	15.8	15.9	16.2	16.6	16.9	17.4	(17.7)
ME		(179)	(179)	179	177	171	152	123	98	(81)		
Nickel 405 (N04405), $\rho = 8.83$												
TC	15.5	19.7	21.4	24.3	27.6	30.6	33.5	36.8	(39.8)	(42.7)	(45.6)	(49.0)
CE	(9.5)	(11.6)	(12.4)	13.7	14.9	15.7	16.1	(16.2)	(16.6)	(16.9)	(17.4)	(17.7)
ME		(179)	(179)	(179)	(177)	(171)	(152)	(123)	(98)	(81)		
Nickel 600 (N06600), $\rho = 8.43$												
TC		13.1	14.5	15.9	17.3	18.9	20.4	22.2	23.9	25.6	27.3	(29.1)
CE	(11.3)	11.9	12.5	13.0	13.6	14.1	14.5	14.9	15.3	15.7	16.1	(16.5)
ME		(228)	(220)	212	207	202	196	188	181	172	163	152

Nickel 800 (N08800), $\rho = 8.02$												
TC	(7.6)	9.4	11.2	12.9	14.5	16.3	17.9	19.6	21.3	23.0	24.7	26.5
CE		(11.5)	(12.9)	14.3	15.6	16.2	16.6	16.7	17.1	17.5	18.0	18.5
ME	212	205	199	191	184	177	170	163	157	150	143	136
Nickel 825 (N08825), $\rho = 8.09$												
TC		(9.7)	(11.4)	13.4	15.1	16.6	18.4	20.0	21.7	23.3	24.8	26.5
CE		(12.0)	(13.0)	14.0	14.8	15.1	15.5	15.7	16.3	16.8	17.3	17.7
ME		(203)	(197)	192	187	181	174	168	162	154	146	131
Alloy X (N06002), $\rho = 8.22$												
TC	6.0	(7.2)	(9.2)	11.1	13.0	14.7	16.8	18.9	20.8	22.7	24.6	(26.5)
CE		(13.1)	(13.5)	13.8	14.1	14.4	14.7	14.9	15.3	15.7	16.0	16.3
ME		(203)	(197)	193	185	179	172	164	158	150	143	134
Alloy C4 (N06455), $\rho = 8.64$												
TC		(7.6)	(9.6)	11.2	13.2	14.9	16.8	18.5	20.4	(22.7)	(24.7)	(26.8)
CE		(8.9)	(9.9)	10.8	11.8	12.5	13.0	13.0	13.4	13.5	14.5	15.1
ME		(219)	(214)	208	202	196	190	183	176	169	160	150
Alloy 625 (N06625), $\rho = 8.44$												
TC		7.9	9.3	10.9	12.3	13.8	15.2	16.8	18.3	19.7	21.5	23.4
CE	9.5	11.3	12.2	12.8	13.1	13.3	13.6	14.0	14.5	15.0	15.5	16.1
ME		(210)	(207)	203	199	193	187	181	174	167	157	145
Alloy G (N06007), $\rho = 8.30$												
TC		(8.1)	(9.7)	11.2	12.6	14.2	15.8	17.5	19.0	(20.6)	(22.3)	(23.9)
CE		(13.3)	(13.3)	13.5	13.8	14.2	14.8	(15.4)	(16.2)			
ME		(197)	193	(189)	(184)	(177)	(171)	(164)	(158)			
Alloy G3 (N06985), $\rho = 8.30$												
TC		(6.9)	(9.3)	11.8	13.8	15.9	17.9	20.0	21.8	(23.6)	(25.4)	(27.0)
CE		(14.5)	(14.5)	(14.5)	14.5	14.6	14.7	15.0	15.3	15.8	16.2	16.7
ME		(204)	(200)	196	191	184	178	171	165	158	152	(145)
Alloy B (N1001), $\rho = 9.25$												
TC		(9.3)	(10.2)	11.1	11.9	13.0	14.5	16.2	18.6	21.2	24.0	25.0
CE			(9.0)	11.0	11.4	11.7	11.8	11.9	12.1	12.2	(12.4)	(12.7)
ME		(217)	212	(208)	(203)	(198)	(192)	(186)	(180)	(174)	(167)	(160)
Alloy B2 (N10665), $\rho = 9.22$												
TC		(10.1)	11.1	12.2	13.4	14.6	16.0	17.3	18.7	(19.9)	(21.2)	(22.5)
CE		(9.5)	9.9	10.4	10.8	11.2	11.4	11.7	(11.9)	(12.0)	(12.1)	(12.2)
ME		(222)	(217)	213	208	203	197	191	185	(179)	(172)	(165)
Alloy 276 (N10276), $\rho = 8.89$												
TC		8.1	9.6	11.2	13.0	14.7	16.4	18.2	20.0	21.8	23.5	25.3
CE		(9.9)	(10.7)	11.3	12.0	12.7	13.1	13.3	13.8	14.4	15.0	15.7
ME		(212)	(207)	201	195	190	184	178	(172)	(167)	(161)	(155)
Alloy C22 (N –), $\rho = 8.69$												
TC		(6.9)	(9.0)	11.1	13.4	15.5	17.5	19.5	21.3	(23.2)	(24.9)	(26.8)
CE		(12.1)	12.2	12.3	12.5	12.6	13.1	13.7	14.3	14.9	15.5	16.1
ME		(212)	(207)	202	197	191	185	179	174	167	160	152

Table 4 Copper and copper alloys

	Temperature											
	−200 73	−100 173	0 273	100 373	200 473	300 573	400 673	500 773	600 873	700 973	800 1 073	900°C 1 173 K
Copper (C10200), ρ = 8.94 g/cm³												
TC	1130	422	405	395	388	381	374	367	360	354	347	341
CE	(13.7)	14.9	15.9	16.7	17.3	17.7	18.1	18.6	18.4			
ME	139	120	113	108	104	99	(93)					
Muntz metal (C36500), ρ = 8.41												
TC			(68)	79	90	100	(111)					
CE			18.8	19.8	20.6	21.6	(22.4)					
ME			(109)	96	83	(69)						
Admiralty brass (C44300), ρ = 8.55												
TC		(99)	110	120	136	152						
CE		(19.4)	(19.6)	19.8	20.0	20.2	(20.3)					
ME		112	110	107	94	79	59					
Naval brass (C46400), ρ = 8.40												
TC	41	69	99	131	142	146	148	150	(151)			
CE		18.0	18.9	19.8	20.7	21.6	22.5	23.4	24.3			
ME	103	101	97	94	90	85	79	(71)				
Phosphor bronze A (C51000), ρ = 8.85												
TC		(65)	(77)	88.6	95.2	(112)						
CE		(11.3)	(14.2)	16.2	17.1	18.0	(18.9)					
ME	115	114	109	103	94	83	67					
Aluminum bronze D (C61400), ρ = 7.89												
TC		(61)	72	83	94	(105)						
CE		(14.8)	(15.3)	15.8	16.4	16.9	(17.4)					
ME	112	112	109	(103)	(94)	(83)	(67)					
Aluminum brass (C68700), ρ = 8.35												
TC		(87)	100	112	125	(138)						
CE		(15.8)	(17.0)	18.0	19.0	20.0	(21.0)					
ME		(112)	110	(107)	(94)	(79)	(59)					
Copper nickel 90/10 (C70600), ρ = 8.90												
TC		(39)	50	57	69	83	88					
CE	12.8	13.6	14.4	15.3	16.1	16.9	(17.7)					
ME	134	129	124	118	112	106	100					
Copper nickel 80/20 (C71000), ρ = 8.95 at 20°C												
TC		(26)	31	36	42	48	55	65	(76)			
CE	12.8	13.5	14.1	14.7	15.3	16.0	(16.6)	(17.2)				
ME	(150)	(149)	146	(142)	(138)	(130)	(122)	(114)				
Copper nickel 70/30 (C71500), ρ = 8.95 at 20°C												
TC		(26)	(29)	32	35	42	49	60	(71)			
CE	11.8	12.7	13.5	14.3	15.1	16.0	(16.8)	(17.6)				
ME	159	156	153	(148)	(143)	(137)	(130)	(121)				

Table 5 Aluminum alloys (annealed condition)

	Temperature											
	−200 73	−100 173	0 273	100 373	200 473	300 573	400 673	500 773	600 873	700 973	800 1 073	900°C 1 173 K
Alloy 1100 (A91100), $\rho = 2.66$ g/cm^3												
TC	327	242	235	240	237	232	226	219				
CE	20.5	21.5	22.5	23.5	24.4	25.4						
ME	(77)	(73)	69.5	(66)	(58)	(47)						
Alloy 3003 (A93003), $\rho = 2.66$												
TC			(202)	180	183	(177)	(171)					
CE		(21.2)	(22.2)	23.2	24.1	(25.1)						
ME	76.5	72.7	69.5	65.5	(57.9)	(46.9)						
Alloy 5052 (A95052), $\rho = 2.66$												
TC			(134)	146	155							
CE		(21.7)	(22.7)	23.8	24.8	(25.7)						
ME	77.9	73.8	71.0	66.9	55.8	(42.7)						
Alloy 5083 (A95083), $\rho = 2.66$												
TC			(113)	126	137							
CE		(22.1)	(23.0)	24.0	25.0	(25.9)						
ME	78.6	74.8	71.4	68.6	60.7	(50.3)						
Alloy 6061 (A96061), $\rho = 2.70$												
TC			(164)	172	176	(179)	(180)					
CE		(21.3)	(22.3)	23.3	24.3	(25.2)						
ME	76.5	72.4	69.6	65.8	60.3	(52.4)						
Alloy 6063 (A96063), $\rho = 2.70$												
TC			(209)	206	202	200	197					
CE		(21.4)	(22.4)	23.4	24.4	(25.4)						
ME	(76.5)	(72.4)	69.6	65.8	57.9	(48.3)						

Table 6 Miscellaneous materials

	Temperature											
	−200 73	−100 173	0 273	100 373	200 473	300 573	400 673	500 773	600 873	700 973	800 1 073	900°C 1 173 K
Unalloyed titanium (grade 2) R50400, $\rho = 4.51$ g/cm³												
TC		26.8	22.0	20.4	19.8	19.1	18.8	18.9	19.7	20.7	21.6	22.4
CE		8.1	8.4	8.6	8.9	9.2	9.4	9.7	9.9	10.0	10.1	(10.1)
ME			(108)	103	97	88	77	(66)				
Zirconium, $\rho = 6.505$												
TC	37	26	23	22	21	21	21	21	22	23	24	26
CE			(5.4)	6.1	6.7	7.1	(7.5)					
ME			(101)	95	86	77	(66)					
Grey cast iron, $\rho = 7.35$												
TC		(58.0)	(56.6)	54.3	51.9	48.8	45.9	42.6	(38.8)			
CE		(9.4)	(9.9)	10.44	11.02	11.59	12.17	12.78	(13.3)			
ME		(92)	(92)	90.7	87.2	81.4	73.4	(62)				

REFERENCES

1. *Koppers Handbuch der Brennstofftechnik*, Heinrich Koppers GmbH, Essen, 1937, pp. 472, 473.
2. Worthing, A. G., and Halliday, D., *Heat*, Wiley & Sons, New York, 1948, p. 136.
3. Kittel, C., *Introduction to Solid State Physics*, 3rd ed., Wiley & Sons, New York, 1966. (*Einführung in die Festkörperphysik*, R. Oldenbourg, München, 1969).
4. Schroedinger, E., Der Energieinhalt der Festkörper im Lichte der Neueren Forschung, *Phys. Z.*, pp. 450–455, 1919.
5. Peierls, R. E., *Quantum Theory of Solids*, Clarendon, Oxford, 1955, p. 233.
6. Kingery, W. D., *Property Measurements at High Temperatures*, Wiley & Sons, New York, 1959, p. 94.
7. D'Ans, J., and Lax, E., *Taschenbuch für Chemiker und Physiker*, Springer-Verlag, Berlin, 1943, p. 1126.
8. Ho, C. Y., Powell, R. W., et al., Thermal Conductivity of the Elements: A Comprehensive Review, *J. Phys. Chem. Ref. Data*, vol. 3 (Suppl. 1), 796 pp., 1974.
9. Childs, G. E., Ericks, L. J., et al., Thermal Conductivity of Solids at Room Temperature and Below, *N.B.S. Monogr.* 131, 624 pp., 1973.
10. Touloukian, Y. S., Powell, R. W., et al., Thermal Conductivity–Nonmetallic Elements and Alloys, in *Thermophysical Properties of Matter*, eds. Y. S. Touloukian and C. Y. Ho, vol. 1, 1595 pp., IFI/Plenum, New York, 1970.
11. Touloukian, Y. S., Powell, R. W., et al., Thermal Conductivity–Nonmetallic Solids, in *Thermophysical Properties of Matter*, eds. Y. S. Touloukian and C. Y. Ho, vol. 2, 1302 pp., IFI/Plenum, New York, 1970.
12. Kowalczyk, L. S., Thermal Conductivity and Its Variability with Temperature and Pressure, *Trans. ASME*, vol. 77, pp. 1021–1035, 1955.
13. V. V. Mirkovich, ed., *Thermal Conductivity 15*, 493 pp., Plenum, New York, 1978. (This volume also contains a list of earlier thermal conductivity conference publications.)
14. Hoyt, S. L., *Metal Data*, Reinhold, New York, 1952, pp. 72–73.
15. Campbell, J. D., and Ferguson, W. G., The Temperature and Strain-rate Dependence of the Shear Strength of Mild Steel, *Philos. Mag.*, vol. 21, no. 169, pp. 63–82, 1970.
16. Hearmon, R. F. S., The Elastic Constants of Anisotropic Materials, *Rev. Mod. Phys.*, vol. 18, pp. 409–440, 1946.
17. Zener, C., *Elasticity and Anelasticity of Metals*, University of Chicago Press, Chicago, 1948.
18. Smithells, C. J., ed., *Metals Reference Book*, 3rd ed., vol. 2, pp. 616–617, Butterworths, London, 1962.
19. Smithells, C. J., ed., *Metals Reference Book*, 4th ed., vol. 3, pp. 714–717, Butterworths, London, 1967.
20. Tang, R., Kratochvil, J., and Conrad, H. Scripta Met. *Three Dimensional Stress Distributions and Degree of Anisotropy in some Hexagonal Metals*, vol. 3, no. 7, pp. 485–488, 1969.
21. Smithells, C. J., ed., *Metals Reference Book*, 4th ed., vol. 3, pp. 709–710, Butterworths, London, 1967.
22. *Standards of Tubular Exchanger Manufacturers Association*, 6th ed., p. 209, TEMA Inc., New York, 1978.
23. Bolz, R. E., and Tuve, G. L., eds., *Handbook of Tables for Applied Engineering Science*, 2d ed., pp. 117–118, CRC Press, Cleveland, 1977. [Gives modulus of elasticity data for pure metals and commercial metals and alloys at room temperature.]
24. Woolman, J., and Mottram, R. A., eds., *The Mechanical and Physical Properties of the British Standard En Steels (BS 970-1955)*, vol. 3, Pergamon Press, Oxford, 1969. [Gives Young's modulus, shear modulus, and some Poisson's ratio data in tabular form for carbon, alloy and stainless steels. Tables include composition, heat treatment, test temperature, and variation with temperature.]
25. Miner, D. F., and Seastone, J. B., eds., *Handbook of Engineering Materials*, pp. 2-05–2-07, Wiley, New York, and Chapman and Hall, London, 1955. [Gives a graph of the effect of temperature on elasticity of stainless steels and states that if the room temperature value is known the value at a higher temperature can be deduced. Figures for copper and copper base alloys are included.]
26. Perry, R. H., and Chilton, C. H., eds., *Chemical Engineers Handbook*, 5th ed., McGraw-Hill, New York, 1973. [Gives properties of various metals including steels, nickel, aluminum, copper, magnesium, titanium, and other nonferrous alloys and includes a figure for tensile (Young's) modulus of elasticity. pp. 23-38–23-53.]
27. Prockter, C. E., ed., *Kempe's Engineers Year Book for 1978*, 83d ed., vol. 1, p. 190, Morgan-Grampian, London, 1978. [A table contains Young's modulus, shear modulus, and Poisson's ratio values for common materials.]
28. Parrish, A., ed., *Mechanical Engineers Reference Book*, pp. 7-85, 7-99, Butterworths, London, 1973. [Data on Young's modulus are given as a function of temperature for carbon, C-Mn and C-Mo steels, and four aluminum alloys. Data are based on BS 1500.]
29. Gray, E. E., ed., *American Institute of Physics Handbook*, 3d ed., McGraw-Hill, New York, 1972. [Contains values of Young's modulus for chemical elements and alloys.]
30. American Society of Mechanical Engineers, *Boiler and Pressure Vessel Code*, Section VIII–Division 2, 9th ed., vols. 1 and 2, ASME, New York, 1977.
31. American Society of Mechanical Engineers, *Boiler and Pressure Vessel Code*, Section VIII, Division 2, ASME, New York, 1983.
32. American Society of Metals, *Metals Reference Book*, 1981.
33. Sandvikens Jernwerks Aktiebolag, Sandviken, Sweden, Sandvik Data Sheets E924/05.
34. British Iron and Steel Research Association, *Physical Constants of Some Commercial Steels at Elevated Temperatures*, Butterworth Scientific Publications, London, 1953.
35. Tubular Exchanger Manufacturers' Association, Standards, 6th ed., pp. 209–211, New York, 1978.
36. Fox, S., Sheffield, England, 1965, Data Sheet for "Esshete CRMS."
37. Draft B. S. Specification for Pressure Vessels for Reactor Primary Circuits. D64/2354, British Standards Institution.
38. Clark, C. L., *High Temperature Alloys*, Pitman, London, 1953.
39. Smith, A. I., et al., Creep, Stress-Relaxation and Metallurgical Properties of Steels for Steam Power Plant, *Proc. Inst. Mech. Eng. London*, vol. 171, no. 34, pp. 918–942, 1957.
40. Gemill, M. G., et al., Study of 7% and 8% Creep-Resisting Steels for Use in Steam Power Plants, *Iron Steel Inst. J.*, vol. 184, no. 3, pp. 122–144, 1956.

REFERENCES

41. *U.S. Atomic Energy Commission Reactor Handbook,* 2d ed., vol. 1, *Materials,* Interscience, New York, 1960.
42. U.K. Atomic Energy Authority, TRG Rept. 840(R), Seddon, B. J., Steels Data Manual, 1965.
43. Wiggin Alloys Ltd., Hereford, England, 1972. Publication Nos. 3269, 3564, 3573, 3664, 3700, 3570.
44. Cabot Corp., Kokomo, Indiana, 1983. Publication Nos. H1019, 1032, 2001, 2002A, 2006, 2007, 2009, 2013, 2015, 2019, 3004, F30037H.

Touloukian, Y. S., Powell, R. W., Ho, C. Y., and Klemens, P. G., Thermal Conductivity of Metallic Elements and Alloys, in *Thermophysical Properties of Matter,* eds. Y. S. Touloukian and C. Y. Ho, vol. 1, IFI/Plenum, New York, 1970.

46. Oak Ridge National Laboratory, Metals and Ceramics Division, Annual Progress Report ORNL 3970, p. 243, 1966.
47. Perry, J. H., *Chemical Engineers' Handbook,* 5th ed., pp. 23–38 to 52, McGraw-Hill, New York, 1973.
48. Liley, P. E., HEDH Section 5.5.6.
49. Copper Development Association, Potters Bar, Hertfordshire, England. Publication Nos. 82, E6, F1, F2, F3, G3, K4, TN27, TN31.
50. Hager, S. F., A Basic Guide to Copper Tubing Alloys for Heat Exchangers and Condensers, *Plant Eng.,* vol. 29, no. 20, pp. 81–89, 1975.
51. Rohsenow, W. M., and Hartnett, J. P., *Handbook of Heat Transfer,* pp. 2–66 to 2–69 and 2–107, McGraw-Hill, New York, 1973.
52. American Society of Metals, *Source Book on Copper and Copper Alloys,* 1979.

INDEX OF GASES AND LIQUIDS

INDEX OF SOLIDS